Sustainability of Natural Resources

Agriculture is the backbone of the economy in most countries and its output can be impacted by climate change effects. India, as well as other countries which are predominantly agricultural are facing various challenges due to increasing population which can be met by technological innovations for sustainable agriculture. Advanced and innovative technologies in agriculture will not only solve the problems of fulfilling the food requirement of the growing population but also sustain agriculture in the future. *Sustainability of Natural Resources Planning and Management* addresses the advancement of innovative techniques to address the issues of water scarcity and agricultural yield. It discusses various aspects of natural resource management, agriculture micro irrigation, AI applications for water management and impacts of climate change on water resources. This book also deals water resource exploration, planning, recent geographic information system-based studies, groundwater modelling, and related applications. It highlights the optimal strategies for sustainable water resource management and development. It also examines precision farming using remote sensing and GIS techniques.

Sustainability of Natural Resources

Planning, Development, and Management

Edited by
Rohitashw Kumar, Kanak N. Moharir, Vijay P. Singh,
Chaitanya B. Pande, and Abhay M. Varade

CRC Press is an imprint of the
Taylor & Francis Group, an informa business

First edition published 2024
by CRC Press
2385 NW Executive Center Drive, Suite 320, Boca Raton FL 33431

and by CRC Press
4 Park Square, Milton Park, Abingdon, Oxon, OX14 4RN

CRC Press is an imprint of Taylor & Francis Group, LLC

Library of Congress Cataloging-in-Publication Data
Names: Kumar, Rohitashw, editor.
Title: Sustainability of natural resources : planning, development, and management /
edited by Rohitashw Kumar, Kanak N. Moharir, Vijay P. Singh, Chaitanya B. Pande, and Abhay M. Varade.
Description: First edition. | Boca Raton : CRC Press, 2024. |
Includes bibliographical references and index. |
Identifiers: LCCN 2023046477 (print) | LCCN 2023046478 (ebook) | ISBN 9781032295312 (hardback) |
ISBN 9781032300559 (paperback) | ISBN 9781003303237 (ebook)
Subjects: LCSH: Watershed management. | Water resources development. |
Renewable natural resources. | Conservation of natural resources. | Environmental management.
Classification: LCC TC413 .S867 2024 (print) | LCC TC413 (ebook) |
DDC 333.91/04–dc23/eng/20240128
LC record available at https://lccn.loc.gov/2023046477
LC ebook record available at https://lccn.loc.gov/2023046478

ISBN: 978-1-032-29531-2 (hbk)
ISBN: 978-1-032-30055-9 (pbk)
ISBN: 978-1-003-30323-7 (ebk)

DOI: 10.1201/9781003303237

Typeset in Times
by codeMantra

Dedicated to:

Rohitashw Kumar: My late parents (Smt. Rajwan Devi and Sh. Chhalu Ram), my wife (Mrs Reshma Devi), daughter Meenu, son in-law Vineet Ravish, and son Vineet Kumar.

Kanak N. Moharir: My parents, husband, and daughter.

Vijay P. Singh: My wife Anita who is no more; son Vinay; daughter Arti; daughter-in-law Sonali; son-in-law Vamsi; and grandsons Ronin, Kayden, and Davin.

Chaitanya B. Pande: My parents, wife and son Ridant.

Abhay M. Varade: My parents, wife, and sons.

Contents

Preface

Over-exploitation of groundwater and marked changes in climate over the past few decades has imposed immense pressure on groundwater resources. As demand for water has increased manifold across the globe for domestic use, agricultural irrigation, and industrial use, the need for groundwater development and management has also increased. This book provides a multidisciplinary overview for academics, administrators, scientists, policymakers, social science professionals, and NGOs involved in departments concerned with sustainable water resources development and management programmes. Growth in agricultural productivity (https://www.sciencedirect.com/topics/economics-econometrics-and-finance/productivity-change) is vital for economic and food security (https://www.sciencedirect.com/topics/economics-econometrics-and-finance/food-security), which are being threatened by climate change. Under the spectre of looming scarcity of water resources in the near future, it has become crucial to quantify and manage available water resources. As demand for water has increased manifold across the globe for domestic use, agricultural irrigation, and industrial use, the need for groundwater development, management, and productivity of aquifers has also increased. Therefore, this book covers recent geographic information system-based studies, advanced methodologies with their applications and groundwater modelling. It highlights groundwater exploration, planning, and designing; and formulating further strategies for sustainable management and development.

Growth in agricultural productivity, which is vital for economic and food security, is being threatened by climate change and other factors. This book focuses on the development and management of groundwater resources in India and other countries, with emphasis on sustainable water resources development and management programmes.

This book examines advanced techniques and themes such as remote sensing and geographical information system (GIS), analytical hierarchy process (AHP), aquifer modelling, groundwater quality analysis, land use and land cover analysis, evapo-transpiration estimation, crop coefficient, groundwater flow modelling, climate, change impact on groundwater and Internet of Things (IoT) and automation. The main objective is to formulate plans for the development and maintenance of groundwater from India and other countries having similar situations regarding natural resources, particularly water. One of the defining topics in this book is climate change as its impacts on people and nature are increasingly evident. This book includes the research work by academics, planners, scientists, and research scholars from various universities, international organizations and institutions from around the world. This book covers water resources exploration, planning recent geographic information system-based studies, and groundwater modelling and applications. It highlights strategies for sustainable water resources management and development and provides a multidisciplinary overview for academics, administrators, scientists, policymakers, social science professionals, and NGOs involved in departments concerned with sustainable water resources development and management programmes.

Chapter 1 presents the systematic survey of the land resources and their mapping for managing the resources in a sustainable way. Four major landforms such as hills, mounds, valleys, and alluvial plains were identified, and detailed soil survey was carried out. The study indicated that clay content and CEC of soils in hills and mounds increased with increasing soil depth, whereas in alluvial plains, the majority of soils showed irregular distribution of OC and clay content and may be attributed to periodic flooding and deposition of materials during different cycles. Considering the major problems and potentials four land management units (LMUs) were identified and an alternate land use was suggested for each LMU of the study area.

Chapter 2 provides a broader picture of sustainable biodiversity conservation in tribal areas. As the steady decline of plant supplies contributes to environmental deterioration, individuals have begun to pay attention to their survival and sustainable usage. This chapter argues that various

Indian tribal groups and their associated information structures will contribute to the sustainable dialogue about biodiversity conservation. The focus is on the insights that aboriginal knowledge can provide, on the principles that govern aboriginal relationships with nature, such as reciprocal recognition and care.

Chapter 3 provides soil bioengineering practices for sustainable ecosystems in landslide-affected areas. Land degradation transforms productive, usable land into uncultivable, unproductive land. Stabilization of landslide-prone slopes by low-impact techniques such as bioengineering is a topic of attention. This chapter focuses on the pros and cons in relation to the implementation of different soil bioengineering work typologies in landslide-affected regions.

Chapter 4, highlighted, the sustainable ecosystem development and landscaping for urban and peri-urban areas. Soil bioengineering is one of the techniques adopted by several countries to protect their slopes by the usage of plant wealth that is not only indigenous but also provides economic and ecosystem values to the regional population. This chapter provides information and a framework for utilizing indigenous plants based on their specific selection criteria focussing on the eco-engineering perspective and the method to utilize them to protect urban and peri-urban slopes.

Chapter 5 illustrates the climate change impacts of fluoride contamination on human health in the dry zone of Sri Lanka. Fluoride in groundwater in arid or dry zones is one of the most pronounced elements in the emerging science of 'medical geology.' Many tropical groundwater sources have high fluoride concentrations resulting in various health deficits. Studies have shown Sri Lanka, particularly in the dry zone, has been affected by excessive quantities of fluoride in groundwater. The spatial distribution of fluoride in groundwater shows that climate and hydrological conditions appear to perform a salient role in the geochemical distribution of fluoride in groundwater. Overall, it is evident that climate change directly or indirectly impacts fluoride contamination thus resulting in human health complications in the dry zone of Sri Lanka.

Chapter 6 presents geospatial technology in catchment modelling using the SCS-CN method for estimating the direct runoff on the Barakar River Basin, Jharkhand. The Damodar basin has two extensive dams built to control the river's flow and store the water in it. The amount of rainfall and the primary runoff caused over the Barakar catchment was thus calculated by the Soil Conservation Service Curve Number (SCS-CN) method.

Chapter 7 provides a hydro-geospatial investigation of water conservation in limestone terrains. Groundwater is a major source of water for more than 2 billion people globally and makes up 30% of the freshwater supplies for the globe. This chapter identifies prospective sites and appropriate artificial recharge structures recommended to recharge by using the geospatial method using overlay analysis and to recommend sites where artificial recharge structures can be built. To enable an increase in aquifer recharge during dry days in the study area and to support sustainable groundwater conservation during the period of low water availability, the overlay operation to create a suitable site map is processed in a GIS environment.

Chapter 8 highlights the rainfall spatio-temporal variability and trends in the semi-arid ecological zone of Nigeria. This chapter describes the rainfall climatology, trends, variability and anomalies of the Sokoto–Rima Basin (SRB) of the north-western axis of Nigeria. Data used were collated from synoptic stations covering 1951–2017 and gridded data spanning 2018–2050 denoting past and future climates accordingly. The Mann-Kendall test was used to detect monotonic trends at the monthly, inter-annual, annual, decadal level including the World Meteorological Organization (WMO) standard.

Chapter 9 highlights the climate change awareness, perception and adaptation strategies for small and marginal farmers in Nigeria. It examines farmers' socio-economic and institutional characteristics, their perceptions of climate change, the factors that influence these perceptions, farmers' climate change and adaptation strategies as well as climate change constraints in the study area. The study used a structured questionnaire and focus group approaches for data collection procedures and utilized inferential statistical techniques for the analysis. It was found that the positive responses from smallholder farmers' awareness and perception of climate change could lead to plans to adopt further climate change initiatives.

Chapter 10 provides the delineation of groundwater prospect based on earth observation data and AHP modelling. The input data for the entire study were derived using satellite data, and maps. The digital layers including geomorphology, slope, geology, drainage density, lineament density, soil, rainfall, groundwater level and LULC were used to delineate the groundwater prospect zones. The accuracy assessment was performed for the LULC with the aid of ground truth data. The results of the study area can be useful to develop regional local sustainable water resources development plan as well as this methodology can be applicable in other geographical regions.

Chapter 11 provides the evaluation of phreatic groundwater for assessing drinking and irrigation appropriateness using the hydro-geochemical method. An attempt has been made, here to understand the hydro-geochemical distinctiveness of groundwater from WRD watershed (Chandrapur district, Central India) where fluoride deposits are being mined out. The dominantly alkaline groundwater from the study area also has high electrical conductivity (EC). The spatial variation in TDS content has been attributed to the distinctive lithology; while the high concentration of Cl^- has been described to anthropogenic activities. The safe limit for fluoride ingestion has been exceeded in the study area and hence such water is unsuitable for drinking purpose and even domestic use.

Chapter 12 deals with the variability of groundwater quality and associated hydro-geochemical processes in quaternary (shallow) aquifers of the Cauvery and Vennar sub-basins in Cauvery delta, Southern India using 200 groundwater samples collected seasonally. The samples were analysed for 12 hydrochemical parameters: EC, pH, TH, TDS, Ca^{2+}, Mg^{2+}, Na^+, K^+, Cl^-, CO_3^{2-}, HCO_3^-, and SO_4^{2-}. These parameters were studied using the spatial interpolation maps prepared with the help of ArcGIS 10.1 using the inverse distance Weighting (IDW) technique for assessing the fitness of groundwater for irrigation and drinking purposes. Groundwater types were determined using the Chadha diagram, and the mechanisms controlling groundwater chemistry were studied using the Gibbs plot. Overall, the groundwater in the Vennar sub-basin was more deteriorated when compared with the Cauvery sub-basin in this Cauvery delta.

Chapter 13 deals with the groundwater potential zones mapping based on the ANN, ML models using AHP and MIF techniques. This chapter describes the advances in all aspects of groundwater, including the importance, exploitation, and exploration of groundwater resources and methods for identifying potential groundwater zones, including conventional, remote sensing, and advanced methods. It focuses specifically on various applications of groundwater mapping and discusses and shows the importance of groundwater estimation. Furthermore, the findings show how the groundwater tools could be important for making proper decisions to ensure water resources management for future planning.

Chapter 14 provides the irrigation water requirement and estimation of evapotranspiration using remote sensing, GIS and other technologies. Irrigation is essential to water resource management and vital to agricultural policy design. Water security in arid environments is very vulnerable to climate change and increasing population activity. This chapter discusses the advances in all aspects of estimating water requirements and reference evapotranspiration (ETo) using remote sensing, GIS, and other approaches, including physical models and artificial methods. Furthermore, we illustrated how the tools applied for estimating ETo and IWR will help the decision-makers and developers achieve better performance with less computational cost and agricultural sustainability.

Chapter 15 provides smart irrigation water using sensors and IoT. Watering crops at the correct times and in the appropriate amounts are the core value propositions for smart irrigation systems, as the IoT and sensor technologies are now widely employed in agriculture. It discusses the advances in all aspects of smart irrigation, including components, different sensors, datasets required, algorithms and benefits of the smart system. Furthermore, it illustrates how the intelligence method tool will help the decision-makers and developers in achieving agricultural water sustainability.

Chapter 16 provides the analysis of climate variability and change impact on rainfall trend patterns in Nigeria. The evaluation of climate change patterns of The Calabar River Basin was necessitated owing to its distinct socio-economic importance to Nigeria.

Chapter 17 deals with the assessment of hydro-chemistry and water quality index for groundwater quality. 105 descriptive groundwater samples were collected and their physicochemical parameters were determined using the standard norms.

Chapter 18 provides the assessment of heavy metal contamination in water and sediments of major industrial streams. Water and sediment were simultaneously assessed and compared to several environmental contamination monitoring parameters, viz. threshold effect level (TEL), probable effect level (PEL) and severe effect lever (SEL) for the sediments, and modified degree of contamination and pollution load indices for the water.

Chapter 19 deals with wastewater treatment and reuse for sustainability. Water waste treatment and reuse are essential for maintaining a sustainable environment and conserving natural resources. Water waste treatment technologies such as biological treatment, chemical treatment, and physical treatment play a significant role in removing contaminants from wastewater, making it safe for reuse. Additionally, reusing treated wastewater reduces the need for freshwater sources, thus conserving energy and reducing greenhouse gas emissions associated with water transportation.

Chapter 20 addresses the watershed development for sustainable agriculture. The sustainable management of watersheds is necessary to maintain their ecological integrity and protect the livelihoods of communities that depend on them. The implementation of effective watershed management strategies requires an understanding of the complex interactions between the physical, biological, and socio-economic components of the watershed. GIS technology enables the integration and analysis of diverse data sources to provide a comprehensive understanding of the watershed's spatial and temporal dynamics. GIS-based models can be used to simulate and predict the impact of various management interventions on the watershed's health and inform decision-making processes.

This book discusses the advanced technologies used to tackle the issue of natural resources management under climate change along with applications in agriculture in different agro-climatic regions across the globe. It is hoped that this book will be useful to increase agricultural production, productivity, profitability and suitability in agriculture while mitigating the impacts of climate change.

Editors
Rohitashw Kumar
Kanak N. Moharir
Vijay P. Singh
Chaitanya B. Pande
Abhay M. Varade

Acknowledgements

I consider it a proud privilege to express my heartfelt gratitude to ICAR - All India Coordinated Research Project on Plastic Engineering in Agriculture Structures & Environment Management, for providing financial assistance to carry out this project at SKUAST-Kashmir, Srinagar. It is highly thankful to Project Coordinated unit, CIPHET, Ludhiana for providing all necessary facilities. I sincerely acknowledge the efforts of Professor Vijay Pal Singh, Distinguished Professor, Department of Biological and Agricultural Engineering, Texas A&M University, College Station, Texas, USA for his support and guidance in bringing this manuscript to final refined shape, constructive criticism and utmost cooperation at every stage during this work.

I am grateful to all my authors Dr. Kanak N. Moharir, Dr. Chaitanya B. Pande, and Dr Abhay M. Varade, who put great efforts and worked day and night to bring this manuscript in a refined shape. I am also highly thankful to all contributors of different chapters of this book.

I am highly obliged to Hon'ble Vice Chancellor Prof. Nazir Ahmad Ganai for his support and encouragement. I am highly thankful to him provide support under NAHEP project. I wish to extend my sincere thanks to the College of Agricultural Engineering and Technology, SKUAST-Kashmir for their support and encouragement.

I am grateful to my students Dr. Saba Parvez, Er. Munjid Maryam, Er. Zeenat Farooq, Er. Tanzeel Khan, Er. Dinesh Vishkarma, Er. Faizan Masoodi, Er, Noureen, Er. Mahrukh, Er Muneeza, Er Nuzhat and Er Khilat Shabir for providing all necessary help to write this book.

I express my regards and reverence to my late parents as their contribution in whatever I have achieved till date is beyond expression. It was their love, affection and blessed care that helped me to move ahead in my difficult times and complete my work successfully. I thank my all family members. I thankfully acknowledge the contribution of all my teachers since schooldays, for showing me the right path at different steps of life.

I sincerely acknowledge with love, the patience and support of my wonderful wife Reshma. She has loved and cared for me without ever asking anything in return and I am thankful to God for blessing me with her. She has spent the best and the worst of times with me but her faith in my decisions and my abilities have never wavered. I would also like to thank my beloved daughter Meenu and son Vineet and my son-in-law Vineet Ravish for making my home lively with their sweet activities.

Finally, I bow my head before the almighty God, whose divine grace gave me the required courage, strength and perseverance to overcome various obstacles that stood in my way.

Rohitashw Kumar
Associate Dean,
College of Agricultural Engineering and Technology,
SKUAST- Kashmir, Srinagar

Editors' Biography

Rohitashw Kumar (B.E., M.E., Ph. D.) is Associate Dean and Professor in the College of Agricultural Engineering and Technology, Sher-e-Kashmir University of Agricultural Sciences and Technology of Kashmir, Srinagar, India. He worked as Professor Water Chair (Sheikkul Alam ShiekhNuruddin Water Chair), Ministry of Jal Shakti, Government of India, at the National Institute of Technology, Srinagar (J&K) for 3 years. He is also Professor and Head, Division of Irrigation and Drainage Engineering. He obtained his Ph.D. degree in Water Resources Engineering from the National Institute of Technology, Hamirpur, and Master of Engineering Degree in Irrigation Water Management Engineering from Maharana Pratap University of Agriculture and Technology, Udaipur. He received a gold medal in 2022, a leadership award in 2020, a Special Research award in 2017, and a Student Incentive Award in 2015 (Ph.D. Research) from the Soil Conservation Society of India, New Delhi. He also got the first prize for the best M. Tech thesis in Agricultural Engineering in 2001 in India. He has published over 130 papers in peer-reviewed journals, more than 25 popular articles, 8 books, 4 practical manuals, and 30 book chapters. He has guided two Ph.D students and 16 M.Tech students in soil and water engineering. He has handled more than 12 research projects as a principal or co-principal investigator. Since 2011, he has been the Principal Investigator of ICAR, All India Coordinated Research Project on Plastic Engineering in Agriculture Structural and Environment Management.

Kanak N. Moharir is working as Assistant Professor at Banasthali University, Jaipur, India. She has more than 5 years of teaching and research experience and has published more than 40 research papers in national and international journals and has presented 15 papers in national and international conferences with more than 870 citations. She is working in remote sensing, watershed management, hydrology, land use and land cover, aquifer mapping, groundwater modeling, hydrogeology, geomorphology, and geology.

Vijay P. Singh is a Distinguished Professor, a Regents Professor, and the inaugural holder of the Caroline and William N. Lehrer Distinguished Chair in Water Engineering at Texas A&M University. His research interests include surface water hydrology, groundwater hydrology, hydraulics, irrigation engineering, environmental quality, water resources, water–food–energy nexus, climate change impacts, entropy theory, copula theory, and mathematical modeling. He graduated with a B.Sc. in Engineering and Technology with emphasis on Soil and Water Conservation Engineering in 1967 from U.P. Agricultural University, India. He earned an MS in Engineering with specialization in Hydrology in 1970 from the University of Guelph, Canada; a Ph.D. in Civil Engineering with specialization in Hydrology and Water Resources in 1974 from Colorado State University, Fort Collins, USA; and a D.Sc. in Environmental and Water Resources Engineering in 1998 from the University of the Witwatersrand, Johannesburg, South Africa. He has published extensively on a wide range of topics. His publications include more than 1,365 journal articles, 32 books, 80 edited books, 305 book chapters, and 315 conference proceedings papers. For his seminar contributions, he has received more than 100 national and international awards, including three honorary doctorates. Currently, he serves as Past President of the American Academy of Water Resources Engineers, the American Society of Civil Engineers (ASCE), and previously, he served as President of the American Institute of Hydrology and Cahir, Watershed Council, ASCE. He is Editor-in-Chief of two book series and three journals and serves on the editorial boards of more than 25 journals. He has served as Editor-in-Chief of three other journals. He is a Distinguished Member of the American Society of Civil Engineers, an Honorary Member of the American Water Resources Association, an Honorary Member of International Water Resource Association, and a Distinguished Fellow of the Association of Global Groundwater Scientists. He is a fellow of five professional societies. He is also a fellow or member of 11 national or international engineering or science academies.

Chaitanya B. Pande has completed his Ph.D. in Environment Science from Sant Gadge Baba Amravati University. He has completed M.Sc. in Geoinformatics from Amravati University in 2011. He has more than 11 years of teaching, research, and industrial experience. He is a reviewer for several scientific journals of the international repute and an editorial board member in the American Journal of Agricultural and Biological Sciences. He has published 57 research papers, 1 textbook, 3 edited books entitled "Groundwater Resources Development and Planning in the Semi-Arid Region," 19 conference papers, and 18 book chapters with more than 1,349 citations. His research interests include remote sensing, GIS, Google Earth Engine, machine learning, watershed management, hydrogeology, hydrological modeling, drought monitoring, land use and land cover analysis, groundwater quality, urban planning, hydro-geochemistry, groundwater modeling, geology, hyperspectral remote sensing, remote sensing and GIS application in natural resources management, watershed management, and environmental monitoring and assessment subjects.

Abhay M. Varade is working as an Associate Professor at the Post Graduate Department of Geology, Rashtrasant Tukadoji Maharaj Nagpur University, Nagpur. He has completed his M.Sc. in Geology from Amravati University, Amravati, and M.Tech. in Petroleum Exploration from the renowned Indian School of Mines, Dhanbad. Before joining the post of Lecturer in the Department of Geology, RTM Nagpur University Nagpur, he obtained experience at the Central Mining Research Institute (CSIR Lab., Dhanbad) and SGS India Pvt. Limited, Mumbai. Presently, he is associated with different Earth science-related societies. Till date, he has published 77 research papers in the journals/ books of international (25 international) and national (52 national) repute and also edited (Editor) three Special Volumes of Gondwana Geological Society (*Journal of Geosciences Research*). His publications mainly cover the topics like sedimentology, coal bed methane technology, coal petrology, Quaternary geology and vertebrate palaentology, hydrochemistry, applications of RS/GIS and geophysical/geostatistical techniques in groundwater and agricultural based studies, pumping test analysis for aquifer characterization, watershed characterization, and prioritization and management. He is a reviewer for high-impact factor journals along with several other Indian referred journals and has worked as a Co-Chairman/Rapporteur in many scientific/ technical sessions. As an Organizing Secretary, he has organized five national/state-level seminars/ conferences. Presently, he is working on an International Book Proposal related to Climate Change Impact – Uzbekistan. His noteworthy contribution includes Advisory Committee Membership of UGC-SAP Programme for the Department of Geology, RTM NU (2016–2021); Core Member of RTM NU Nagpur for NAAC (2020 NAAC); Member of the Task Force Committee-Board of Studies in Geology (2016–2018); Ex-Executive Member of the Alumni Association of the Department of Geology, RTM NU (2016–2018); Ex-Executive Member (Joint Secretary) of the Gondwana Geological Society, Nagpur; Deputy Chief Officer for the Examination of RTM NU (2012–2013); and Training and Placement In-charge (2004–2009; 2016–2017; 2017–2021).

List of Contributors

Abhay M. Varade
Post Graduate Department of Geology
RTM Nagpur University
Nagpur, India

K. R. Aher
Groundwater Surveys and Development Agency
Aurangabad, India

Ahmed Elbeltagi
Agricultural Engineering Department, Faculty of Agriculture
Mansoura University
Mansoura Egypt

Ali Mokhtar
Department of Agricultural Engineering, Faculty of Agriculture
Cairo University
Giza, Egypt
State Key Laboratory of Soil Erosion and Dryland Farming on the Loess Plateau, Institute of Soil and Water Conservation
Northwest Agriculture and Forestry University, Chinese Academy of Sciences and Ministry of Water Resources
Yangling, China

Amal Mohamed
Department of Agricultural Engineering, Faculty of Agriculture
Cairo University
Giza, Egypt

Amit Biswas
Department of Agricultural Engineering, School of Agriculture & Bio-Engineering
Centurion University of Technology and Management
Odisha, India

Amit P. Multaniya
Department of Environmental & Water Resources Engineering
University Teaching Department, CSVTU Bhilai
Chhattisgarh, India

André Gustavo de Sousa Galdino
Federal Institute of Education, Sciences and Technology of Espírito Santo
Jucutuquara, Vitória, ES, Brazil.

Arishmita Ghosh
Symbiosis Institute of Geoinformatics (SIG)
Symbiosis International (Deemed University) (SIU)
Pune, Maharashtra, India.

Aswin Kokkat
Earth Process Modeling Group, CSIR-National Geophysical Research Institute
Hyderabad, India

S. M. Deshpande
Post Graduate Department of Geology, Institute of Science
Aurangabad, India

Dimple
Soil and Water Engineering, College of Engineering and Technology
Maharana Partap University of Technology
Udaipur, Rajasthan, India

M. L. Dhumal
Department of Mathematics, Deogiri College
Aurangabad, India

Falaq Firdous
College of Agricultural Engineering and Technology
SKUAST-Kashmir, Srinagar, India

G. P. Ganapathy
VIT University
Vellore, Tamil Nadu

E. Gayathiri
Guru Nanak College
Tamil Nadu, India

Gayathiri Ekambaram
Department of Plant Biology and Plant Biotechnology, Guru Nanak College
Chennai, Tamil Nadu, India

Gobinath R
SR University
Telangana, India

M. D. K. L. Gunathilaka
Department of Geography, University of Colombo
Colombo, Sri Lanka

J. Jayanthi J
Hindusthan College of Arts and Science
Coimbatore, India

E. J. James
Water Institute, Karunya Institute of Technology and Sciences
Coimbatore, Tamil Nadu, India

Jamal Khatib
Department of Civil and Environmental Engineering, Faculty of Engineering
Beirut Arab University
Lebanon

R. K. Jena
Indian Council for Agricultural Research, Indian Institute of Water Management
Bhubaneswar, India

Jitendra Rajput
Water Technology Center, Indian Council for Agricultural Research, IARI
New Delhi, India

Kamal Kishor Sahu
Chhattisgarh Council of Science and Technology
Vigyan Bhwan, Vidhansabha Road Raipur
Chhattisgarh, India

Kanak N. Moharir
Sant Gadge Baba Amravati University
Amravati (MS), India

Kalyan Kumar Bhar
Department of Civil Engineering, Indian Institute of Engineering Science and Technology
Shibpur

Khilat Shabir
College of Agricultural Engineering and Technology
SKUAST-Kashmir, Srinagar, India

Kumaravel Priya
Department of Biotechnology, St. Joseph College (Arts & Science)
Kovur, Chennai, Tamil Nadu, India

Lasantha Manawadu
Department of Geography
University of Colombo
Colombo, Sri Lanka

Manish Kumar Sinha
Department of Environmental & Water Resources Engineering
University Teaching Department
CSVTU Bhilai, Chhattisgarh, India

Mayowa Fasona
Department of Geography
University of Lagos
Akoka-Yaba, Lagos, Nigeria

Mehmet Serkan Kırgız
Senior Editor, Middle East & North Africa (MENA) region in Springer
Department of Architecture, Faculty of Engineering-Architecture
Nisantasi University, İstanbul, Turkey

Mohammed Bashir Umar
Department of Agricultural Economics and Extension, Faculty of Agriculture
Federal University
Gashua, Yobe State, Nigeria

N. C. Mondal
Earth Process Modeling Group, CSIR-National Geophysical Research Institute
Hyderabad, India

S. Mukhopadhyay
Indian Council for Agricultural Research, NBSS&LUP
West Bengal, India

Munjid Maryam
College of Agricultural Engineering and Technology
SKUAST-Kashmir, Srinagar, India

Y. A. Murkute
P. G. Department of Geology
R.T.M. Nagpur University, Law College Square
Nagpur, India

D. C. Nayak
Indian Council for Agricultural Research, NBSS&LUP
West Bengal, India

Nihan Naiboğlu
Department of Architecture
Nisantasi University
İstanbul, Turkey

Pandurang Choudhari
Department of Geography
University of Mumbai
Mumbai (MS), India

Pankaj Bakshe
Department of Chemistry
RTM Nagpur University
Nagpur, India

Paniswamy Prakash
Department of Botany
Periyar University
PeriyarPalkalai Nagar, Salem, Tamil Nadu, India

Papri Mukherjee
Department of Agricultural Engineering, School of Agriculture & Bio-Engineering
Centurion University of Technology and Management
Odisha, India

M. G. Ragunathan
Guru Nanak College
Chennai, Tamil Nadu, India

S. Ramachandran
Indian Council for Agricultural Research, Indian Institute for Horticulture Research
Karnataka, India

Ravin Jugade
Department of Chemistry
RTM Nagpur University
Nagpur-330033, India

P. Ray
Indian Agricultural Research Institute
Pusa, New Delhi, India

S. K. Ray
Indian Council for Agricultural Research, NBSS&LUP
West Bengal, India

Renata Graf
Department of Hydrology and Water Management, Institute of Physical Geography and Environmental Planning
Adam Mickiewicz University
Poznań, Poland

S. K. Reza
Indian Council for Agricultural Research, NBSS&LUP
West Bengal, India

Rituparna Saha
Advanced Technology Development Centre, Indian Institute of Technology Kharagpur
West Bengal, India

Rohitashw Kumar
College of Agricultural Engineering and Technology
SKUAST-Kashmir, Srinagar, India

Saheed Adekunle Raji
Department of Environmental Management and Toxicology
Federal University of Petroleum Resources
Effurun, Nigeria

P. J. Sajil Kumar
Freie Universit ät Berlin, Institute of Geological Sciences, Hydrogeology Group
Berlin, Germany

S. Saha
Indian Council for Agricultural Research, NBSS&LUP
West Bengal, India

Sandipan Das
Symbiosis Institute of Geoinformatics (SIG)
Symbiosis International (Deemed University) (SIU)
Pune, Maharashtra, India

Shakirudeen Odunuga
Department of Geography
University of Lagos
Akoka-Yaba, Lagos, Nigeria

S. K. Singh
Indian Council for Agricultural Research, Indian Agricultural Research Institute Goa
India

Sobhy M. Mahmoud
Agricultural Engineering Department, Faculty of Agriculture
Ain Shams University
Cairo, Egypt

Sudhir Kumar Singh
K Banerjee Centre of Atmospheric and Ocean Studies, IIDS, Nehru Science Centre
University of Allahabad
Allahabad, India

Abali, Temple Probyne
Department of Geography and Environmental Management
Rivers State University, P. M. B., Nkpolu-Oroworukwo Port Harcourt
Rivers State, Nigeria

Venkata S. S. R. Marella
S R University
Warangal, Telangana, India

Wessam El-Ssawy
Department of Agricultural Engineering, Faculty of Agriculture
Cairo University, Giza, Egypt
Irrigation and Drainage Department, Agricultural Engineering Research Institute
Giza, Egypt

V. P. I. S. Wijeratne
Department of Geography
University of Colombo
Colombo, Sri Lanka

R. O. Yenkie
Department of Geology
RTM Nagpur University
Nagpur, India

1 Characterization and Mapping of Soils for Sustainable Management Using Geospatial Techniques

S.K. Reza, P. Ray, S. Mukhopadhyay, S. Ramachandran, R.K. Jena, D.C. Nayak, S.K. Singh, S.K. Ray, and S. Saha

1.1 INTRODUCTION

The Purvanchal range is a sub-mountain range of the Himalayas, covering an area of about 94,800 km^2 with a population of over 4 million residing in Tripura, Nagaland, Manipur, Mizoram hills, and part of Arunachal Pradesh of Northeastern, India. In Tripura, these hills are a series of parallel north-south folds, decreasing in elevation to the south until they merge into the greater Ganges-Brahmaputra lowlands (also called the eastern plains) (Singh, 2006) and display a variety of landforms with varying types of soil (Bhattacharyya et al., 1996, 2003, 2010; Gangopadhyay et al., 2001, 2008; Reza et al., 2019a). Land is a delineable area on the earth's surface and the basic unit of all material production. The limited and inexpansible land resource has to be used very judiciously to meet the expectations of the people and their competing demands. Though, India represents only 2.4% of the geographical area it supports 17.5% of the total world's population (Mythili and Goedecke, 2016; Jangir et al., 2020). Globally, the present-day crisis on food, fuel and energy, increasing food prices in the international market, and conversion of good quality arable lands to several non-agricultural uses, like industry and urbanization, etc., is putting immense pressure to feed the growing population with shrinking and deteriorating land and water resources (Patode et al., 2021; Moharir et al., 2021). Therefore, a systematic survey of land resources and their mapping is essential for managing the resources in a sustainable way (Sarkar, 2011; Supriya et al., 2019).

Soil mapping is basically an inference process, where the soil is described as a function of climate, organisms, relief, parent material, and time, referred to as CLORPT (Jenny, 1941) and recognition of interactions between soil-forming factors is potentially important because it is one possible source of detailed soil patterns (McBratney et al., 2003). In small areas where climate, parent material, and time are almost similar, the major factors influencing the soil properties can be attributed to variation in relief and flora and fauna (Dobos et al., 2000; Srivastava and Saxena, 2004). Soil maps can be produced on different scales, such as 1:250,000, 1:1,000,000, or smaller, medium scales like 1:100,000, 1:50,000, and large scales like 1:25,000, 1:10,000, or larger, depending upon the purpose and requirement of user agencies (Srivastava and Saxena, 2004).

The application of satellite remote-sensing data products for small (country-level) and medium-scale (district-level) soil mapping is widely accepted (Soil Survey Division Staff, 2000). However, their utility is limited for large-scale soil mapping due to the coarse resolution of satellite data. Earlier large-scale soil mapping was mostly done with conventional methods. These were difficult, time-consuming, and expensive with low repetitive value especially, in hilly and mountainous regions, wetlands, and other problematic areas (Adam et al., 2010). With the advancement in terms of spatial, spectral, and radiometric resolutions of the sensors with stereo capabilities, studies have

DOI: 10.1201/9781003303237-1

been initiated to characterize soils at a large scale through the physiography-land-use soil relationship. Srivastava and Saxena (2004) discussed the technique of large-scale soil mapping (1:12,500 scale) in a basaltic terrain with a physiography-land-use (PLU) approach and differentiated soil types using topographic information available in the Survey of India toposheet and land-use/land cover information from IRS-1C PAN merged data (Pande et al., 2018; Pande et al., 2021). Nagaraju et al. (2014) also prepared a large-scale soil map (1:5,000 scale) in a basaltic terrain with a PLU approach using landform, slope, and land-use/land cover maps (Pande and Moharir, 2017).

Soil resource inventory (SRM) at 1:7 million, 1:5 million, 1:4 million, and 1:1 million enumerates major physiography, AER (agro-ecological region), AESR (agro-ecological sub-region), etc. in Tripura. But it provided information on climate such as temperature and rainfall and a few selected soil properties. The climate and soil data were used to estimate the length of the growing period in each region for selected crops (Sehgal et al., 1992; Velayutham et al., 1999). The entire state of Tripura has been mapped at a 1:50,000 scale with soil series association as soil mapping units (Bhattacharyya et al., 1996, 2002). Though antecedent data on SRM at a 1:50,000 scale provides physiographic units and soil information at a smaller scale, the same at a larger scale is lacking in Tripura, particularly in the Sepahijala district, for planning land-use systems at a larger scale or at a village level. Land resource inventory (LRI) on 1:10,000 scales provides adequate information on characteristics and spatial distribution soils and properties of soils that support sustainable land management, which, among others, includes erosion control, fertility management, crop choice, and possibilities for irrigation (van de Wauw et al., 2008; Seid et al., 2013; Singh et al., 2016). After the characterization of soil resources land evaluation is essential to know the suitability of a particular crop or a group of crops. For the evaluation of the capability and suitability of the soils for particular land use, detailed studies on specific soil-related constraints like soil fertility, available water content, degradation hazards, and soil erosion are necessary (Abdel Rahman et al., 2016; Fekadu et al., 2020; Mandal et al., 2020).

Further, this information is a prerequisite for developing a land-use plan at the block level. Land-use planning consisting of the right land use and the right technology in a site-specific mode may be one of the options that may help in meeting the demand for food as well as in preserving the quality of land for the future. LRI on 1:10,000 scales will help in developing such site-specific information, which paves the way for applying the right land-use, right technology at the right place. Hence, the present study is an attempt to supplement the information gap in the Nachar block of Sepahijala district in the Purvanchal range of Tripura, especially with an emphasis on characterization and mapping of soil resources at 1:10,000 scale, modeling soil physiographic relations, finding crop suitability, land-use options, and conservation of natural resources.

1.2 MATERIALS AND METHODS

1.2.1 Study Area

The area under investigation belongs to the Nalchar block of Sepahijala district (23°26′58″–23°35′40″N, 91°14′54″–91°30′23″E) covering an area of 193.84 km^2 (Figure 1.1) in Tripura, India. It is bounded by the Charilam block in the north, the Mohanbhog block in the south, the Boxanagar block in the west of the Sepahijala district, and the Gomati district of Tripura in the east. The area is characterized by a humid subtropical climate with an annual mean maximum temperature being 36°C and an annual mean minimum temperature of 7°C. Mean annual rainfall is 2,340 mm and about 85% of the rainfall is from the south-west monsoon. The difference between the mean summer and mean winter soil temperature is more than 6°C (Reza et al., 2019a, 2019b). The soil temperature class is 'hyperthermic.' The soil moisture regime is 'aquic' in the valleys/plains and 'ustic' in the hills/mounds (Soil Survey Staff, 2003). Geomorphologically, the area represents an undulating topography (Bhattacharyya et al., 1996, 2010). The cropping pattern in the block is characterized by two distinct farming systems, i.e., field crops are cultivated in the plains or valleys and rubber

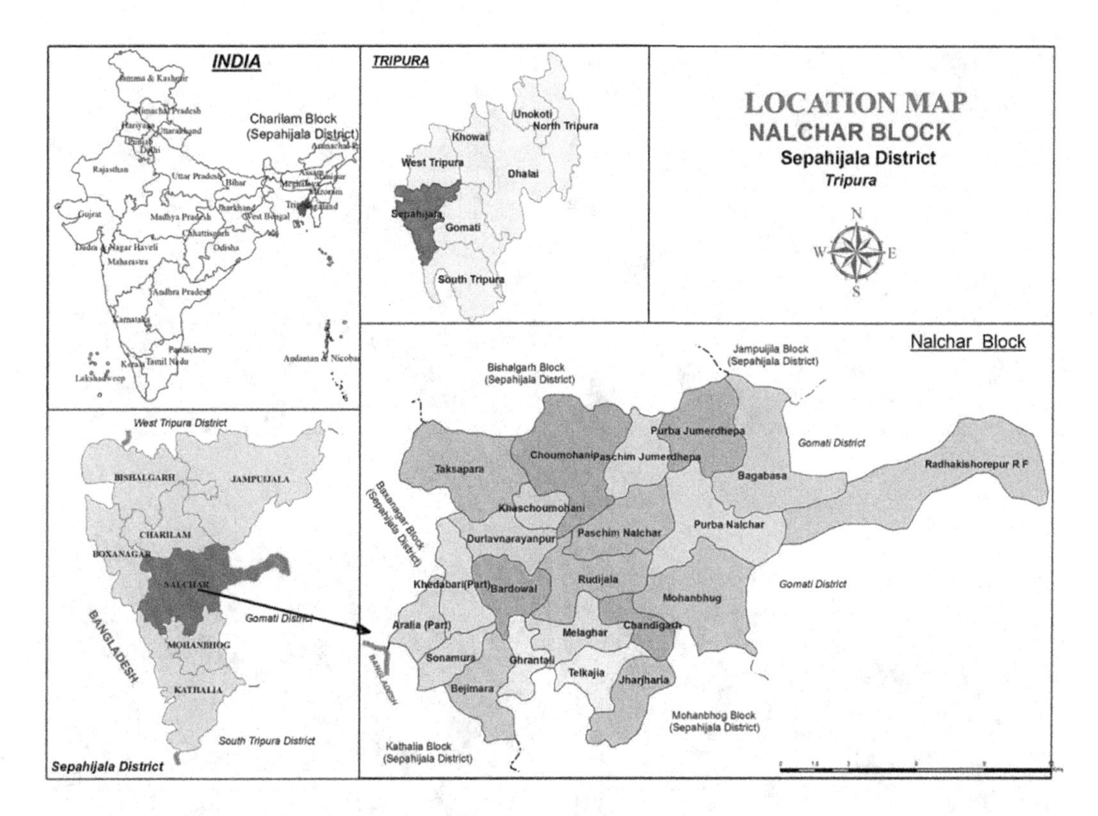

FIGURE 1.1 Location map of the study area.

plantations and pineapples in the hills and mounds. Paddy is the major crop grown in the block. Paddy is cultivated in three seasons viz. Aush (*pre-kharif*), Aman (*kharif*), and Boro (summer). The major *kharif* crops are rice, pigeon pea, black gram, green gram and cowpea, and *kharif* vegetables. Different crops during the *rabi* season are rice, peas, green gram, lentil, rapeseed-mustard, potato, and *rabi* vegetables (Figure 1.2). Natural vegetation comprises trees, shrubs, grasses, and weeds. Most tree species are moist mixed deciduous in nature. The major tree species are mango (*Mangifera indica*), jamun (*Syzigium cumini*), sal (*Shorea robusta*), teak (*Tectona grandis*), patches of acacia or wattle (*Acacia auriculiformis*), and bamboo (*Bambusa sp.*).

1.2.2 Geology

The Geology of the study area is represented by sedimentary rocks which range in age from Miocene to loosely consolidated sediments of recent age. The rocks are sandstone, siltstone, and shale grading into clay. These rock types are repeated as layers one above the other. Depending on their characters and the presence of fossils, these sedimentary rock sequences are divided into the Surma group (the oldest), the Tipam group, and the Dupitila group (the youngest). Geologically the hills consist of tertiary rocks formed by the force of the tectonic movement that caused the sub-stratum to rise up during the upliftment of the Himalayas during the Eocene period (GSI, 2011). Tipam and Dupitila are the two sedimentary rock formations found here lying one above the other. These formations consist of mainly clay and silt with some intercalations of gritty and ferruginous sandstones. The mounds consist of the Tipam and Surma groups having shale, siltstone, and mudstone with yellow to buff-colored fine-grained sandstone, while valleys and alluvial plains were developed from unstabilized alluviums carried out with the flow of water. Accordingly, the soils in the study area have developed on the above-mentioned geology under representative landforms.

FIGURE 1.2 Vegetable cultivation in the alluvial plains and valleys.

1.2.3 Preparation of Base Maps

Survey of India toposheets on a 1:50,000 scale, IRS-R2 LISS-IV data of January 6, 2018 (5.8 m resolution) were georeferenced using WGS 84 datum, Universal Transverse Mercator (UTM) projections, and ground control points (GCPs). After the geo-referencing process, the different band combinations of satellite data were used to generate a false color composite (FCC) for image interpretation and on screen mapping. The delineation of landforms was done using the SRTM (Shuttle Radar Topography Mission) DEM (Digital Elevation Model) of 30 m resolution and on screen image visual interpretation techniques in a GIS environment (ArcGIS ver. 10.5). Geomorphic features were interpreted based on key image elements such as shape, tone or color, pattern, shadow, association, and texture. The landform, slope, and land-use/land cover layers were integrated into ArcGIS and an LEU layer was prepared. These LEU units are relatively homogeneous in terms of the main factors of soil formation and typical predictors of soil characteristics and are used as a base map for ground-truth verification.

1.2.4 Ground-Truth Verification

The area was traversed for the identification of different landform units, slopes, and present land-use/land cover classes, and correlated with image interpretation units. The boundaries that were originally derived during the base map preparation were verified and corrected wherever necessary. To understand the soil variability in the study area, representative sites on each physiographic unit were selected, the exact location was determined using a handheld Global Positioning System (GPS), and profile observations were taken as per variation in phases and were described for site and soil characteristics such as depth, color (matrix and mottle), boundary, structure, texture, cutans, etc. following the guidelines for field soil descriptions (Soil Survey Staff, 1995).

1.2.5 Soil Sampling and Analysis

The soil samples collected during fieldwork were initially air-dried in the laboratory at room temperature, ground using a wooden pestle and mortar, screened through a 2 mm sieve, properly labeled, and stored in polythene bags for laboratory analysis. The soil samples were analyzed in the laboratory for physical and chemical parameters using standard procedures. The particle size analysis was done by the International pipette method. A combined glass-calomel electrode was used to determine the pH of aqueous suspensions (1:2.5 soil/solution ratio). Organic carbon (OC) was determined using the wet digestion method of Walkley and Black (1934). Available nitrogen (N) was measured by the alkaline permanganate method as described by Subbiah and Asija (1956). Available phosphorus (P) was determined by the Bray II method (Bray and Kurtz, 1945). The cation exchange capacity (CEC) of soil was determined as per the procedure outlined by Jackson (1976). Exchangeable cations [calcium (Ca), potassium (K), and magnesium (Mg)] were extracted with 1 M ammonium acetate (NH_4OAc) (pH 7.0). Potassium content was determined by flame photometry (Rich, 1965), while Ca and Mg were determined in ethylene diamine tetra acetic acid (EDTA) titration. Exchangeable Al was extracted with 1 N potassium chloride (KCl) solution and titrated with a 0.1 N sodium hydroxide (NaOH) solution. Soils were classified according to Keys to Soil Taxonomy (Soil Survey Staff, 2014).

1.2.6 Soil Classification

As per Soil Survey Staff (2014) the following criteria are used to classify the soils of the study area. Orders: Ultisols – the presence of argillic (Bt) horizon that showed clay alleviation with more than 1.2 times over surface horizon and base saturation of <35%, Alfisols – the presence of argillic (Bt) horizon that showed clay alleviation with more than 1.2 times over surface horizon and base saturation of >35%, Inceptisols – the presence of cambic (Bw) horizon and structural development in the subsurface horizon, and Entisols – soils that do not show any profile development and no diagnostic horizons, and most are basically unconsolidated sediments with little or no alteration from their parent materials.

Suborders: Udults–Ultisols that have an udic soil moisture regime, Aqualfs–Alfisols that have an aquic soil moisture regime, Udepts–Inceptisols that have an udic soil moisture regime, and Aquents–Entisols that have an aquic moisture regime.

Great groups: Hapludults–Udults, which doesn't meet the requirement of other great groups of the Udults suborder; Endoaqulfs–Aqualfs with endo-saturation; Dystrudepts–Udepts, which doesn't meet the requirement of other great groups of the Udepts suborder and Fluvaquents–Aquents with an irregular decrease in organic carbon content between a depth of 25 cm and a depth of 125 cm below the mineral soil surface.

Subgroups: Typic Hapludults– Hapludults, which doesn't meet the requirement of other subgroups of the Hapludults great group; Typic Endoaqulafs– Endoaqulafs, which doesn't meet the requirement of other subgroups of the Endoaqulafs great group; Typic Dystrudepts– Dystrudepts, which doesn't meet the requirement of other subgroups of the Dystrudepts great group and Typic Fluvaquents–Fluvaquents, which doesn't meet the requirement of other subgroups of the Fluvaquents great group.

1.2.7 Development of Soil Mapping Legend

The map unit considered in the present study is the phases of the soil series. The soil series is a group of soils or polypedons that have horizons similar in arrangement and in differentiating characteristics with a relatively narrow range in sets of properties (Soil Survey Division Staff, 2000).

The soil phases considered were soil depths, surface texture, slope, and erosion. The soil profiles described during the ground truth were correlated in each major landform and soil series were identified. The soil series information was extended to sub-units of major landforms using the diagnostic soil characteristics from the soil profile and augur observations. A soil map showing the soil series and their phases was prepared at a 1:10,000 scale. The soil legend code developed depicts the name of the series followed by surface texture, slope, and erosion (Singh et al., 2016).

1.2.8 Land Evaluation

Land capability classification (LCC) is carried out to find out the general capability of the resources of an area for agricultural, forestry, and other uses. In this classification, the mapping units are grouped according to their limitations to field crops and the way they respond to management into various capability units. The classification is based on the inherent soil characteristics, external land features, and environmental factors that limit the use of land (AIS & LUS, 1970). The characteristics used to group the land resources identified in the study area are: texture, slope, erosion, and drainage. In the capability system, mapping units are generally grouped at three levels – capability class, sub-class, and unit. Capability classes, the broadest group, are designed by Roman numerals I–VIII. The numerals indicate progressively greater limitations and narrow choices for practical use. The eight classes used in the system are defined as: Class I–Class IV are cultivable. They are capable of producing commonly cultivated crops of the region under good management. Classes V–VII are suited to adopted native plants, pasture, or forestry. Class VIII is neither suitable for agriculture nor for silvi-culture.

Capability sub-classes are formed based on the limitations observed within the capability classes. These are designed by adding a lowercase letter like e, s, w, or c to the class numeral. For example, in sub-class IVe, the letter 'e' shows that the main hazard in class IV land is the risk of erosion. Similarly, the symbol 'w' indicates drainage or wetness as a limitation for plant growth; the symbol 's' indicates root zone limitations, and 'c' indicates climate or rainfall with short growing periods. Soil site suitability has been evaluated by the maximum likelihood method (Sys et al., 1993; Naidu et al., 2006). The suitability criteria included climatic attributes (c) viz. rainfall and temperature; wetness aspect (w) viz. drainage and flooding; physical condition (s) viz. surface texture, rooting condition as soil depth (d), and soil fertility factor (f) viz. pH, OC, apparent CEC, base saturation, and sum of cations. Soils have been evaluated as highly suitable (S1), moderately suitable (S2), marginally suitable (S3), temporarily not suitable (N1) and permanently not suitable (N2) classes.

1.3 RESULTS AND DISCUSSION

1.3.1 Landform Delineation

Visual interpretation of LISS-IV data indicated that the block was characterized by hills, mounds, uplands, alluvial plains, and valleys. The major landforms were further subdivided based on elevation, land-uses, and other local features. Hills were subdivided into steeply sloping moderately dissected medium-relief hills & narrow valleys and moderately steeply sloping moderately dissected low-relief hills & narrow valleys, moderately steeply sloping low mounds interspersed with narrow valleys. Other landforms were moderately sloping undulating uplands with mounds, very gently sloping alluvial plains, and very gently sloping narrow valleys.

1.3.2 Land-Use/Land Cover

Based on the IRS-R2 LISS-IV data (January 6, 2018), six land-use/land-cover classes were identified. The land-use data indicates that about 73.20% TGA of the block is under agriculture, 5.0%

is under forest and open scrub, and 21.98% TGA is under miscellaneous viz., habitation, water body, and river system. The land-use/land-cover map indicated that double-cropped area is mainly practiced in the landform, which is dominant by depositional processes such as along the banks of Gomti and Noa Cherra rivers (Table 1.1). Single crops are practiced in alluvial plains and the narrow and broad valley whereas rubber plantation is dominant in the moderately dissected medium relief hill to low mounds in the study area.

1.3.3 Landform and Landscape Ecological Units (LEUs)

The landform, slope, and land-use/land-cover maps were integrated into ArcGIS and an LEU map was prepared. The land-use and landform map of the study area indicates that landforms influence the land-uses. Based on integration, 21 LEU units were delineated in the study area and the characteristics of each LEU unit were described (Table 1.1). On hills, 12 LEU units were identified based on two slope classes (10%–15% and 15%–25%) and four land-use/land-cover classes (plantation, single crop, double-crop, plantation/forest, and forest). Three LEU units were identified on the upland with one slope classes (5%–10%) and three land-use/land-cover classes (single crop, plantation, and plantation/forest). Five LEU units were identified with five land-use/land cover classes (single crop, double-crop, plantation, plantation/forest, and open scrub) in the alluvial plain and one LEU unit in the valley.

1.3.4 Soil-Landform Relationship

Soils of hills and mounds were deep to very deep and had two genetic horizons A and B with clear smooth boundaries on the surface and gradual smooth boundaries on the subsurface horizons (Table 1.2). The pedons showed a difference in surface and subsurface matrix color. The surface horizon colors are brown (7.5YR 4/2) (Jharjharia series) and dark yellowish brown (10YR 4/4) (Radhakishorepur RF and Bagabasa series) in hills, and very dark greyish (10YR 3/2) (Khaschowmuhani series) in mounds. Whereas, the subsurface color for hills and mounds varied from brown (7.5YR 4/3) to strongly brown (7.5YR 5/8) and red (2.5YR 4/8) to dark reddish grey (2.5YR 4/1), respectively. The darker color of surface horizons indicates rich in organic matter (OM) whereas subsurface horizons low in OM showed lighter color with an increasing depth of soils. The results showed the matrix color is highly influenced by OM. Similar observations were also reported by Mulugeta and Sheleme (2010) and Dessalegn et al. (2014). The soils of hills were slightly restricted in development due to erosion by a surficial downward flow of water. This leads to the removal of fine soil particles from the upper slope and their deposition on the lower slope (Deka et al., 2009; Reza et al., 2011, 2014a, 2018a). Many researchers (Moore et al., 1993; Dessalegn et al., 2014) also reported the transport of fine and medium size particles was higher than coarse-size particles (sand) with the movement of water. Hence, the subsurface horizons of most of the profiles were finer than their respective surface horizons.

Soils of alluvial plains and valleys are deep to very deep and have an A–B horizon except for the Sonamura series of alluvial plains (Table 1.2) with a clear smooth and gradual smooth boundary in surface and subsurface horizons, respectively indicating that these soils were developed under a fluvial process. The soils of alluvial plains showed differences in surface and subsurface matrix color. The surface horizons are brown (10YR 5/3) (Bardowal series) and dark greyish brown (10YR 4/2) (Sonamura series) with subsurface color of brown (10YR 4/3) to dark yellowish brown (10YR 4/4) and greenish grey (GLEY1 5/5GY) to dark greyish grey (GLEY1 4/10GY), respectively. The soils of valleys (Choumohani series) are dark greyish brown (10YR 4/2) (surface) and grey (10YR 5/1) (subsurface) color with red mottles. The Sonamura series of alluvial plains and the Choumohani series of valleys showed a grey matrix, which indicates that these soils were under prolonged submergence and subsequently developed under reducing conditions during flooding. The low chroma of soils judged the severity of gleying due to poor drainage conditions in a high ground water table

TABLE 1.1
Landscape Ecological Units (LEU) of Nalchar Block

Landscape Ecological Unit (LEU)	Area (ha)	TGA (%)
Hills		
Moderately steep sloping moderately dissected medium relief hills and narrow valleys under single crop (NcMH^{Md}Nv6s)	13	0.07
Moderately steep sloping moderately dissected medium relief hills and narrow valleys under plantation (NcLH^{Md}Nv6p)	59	0.30
Moderately steep sloping moderately dissected medium relief hills and narrow valleys under plantation with scrubs (NcLH^{Md}Nv6p/s)	465	2.40
Moderately steep sloping moderately dissected low-relief hills and narrow valleys under double crop (NcLH^{Md}Nv5d)	79	0.41
Moderately steep sloping moderately dissected low-relief hills and narrow valleys under single crop (NcLH^{Md}Nv5s)	253	1.31
Moderately steep sloping moderately dissected low-relief hills and narrow valleys under plantation (NcLH^{Md}Nv5p)	3,122	16.11
Moderately steep sloping moderately dissected low-relief hills and narrow valleys under plantation/forest/scrubs (NcLH^{Md}Nv5p/f/s)	2,079	10.73
Moderately steep sloping moderately dissected low-relief hills and narrow valleys under forest (NcLH^{Md}Nv5f)	794	4.10
Moderately steeply sloping low mounds interspersed with narrow valleys under single crop (NcL^{mo}Nv5s)	169	0.87
Moderately steeply sloping low mounds interspersed with narrow valleys under plantation (NcL^{mo}Nv5p)	2,694	13.90
Moderately steeply sloping low mounds interspersed with narrow valleys under plantation/forest/scrubs (NcL^{mo}Nv5p/f/s)	64	0.33
Moderately steeply sloping low mounds interspersed with narrow valleys under forest (NcL^{mo}Nv5f)	147	0.76
Mounds		
Moderately sloping undulating upland with mounds under single crops (NcUu^{mo}4s)	256	1.32
Moderately sloping undulating upland with mounds under plantation (NcUu^{mo}4p)	1,461	7.54
Moderately sloping undulating upland with mounds under plantation/forest (NcUu^{mo}4p/f)	29	0.15
Alluvial Plains		
Very gently sloping lower alluvial plain under double crops (NcAl3d)	657	3.39
Very gently sloping lower alluvial plain under single crop (NcAl3s)	2,063	10.64
Very gently sloping lower alluvial plain under plantation (NcAl2p)	19	0.10
Very gently sloping lower alluvial plain under plantation/forest (NcAl3p/f)	5	0.02
Very gently sloping lower alluvial plain under open scrub (NcAl3os)	23	0.12
Valley		
Gently sloping narrow valleys under single crop (NcNv2s)	673	3.47
Miscellaneous (habitation/river/water body)	4,260	21.98
Total area	19,384	100.00

TABLE 1.2
Morphological Properties of Soils

Depth (cm)	Horizon	Boundary	Matrix color (moist)	Texture	Structure	Cutans	Roots
			Hills				
Jharjharia: Coarse-Loamy, Mixed, Hyperthermic Typic Dystrudepts							
0–18	A	cs	7.5YR 4/2	Sandy clay	m2sbk	–	cm
18–56	Bw1	gs	7.5YR 4/3	Loam	m2sbk	–	cm
56–93	Bw2	gs	7.5YR 4/4	Sandy loam	f1sbk	–	ff
93–122	BC1	gs	7.5YR 4/6	Sandy clay loam	m2sbk	–	fvf
122–150	BC2	–	7.5YR 5/8	Sandy loam	f1sbk	–	–
Radhakishorepur RF Series: Fine, Mixed, Hyperthermic Typic Dystrudepts							
0–30	A	cs	10YR 4/4	Sandy clay	m2sbk	–	cm
30–50	Bw1	gs	7.5RY 4/6	Sandy clay	m2sbk	–	cm
50–88	Bw2	gs	7.5RY 5/6	Sandy clay loam	f1sbk	–	cm
88–130	Bw3	gs	7.5RY 4/6	Sandy clay loam	f1sbk	–	ff
130–150	Bw4	–	7.5RY 5/8	Sandy loam	f1sbk	–	–
Bagabasa Series: Fine-Loamy, Mixed, Hyperthermic Typic Hapludults							
0–14	A	cs	10YR 4/4	Loam	m2sbk	–	cm
14–36	Bt1	gs	7.5YR 4/6	Clay loam	m2sbk	T tn p	cm
36–80	Bt2	gs	7.5YR 4/4	Loam	f1sbk	T m c	ff
80–130	Bt3	gs	7.5YR 5/6	Loam	f1sbk	T m c	ff
130–150	Bt4	–	7.5YR 4/6	Sandy loam	f1sbk	–	–
			Mounds				
Khaschowmuhani Series: Very-Fine, Mixed, Hyperthermic Typic Hapludults							
0–17	A	cs	10YR 3/2	Clay	m2sbk		cm
17–46	Bt1	gs	2.5YR 4/6	Clay	m2sbk	T tk c	cm
46–74	Bt2	gs	2.5YR 4/1	Clay	m2sbk	T tn p	ff
74–105	Bt3	gs	2.5YR 4/8	Clay	m2sbk	T m c	ff
105–150	Bt4	–	2.5YR 4/6	Clay	m2sbk	T m c	–
			Alluvial Plains				
Bardowal Series: Fine-Loamy, Mixed, Hyperthermic Typic Dystrudepts							
0–16	Ap	cs	10YR 5/3	Clay loam	m2sbk	–	cm
16–33	Bw1	gs	10YR 4/4	Clay loam	m2sbk	–	ff
33–55	Bw2	gs	10YR 4/4	Sandy clay loam	f1sbk	–	fvf
55–78	Bw3	cs	10YR 4/3	sandy clay	f1sbk	–	–
78+			Sand				
Sonamura Series: Fine, Mixed, Hyperthermic Family of Typic Fluvaquents							
0–12	Ap	cs	10YR 4/2	Clay	m2sbk	–	cf
12–38	AC1	gs	GLEY1 5/5GY	Sandy clay loam	m2sbk	–	ff
38–59	AC2	gs	GLEY1 4/5GY	Sandy clay loam	m2sbk	–	–
59–93	2AC1	–	GLEY1 4/10GY	Clay	f1sbk	–	–
93+			Ground water table				

(Continued)

TABLE 1.2 (*Continued*)
Morphological Properties of Soils

Depth (cm)	Horizon	Boundary	Matrix color (moist)	Texture	Structure	Cutans	Roots
			Valleys				
Choumohani Series: Fine, Mixed, Hyperthermic Typic Endoaqualfs							
0–15	Ap	cs	10YR 4/2	Clay	m2sbk	–	ff
15–36	Bt1	gs	10YR 5/1	Clay	m2sbk	–	ff
36–70	Bt2	–	10YR 4/1	Clay	m2sbk	–	–
70+			Ground water table				

Boundary: cs, clear smooth; gs, gradual smooth. Structure: f1 sbk, fine weak subangular blocky; m2 sbk, medium moderate subangular blocky. Cutans: T tn p, argillan thin patchy; T m c, argillan moderate continuous; T tk c, argillan thick continuous. Roots: f vf, few very fine; f f, few fine; c m, common medium; m vf, many very fine; f c, few coarse.

in a lower topographical position (Stoop and Eswaran, 1985). When these soils become dry, the reduced iron (Fe^{3+}) is oxidized and precipitates by releasing H^+ ions to acidify and disintegrate the clay. Under saturated conditions for a long time, these soils developed distinctive gley horizons resulting from the oxidation and reduction process and have iron and manganese mottles or streaks in the B horizons due to a slow diffusion process (Ponnamperuma, 1972, 1985). These soils are also known as hydromorphic soils and gleization is the major pedogenic process operating for their development (Khan et al., 2012). The textural variations in alluvial soils are mainly due to the deposition of alluvial brought by water from different landforms. However, the soils of the valleys were clay in texture.

The variation of soil properties with depths indicates the dominant soil processes operating over the course of profile development. In the initial stage, the OM input and mineral weathering occurs in weakly developed soils (Entisols and Inceptisols). With more time, the continued weathering of primary minerals, and synthesis of secondary clay minerals lead to developed matured soil (Ultisols) (Bhattacharyya et al., 2007; Lindeburg et al., 2013). Hence, the depth-wise variations of important soil properties namely; OC, clay content, and CEC in the identified soil series are presented in Table 1.3. Soils of the Jharjharia series (*Coarse-loamy, mixed, hyperthermic Typic Dystrudepts*), the Radhakishorepur RF series (*Fine, mixed, hyperthermic Typic Dystrudepts*) and the Bagabasa series (*Fine-loamy, mixed, hyperthermic Typic Hapludults*) in hills, the Khaschowmuhani series (*Very-fine, mixed, hyperthermic Typic Hapludults*) in mounds showed variation of clay content with depths indicating the vertical movement of clay from the surface to subsurface horizons. This showed that the pedogenic processes of eluviation and illuviation are prevalent in these landforms (Mulugeta and Sheleme, 2010). Clay cutans were also observed in the soils of the mounds indicating clay migration (Buol et al., 2003). Besides this, the in-situ synthesis of secondary clays and the weathering of primary minerals in the B-horizon could have also contributed to the accumulation of clay in the subsurface horizons (Chadwick and Grahm, 2000; Buol et al., 2003). The content of exch. cations (Ca, Mg, Na, and K) increased with increasing depth, which could be attributed to their leaching from the surface to the subsurface horizon and increased the CEC with depth as the exchange complex of the soils was dominated by Ca followed by Mg, Na, and K (Table 1.3). This result is in agreement with that of Gangopadhyay et al. (2001) who showed similar results in rubber growing soils of Tripura. Therefore, the variability in soil properties under hills and mounds may be attributed to the existing pedogenic processes of eluviation and illuviation due to the downward movement of water in the study area (Hall, 1983). The Bardowal series (*Fine-loamy, mixed, hyperthermic Typic Dystrudepts*) in alluvial plains and the Choumohani series (*Fine, mixed,*

TABLE 1.3
Physical and Chemical Properties of Soils

Depth (cm)	Horizon	Sand (%)	Silt (%)	Clay (%)	pH H_2O (1:2.5)	OC (%)	Exchangeable cations Ca	Mg	Na	K	CEC
							cmol(p+)kg^{-1}				
Hills											
Jharjharia: Coarse-Loamy, Mixed, Hyperthermic Typic Dystrudepts											
0–18	Ap	45.3	7.5	47.2	5.3	0.73	3.57	0.87	0.06	0.03	6.1
18–56	Bw1	51.8	35.7	12.5	5.3	0.68	2.9	0.43	0.06	0.03	7.6
56–93	Bw2	59.5	30.7	9.8	5.2	0.52	2.68	0.65	0.06	0.03	4.5
93–122	BC1	72.5	4.6	23.0	5.4	0.40	3.12	0.95	0.06	0.02	6.1
122–150	BC2	79.2	3.7	17.1	5.7	0.15	4.24	0.39	0.28	0.06	8.6
Radhakishorepur RF series: Fine, mixed, hyperthermic Typic Dystrudepts											
0–30	A	52.9	2.9	44.2	4.7	0.77	0.89	0.40	0.06	0.05	3.4
30–50	Bw1	52.2	2.3	45.6	4.7	0.70	0.89	0.40	0.05	0.04	5.3
50–88	Bw2	54.2	10.6	35.3	5.0	0.48	1.12	0.36	0.06	0.03	3.2
88–130	Bw3	57.9	18.8	23.3	5.0	0.45	1.34	0.51	0.05	0.02	4.5
130–150	Bw4	59.1	24.0	17.0	5.4	0.20	1.12	0.36	0.05	0.02	2.5
Bagabasa Series: Fine-Loamy, Mixed, Hyperthermic Typic Hapludults											
0–14	A	42.6	38.2	19.3	5.0	1.06	0.59	0.50	0.05	0.04	4.3
14–36	Bt1	40.1	22.3	37.6	4.9	0.86	0.61	0.41	0.06	0.04	4.5
36–80	Bt2	46.5	22.9	42.6	5.2	0.50	0.62	0.29	0.06	0.03	4.3
80–130	Bt3	48.0	42.9	9.1	5.3	0.27	0.32	0.24	0.05	0.03	3.8
130–150	Bt4	54.6	33.9	11.5	5.5	0.09	0.42	0.44	0.07	0.03	4.0
Mounds											
Khaschowmuhani Series: Very-Fine, Mixed, Hyperthermic Typic Hapludults											
0–17	A	35.9	17.5	46.7	4.4	1.02	0.89	0.59	0.06	0.06	6.2
17–46	Bt1	28.8	3.9	67.4	4.8	0.50	0.67	0.44	0.06	0.04	4.5
46–74	Bt2	29.7	9.6	60.7	4.9	0.30	0.89	0.59	0.06	0.03	4.9
74–105	Bt3	32.5	7.6	59.9	4.9	0.30	0.67	0.44	0.05	0.03	4.8
105–150	Bt4	33.7	12.3	54.0	4.8	0.18	0.67	0.44	0.07	0.03	4.0
Alluvial plains											
Bardowal Series: Fine-Loamy, Mixed, hyperthermic Typic Dystrudepts											
0–16	Ap	29.9	34.9	35.3	4.6	0.45	5.58	2.38	0.22	0.04	9.7
16–33	Bw1	29.3	35.8	35.0	4.9	0.42	4.92	1.00	0.24	0.03	6.9
33–55	Bw2	60.0	8.0	32.0	5.3	0.17	4.02	2.08	0.25	0.03	6.9
55–78	Bw3	57.4	3.9	38.8	5.7	0.06	3.80	1.75	0.24	0.02	8.1
Sonamura Series: Fine, Mixed, Hyperthermic Family of Typic Fluvaquents											
0–12	Ap	31.4	20.4	48.2	4.9	0.71	3.08	1.28	0.08	0.04	7.6
12–38	AC1	53.0	15.1	31.9	6.4	0.21	3.52	1.83	0.06	0.02	4.8
38–59	AC2	64.6	7.8	27.6	6.5	0.18	2.64	1.72	0.06	0.02	4.3
59–93	2AC1	30.1	26.0	43.9	6.5	0.27	4.18	2.55	0.081	0.03	6.9
93+		Ground water table									

(Continued)

TABLE 1.3 (*Continued*)
Physical and Chemical Properties of Soils

Depth (cm)	Horizon	Sand (%)	Silt (%)	Clay (%)	pH H_2O (1:2.5)	OC (%)	Exchangeable cations Ca	Mg	Na	K	CEC
							cmol(p+)kg^{-1}				
Valleys											
Choumohani Series: Fine, Mixed, Hyperthermic Typic Endoaqualfs											
0–15	Ap	22.0	49.0	29.0	6.0	0.67	4.9	1.0	0.3	0.3	8.5
15–36	Bw1	16.1	55.2	28.7	7.0	0.30	4.7	1.1	0.2	0.3	7.4
36–70	Bw2	10.5	58.9	30.6	7.1	0.28	4.6	1.5	0.1	0.2	7.4

hyperthermic Typic Endoaqualfs) in valleys show systematic variation with depth which may be due to the presence of uniform parent materials from where soil profiles developed and reflect pedogenesis (Table 1.3). However, the Sonamura series (*Fine, mixed, hyperthermic Typic Fluvaquents*) in alluvial plains showed an irregular distribution of OC, clay content, and CEC (Table 1.3). Such irregular distribution could be attributed to the pedogenic processes namely, mass movement, periodic flooding, and deposition of alluvium brought down by water during different fluvial cycles (Huggett, 1975, 1976).

1.3.5 Soil Mapping

Soil is an open system and its properties are related to the functions operating in the system (Jenny, 1941). With changes in the system, the soil properties change and directly depend on soil formation factors expressed as follows:

$$S = (cl,\ o,\ r,\ p,\ t,\ \ldots)$$

where S denotes soil property; cl, climate (rainfall and temperature); o, organisms (flora and fauna); r, relief; p, parent material; and t, time or age.

The present study area is almost similar in climate, parent material, and time or age, the soil properties vary mainly on variation in relief or topography (r) and flora and fauna biosphere organisms (o). Hence, the LEU concept was used in the study area for mapping soils. The morphological characteristics observed during the soil survey and with analyzed soil properties, the soils were classified up to family level as per the Keys to Soil Taxonomy (Soil Survey Staff, 2014). Seven soil series have been identified and mapped on a 1:10,000 scale with ten soil mapping units (phases of series). A brief description of the soil series identified along with their taxonomic classification is given in the mapping legend (Table 1.4).

1.3.6 Soil Survey Interpretation

Soil maps and other interpretative maps are the ultimate products of a soil survey. They provide valuable information on various aspects like physiography/landform, geology, vegetation, soils, drainage, etc., and are useful to planners, administrators, and other user agencies. Land-use/agricultural planning of any particular area is largely based on soil resource interpretations (site characteristics and soil properties).

TABLE 1.4
Soil Series and Phases of Nalchar Block

Landform	LEU Map Unit	Soil Series	Soil Map Unit	Mapping Legend	Brief Description of Soil Series	Area (ha)	TGA (%)
Steep sloping moderately dissected medium relief hills & narrow valleys ($NcMH^{Md}Nv6$)	$NcMH^{Md}Nv6s$ $NcMH^{Md}Nv6p$ $NcMH^{Md}Nv6p/f/s$	Jharjharia	1	Jha6eH3	Very deep, well-drained, brown to strong brown, loamy sand to sandy clay soils on strongly sloping moderately dissected medium relief hills with narrow valleys having sandy clay surfaces and severe erosion (*Coarse-loamy, mixed, hyperthermic Typic Dystrudepts*)	537	2.77
Moderately steeply sloping moderately dissected low-relief hills & narrow valleys ($NcLH^{Md}Nv5$)	$NcLH^{Md}Nv5d$	Radhakishorepur RF	2	Rad6cE2	Very deep, well-drained, dark yellowish brown to strong brown, sandy loam to sandy clay soils on moderately steeply sloping moderately dissected low-relief hills with narrow valleys having sandy clay surfaces and moderate erosion (*Fine, mixed, hyperthermic Typic Dystrudepts*)	794	4.10
	$NcLH^{Md}Nv5p$		3	Rad6cE3	Same as Radhakishorepur RF with severe erosion	2,079	10.73
	$NcLH^{Md}Nv5s$	Bagabasa	4	Bag6hE2	Very deep, well-drained, dark yellowish brown to brown, sandy loam to clay loam soils on moderately steeply sloping moderately dissected low-relief hills with narrow valleys having loam surface and moderate erosion (*Fine-loamy, mixed, hyperthermic Typic Hapludults*)	3,122	16.11
	$NcLH^{Md}Nv5p/f/s$ $NcLH^{Md}Nv5f$		5	Bag6kE2	Same as Bagabasa with a silty clay loam surface	332	1.71

(Continued)

TABLE 1.4 (*Continued*)
Soil Series and Phases of Nalchar Block

Landform	LEU Map Unit	Soil Series	Soil Map Unit	Mapping Legend	Brief Description of Soil Series	Area (ha)	TGA (%)
Moderately steeply sloping low mounds interspersed with narrow valleys ($NcL^{mo}Nv5$) and moderately sloping undulating upland with mounds ($NcUu^{mo}4$)	NcLmoNv5s NcLmoNv5p NcLmoNv5p/f/s/t NcLmoNv5f NcUumo4s NcUumo4p	Khaschowmuhani	6	Kha6hD3	Very deep, well-drained, very dark greyish brown to red, clay soils on moderately sloping low mounds interspersed with narrow valleys and moderately sloping undulating upland with mounds having clay surfaces texture and severe erosion (*Very-fine, mixed, hyperthermic Typic Hapludults*)	4,564	23.55
	NcUumo4p/f/s/t		7	Kha6hE2	Same as Khaschowmuhani with moderate erosion	256	1.32
Very gently sloping lower alluvial plain (NcAl2)	NcAl3s NcAl3p	Bardowal	8	Bar6cB2	Very deep, moderately well-drained, yellowish brown to dark yellowish brown, sandy clay loam to clay loam soils on gently sloping lower alluvial plain having clay loam surface and moderate erosion (*Fine-loamy, mixed, hyperthermic Typic Dystrudepts*)	7,040	3.63
	NcAl3d	Sonamura	9	Son6eB1	Very deep, somewhat poorly drained, dark greyish brown to dark greenish grey, loamy sand clay loam to clay soils on gently sloping lower alluvial plain having clay surfaces and slight erosion (*Fine, mixed, hyperthermic Typic Fluvaquents*)	2,063	10.64
Very gently sloping narrow valleys (NcNv2)	NcNv2s	Choumohani	10	Cho6eB1	Very deep, somewhat poorly drained, grey to dark grey, clay soils on very gently sloping narrow valleys having clay surfaces and moderate erosion (*Fine, mixed, hyperthermic Typic Endoaqualfs*)	673	3.47
Miscellaneous (habitation/river/water body)						4,260	21.98
Total area						19,384	100.00

Following the criteria outlined in the Field Manual (Sehgal et al., 1987) and *Hand Book of Agriculture* (Takkar, 2009), various thematic maps such as surface texture, slope, drainage, soil reaction (pH), OC, and available N, P, and K have been prepared. The site characteristics and the soil properties of the surface soils of each of the soil phases have been considered for the preparation of different thematic maps.

1.3.7 Surface Texture

Soil texture, the relative proportion of sand, silt, and clay, is one of the most important physical soil properties that govern nearly all of the other attributes of soils (Zhai et al., 2006; Adhikari et al., 2009). Considering the effects of the soil texture on soil-water retention, its availability (Katerji and Mastrorilli, 2009; Reza et al., 2016), leaching and erosion potential (Adhikari et al., 2009), plant nutrient storage (Kettler et al., 2001), organic matter dynamics (Kong et al., 2009), and energy balance of the soil-plant system and pedogenesis (Western et al., 2003), plays a key role in total behavior of soil. Soils of the block have been grouped into four textural classes. Soils are dominantly clay (38.98% TGA) followed by loam (17.82% TGA), sandy clay (17.59% TGA), and clay loam (3.63% TGA).

1.3.8 Drainage

Soil drainage, landscape position, and the depth of the ground water table directly influence the internal drainage of the soils. The soils of the block have been grouped into three soil drainage classes. Well-drained soils occupied 60.28% TGA of the block followed by somewhat poorly drained (14.11% TGA), and moderately well-drained (3.63% TGA), respectively.

1.3.9 Soil Reaction (pH)

Soil reaction (pH) is a measure of the intensity of soil acidity or alkalinity. It acts as an indicator to assess the availability of different plant nutrients and also the percentage base saturation (Black, 1968). The pH value also helps to determine the amount of various amendments to be added to the soils for acidity or alkalinity. Soils of the block have been grouped into three soil reaction classes. It was observed that very strongly acidic soils occupy 50.39% TGA followed by extremely acidic (24.87% TGA) and strongly acidic (2.77% TGA). In the study area, the humid tropical climate poses the problem of soil acidity which is the major limiting factor for low productivity potential. The conditions conducive to the formation of acid soils are high rainfall, high temperature, and hilly topography. The rapid weathering and intense leaching under high rainfall conditions favor the development of soil acidity. This is also possibly due to the high content of organic matter, clay, and free iron oxide (Reza et al., 2018b).

1.3.10 Organic Carbon

Organic matter serves as a reservoir of soil nutrients that are essential for plant growth and is, therefore, considered as the vital and essential soil attribute controlling productivity. Soils of the block have been grouped into three organic carbon classes. Organic carbon status in soils was low to high. Data indicated that the high (>0.75%) to medium (0.5%–0.75%) level of organic carbon occupies 74.39% TGA and only 3.63% TGA is low (<0.5%) in organic carbon. High OC in the hills and mounds is due to vegetation cover (rubber plantation, plantation/forest, and forest) and is most likely caused by the balance between organic matter input and decomposition over time. It is reported that during the mature phase, an amount of 5–7 Mg/ha of plant residue is added annually to the soil floor which on subsequent decomposition accumulates a considerable amount of C (Hutchinson et al., 2007). The low values of organic carbon in the soils of alluvial plains and valleys may be due to the active fluvial process operating in that area (Reza et al., 2020a).

1.3.11 Available Nitrogen

Soil nitrogen (N) is an important macronutrient and plays an important role in crop growth and development. The presence of N in the soil also governs the yield of the crop and its yield attributes. However, its adverse effects on crop production and productivity may also be observed due to imbalance use in soil (Reza et al., 2017, 2019a, 2019c). In the study area soils were grouped into two classes. It was observed that 59.57% of TGA are medium in N, whereas 18.45% of TGA are in the low category.

1.3.12 Available Phosphorous

Phosphorus (P) is one of the three major nutrients and plays a vital role in the life cycle of plants, right from the stimulation of root growth to proper seed filling and seed setting, in addition to being an indispensable constituent of genetic material (Khasawneh et al., 1986). It also plays a role in photosynthesis, the breakdown of carbohydrates, and the transfer of energy through ATP and ADP compounds in various metabolic transformations. The available phosphorous content (kg $P_2O_5ha^{-1}$) of surface soils of the block has been grouped into two classes viz. low and medium. It was observed that 60.28% TGA of the block is medium in phosphorous, whereas 17.75% of TGA comes under the low category. There is extensive literature on the variation in soil P concentrations (Reza et al., 2012), and nearly all researchers, reported that P is among the most variable soil nutrient. Dobermann et al. (1995) suggested that P concentrations are variable because P is less mobile in soil than nearly all other solutes, tending to concentrate in patches and resist homogenization in water flow across the landscape.

1.4 AVAILABLE POTASSIUM

The importance of potassium (K) is well recognized in agriculture (Krauss and Johnson, 2002) and it is an essential nutrient for plant growth. Exchangeable K i.e., available K is widely used to evaluate the soil K status and to predict the crop K requirements (Askegaard and Jørgen, 2002; Reza et al., 2014b, 2014c). K content of soils was grouped into medium and high and is depicted. It was observed that the whole area of the block has a low content of available potassium.

1.5 LAND CAPABILITY CLASSIFICATION

Land capability classification is an interpretative grouping made primarily for broad agricultural and non-agricultural uses. The United States Department of Agriculture (USDA) has placed the arable lands into I–IV classes according to their limitations, grazing, and forestry, into class V–VII and class VIII lands for recreation having maximum limitations, wildlife, and quarrying. The capability classes were further sub-divided into sub-classes based on dominant limiting factors, such as erosion (e), soil (s), climate (c), and wetness (w). It was observed that the soils of the study area were divided into three land capability classes viz. II, III, and IV. The major limiting factors in the block are topography, soil, and erosion. Six land capability sub-classes were recognized viz. IIs2w2 (17.75% TGA), IIIs3t1 (1.32% TGA), IVe2s2t1 (23.55% TGA), IVe3s2t2 (12.44% TGA), IVe3s3t2 (20.20% TGA), and IVe3s3t3 (2.77% TGA) in the study area.

1.6 SOIL SUITABILITY FOR CROPS

Soil and climatic conditions play a vital role in optimal crop growth. The physico-chemical characteristics and micro-environments of soils were largely influenced by water and plant nutrient availability.

As such, soil depth, subsoil texture, fertility and drainage conditions, etc. are taken into account for soil site evaluation, so that soil maps can be interpreted in terms of suitability for agricultural crops for better socio-economic upliftment. The results showed that soils of the study area were moderately suitable for paddy (7.10% TGA), maize (7.10% TGA), mustard (17.75% TGA), rabi vegetables 17.75(% TGA), and rabi pulses (17.75% TGA) due to topography from steeply sloping to moderately sloping and non-availability of moisture during growing period whereas, among plantation crops rubber (57.50% TGA), tea (35.20% TGA), and citrus (57.51%) were moderately suitable in moderately sloping (5%–10% slope) to steeply sloping (15%–25% slope) with high KCl-Al. A similar result was also reported by Bhattacharyya et al. (2010) in the Tripura state, plantation and horticultural crops are suitable for those soils where KCl-Al is very high. However, 20.51%, 14.27%, and 20.51% of areas of TGA were marginally suitable for rubber, tea, and citrus due to fertility and drainage condition limitations (poor drainage and reduced matrix color). Furthermore, the suitability evaluation of tea showed that 28.54% TGA of the block was highly suitable due to favorable climate, soil depth, and no rockiness and gravel content in the soil.

1.7 RECOMMENDATION

The low productivity of agricultural crops in the block is the combined effect of problems of soil, water, and climate. Major soil problems are soil acidity, soil erosion, light texture soil, and low fertility status (Reza et al., 2018b, 2019b, and 2019d). Based on the above-mentioned characteristics four land management units (LMUs) were identified and mapped after merging seven soil series (Figure 1.3) in the study area. Hence, after carefully merging the soil series with a similar range of soil characteristics like soil texture, soil pH, internal soil drainage conditions, and status of fertility and erosion the LMUs were mapped (Ghosh et al., 2018; Reza et al., 2020b). In each LMU the present and alternate land-use option was proposed in Table 1.5 for the study area. The adaptation of an LRI-based land-use plan will help the farmers to increase productivity and profitability as compared to the traditional-based land-use system.

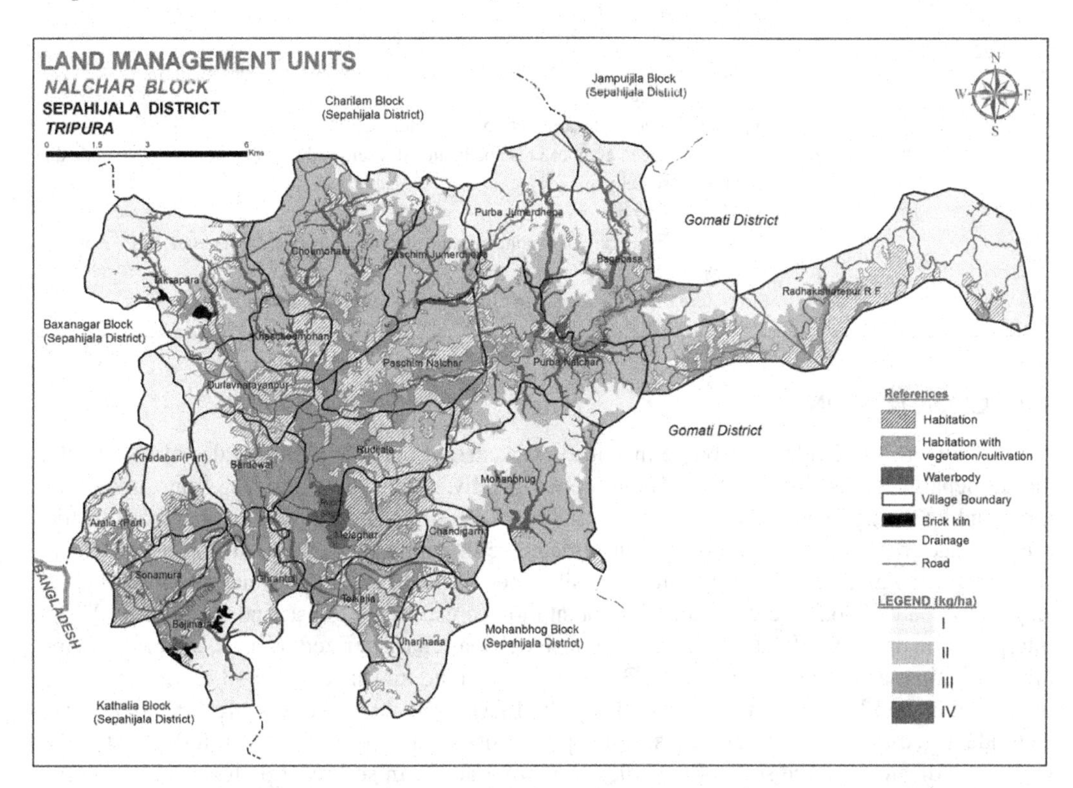

FIGURE 1.3 Land management units map.

TABLE 1.5
Present and Suggested Land-Use of the Block

LMU	Present Land-Use	Suggested Land-Use Options

I	Rubber plantation/ mixed (rubber plantation and forest)	• Rubber plantation with the application of a recommended dose of NPK fertilizer in two equal splits.
		• Inter-cropping with pineapple, ginger, and turmeric for the first 3 years.
		• Managements:
		  • Amelioration of soil acidity with the application of 200–250 kg lime/ha in furrows.
		  • Use of biofertilizers particularly phosphate solubilizing micro-organisms to improve the availability of phosphorus.

II	Rubber plantation	• Rubber plantation with the application of a recommended dose of NPK fertilizer in two equal splits.
		• Managements:
		  • Amelioration of soil acidity with the application of 200–250 kg lime/ha in furrows.
		  • Use of biofertilizers particularly phosphate solubilizing micro-organisms to improve the availability of phosphorus.
		  • Use of manures to reduce the adverse effect of soil acidity (particularly high Al) and increase soil fertility.
		  • Use of specific management practices like mulching, ridge and furrow systems, etc. which can help in conserving soil moisture.

III	Paddy – mustard/ pea/lentil/winter vegetables/ groundnut/paddy/ fallow	• No change, prefer medium duration HYV paddy varieties.
		• Managements:
		  • Timely land preparation, sowing & transplanting.
		  • Rain water harvesting by 30 cm high bunding.
		  • Utilization of waters for irrigation from nearby beels, ponds, rivers, natural depressions, etc.
		  • SRI technology should be properly adopted.
		  • Timely weeding, at critical growth stages and short-duration drought tolerant crops should be grown.

VI	Kharif paddy/ fallow	• Paddy/maize – mustard/lentil/pea/groundnut.
		• Summer green gram can also be included where paddy cultivation is problematic due to water scarcity.
		• Managements:
		  • Green manuring of Dhaincha can be included after medium duration HYV of paddy.
		  • Zero tillage or minimum tillage for mustard/lentil cultivation.

1.8 CONCLUSION

In this study, a detailed land resource inventory was carried out at 1:10,000 scales of the Nalchar block, Sepahijala district, Tripura. Geomorphologically, the study area represents undulating topography (3%–25% slope). The geology of the block is represented by sedimentary rocks which range in age from Miocene to loosely consolidated sediments of recent age. The rocks are sandstone, siltstone, and shale grading into clay. Six types of land-use/land cover were observed viz. single crop paddy, double crop, plantation, plantation/forest/scrub, forest, and open scrub. Visual interpretation of LISS-IV data indicated that the block was characterized by hills, mounds, upland, alluvial plains, and valleys. In the ArcGIS landform, slope and land-use/land cover maps were integrated and an LEU map was prepared with 21 LEU units. At 1:10,000 scale, seven soil series were identified and mapped into ten soil mapping units (phases of series). The fertility status of the soils indicates that the soils of the study area were acidic in soil reaction (extremely acidic to strongly acidic), low to high in organic carbon, low to medium in available nitrogen and available phosphorus, and low in available potassium. Soil survey interpretation showed that the study area was divided into three land capability classes viz. II, III, and IV which are further sub-divided into six sub-classes based on the limitation of soils, erosion, climate, and wetness. The suitability for different crops grown in the study area showed moderate to marginally suitable due to topography, fertility, and wetness limitations. Finally, based on major problems and potential the study area was divided into four soil management units and suggested LMU-wise alternative land-use options for the study area.

ACKNOWLEDGMENTS

This research was financially supported by the Indian Council of Agricultural Research (ICAR), New Delhi, India. We acknowledge all our colleagues who helped directly or indirectly in carrying out the study.

REFERENCES

Abdel Rahman, M.A.E., Natarajan, A., and Hegde, R. (2016). Assessment of land suitability and capability by integrating remote sensing and GIS for agriculture in Chamarajanagar district, Karnataka, India. *Egypt. J. Remote Sens. Space Sci.* 19:125–141. https://doi.org/10.1016/j.ejrs.2016.02.001

Adam, E., Mutanga, O., and Rugege, D. (2010). Multispectral and hyperspectral remote sensing for identification and mapping of wetland vegetation: a review. *Wetlands Ecol. Manage.* 18:281–296. https://doi.org/10.1007/s11273-009-9169-z

Adhikari, K., Guadagnini, A., Toth, G., and Hermann, T. (2009). Geostatistical analysis of surface soil texture from Zala County in western Hungary. *Proceedings of International Symposium on Environment, Energy and Water in Nepal: Recent Researches and Direction for Future*. March 31–April 1, Kathmandu, pp. 219–224.

AIS & LUS (All India Soil and Land Use Survey Organization). (1970). *Soil Survey Manual*. IARI, New Delhi.

Askegaard, M. and Jørgen, E. (2002). Exchangeable potassium in soil as indicator of potassium status in an organic crop rotation on loamy sand. *Soil Use Manage.* 18:84–90. https://doi.org/10.1111/j.1475-2743.2002.tb00224.x

Bhattacharyya, T., Sehgal, J.L., and Sarkar, D. (1996). Soils of Tripura for optimizing land use: their kinds, distribution and suitability for major field crops and rubber. NBSS Publication No. 65, Nagpur, pp. 154.

Bhattacharyya, T., Ram, B., Sarkar, D., Mandal, C., and Nagar, A.P. (2002). Soil erosion of Tripura, a model for soil conservation and crop performance. NBSS Publ No. 97, NBSS&LUP, Nagpur, pp. 80+1 sheet of map (1:250,000 scale).

Bhattacharyya, T., Pal, D.K., and Vaidya, P.H. (2003). Soil landscape model for suitable cropping pattern in Tripura, part I. Soil resources in Tripura–their extent, nature and characteristics. Final DST Project report, NBSS&LUP (ICAR), Nagpur, pp. 114.

Bhattacharyya, T., Chandran, P., Ray, S.K., Mandal, C., Pal, D.K., Venugopalan, M.V., Durge, S.L., Srivastava, P., Dubey, P.N., Kamble, G.K., Sharma, R.P., Wani, S.P., Rego, T.J., Pathak, P., Ramesh, V., Manna, M.C., and Sahrawat, K.L. (2007). Physical and chemical properties of selected benchmark spots for carbon sequestration studies in semi-arid tropics of India. Global Theme on Agro-ecosystems Report No. 35. Patancheru 502 324, Andhra Pradesh, India: International Crops Research Institute for the Semi-Arid Tropics (ICRISAT), and New Delhi. Indian Council of Agricultural Research (ICAR), India, pp. 236.

Bhattacharyya, T., Sarkar, D., Pal, D.K., Mandal, C., Baruah, U., Telpandey, B., and Vaidya, P.H. (2010). Soil information system for resource management-Tripura as a case study. *Curr. Sci.* 99:1208–1217.

Black, A.A. (1968). *Soil-Plant Relationships*. Wiley, New York.

Bray, H.R. and Kurtz, L.T. (1945). Determination of total organic and available forms of phosphorus in soil. *Soil Sci.* 59:39–46. https://doi.org/10.1097/00010694-194501000-00006

Buol, S.W., Southard, R.J., Grahm, R.C., and McDaniel, P.A. (2003). *Soil Genesis and Classification*, (5th ed.). Iowa State University Press, Ames, USA.

Chadwick, O.A. and Grahm, R.C. (2000). Pedogenic processes. In: Sumner, M.E. (ed.) *Handbook of Soil Science*. CRC Press, Boca Raton, FL, pp. 41–75.

Deka, B., Baruah, T.C., Dutta, M., and Karmakar, R.M. (2009). Landscape soil relationships and pedonegic evaluation of soils in Ghiladhari watershed of the Brahmaputra valleys of Assam. *J. Indian Soc. Soil Sci.* 57:245–252.

Dessalegn, D., Beyene, S., Ram, N., Walley, F., and Gala, T.S. (2014). Effects of topography and land use on soil characteristics along the toposequence of Ele watershed in southern Ethiopia. *Catena.* 115:47–54. https://doi.org/10.1016/j.catena.2013.11.007

Dobermann, A., Goovaerts, P., and George. T. (1995). Sources of soil variation in an acid Ultisol of the Philippines. *Geoderma.* 68:173–191. https://doi.org/10.1016/0016-7061(95)00035-M

Dobos, E., Micheli, E., Baumgardner, M.F., Biehl, L., and Helt, T. (2000). Use of combined digital elevation model and satellite radiometric data for regional soil mapping. *Geoderma.* 97:367–391. https://doi.org/10.1016/S0016-7061(00)00046-X

Fekadu, E., Negese, A., and Yildiz, F. (2020). GIS assisted suitability analysis for wheat and barley crops through AHP approach at Yikalo sub-watershed, Ethiopia. *Cogent Food Agric.* 6:1743623. https://doi.org/10.1080/23311932.2020.1743623

Gangopadhyay, S.K., Bhattacharyya, T., and Sarkar, D. (2001). Characterization and classification of some rubber growing soils of Tripura. *J. Indian Soc. Soil Sci.* 49:164–170.

Gangopadhyay, S.K., Bhattacharyya, T., and Sarkar, D. (2008). Nature of acidity in some soils of south Tripura. *Agropedology.* 18:12–20.

Ghosh, B.N., Das, K., Bandyopadhyay, S., Mukhopadhyay, S., Nayak, D.C., and Singh, S.K. (2018). Impact assessment of GIS based land resource inventory towards optimizing agricultural land use plan in Dandakaranya and Easternghats physiographic confluence of India. *J. Indian Soc. Remote Sens.* 46(4):641–654. https://doi.org/10.1007/s12524-017-0743-1

GSI (Geological Survey of India). (2011). Geology and mineral resources of Manipur, Mizoram, Nagaland and Tripura, Misc. Publ. No. 30, Part IV, 1 (Part-2), pp. 1–104.

Hall, G.F. (1983). Pedology and geomorphology. In: Wilding, L.P., Smeck, N.E., and Hall, G.F., (eds.) *Pedogenesis and Soil Taxonomy. I. Concepts and Interactions.* Elsevier Science Publishers BV, Amsterdam, the Netherlands, pp. 117–140.

Huggett, R.J. (1975). Soil landscape systems: a model of soil genesis. *Geoderma.* 13:1–22. https://doi.org/10.1016/0016-7061(75)90035-X

Huggett, R.J. (1976). Lateral translocation of soil plasma through a small valley basin in the Northway Great Wood Hertfordshire. *Earth Surf. Process.* 1:99–109. https://doi.org/10.1002/ESP.3290010202

Hutchinson, J.J., Campbell, C.A., and Desjardins, R.L. (2007). Some perspectives on carbon sequestration in agriculture. *Agric. Meteorol.* 142:288–302. https://doi.org/10.1016/j.agrformet.2006.03.030

Jackson, M.L. (1976). *Soil Chemical Analysis.* Prentice Hall, Englewood Cliffs, NJ.

Jangir, A., Tiwari, G., Sharma, G., Dash, B., Paul, R., Vasu, D., Malav, L.C., Tiwary, P., and Chandran, P. (2020). Characterization, classification and evaluation of soils of Kamrej Taluka in Surat district, Gujarat for sustainable land use planning. *J. Soil Water Conserv.* 17:15–24. https://doi.org/10.5958/2455-7145.2020.00046.6

Jenny, H. (1941). *Factors of Soil Formation. A System of Quantitative Pedology.* McGraw-Hill, New York.

Katerji, N. and Mastrorilli, M. (2009). The effect of soil texture on the water use efficiency of irrigated crops: results of multi-year experiment carried out in the Mediterranean region. *European J. Agron.* 30:95–100. https://doi.org/10.1016/j.eja.2008.07.009

Kettler, T.A., Doran, J.W., and Gilbert, T.L. (2001). Simplified method for soil particle-size determination to accompany soil-quality analyses. *Soil Sci. Soc. Am. J.* 65:849–852. https://doi.org/10.2136/sssaj2001.653849x

Khan, Z.M., Hussain, M.S., and Otiner, F. (2012). Morphogenesis of three surface-water gley soils from the Meghna floodplain of Bangladesh. *J. Biol. Sci.* 21:17–27. https://doi.org/10.3329/DUJBS.V21I1.9741

Khasawneh, F.E., Sample, E.C., and Kamprath, E.J. (1986). The role of phosphorus in agriculture. In: *Proceeding of Symposium of Phosphorus.* NEDC, Muscle Shoals, Albana, USA, pp. 48–56.

Kong, X., Dao, T.H., Qin, J., Qin, H., Li, C., and Zhang, F. (2009). Effects of soil texture and land use interactions on organic carbon in soils in North China cities urban fringe. *Geoderma.* 154:86–92. https://doi.org/10.1016/j.geoderma.2009.09.016

Krauss, A. and Johnson, A.E. (2002). Assessing soil potassium; can we do better? 9th International Congress of Soil Science. March 18–20, Faisalabad.

Lindeburg, K.S., Almond, P., Roering, J.J., and Chadwick, O.A. (2013). Pathways of soil genesis in the Coast Range of Oregon, USA. *Plant Soil.* 367:57–75. https://doi.org/10.1007/s11104-012-1566-z

Mandal, V.P., Rehman, S., Ahmed, R., Masroor, M., Kumar, P., and Sajjad, H. (2020). Land suitability assessment for optimal cropping sequences in Katihar district of Bihar, India using GIS and AHP. *Spat. Inf. Res.* 28:589–599. https://doi.org/10.1007/s41324-020-00315-z

McBratney, A.B., Santos, M.L.M., and Minasny, B. (2003). On digital soil mapping. *Geoderma*. 117:3–52. https://doi.org/10.1016/S0016-7061(03)00223-4

Moharir, K., Pande, C., Patode, R.S., Nagdeve, M.B., and Varade, A.M. (2021). Prioritization of sub-watersheds based on morphometric parameter analysis using geospatial technology. In: Pandey, A., Mishra, S., Kansal, M., Singh, R., and Singh, V. (eds.) *Water Management and Water Governance. Water Science and Technology Library*, vol 96. Springer, Cham. https://doi.org/10.1007/978-3-030-58051-3_2

Moore, I. D., Gessler, P. E., Nielsen, G. A., and Peterson, G. A. (1993). Soil attributes prediction using terrain analysis. *Soil Sci. Soc. Am. J.* 57:443–452. https://doi.org/10.2136/sssaj1993.03615995005700020026x

Mulugeta, D. and Sheleme, B. (2010). Characterization and classification of soils along topo-sequence in Kindo Koye watershed in southern Ethiopia. *East African J. Sci.* 4:65–77.

Mythili, G. and Goedecke, J. (2016). Economics of land degradation in India. In: Nkonya, E., Mirzabaev, A., and von Braun, J., (eds.) *Economics of Land Degradation and Improvement-A Global Assessment for Sustainable Development*. Springer, Cham, pp. 431–469. https://doi.org/10.1007/978-3-319-19168-3_15

Nagaraju, M.S.S., Kumar, N., Srivastava, R., and Das, S.N. (2014). Cadastral-level soil mapping in basaltic terrain using Cartosat-1-derived products. *Int. J. Remote Sens.* 35:3764–3781. https://doi.org/10.1080/01431161.2014.919675

Naidu, L.G.K., Ramamurthy, V., Challa, O., Hegde, R., and Krishnan, P. (2006). Soil Site Suitability Criteria for Major Crops. NBSS Publ. No.129, ICAR-NBSS&LUP, Nagpur.

Pande, C.B. and Moharir, K. (2017). GIS based quantitative morphometric analysis and its consequences: a case study from Shanur River Basin, Maharashtra, India. *Appl. Water Sci.* 7:861–871. https://doi.org/10.1007/s13201-015-0298-7

Pande, C.B., Moharir, K.N., Khadri, S.F.R., et al. (2018). Study of land use classification in an arid region using multispectral satellite images. *Appl. Water Sci.* 8:123. https://doi.org/10.1007/s13201-018-0764-0

Pande, C.B., Moharir, K.N., and Khadri, S.F.R. (2021). Assessment of land-use and land-cover changes in Pangari watershed area (MS), India, based on the remote sensing and GIS techniques. *Appl. Water Sci.* 11:96. https://doi.org/10.1007/s13201-021-01425-1

Patode, R.S., et al. (2021). Groundwater development and planning through rainwater harvesting structures: a case study of semi-arid micro-watershed of Vidharbha Region in Maharashtra, India. In: Pande, C.B. and Moharir, K.N. (eds.) *Groundwater Resources Development and Planning in the Semi-Arid Region*. Springer, Cham. https://doi.org/10.1007/978-3-030-68124-1_26

Ponnamperuma, F.N. (1972). The chemistry of submerged soils. *Adv. Agron.* 24:29–96. https://doi.org/10.1016/S0065-2113(08)60633-1

Ponnamperuma, F.N. (1985). Chemical kinetics of wetland rice soil relative to fertility. In *Wetland Soils-Characterization, Classification and Utilization*. International Rice Research Institute (IRRI), Philippines, pp. 71–90.

Reza, S.K., Baruah, U., Sarkar, D., and Dutta, D.P. (2011). Influence of slope positions on soil fertility index, soil evaluation factor and microbial indices in acid soil of Humid Subtropical India. *Indian J. Soil Conserv.* 39:44–49.

Reza, S.K., Baruah, U., and Sarkar, D. (2012). Mapping risk of soil phosphorus deficiency using geostatistical approach: a case study of Brahmaputra plain, Assam, India. *Indian J. Soil Conserv.* 40: 65–69.

Reza, S.K., Baruah, U., Nath, D.J., Sarkar, D., and Gogoi, D. (2014a). Microbial biomass and enzyme activity in relation to shifting cultivation and horticultural practices in humid subtropical North-Eastern India. *Range Mgmt. Agroforestry*. 35:78–84.

Reza, S.K., Baruah, U., Dutta, D., Sarkar. D., and Dutta, D.P. (2014b). Distribution of forms of potassium in Lesser Himalayas of Sikkim, India. *Agropedology*. 24:106–110.

Reza, S.K., Baruah, U., Chattopadhyay, T., and Sarkar, D. (2014c). Distribution of forms of potassium in relation to different agroecological regions of North-Eastern India. *Arch. Agron Soil Sci.* 60:507–517. https://doi.org/10.1080/03650340.2013.800943

Reza, S.K., Nayak, D.C., Chattopadhyay, T., Mukhopadhyay, S., Singh, S.K., and Srinivasan, R. (2016). Spatial distribution of soil physical properties of alluvial soils: a geostatistical approach. *Arch. Agron Soil Sci.* 62:972–981. https://doi.org/10.1080/03650340.2015.1107678

Reza, S.K., Nayak, D.C., Mukhopadhyay, S., Chattopadhyay, T., and Singh, S.K. (2017). Characterizing spatial variability of soil properties in alluvial soils of India using geostatistics and geographical information system. *Arch. Agron Soil Sci.* 63:1489–1498. https://doi.org/10.1080/03650340.2017.1296134

Reza, S.K., Baruah, U., Nayak, D.C., Dutta, D., and Singh, S.K. (2018a). Effects of land-use on soil physical, chemical and microbial properties in humid subtropical northeastern India. *Natl. Acad. Sci. Lett.* 41:141–145. https://doi.org/10.1007/s40009-018-0634-1

Reza, S.K., Bandyopadhyay, S., Ray, P., Ramachandran, S., Mukhopadhyay, S., Sah, K.D., Nayak, D.C., Singh, S.K., and Ray, S.K. (2018b). Rubber growing soils of Bishalgarh block, Sepahijala district, Tripura: their characteristics, suitability and management. *Field Forester*. 3:180–184.

Reza, S.K., Ray, P., Ramachandran, S., Bandyopadhyay, S., Mukhopadhyay, S., Sah, K.D., Nayak, D.C., Singh, S.K., and Ray, S.K. (2019a). Spatial distribution of soil nitrogen, phosphorus and potassium contents and stocks in humid subtropical North-eastern India. *J. Indian Soc. Soil Sci.* 67:12–20. https://doi.org/10.5958/0974-0228.2019.00002.1

Reza, S.K., Bandyopadhyay, S., Ramachandran, S., Ray, P., Mukhopadhyay, S., Sah, K.D., Nayak, D.C., Singh, S.K., and Ray, S.K. (2019b). Land resource inventory of Bishalgarh block of Sepahijala district, Tripura at 1:10000 scale for farm planning. NBSS Publ. No. 1116, ICAR-NBSS&LUP, Nagpur, pp. 76.

Reza, S.K., Dutta, D., Bandyopadhyay, S., Singh, and S.K. (2019c). Spatial variability analysis of soil properties of Tinsukia district, Assam, India, using geostatistics. *Agric. Res.* 8:231–238. https://doi.org/10.1007/s40003-018-0365-z

Reza, S.K., Ray, P., Ramachandran, S., Bandyopadhyay, S., Mukhopadhyay, S., Sah, K.D., Nayak, D.C., Singh, S.K., and Ray, S.K. (2019d). Profile distribution of soil organic carbon in major land use systems in Bishalgarh block, Tripura. *J. Indian Soc. Soil Sci.* 67:236–239. https://doi.org/10.5958/0974-0228.2019.00026.4

Reza, S.K., Ray, P., Ramachandran, S., Jena, R.K., Mukhopadhyay, S., and Ray, S.K. (2020a). Soil organic carbon fractions in major land use systems in Charilam block of Tripura. *J. Indian Soc. Soil Sci.* 68:458–461. https://doi.org/10.5958/0974-0228.2020.00037.7

Reza, S.K., Bandyopadhyay, S., Ray, P., Ramachandran, S., Mukhopadhyay, S., and Ray, S.K. (2020b). Delineation of land management units for alternate land use options: a case study of Bishalgarh block in Sepahijala district of Tripura. *Agric. Observ.* 1:118–123.

Rich, C.I. (1965). Elemental analysis by flame photometry. In: Black, C.A. (ed.) *Methods of Soil Analysis, Part 2: Chemical and Microbiological Properties*. American Society of Agronomy, Madison, WI, pp. 849–864.

Sarkar, D. (2011). Geo-informatics for appraisal and management of land resources towards optimizing agricultural production in the country-issues and strategies. *J. Indian Soc. Soil Sci.* 59(suppl.):35–48.

Sehgal, J.L., Saxena, R.K., and Vadivelu, S. (1987). Field Manual. Soil resource mapping of different states of India. NBSS Publ. No. 13, ICAR-NBSS&LUP, Nagpur.

Sehgal, J., Mandal, D.K., Mandal, C., and Vadivelu, S. (1992). Agro-ecological regions of India. Technical Bulletin No. 24, NBSS&LUP, Nagpur, pp. 130.

Seid, N.M., Yitaferu, B., Kibret, K., and Ziadat, F. (2013). Soil-landscape modeling and remote sensing to provide spatial representation of soil attributes for an Ethiopian watershed. *Appl. Environ. Soil Sci.* 798094:1–11. https://doi.org/10.1155/2013/798094

Singh, R.L. (2006). *India: A Regional Geography*. National Geographical Society of India, Varanasi.

Singh, S.K., Chatterji, S., Chattaraj, S., and Butte, P.S. (2016). Land Resource Inventory on 1:10000 scale, Why and How? NBSS Publ. No. 172, ICAR-NBSS&LUP, Nagpur, pp. 110.

Soil Survey Division Staff. (2000). *Soil Survey Manual. Soil Conservation Service. USDA Handbook 18, Revised.* Scientific Publishers, Jodhpur.

Soil Survey Staff. (1995). *Soil Survey Manual, USDA, Agricultural Handbook No. 18, New Revised Edition.* Scientific Publishers, Jodhpur.

Soil Survey Staff. (2003). *Keys to Soil Taxonomy* (9th ed.). USDA-Natural Resources Conservation Service, Washington, DC.

Soil Survey Staff. (2014). *Keys to Soil Taxonomy* (12th ed.). USDA-Natural Resources Conservation Service, Washington, DC.

Srivastava, R. and Saxena, R.K. (2004). Technique of large-scale soil mapping in Basaltic terrain using satellite remote sensing data. *Int. J. Remote Sens.* 25:679–688. https://doi.org/10.1080/0143116031000068448

Stoops, G. and Eswaran, H. (1985). Morphological characteristics of wet soils. In: *Wetland Soils-Characterization, Classification and Utilization*. International Rice Research Institute (IRRI), Philippines, pp. 177–189.

Subbiah, B.V. and Asija, G.L. (1956). A rapid procedure for estimation of available nitrogen in soils. *Curr. Sci.* 25:259–260.

Supriya, K., Naidu, M.V.S., Kavitha, P., and Reddy, M.S. (2019). Characterization, classification and evaluation of soils in semi-arid region of Mahanandi mandal in Kurnool district of Andhra Pradesh. *J. Indian Soc. Soil Sci.* 67:125–136. https://doi.org/10.5958/0974-0228.2019.00014.8

Sys, C., Van Ranst, E., Debaveye, J., and Beernaert, F. (1993). Land Evaluation, Part-III. Crop Requirements. Agricultural Publications No.7, GADC, Brussels, Belgium.

Takkar, P.N. (2009). Soil fertility, fertilizer and integrated nutrient use. In: Rai et al. (eds.) *Hand Book of Agriculture* (6th revised ed.). Indian Council of Agriculture, New Delhi, pp. 516.

Van, W.J., Baert, G., Moeyersons, J., Nyssen, J., De Geyndt, K., Taha, N., Zenebe, A., Poesen, J., and Deckers, J. (2008). Soil-landscape relationships in the Basalt-dominated highlands of Tigray, Ethiopia. *Catena.* 75:117–127. https://doi.org/10.1016/j.catena.2008.04.006

Velayutham, M., Mandal, D.K., Mandal, C., and Sehgal, J. (1999). Agro-ecological subregion of India for planning and development. NBSS Publ. 35, NBSS&LUP, Nagpur, pp. 372.

Walkley A. and Black, I.A. (1934). An examination of the digtjareff method for determining soil organic matter and a proposed modification of the chromic acid titration method. *Soil Sci.* 37:29–38. https://doi.org/10.1097/00010694-193401000-00003

Western, A.W., Grayson, R.B., Blöschl, G., and Wilson, D.J. (2003). Spatial variability of soil moisture and its implication for scaling. In: Pachepsky, Y., Radcliffe, D.E., Selim, H.M. (eds.) *Scaling Methods in Soil Physics.* CRC Press, Boca Raton, FL, pp. 119–142.

Zhai, Y., Thomasson, J.A., Boggess, J.E., and Sui, R. (2006). Soil texture classification with artificial neural networks operating on remote sensing data. *Comput. Electron Agric.* 54:53–68. https://doi.org/10.1016/j.compag.2006.08.001

2 Sustainable Biodiversity Conservation in Tribal Area

E. Gayathiri, R. Gobinath, J. Jayanthi J, Paniswamy Prakash, and M.G. Ragunathan

2.1 INTRODUCTION

Around 100 million indigenous tribes makeup India's communities. There are 705 Scheduled Castes and Tribes groups in India (MOTA and GoI, 2018; Das et al., 2010; Government Census, 2011). The main areas of tribal occupation are the countries of Burma and North-East China, as well as the highlands and plains of Peninsula India. In central India, the Scheduled Tribes were referred to as Adivasis or indigenous peoples. Many other indigenous groups were not legally recognized as Scheduled Tribes, which is creating a cause of concern. Racial groups worldwide are real keepers of Nature's bounty and have been recognized as real treasurers of natural medicines. Traditional native knowledge has been passed down orally for several decades and is gradually fading due to technological advancements resulting in the culture of indigenous groups (Ganesan et al. 2004; Suranjit, 1996). In India, tribal people were classified as Adivasi (original settlers), Vanyajati (forest caste), Aboriginal (indigenous), Girijan (hillsmen), Adimjati (primitive caste), Vanavasi (forest inhabitants), Adam Niwasi (The population's earliest ethnological subgroup), and Janjati (forest inhabitants). Adivasi is the most well-known of these terms, though AnusuchitJanjati (Scheduled Tribes) is the legislative term. Tribes or ethnic communities were well known across the globe and have their very own distinctive cultures, traditions, social and religious status, folklore, folktales, legends, taboos, totems, poetry, rituals, and food and medicine theories. Trees and plants play a major role in the tribe's life (Figure 2.1).

Most of the Indian tribes are believed to be the most primitive human societies, totally dependent on the forest for life and living in absolute harmony with nature. The Andaman and Nicobar Islands, Jarawas, Sentinels, Shompens, Onges, and Kobo are among these native populations. Abhujhmaria, Baiga, Kamar, Birhors, the five tribes of Chhattisgarh, Kani, Kurumba, Malayali, Vadugar, and Irular of Tamil Nadu have also been identified by the Indian Government. Tribal populations like these are on the edge of extinction. The Indian Constitution guarantees indigenous people equal rights under Article 466. Native peoples have their history, worship ceremonies, eating habits, and a strong understanding of traditional medicine (Harsha et al., 2003). Native phytocure methods have been maintained by the tribe as part of their culture in the midst of many of these challenges. Furthermore, these people are a little hesitant to change their way of living. However, in the present scenario, traditional knowledge of medicinal plants is fading; it should be preserved and passed on to further generations (Burmol and Naidu, 2007).

The states of Sikkim, Bihar, Uttar Pradesh, Assam, Kerala, West Bengal, Himachal Pradesh, and Bihar have the maximum tribal groups. Ethnic peoples' livelihoods depend on how well they manage their ecosystem because their survival depends on its sustainability. These organizations have been designated as "true ecologists" because they take care of their land and habitat more than everyone else (Table 2.1).

The tribes have similar traits and are far more homogeneous and self-contained than other non-tribal social groups, who were particularly given regional variations. Consequently, inside tribal links and the relationship between tribes and the state, numerous inconsistencies (both visible and invisible) emerge. The conventional and widely accepted solution is to maintain tribal communities'

 DOI: 10.1201/9781003303237-2

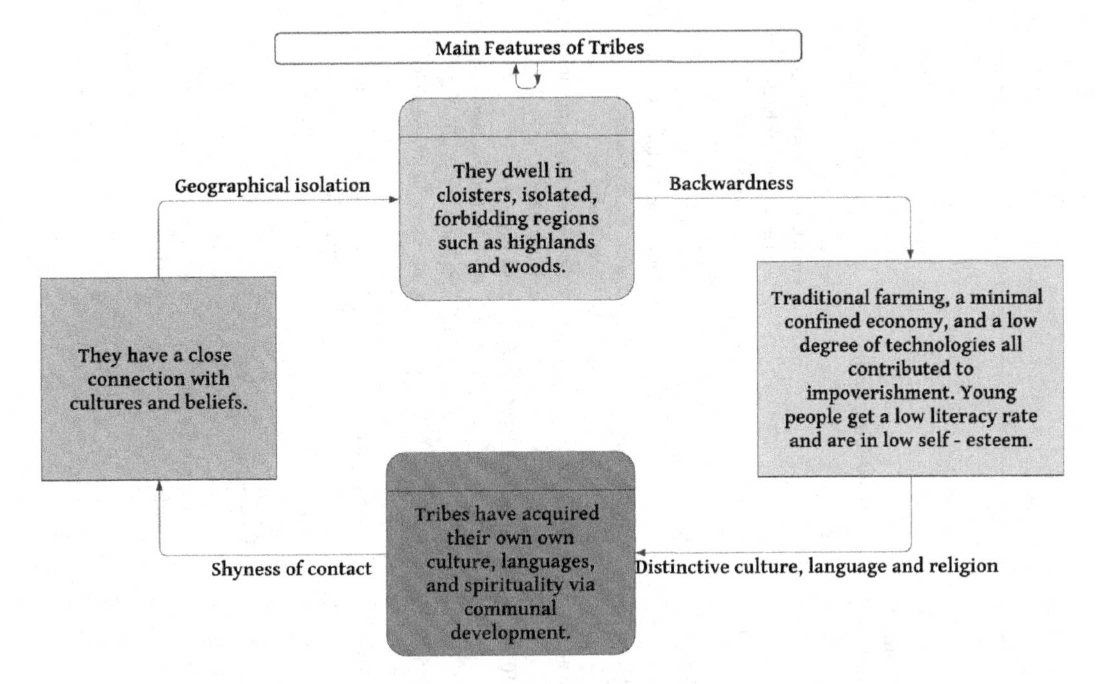

FIGURE 2.1 Main distinctive features of tribes.

rights and their relation to assimilation while defining the contours of a national policy that allows them to sustain their way of life without jeopardizing development.

2.2 TRIBES ROLE IN INDIGENOUS BOTANICAL KNOWLEDGE (IBK)

Tribal communities play a major role in preserving the biodiversity of a diverse range of natural forests and preserving various plants and animals in protected tribal groves, which would otherwise have perished from the natural ecosystem (Vartak, 1996). Indigenous intelligence and neighborhood knowledge obtained by indigenous communities are unique to their ethnic group (Panghal et al., 2010). Scientific investigation and recording of aboriginal information about plant products are critical strategies for gaining a better understanding of the cultural lifestyles of indigenous people (Gurib-Fakim, 2006). An orthodox herbal formulation based on indigenous wisdom is regarded as the world's oldest and greatest method of health treatment. They utilize plant species as primary sources of medication and is also believed that they have a major role in curing all ailments. Common plants with little harmful effects are a part of our everyday lives, and their toxic metabolites have a strong correlation with human health. Over centuries, traditional people have used toxic seeds, either refined or untreated, to cure diseases (Mukherjee and Wahile, 2006; Sevugaperumal, 2004).

2.3 TRIBES ETHNO BOTANY AND TRADITIONAL PRACTICES

Ethnobotany is the study of the "interrelationships between flora and fauna, plants and humans, and plants and plants." Over the last few decades, ethnobotany has gained much importance as a study area focusing on supporting the protection of knowledge at the tribal level; historical and cultural information has been used for resource management and biological diversity protection, ecology, habitat, forest form, and regional levels (Lewis and Elvin-Lewis, 2003). Ethnobotanical material, on the other hand, has been established to the extent that different terms may be represented in different contexts. Indigenous Botanical Knowledge (IBK) and Indigenous Traditional Knowledge (ITK) are two terms that have been recently used to describe the material of ethnobiology (Figure 2.2).

TABLE 2.1
List of Tribal Communities

S.No	State	Region	Tribal Communities	Total Population (In Lakhs)	Worshiping God	Work Done to Conserve Forest
1	Andaman and Nicobar Island	Andaman	Andamanese, Nicobarese, Onge, Jarawa, Shompen, Sentenalese	3.81	Sun	Herbal science and traditional knowledge of practitioners use leaf extract for healthcare
2	Andhra Pradesh	Nallamala Forest	Chenchu, Yandai, Kurumba, Khond, Bagdaz	493.87	Sun	Roots of aloevera as food during famine
3	Arunachal Pradesh	Central Region	Aptani, Mishmi, Daffla, Miri, Aka, Sinpho, Khamti etc.	13.84	Sun	Leaves of *Ageratum conyzoides* for blood clotting
4	Assam	South-Central Region of Asiatic descent	Chakma, Mikir, Kachari, Bora etc.	312.06	Goddess Tusu	Tea garden community (Tea-Tribes)
5	Bihar	Jharkhand	Santhal, Oraon, Munda, Kharwar, Kharia, Bhumji, Ho	1,040.99	Sun	*Leonotisnepetifolia* for curing skin disease
6	Chandigarh	Border of Both States	Ad Dharmi, Bangali, Chanal, Darain, Dagi, Garga, Kori, Nat, Pasi, Sanhal	10.55	Hindu Goddess Chandi	Leaf paste of *Pyracanthacrenulata* is used for burns
7	Chhatisgarh	South of Kanker, Forested KeshkelGhat	Parja, Bhattra, Agaria, Bhil, Saharia, Kowra, Halba	255.45	Rain	Conserve wild fruits and seeds
8	Dadra and Nagar Haveli	Silvassa	Dhodia, Dubla, Kathodi, Kokna, KoliDhor, Naikda, Varli	3.44	Sun and Moon	Regulated and planned the cutting of trees
9	Daman and Diu	Diu	Dubla, Dhodia, Varli, Naikda, Siddi	1.58	Sun	Protects green forests and winding rivers
10	Goa	Interior of Goa	Gowdas, Kunbis, Velips, Dhangars	14.59	Lord Shiva	Soil conservation and water harvesting
11	Gujarat	Valasd, Surat, Dangs	Bavacha, Charan, Bharwad, DholiBhil, Vasava, Chaudhari	604.4	Tigar crocodile and Snake	Conserve teak and bamboo widely

(*Continued*)

TABLE 2.1 (*Continued*)
List of Tribal Communities

S.No	State	Region	Tribal Communities	Total Population (In Lakhs)	Worshiping God	Work Done to Conserve Forest
12	Haryana	Hisar, Bhiwani, Faridabad	Bazigar, Mirasi, Sikligar, Spera	253.31	Kali	Land reclamation and soil conservation
13	Himachel Pradesh	Hamirpur, Kullu, Kangar, Spiti	Gaddi, Gujjar, Kinnar	68.65	Phoomi	Protection of forests by reforestation and afforestation
14	Jammu and Kashmir	Doda, Ganderbal, Anantnag	Gaddi, Bakarwal	125.41	Sun	Conserve forest ecology
15	Jharkand	Kol, Savar, Oraon	Kharia, Oraon, Santhal, Munda Paharia, Ho, Birhor, Tamaria	329.88	Peace	Thrown out timber mafia and conserved forest
16	Karnataka	Chikmangalur, Udupi, Hassan	KoliDhor, Gond, Naikda, Marati	610.95	Ancestors	Adopted the policy of no harm to flora and fauna
17	Kerala	Malabar District, Malappuram, Hosedrug	Irula, Kurumba, Kadar, Puliyan	334.06	Ancestors	Work to avoid forest fire
18	Ladakh	Leh, Kargil	Mon, Bot, Beda, Garra, Changpa		Monasteries	Protect landforms and wild fruits
19	Lakshadweep	laccadive, Minicoy, Amindivi	Andrott, Amini, Bitra, Minicoy, Kavaratti	0.64	Mandaps	Conserve the species of sponges and protect their ecosystem
20	Madhya Pradesh	Bhopal, Schore, Vidisha, Tikamgarh	Hill Maria, Muria, Dandami, Gond, Baiga	726.27	Lord Shiva	Guardians of wildlife especially tigers
21	Maharastra	Chandrapur, Hiangoli, Jalni	Bhil, Garasia, Kokni, Kawar, Pardhan	1,123.74	Vishnu	Conserve their dense forest, protection of bamboo
22	Manipur	Churachandur, Kangpokpi, Tamenglong, Thoubal	Kuki, Lepcha, Mugh	28.56	Sun	Conserve wildlife
23	Meghalaya	Khasi Hills, Jaintia Hills, Garo Hills	Garo, Khasi, Janitia, Hamar	29.67	Nature	Conserve forest with sacred grooves
24	Mizoram	Siaha, Lungeli, Aizawal	Mizo, Lakher	10.97	Sun	Conservation of the environment and wildlife

(*Continued*)

TABLE 2.1 (*Continued*)
List of Tribal Communities

S.No	State	Region	Tribal Communities	Total Population (In Lakhs)	Worshiping God	Work Done to Conserve Forest
25	Nagaland	Wokha, Phek, Mon, Kohima, Dimapur	Angami, Konyak, Lotha	19.79	Natural World	Conservation of nature and protection of forest
26	Odisha	Nabarangpur, Koraput, Gajapati, Kadhamal	Zuang, Sawara, Karia, Khond, Kandh	419.74	Animal and Nature	Worship and protect the forest as god
27	Puducherry	Yanam, Karaikal	Adivasi	12.48	Sun	Protect land and water resources
28	Punjab	Bhagat Singh Nagar, Sri Muktsar Sahib, Jalandhar, Moga	Gurjar, Jats, Labana	277.43	Sun	Land reclamation and soil conservation
29	Rajasthan	Banswara, Dungarpur, Chittorgarh	Bhil, Meena, Kathoria, Garasia	685.48	Sun, Lord Shiva, Vishnu	The religious act of these communities is to conserve forest
30	Sikkim	Gangtok, Mangan	Lepchas, Bhutias, Nepalese	6.11	Sun, Shiva	Preserve nature through tradition
31	Tamil Nadu	Nilgari, Pudukotti, Karur, Salam	Toda, Kota, Kurumba, Bagada	721.47	Birds, Animals, Snakes, Rock Hillocks	Conserve ecology and protect the forest
32	Telangana	Khammam, Warangal, Adilabad	Koya, Bagata, Gadaba	351.94	Snake and Tiger	Conserve wildlife and social forestry
33	Tripura	Sipahijala, Khowai	Bhutia, Chakma, Garo, Kuki	36.74	Goddess Lakshmi	Conservation of biodiversity by cultural heritage
34	Uttarakhand	Nainatal, Udham Singh Nagar, Uttarkashi	Raji, Khasa, Bhuia, Kharwar, Manjhi	100.86	Lord Rama and Krishna	Conservation of forest and wildlife
35	Uttar Pradesh	Agariya, Bhotia, Buksa, Chero	Tharu, Bhatia, Jaunsari, Bhoksha, Kol	1,998.12	Sun	Conservation of biodiversity by plantations
36	West Bengal	Hooghly, Bankur, Alipurduar	Asur, Bhumji, Birhor, Lodha, Lepcha, Mag, Mahali, Malpaharia, Polia	912.76	Sun	Conservation and preservation of forest resources

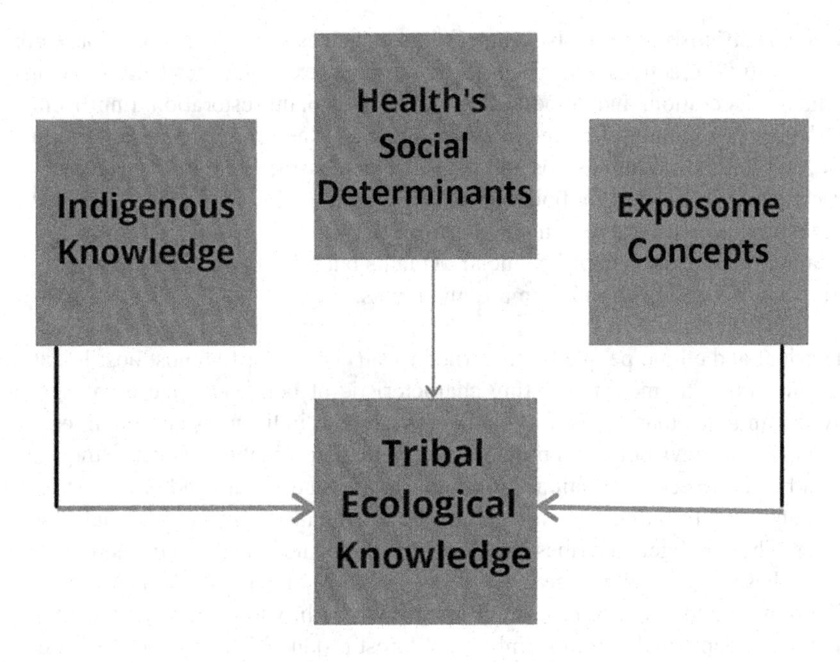

FIGURE 2.2 Flow of tribal ecological knowledge.

Also, it is very evident that documenting information is absolutely essential in ethnobiology since; it offers evidence for subsequent research that could be motivated by metaphysical or primarily utilitarian concerns (Berlin, 1973, 1992). We can only appreciate the interaction between plants and human cultures through an interdisciplinary approach that incorporates botany, chemistry, anthropology, ecology, archaeology, psychology and pharmacology, Ethnopharmacology has historically taken a utilitarian view, aiming to investigate experimentally and validate medical plant and bioactive natural medicines (Balick and Cox, 1996). Ethnopharmacology's extensive approach contextualizes nature and examines how plants are perceived, used, pharmacology, and physiology in human cultures (Etkin, 1988; Etkin and Elisabetsky, 2005). Cultural anthropology stresses the influence of thinking and emotion on cultural behavior, while ethnobotany seeks to improve comprehension of both the reason for plant usage and the classification of the natural environment (Berlin et al., 1992; Endicott and Welsch, 2003). Ethnopharmacology is particularly pertinent when it comes to further evolving and testing indigenous pharmacopoeias (Robineau and Soejarto, 1996; Frei et al., 1998; Leonti et al., 2001). In either event, will the specificity of study goals for "ethno-pharmacologists of all backgrounds" be improved "by projecting pharmacologic evidence against a context of medical ethnography, as well as by extending information resource of medical practice by an exploration of plant physiologic potentials" (Etkin, 2001).

2.4 TRIBES RELATIONSHIP WITH FLORA PROTECTION

Tribals have protected a long list of flora through their indigenous practices that have been available in practice for centuries, it happens through their regular interaction between them and the plants as follows

The tribals-plant interaction is classified into two categories, i.e. (1) Abstract relationship (2) Concrete relationship (Jain et al., 1989).

a. **Abstract relationship**: It involves belief in the beneficial or harmful properties of plants, taboos, sacred plants, spirituality, and mythology. Plant-based similes and descriptions are used in folklore, as are fables or scriptures concerning, or references to, plants.

b. **Concrete relationship**: It involves material uses such as food, medicines, house construction, agricultural practices, other household activities, exchange, aesthetic drawings, carvings, house decoration, and domestication, as well as plant restoration, improvement, and degradation. All human-plant relationships are first classified as material, cultural, economic, and religious relationships and then further classified into four categories:
   a. Interactions beneficial for both humans and plants.
   b. Interactions beneficial to man and harmful to plants.
   c. Plant-human relationships beneficial to plants but detrimental to humans.
   d. Interactions unsafe to both humans and plants.

The world's tribal and ethnic people have learned to survive under the most hostile natural conditions in the universe. The most interesting characteristic of both local and ethnic groups is that they mostly live in areas that are rich in biodiversity. These indigenous communities have always played a pivotal role in environmental protection as well as sustainability because they have cultural expertise that helped in eco-restoration. Moreover, people understand and practice the way nature intrinsically functions. In India, there are 68 million inhabitants from 227 national groups and 573 tribal societies. These indigenous tribes have long protected and conserved the habitats of their own communities. These tribes seek refuge in the forest and consume native plant species both raw and cooked. The flowers and fruits were usually consumed fresh, while the tubers, leaves, and seeds were fried. Tribal people make use of timber and forest produce. These tribes have been living in the forest for a long time and have formed a special bond in protecting the flora prevailing in their vicinities.

It has been observed that there are 45,000 species of wild plants, with 9,500 considered to be vital. Of these 9,500 species, 7,500 are used medicinally in traditional health practices. Tribes consume approximately 3,900 plant species as food, 525 species have been used for fiber, 400 for fodder, 300 for the preparation and extraction of bioactive compounds used as natural sources of insecticides and pesticides, and a further 300 for the extraction of resins, gum, dyes, and perfume. Aside from these, numerous plants were used as timber and building materials, and approximately 700 plant species are significant in terms of religious, cultural, aesthetic, moral, and social significance. The Indian subcontinent is one of twelve mega-centers of flora and fauna, indicating two of the eighteen hotspots of biodiversity, one in the North-Eastern Himalayas and the other in the Western Ghats (Rai, 2012) (Table 2.2).

TABLE 2.2
Tribes Relationship with Flora Across Various Regions in India

S.No	Botanical Name of the Flora	Family & Local Name	Region of the Tribe Where Consumed
1	*Achyranthes Aspera*	Amarathaceae & Apamarg	Andhra Pradesh
2	*Basellarubra*	Basellacaeae & Poi	Uttar Pradesh
3	*Chenopodium album*	Chenopodiaceae & Chaulai	Uttar Pradesh & Madhya Pradesh
4	*Dipsacusinermis*	Caprifoliaceae & Wopal Haakh	Jammu and Kashmir
5	*Stellaria media*	Caryophyllaceae & Koku, Kokuwa	Himachal Pradesh, Jammu and Kashmir
6	*Malvaneglecta*	Malvaceae & Sonchal	Himachal Pradesh, Jammu and Kashmir
7	*Hydrocotyle Javanica*	Araliaceae & Manimuni	Assam
8	*Alangiumsalvifolium*	Alangiaceae	Uttar Pradesh & Andhra Pradesh
9	*Bauhinia purpurea*	Caesalpiniaceae	Madhya Pradesh
10	*Phoenix sylvestris*	Arecaceae	Madhya Pradesh & Andhra Pradesh

2.5 ETHNIC AND ABORIGINAL PEOPLES' ROLE IN CONSERVATION

As per the Indian Ethno-biology Survey undertaken by the MoEF India, over 7,500 plant species can be used for human and veterinary health care which is protected and used by 4,635 ethnic communities throughout the country. 90% of flora and fauna are protected in their natural environment by tribes, owing to tribal beliefs, the habitats' were considered as living God. The tribal cultures are widespread in the parts of Central India such as Madhya Pradesh's Dindori, Mandala and Balaghat districts, and Chattisgarh's Bilaspur and Kawardha districts. Moreover, natural forests in Central, Peninsular India, and NorthEast are known as holy groves. Many wild varieties of classical seed varieties of rice, fruits, millets, maize, legumes, beans, and vegetables have been conserved by indigenous people in the central, peninsular regions, and northeast of India. These varieties are superior to currently cultivated rice varieties in characteristics such as scent, grain consistency, digestibility, protein content, pest resistance and disease tolerance. Forty percent of herbs and shrubs in the forest are used as anti-venom for scorpion-stings and snake bites by tribal herbal practitioners. Tribes use the stem and leaves of *Moringaoleifera*, rhizomes of *Acoruscalamus*, *Achyrnthusaspera*, and *Gynandropsisgynandra* for fever, cough, and digestion. Likewise, the stem bark of *B. lanzan* and *B. ceiba* are used as antidotes for scorpion stings and snake bites. Similarly, Rhizome paste is made and applied to wounded areas (Nelson et al., 2014).

Several endangered types of plants are prevented in their natural environment and used by native medicinal healers to set bone fractures and in orthopedic care. The paste made from the stems and leaves of plants such as *V. tessala* and *A. sessiles*, as well as the roots of *C. adnata*, *B. purpurea*, and *S. cordata*, is tied over the wound and allowed to heal for more than 2 weeks on broken bones. Tribal herbal healers preserve these plants in natural forests for orthopedic treatments. For several decades, primitive and tribal societies used several parts of the plants to treat illness, and these plants have gained widespread popularity in conventional medical usage. Plants such as *E. ramosissimum* and *A. maxicana* are processed, ground, and applied as a paste to infested skin and wounds. Plants such as *Albizzialebbeck*, *Sidaacuta*, *Bauhiniapurpurea*, *Grewiahirsutum*, *Jatrophacurcus*, and *Capparisdeciduas* are protected since they are used to treat muscle discomfort, headache, fever, and body inflammation. Jhum cultivation performed by tribes in India's northeast and southern states is the best practice, they harvest and shift the yielded crops and store them in special vats/pots or vessels made of special metals, which hold and preserve the grains for more than 20 decades. In abandoned areas, valuable plants such as *Ardisiapolycephala*, *Phoenisx*, *Meliosmaipata*, *Ardisiacripsa*, *Caseariaglomerata* and *Rhus sp.* are planted. The Indigenous tribes play a pivotal role in sustaining the biodiversity of many forest reserves and help in conserving a wide range of flora and fauna in tribal sacred groves. This flora and fauna would have perished from the natural ecosystem if they hadn't been saved, so maximum human contact with flora was carefully monitored in sacred groves (Zubair et al., 2017). There are many reports of aboriginal groups acting as the greatest conservationists and defenders of the natural environment. Aerial imagery and academic analyses have shown that indigenous peoples maintain and process the soil in such a way that, they can be prevented from erosion (Arora, 1991a, 1991b, 1997).

2.6 TRIBAL RELATIONSHIP WITH FAUNA

Animals were imbued with immense symbolic meaning in many Indigenous communities. The natural world taught man the following aspects

i. How to live close to the earth,
ii. A link that has been formed between the animals and man
iii. Instilled the concept of practicing the conventional way of healing
iv. Respect for all life which directly or indirectly benefits us (Bruchac, 2014).

Despite the interconnectedness of animals and humans, Aboriginal Peoples were vigilant as well as conscientious of the natural life cycles of the animals with which they shared the Earth and made attempts not to overfish, overhunt, or overharvest. However, they also hunted, fished, and gathered everything they needed to feed their relatives, tribes, or clans, but not made the whole habitat extinct. In any situation, if the animal were found dead due to ill-health, they follow subsequent ceremonies and practices and show their gratitude and respect toward them. Overharvesting has little impact on stock or species decline (Nasi et al., 2008). To ensure that humans and animals coexist comfortably, India's biodiversity conservationists need the intervention of tribal groups rather than symbolic advocacy from the metropolitan (Biplab et al., 2010).

Human-animal relationship is fundamental to most tribal value structures since it is providing sacred and mutual bond. Tribal lands provide a safe refuge for the most vulnerable native species and plants. Tribes have long maintained their lands and native habitat. Tribal folklore aids in the understanding and protection of different animal types. Their anti-hunting and anti-cow-killing legislation will assist in the survival of animal habitats in modern times (Philip, 2007). For example, in the year 2009, the Baiga tribe in India launched their initiative to "save the forest from the forest department" outlining guidelines for both their people and visitors to follow and to conserve the forest and its biodiversity. Consequently, plant wealth has improved, and they have been able to harvest more herbs and medicines from the trees (Vikram and Shailly, 2018). Baiga do not hunt tigers; also on the alternative, they refer to the tiger as their little brother. Nonetheless, thousands of Baiga, like many other tribal populations in India, have been unlawfully and forcefully evicted from their ancestral homeland in the name of "tiger conservation" (Bijoy et al., 2010).

The Monpa and Shertukpen tribes of Arunachal Pradesh have followed a tradition where they capture cattle, birds, and fish and release them in a safer place. Tsedar is the name assigned to this exercise. People have a deep conviction that liberating horses, insects, and fishes would bring bad curse to them. Generally, it is strictly regulated that, people should not slaughter freed creatures, thus saving them. The majority of animal liberation occurs in nature reserves and monasteries that are lawfully or culturally secured, guaranteeing the welfare of those species. Buddhists believe the lakes to be holy and prohibit fishing and any other kind of extraction. As a result, fish are mostly covered in such holy lakes. They often release fish into holy lakes and waterways to maintain a healthy harbor (Namsa et al., 2011). In all the above-mentioned examples it is very clear and evident that tribals not only worship and manage them but also help in protecting their availability.

2.7 TRIBES ROLE IN CONSERVATION OF INSECTS

Insects are the most diverse species on the planet, accounting for 70% of all creatures and they contribute significantly to biological diversity (Sattler et al., 2011; New, 2015; Alessa et al., 2016). They rule all conditions except the snowiest and most salty. They are found in a variety of habitats, from desert to tropical woodland, and from pools to cascading waterfalls. Insects are associated with both terrestrial and freshwater, including mosses and liverworts. The majority of plants with flowers, and their reproduction is reliant on insect visitors pollinating and thereby reproducing them. Almost all frogs and lizards need insects to survive. About one-third of fish and animals eat insects whereas, about one-third of insects consume other insects. In a nutshell, insects are the bedrock of all land-associated habitats. Additionally, humans cannot exist without them, since almost a third of our food, particularly the most nutritional components, such as fruit and nuts, are substantially or fully pollinated by insects. From the dawn of civilization, tribal communities have played a critical role in conserving insect biodiversity in and around areas of their natural environment (Michael and Sam, 2018).

The biological roles they serve could bring various advantages to tribal inhabitants (Prather, 2013; Bennett et al., 2014; Baldock, 2015), as well as several negative consequences (Rust and Su, 2012; Dunn, 2010). Flora and insects often co-evolved in near proximity, with certain insect organisms exhibiting a strong degree of specialization (Forister et al., 2012). Plant species also provide tools for insect detritivores, pests, and parasitoids such as fruit, foraging, spawning, oviposition, shelter, and overwintering; indeed, activities that encourage these resources in agroecosystems through promoting plant diversity and systemic complexity are critical components of pest management (Landis et al., 2000). Thus, the existence of tribal people maintains the biodiversity of appropriate plant species which can sustain the vital element affecting insect diversity (Aronson, 2016). To introduce value-based insect protection, tribal communities follow certain traditional principles like insect conservation practices, which seek to recognize and encourage human local community concern for insects, thus guaranteeing that these insects serve as an essential source fostering human and insect well-being (Simaika and Samways, 2018). Since time immemorial, the bee product honey has been used by tribal communities in many Ayurveda formulations. It was also demonstrated by many tribal informants that insects, in particular could be used to produce drugs with analgesic, anti-rheumatic, anti-bacterial, immunological, anesthetic, and diuretic properties (Chakravorty et al., 2011). Cultural ethno-biological understanding and the acceptance of insects as food is an important part of indigenous therapies that are still known only by native people. Moreover, the medicinal properties of insects are often a tightly held secret that is often handed on orally from generation to generation. This method of information transfer is an age-old tradition and a widely recognized socio-cultural characteristic of the ethnic communities of India (Yamakawa, 1998).

While insects are most abundant in India, with approximately 645 district tribes and a large percentage of the population living in rural and semi-urban areas, the diversity of insects consumed as food is comparatively lesser, especially in the south and central parts of the country. It's difficult to explain why the condition is as it is; either insects as food are underappreciated, or ethnic people are sacrificing their rituals as a result of legislation requiring forest cover conservation and eradicating their extensive indigenous knowledge of utilizing insects in general, and as food in particular, before any scientific documentation. This tribal cultural structure has often fulfilled the dual function of using insects as food (i.e. food items and feed) and conservation (Chakravorty, 2014). While insects (species and individuals) are abundant in Arunachal Pradesh, groups of tribal people never capture and eat insects hastily, randomly, or without selection, but rather according to unwritten rules and traditions. Not only do the rituals dictate, which organisms may be harvested and used as food, but they often address the insects' medicinal properties (Meyer-Rochow et al., 2008). Some of the taboos among the tribal people such as tiger killing are unlucky; sparrows are symbols of good fortune, etc., also helped in their conservation.

2.8 TRIBES OF SOUTHERN INDIAN

The tribal people of South India are remarkable in that they remain uninhabited by modernity; consequently, people possess simplistic preferences yet are devoted to their traditions and beliefs. Dynamic and vibrant, tribal groups are unique in their customs and history. The biological diversity we see at present is the gift of our descent from thousands of years of evolution. Genetic diversity (diversity between habitats), species diversity (diversity at the species level), and ecological diversity are all used to quantify diversity. The preservation of biological diversity is critical for the survival of humans and other species of life. From their inception, plants have provided both physical and moral requirements for humans. The relationship has culminated in creating a one-of-a-kind information framework for the use and protection of plant genetic capital (IISE, 2011). The value of plant resources has been taught by cultural diversity in terms of ethnic populations. Understanding native communities about the cultural, moral, social, and economic qualities of plants can be extremely beneficial to humanity as a whole. It can have several dominant genes for the production of crop plants that are widely grown today (Armstrong and Botzler, 2003). It has the power to supply humans with a slew of new chemicals to tackle a range of human illnesses. Many of these stories from around the globe show how the ethnic experience has aided in the advancement of

Western society. New medicine for memory retention has been produced and sold from *B. monnieri*, which has historically been used in India to improve memory capacity (Jogender et al., 2020). Both of these references specifically demonstrate that cultural complexity is the primary basis of plant utilitarianism. The presence of cultural diversity is inextricably related to biological diversity. Since tribal cultures depend on biological resources for spiritual, religious, and cultural needs, this traditional biodiversity interpretation of native communities is not limited to sustenance. Both of these were viewed as life-sustaining services by tribal groups. As a result, they not only use them but still conserve them. The loss of any of these types of diversity will have a significant impact on humanity. Therefore, for meaningful conservation, all biological and cultural diversity can be viewed as a unit (Michelle, 2006). Ravishankar (1990, 1992, 1995) observed the conservation and sustainable use activities of some southern Indian tribal groups in his study, including the Kolams, Paliyars, Gonds, Koyas, KondaReddys, Malasar, Pardhans, Naikpods, Pulay, Irulas, Malaimalasar, Malayalis, Kadars, Lambadis of Andhra Pradesh, and Muthuvans. Southern India is home to a rather diverse ecosystem of fascinating species, due to its tribal residents' use of plants. Most of the plants are used only for one purpose, while there might be future applications in other parts of plants. *Brideliaretusa*, *Canthiumdoccum*, *Ficusracemosa*, *Longifoliaficus*, *Phyllaneralatifolia*, and *Terminaliabellerubarbila* were predominantly used for their edible seeds and fruits. Plant usage may be well defined by the selection. Seventy percent of humanity draws much of their nourishment from only a handful of animals. The tribals in southern India use between 1,000 and 1,500 plant varieties. Because it is hard to name just a few kinds of fruits, they usually consume fruits of various plant species, including Carissa, Cordia, and Memecylon (Xu, 2019; Ravichandran and Antony, 2019). Tribal groups use resources with caution and ecological competence. For e.g., in Tamil Nadu, "Kadars" pick only matured plants of the Dioscorea for tuber harvesting (Thamizoli and Balakrishna, 2015).

The production is shared by the whole population, preventing overexploitation. In the off-season, a part of the collection is prepared for consumption. This is a one-of-a-kind demonstration of neighborhood cooperation in plant utilities and resource management (Ravishankar, 2003). Tribal citizens in Sathyamangalam, which lies at the crossroads of Tamil Nadu, Karnataka, and Kerala, live a comfortable life with few everyday necessities. For decades, tribal communities residing in and near the forest have participated in hunting and collecting, cultivation, and evolving agriculture. They gathered food from the forest without endangering it (Arul and Kumar, 2019). Agriculture is the dominant source of production in all of the tribal settlements surveyed. Hill rice and millets such as Tenay, Ragi, Maize, Samai, and Kambu are among the subsistence crops grown by others. Farming is the main source of production in all of the tribal settlements surveyed. Cash crops like pepper and banana are increasingly making their way through several settlements (Janetius, 2017).

However, the genetic material conserved by the tribes up to this stage is under pressure due to the rising population in tribal region. This leads to less interaction among them which directly or indirectly inculcates unhealthy lifestyles among themselves. Outsiders' entrepreneurial behaviors contributed to the overexploitation of wealth that the tribes relied on for survival. Thus, it leads to depleting tribal capital which makes them do unnecessary practices for their income, finally resulting in unhealthy tribal attitudes. Traditional cultivars must be conserved by gene bank conservation to avoid genetic erosion since they provide the essential raw material for future crop improvement (Koen and Brain, 2004).

2.9 NORTH EASTERN TRIBES

The Monpa and Shertukpen tribes of Arunachal Pradesh's hilly western regions have ancient religious traditions and a wealth of indigenous knowledge. Their culture is identical to that of the other tribes in Arunachal Pradesh, but it is unique and strongly related to the climate. Their society is, in reality, strongly dominated by religious beliefs. Traditional cultures' rich information and traditions connected with the usage and protection of livestock, plants, and the environment as a whole have been the social pillars for shared coordination and long-term sustainability of the inhabitants (Ranjaysingh, 2013).

2.10 TRIBES ASSOCIATION IN PROTECTING NATURE

India has its own concerns regarding trees, which have a major role in the nation since ancient times. After the African continent, the second-largest tribal people reside in India (Rao et al., 2019). Realizing the tribes' privileges to obtain and distribute small forest products has immense potential for growing the tribes' revenue and raising their quality of life. This direct or indirect means of dependency on the forest for the economy would enable them to conserve it. In this way, tribes would be released from the clutches of money lenders and small business people who purchase lucrative minor produce from tribes at a discount and earn handsomely. The tribes' well being and welfare metrics would certainly increase as a consequence of the discretionary income. The tribals are widely regarded as the forests and resources guardians (Figure 2.3). The tribal people's present way of life is focused on subsistence farming. They also serve as wage workers in the forestry and sylviculture activities of state governments' forest departments (National Portal of India, 1951). Their needs and practices have to be documented so that herbal practitioners will have a greater understanding of how these tribes are related to the forest and what their unique

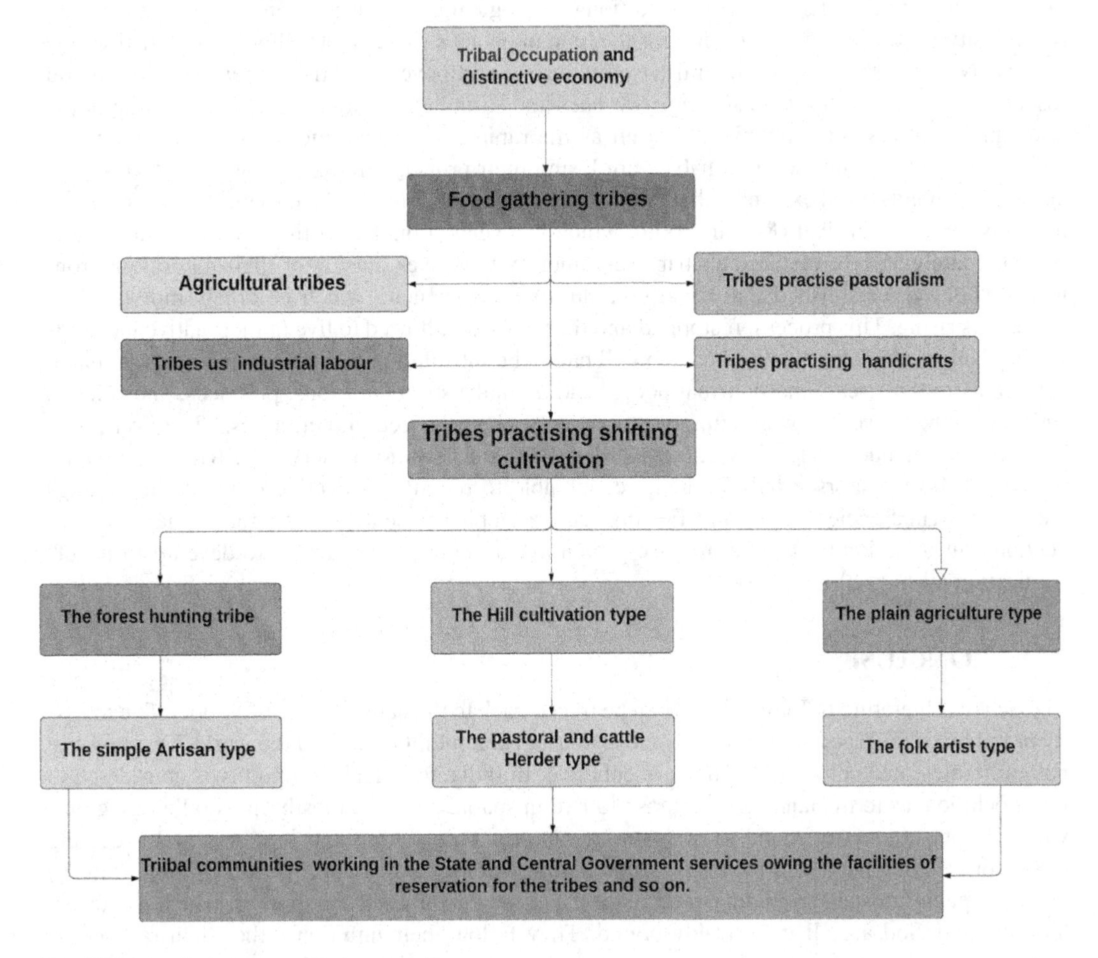

FIGURE 2.3 Tribal occupation and distinctive economy.

practices are. The social and legal facets of the partnership between tribals and the forest should have a strict implication, which further establishes that such a partnership has a solid legal framework because it is largely focused on socio-cultural influences and customs (Wackernagel et al., 2005; Philip, 2007).

The current big concern faced by the Aboriginal people is a lack of recognition of their rights. They are unaware of the specific privileges that have been bestowed upon them. The absence of any substantive law that defines tribal interests in environmental protection is the source of this shambles. Tribal people have been worshipping the forest and hilly regions as though it were a deity since the dawn of time and this is one of the key explanations why they defend it from damage as part of their religious obligations. But, owing to the evolving terms, the tribals have been suppressed in numerous forms by many influential factors of society. Consequently, it is important to identify their privileges and pass laws to assist tribal communities in protecting their rights (Narwani, 2004). Incentives should also be offered to tribal groups that grow conventional cultivars, which produce lower yields than high-yielding varieties. This allows tribals to plant genetically diverse, drought-resistant, and pest-tolerant crops, which will compensate for their poor production (Röös et al., 2018). Policy choices that could affect the natural equilibrium and nature can be undertaken with the help of indigenous people who reside in the region and may provide realistic feedback. Ethnic people are the ones who struggle the most in any environmental disaster when they are at the bottom of the socioeconomic ladder. Biodiversity depletion leads to cultural diversity loss, which is the cradle of awareness about plant beliefs. Ignoring cultural diversity security and the associated traditional ecological consciousness endangers the cause of biodiversity restoration, which is vital for the human planet's survival. The bourgeoisie insists that major MNCs investing in the mineral-based industry in impoverished tribal areas rich in mineral resources would raise foreign exchange revenues and stimulate economic growth, resulting in the development of essential amenities as well as the rapid socio-economic transition of the native community. It was identified that tribals are losing their property in big mining, iron companies, and also with non-tribal externals that coexist in such places. This leads to the migration of tribes to slums for their survival (Mehar, 2009), which is a generational shift that no one wants to see. Approximately 55% of the nation's tribal community now lives outside of their natural environments. It is well established that tribal community displacement, which is largely motivated by distress, is rising. This process, if stopped and the tribes are allowed to live in their native locations will not only make them survive but also will have a bountiful impact in improving environmental conditions in that area. Finally, tribal people should make strict decisions, practices, and rules in such a way they never allow any immoral practices to be practiced in their areas. The government has already amended strict Government order (GO) and laws to preserve the tribal communal properties. Investigators should be held accountable to record prospective use of their findings and the inherited ancient traditions. For instance, a biopharmaceutical corporation may use conventional information unearthed by a well-intentioned botanical researcher to develop a "novel" medication (Whitt, 2009).

2.11 CONCLUSION

The rich oral literature of India's tribes can be traced back to the nation's literature roots. The majority of Indian tribal groups are culturally close to tribal communities around the world. They exist in naturally integrated communities that are cohesive. In India, the tribal community is considered to be autochthonous to the land. They choose to live in spatial solitude, mostly in woodland regions, which do not often correspond to mainstream culture. They don't seem to care about acquiring money or utilizing labor as a means of accumulating interest and capital. Also, they believe in the human capacity to spell and understand reality, and they embrace a paradigm in which existence, humans, and God are all inextricably bound. They follow their intuition rather than rationality, regard the room around them as religious rather than secular, and have a subjective rather than logical sense of time. As a result, the world of tribal fantasy is opposed to contemporary Indian society. Indigenous peoples are an essential part of cultural complexity and ecological equilibrium, as well as an evolutionary treasure for scientists researching evolutionary and migratory history. The world owes them the courtesy of allowing them to remain in harmony on sacred lands that they have revered and protected for generations. Although Indigenous peoples strive to advocate for the environment. "We're all in this together". We need the rest of the planet to understand this, "This isn't about Indigenous communities heroically battling and losing our life to defend the earth". This is about all of us coming together through nations, races, and social backgrounds to transform the way our global economy operates.

REFERENCES

Alessa, L., Kliskey, A., and Gamble, J. (2016). The role of Indigenous science and local knowledge in integrated observing systems: moving toward adaptive capacity indices and early warning systems. *Sustain Science, 11*, 91–102. https://doi.org/10.1007/s11625-015-0295-7

Armstrong, S.J. and Botzler, R.G. (2003). *Environmental Ethics: Divergence and Convergence*. McGraw-Hill, Columbus, OH.

Aronson, J.K. (2016) The Hitchhiker's Guide to Clinical Pharmacology, Pharmacodynamics: How Drugs Work. https://www.cebm.net/wp-content/uploads/2016/05/Pharmacodynamics-How-drugs-work.pdf

Aronson, M.F.J., Lepczyk, C.A., Evans, K.L., Goddard, M.A., Lerman, S.B., MacIvor, J.S., Nilon, C.H., and Vargo. T. (2017). Biodiversity in the city: key challenges for urban green space management. *Frontiers in Ecology and the Environment, 15*, 189–196.

Arora, R.K. (1991a). Conservation and management concept and approach in plant genetic resources. In: Paroda, R.S. and Arora, R.K. (eds.) *IBPGR*. Regional office South and Southeast Asia, New Delhi, pp. 25.

Arora, R.K. (1991b). Plant diversity in Indian gene centre. In: Paroda, R.S. and Arora, R.K. (eds.) *Plant Genetic Resources Conservation and Management*. International Board for Plant Genetic Resources, New Delhi, India, pp. 25–54.

Arora, R.K. (1997). Ethnobotany and its role in conservation and use of plant genetic resources in India. *Journal of Ethnobotany, 9*(2), 6–15.

Arul, A.C. and Kumar, S.C. (2019). Livelihood status of uraly tribes in sathyamangalam block, Tamil Nadu. *International Journal of Scientific & Technology Research, 8*(12), 117–120.

Baldock, K. (2015). Where is the UK's pollinator biodiversity? The importance of urban areas for flower-visiting insects. *Proceedings of the Royal Society of London B, 282*, 20142849.

Balick, M.J. and Cox, P.A. (1996). *Plants, People, and Culture: The Science of Ethnobotany*. Scientific American Library, New York.

Bennett, T.M.B., Maynard, N.G., Cochran, P., Gough, R., Lynn, K., Maldonado, J., Vogesser, G., Wotkyns, S., and Cozzetto, K. (2014). Indigenous peoples, lands, and resources. In: Melillo, J.M., Richmond, T.C., and Yohe, G.W. (eds.) *Climate Change Impacts in the United States. The Third National Climate Assessment*. U.S. Global Change Research Program, [Place of Publication Unknown], pp. 297–317, Chapter 12.

Berlin, B. (1992). On the making of a comparative ethnobiology. In: Berlin, B. (ed.) *Ethnobiological Classification: Principles of Categorization of Plants and Animals in Traditional Societies*. Princeton University Press, Princeton, NJ, pp. 3–51.

Berlin, B., Breedlove, D.E., and Raven, P.H. (1973). General principles of classification and nomenclature in Folk biology. *American Anthropologist, 75*, 214–242.

Bijoy, C.R., Gopalakrishnan, S., and Khanna, S. (2010). India and the Rights of Indigenous Peoples. Constitutional, Legislative and Administrative Provisions Concerning Indigenous and Tribal Peoples in India and their Relation to International Law on Indigenous Peoples. Asia Indigenous Peoples Pact (AIPP) Foundation.

Biplab, D., Debbarma, T., Sen, S., and Chakraborty, R. (2010). Tribal life in the environment and biodiversity of Tripura, India. *Current World Environment, 5*(1), 59–66. http://doi.org/10.12944/CWE.5.1.08

Bruchac, M.M. (2014). Indigenous knowledge and traditional knowledge. In Claire S. (ed.) *Encyclopedia of Global Archaeology* (Vol. 10, pp. 3814–3824). Springer Science and Business Media, New York.

Burmol, K.S. and Naidu, T.S. (2007). National seminar on Tribal medicinal system and its contemporary relevance –Alluri Seetharama Raju centre for tribal studies and Research.

Chakravorty, J. (2014). Diversity of edible insects and practices of entomophagy in India: an overview. *Journal of Biodiversity, Bioprospecting and Development*, *1*, 124. http://doi.org/10.4172/2376-0214.1000124

Chakravorty, J., Ghosh, S., and Meyer-Rochow, V.B. (2011). Practices of entomophagy and entomotherapy by members of the Nyishi and Galo tribes, two ethnic groups of the state of Arunachal Pradesh (North-East India). *Journal of Ethnobiology and Ethnomedicine*, *7*, 5. https://doi.org/10.1186/1746-4269-7-5

Das, M.B., Kapoor, S., and Nikitin, D. (2010). A closer look at child mortality among adivasis in India. In World Bank (eds.) *Policy Research Working Paper*. Available at: http://documents.worldbank.org/curated/en/955711468044086021/pdf/WPS5231.pdf

Dunn, R.R. (2010). Global mapping of ecosystem disservices: the unspoken reality that nature sometimes kills us. *Biotropica*, *42*, 555–557.

Endicott, K.M. and Welsch, R.L. (2003). *Taking Sides. Clashing Views on Controversial Issues in Anthropology* (2nd ed.). McGraw-Hill/Dushkin, USA.

Etkin, N. (1988). Ethnopharmacology: biobehavioral approaches in the anthropological study of indigenous medicines. *Annual Reviews in Anthropology*, *17*, 23–42.

Etkin, N.L. (2001). Perspectives in ethnopharmacology: forging a closer link between bioscience and traditional empirical knowledge. *Journal of Ethnopharmacology*, *76*, 177–182.

Etkin, N.L. and Elisabetsky, E. (2005). Seeking a transdisciplinary and culturally germane science: the future of ethnopharmacology. *Journal of Ethnopharmacology*, *100*, 23–26.

Forister, M.L., Dyer, L.A., Singer, M.S., Stireman, J.O., and Lill, J.T. (2012). Revisiting the evolution of ecological specialization, with emphasis on insect–plant interactions. *Ecology*, *93*, 981–991.

Frei, B., Baltisberger, M., Sticher, O., and Heinrich, M. (1998). Medical ethnobotany of the Zapotecs of the Isthmus-Sierra (Oaxaca, Mexico): documentation and assessment of indigenous uses. *Journal of Ethnopharmacology*, *62*, 149–165.

Ganesan, S., Suresh, N., and Kesavan, L. (2004). Ethnomedicinal survey of lower palani of Tamilnadu. *Indian Journal of Traditional Knowlrdge*, *3*(3), 299–304.

Gurib Fakim, A. (2006). Medicinal plants: traditions of yesterday and drugs of tomorrow. *Molecular Aspect Medicine*, *27*(1), 1–93.

Harsha, V.H., Hebbar, S.S., Sripathi, V., and Hedge, G.R. (2003). Ethnomedicobotany of Uttarakhand District of Karnataka, India-plants in treatment of skin diseases. *Journal of Ethnopharmacology*, *84*, 37–40.

International Institute for Species Exploration (IISE). (2011). 2011 State of Observed Species (SOS). IISE, Tempe, AZ. Accessed May, 20, 2012. http://species.asu.edu/SOS

Jain, S.K., Singh, B.K., and Saklani, A. (1989). Some interesting medicinal plants known among several tribal societies of India. *Ethnobotany*, 1989, 89–100.

Janetius, S.T. (2017). Sathyamangalam dilemma: tribal relocation plan for tiger reserve expansion and the associated psychosocial problems. *International Research Journal* of *Social Sciences*, *6*(12), 1–4.

Jogender, M., Pooja, G., Monika, P., Deepti, D., and Diksha, D. (2020). Indian medicinal herbs and formulations for Alzheimer 's disease, from traditional knowledge to scientific assessment. *Journal of Brain Sciences*, *10*, 964. http://doi.org/10.3390/brainsci10120964

Koen, K. and Brain, B. (2004). Forest Products, Livelihood and Conservation: Case Studies of Non- Timber Forest Product Systems. Center for International Forestry Research, Bogor.

Landis, D.A., Wratten, S.D., and Gurr, G.M. (2000). Habitat management to conserve natural enemies of arthropod pests in agriculture. *Annual Review of Entomology*, *45*,175–201.

Leonti, M., Vibrans, H., Sticher, O., and Heinrich, M. (2001). Ethnopharmacology of the Popoluca, México: an evaluation. *Journal of Pharmacy and Pharmacology*, *53*, 1653–1669.

Lewis, W.H. and Elvin-Lewis, M.P. (2003*). Medical Botany: Plants Affecting Human Health*. John Wiley & Sons, New York.

Mehar, S.K. (2009) *Weathering Behaviour of Bagasse Fibre Reinforced Polymer Composite*. Thesis, National Institute of Technology, Rourkela

Meyer-Rochow, V.B., Nonaka, K., and Boulidam, S. (2008). More feared than revered: Insects and their impacts on human societies (with some specific data on the importance of entomophagy in a Laotian Setting). *Entomologieheute*, *20*, 3–25.

Michelle, C. (2006). Biocultural diversity: moving beyond the realm of 'indigenous' and 'local' people. *Human Ecology*, *34*(2), 185–200. http://doi.org/10.1007/s10745-006-9013-5

Michael, J. and Sam, W. (2018). Insect conservation for the twenty-first century. In Mohammad, M.S. and Umar, S. (eds.) *Insect Science-Diversity, Conservation and Nutrition*. IntechOpen, London. http://doi.org/10.5772/intechopen.73864

MOTA and GoI. (2018). Ministry of Tribal Affairs: Annual Report 2017–18. Retrieved from https://tribal.nic.in/writereaddata/AnnualReport/AR2017-18.pdf

Mukherjee, P.K. and Wahile, A. (2006). Integrated approaches towards drug development from Ayurveda and other Indian system of medicines. *The Journal of Ethnopharmacology*, *103*(1), 25–35.

Namsa, N.D., Mandal, M., and Tangjang, S. (2011). Ethnobotany of the Monpa ethnic group at Arunachal Pradesh, India. *Journal of Ethnobiology and Ethnomedicine*, *7*, 31. https://doi.org/10.1186/1746-4269-7-31

Narwani, G.S. (2004). *Tribal Law in India* (1st ed.). Rawat Publications, Jaipur, India.

Nasi, R., Brown, D., Wilkie, D., Bennett, E., Tutin, C., van Tol, G., and Christophersen, T. (2008). Conservation and use of wildlife-based resources: the bushmeat crisis. Secretariat of the Convention on Biological Diversity, Montreal, and Center for International Forestry Research (CIFOR), Bogor. Technical Series no. 33, 50.

Nelson, E.A.R., Benett, A.B., and Lovell, S.T. (2014). A comparison of arthropod abundance and arthropod mediated predation services in urban green spaces. *Insect Conservation and Diversity*, *7*, 405–412.

New, T.R. (2015). *Insect Conservation and Urban Environments*. Springer, New York.

Panghal, M., Arya, V., Yadav, S., Kumar, S., and Yadav, J.P. (2010). Indigenous knowledge of medicinal plants used by Saperas community of Khetawas, Jhajjar District, Haryana, India. *Journal of Ethnobiology and Ethnomedicine*, *6*(1), 4.

Philip, H. (2007). The human impact on biological diversity. How species adapt to urban challenges sheds light on evolution and provides clues about conservation. *EMBO Reports*, *8*(4), 316–318. https://doi.org/10.1038/sj.embor.7400951

Prather, C.M. (2013). Invertebrates, ecosystem services and climate change. *Biological Reviews*, *88*, 327–348.

Rajiv Rai, V.N. (2012). The Role of Ethnic and Indigenous People of India and Their Culture in the Conservation of Biodiversity. http://www.fao.org/3/xii/0186-a1.htm#fn1

Ranjaysingh, K. (2013). Ecoculture and subsistence living of Monpa community in the eastern Himalayas: an ethnoecological study in Arunachal Pradesh. *Indian Journal of Traditional Knowledge*, *12*(3), 441–453.

Ravichandran, L.K. and Antony, U. (2019). The impact of the green revolution on indigenous crops of India. *Journals of Ethnic Food*, *6*, 8. https://doi.org/10.1186/s42779-019-0011-9

Ravishankar, T. (1990). Ethnobotanical studies in Adilabad and Karimnagar districts of Andhra Pradesh, India. Ph. D. Thesis. Bharathiar University, Coimbatore.

Ravishankar, T. (2003). Traditional Knowledge and Conservation of Biodiversity for Sustainable Livelihoods by Tribal Communities in Southern India. XII world Forest congress.

Ravishankar, T. and Henry, A.N. (1992). Ethnobotany of Adilabad district, Andhra Pradesh, India. *Ethnobotany*, *4*, 45–52.

Ravishankar, T. and Hosagouda, V.B. (1995). *Centres of Plant Diversity, A guide and Strategy for their Conservation*. IUCN Publications Unit, Cambridge, UK.

Robineau, L. and Soejarto, D.D. (1996). Tramil: a research project on the medicinal plant resources of the Caribbean. In Balick, M.J., Elisabetsky, E., and Laird, S.A. (eds.) *Medicinal Resources of the Tropical Forests*. Columbia University Press, New York, pp. 317–325.

Röös, E., Mie, A., and Wivstad, M. (2018). Risks and opportunities of increasing yields in organic farming. A review. *Agronomy for Sustainable Development*, *38*, 14. https://doi.org/10.1007/s13593-018-0489-3

Rust, M.K. and Su, N.Y. (2012). Managing social insects of urban importance. *Annual Review of Entomology*, *57*, 355–375.

Sattler, T., Obrist, M., Duelli, P., and Moretti, M. (2011). Urban arthropod communities: added value or just a blend of surrounding biodiversity? *Landscape and Urban Planning*, *103*, 347–361.

Sevugaperumal, G. (2004). Ethnomedicinal survey of lower Palani Hills of Tamil Nadu. *Indian Journal of Traditional Knowledge*, *3*(3), 299–304.

Simaika, J.P. and Samways, M.J. (2018). Insect conservation psychology. *Journal of Insect Conservation*, *22*, 635–642. https://doi.org/10.1007/s10841-018-0047-y

Sreenivasa Rao, J., Shivudu, G., Hrusikesh, P., and Kalyan, R.P. (2019). Livelihood strategies resource and nutritional status of forest dependent primitive tribes Chenchu in Andhra Pradesh and Telangana States. *Nutrition and Food Science International Journal*, *8*(2), 555735. https://doi.org/10.19080/NFSIJ.2019.08.555735

Suranjit, S.K. (1996). Early state formation in tribal areas of east-central India. *Economic and Political Weekly*, *31*(13), 824–834.

Thamizoli, P. and Balakrishna, P. (2015). *Sustainable Development – Stories from Those Making It Possible*. Fledge, India.

The 2011 Government of India Census recorded 8.2 per cent of India's population as tribal.
The Constitution (First Amendment) ACT (1951). National Portal of India.
Vartak, K.V.D. (1996). Sacred groved of tribals for in-situ conservation of biodiversity. In Jain, S.K. (eds.) *Ethnobiology in Human Welfare*. Deep Publications, New Delhi, pp. 300–302.
Vikram, S. and Shailly, D. (2018). Ethnomedicine and tribes: a case study of the Baiga's traditional treatment. *Research & Reviews: A Journal of Health Professions*, *8*(2), 62–67.
Wackernagel, M., Kitzes, J., Cheng, D., Goldfinger, S., Espinas, J., Moran, D., Monfreda, C., Loh, J., O'Gorman, D., and Wong, I. (2005). Asia-Pacific 2005 - The Ecological Footprint and Natural Wealth. WWF, Global Footprint Network, Kadoorie Farm and Botanic Garden.
Whitt, L. (2009). *Science, Colonialism, and Indigenous Peoples: The Cultural Politics of Law and Knowledge*. Cambridge University Press, Cambridge, UK, pp. 284.
Xu, J. (2019). Sustaining biodiversity and ecosystem services in the Hindu Kush Himalaya. In Wester, P., Mishra, A., Mukherji, A., and Shrestha, A. (eds.) *The Hindu Kush Himalaya Assessment*. Springer, Cham. https://doi.org/10.1007/978-3-319-92288-1-5
Yamakawa, M. (1998). Insect antibacterial proteins: regulatory mechanisms of their synthesis and a possibility as new antibiotics. *The Journal of Sericultural Science of Japan*, *67*, 163–182.
Zubair, A., Sushma, S., Mudasir, Y., Malik, R., and Bussmann, W. (2017). Sacred groves: myths, beliefs, and biodiversity conservation—a case study from Western Himalaya, India. *International Journal of Ecology*, 9, 12. https://doi.org/10.1155/2017/3828609

3 Soil Bioengineering Practices for Sustainable Ecosystem Restoration in Landslide-Affected Areas

Gobinath R, Gayathiri Ekambaram, Paniswamy Prakash, Kumaravel Priya, and Venkata SSR Marella

3.1 INTRODUCTION

Soil bioengineering stabilizes landslides and reduces habitat damage and it's becoming more common across the world, but its efficacy needs to be further assessed by post-intervention environmental testing. To shield slopes from erosion and shallow mass migration, soil bioengineering steps incorporate the usage of living plants and inert mechanical constructions. However, considering the reality, performing such experiments is vital in developing potential remedial action, which is seldom carried out. In particular, instilling soil bioengineering studies and their biotechnical characteristics into native people will be essential and highly beneficial, as these techniques, when combined with conventional practices, would aid in soil stabilization, enhancing the effectiveness of such projects. According to a recent survey, Hong Kong, the Philippines, and Malaysia have used soil bioengineering technologies rather than many of the regions, and bioengineering alteration has a strong degree of understanding, especially for improving soil stability (Vaughn et al., 2010).

3.2 ECOSYSTEM DETERIORATION AND NEED TO REPLENISH

Ecological restoration seeks to re-establish, initiate, or accelerate the development of a disturbed environment. Disturbances are environmental modifications that disrupt the structure and function of an ecosystem. Landslides, deforestation, damming waterways, heavy grazing, earthquakes, flooding, and fires are also common disturbances. Restoration efforts can be intended to replicate an environment that existed prior to the disturbance or to establish a new ecosystem where none existed previously. The experimental study of rehabilitating disturbed habitats by human activity is called restoration ecology (Figure 3.1).

Restoration is focused on a number of landscape ecology principles. Since restored areas are frequently limited and remote, they are particularly vulnerable to issues associated with habitat degradation. The term "habitat fragmentation" refers to the process by which contiguous region of habitats were isolated due to ecological or anthropogenic causes such as constructing roadways in forests, for instance. Generally, annihilation results in tiny, fragmented areas of hospitable habitat. Smaller ecosystems host fewer plants and communities that are more susceptible to inbreeding and local extinction. According to island biogeography theory, species are more likely to survive in vast and/or well-connected habitat patches that are connected to ecosystems in other hospitable environments. This hypothesis presupposes a uniform and inhospitable matrix—the area between habitat patches. Oceanic islands are the most prominent manifestations of this concept; they are clusters of marine species surrounded by unlivable land. The existing distinction between accommodating

DOI: 10.1201/9781003303237-3

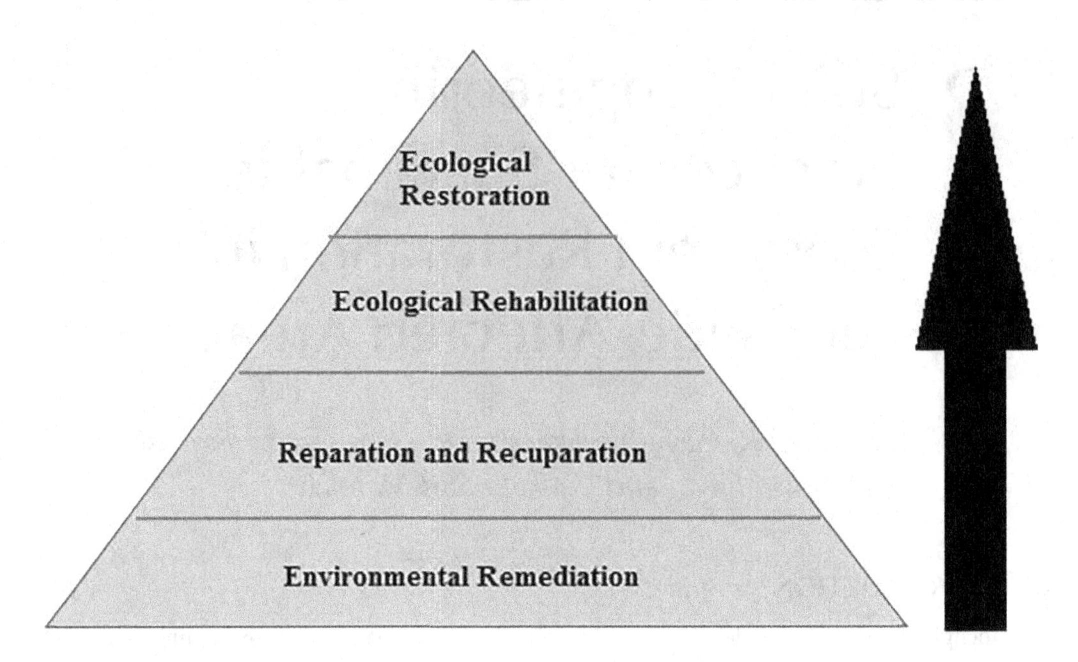

FIGURE 3.1 Ecological restoration process.

versus uninhabitable habitats also has recently been modified to allow only for the existence of different types of habitat patches that are combined to form a patch mosaic. The mosaic's various patches can be more or less conducive to the plants, populations, and ecosystem functions targeted by conservation activities.

Fragmentation will also increase detrimental edge effects by:

- The effects of one ecosystem on an adjacent habitat and
- By increasing the volume of edge habitat and decreasing the distances between edges.

For example, due to the increased prevalence of native weeds around forest edges, small forest fragments (with more edge habitat) are more prone to invasion. By establishing or restoring linkages between habitat patches in fractured ecosystems, restoration efforts often aim to improve communication within habitat patches (Figure 3.2). Corridors and stepping stones are two types of linkages that are often used to increase communication. Corridors are longitudinal swaths of forest that connect habitat patches that were isolated. Isolated habitat areas that are well enough to allow for better movement across the landscape (Vaughn et al., 2010).

3.2.1 Evaluating the Area

A critical understanding of the restoration site's current condition is important in determining the safest option. This move identifies the sources of habitat disruption and strategies for halting or restoring them.

3.2.2 Identifying Project Goals

Practitioners can visit reference sites (similar, neighboring natural environments) or its historic place and record the documents of community in the pre-disaster duration in order to identify goals

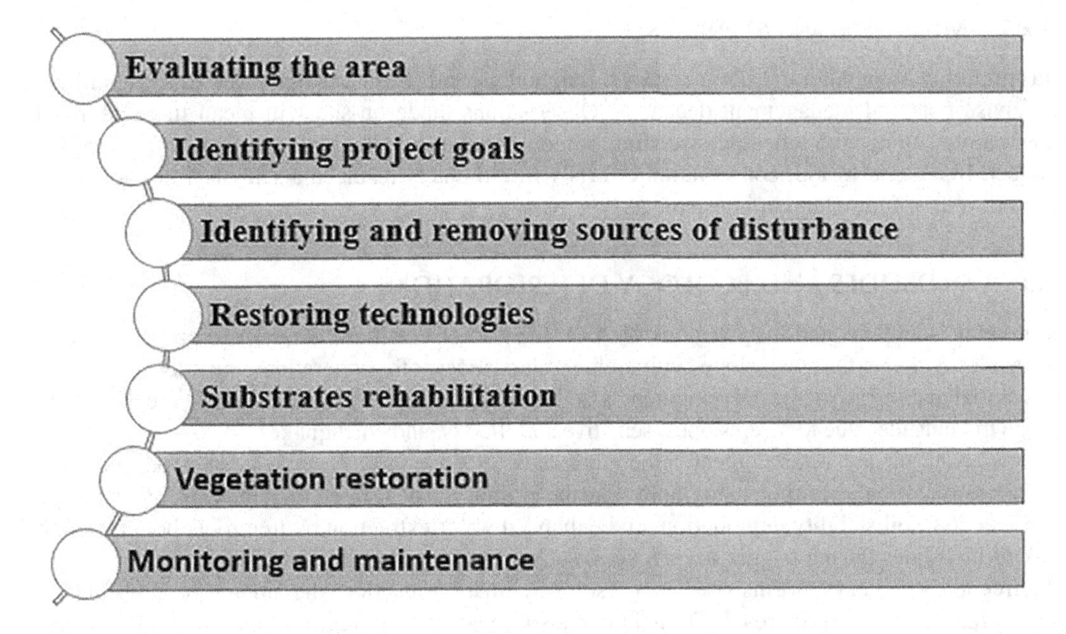

FIGURE 3.2 A schematic diagram explaining the steps involved in restoration project goals.

for the revived community. Additionally, goals can require an assessment of the organisms that are ideally adapted to current or potential environmental conditions.

3.2.3 Identifying and Removing Sources of Disturbance

Before reconstruction can be effective, it will be necessary to identify and eliminate sources of disturbance. For example, cessation of logging, forestry, or other erosion-causing activities; restriction of livestock in riparian areas; removal of radioactive substances from land or sediment; and eradication of invading tropical habitats.

3.2.4 Restoring Technologies

Sometimes, restoring critical ecological catastrophic mechanisms, such as natural flood and fire regimes, is sufficient to maintain ecosystem integrity. Native plants and animals that have adapted to tolerate or need natural disturbance regimes may reintroduce themselves in these instances without practitioner interference.

3.2.5 Substrates Rehabilitation

This term refers to any practice that aims to restore altered surface texture or chemical composition, as well as altered hydrogeological frameworks or water management.

3.2.6 Vegetation Restoration

Conservation efforts often include the direct revegetation of a property. Typically, natural plants that are adapted to the local environmental conditions are planted. To maintain genetic diversity, cuttings or seeds are typically harvested from a range of sources in a particular area. Planting vegetation using such seeds or seedlings is possible.

3.2.7 Monitoring and Maintenance

It is important to monitor the site recovery throughout periods to ensure that objectives are fulfilled to advise potential management decisions. Observations made on-site will mean that additional intervention, along with scheduled weeding, is required to ensure the long-term performance of the project. In an ideal world, conservation efforts will ultimately result in a self-sustaining environment devoid of future human interference.

3.3 LANDSLIDES AND ECOLOGY DETERIORATION

In several mountains and hilly regions of Asia, the risk of death and damage to property due to landslides is rising. On steep slopes vulnerable to landslides, other operations, such as mining and residential and infrastructure development, also began to expand. Wide areas of Asia are especially prone to landslides due to steep slopes, sensitive soil, floods, and earthquakes.

Increasing surface water content is the main cause of slope failure, while steep slopes, unstable soils, or water-concentrating topography are the primary risk factors for landslide. Poor bridge construction, soil stability degradation, and enhanced water extraction by tree roots both raise the risk of landslides during trigger occurrences, such as prolonged heavy rainfall or earthquakes. If the frequency of heavy storms rises as a result of climatic transition, the amount of landslides in some areas of Asia will increase. Drought can also impact certain regions, resulting in root death and decay and forest fire, which would possibly decrease the soil reinforcement offered by trees and raise the occurrence of landslides. Various forms of landslides perform differently, create high challenges, and have varying environmental consequences. To manage landslides and massive landslide terrain intelligently and efficiently, the identification and classification were needed. Today, the most often employed classifications are those given by Hungr et al. (2001) and Varnes (1996).

3.4 LANDSLIDE RESTORATION

Ecosystems have a characteristic propensity to self-alleviate after either natural or human-induced disruption. Even so, innovations are used to improve the recovery phase in heavily damaged areas, including the Highlands reducing destruction incurred by landslides (Aronson et al., 1993 Vidya et al., 2021). A cornerstone of the concept of ecological regeneration is that the manipulations of succession must work against any strategic measures, like that of keeping some species or vegetation patterns intact or maintaining the ecosystem in place (Walker et al., 2007).

3.5 LANDSLIDE RESTORATION: LOSSES INCURRED DIRECTLY VS. INDIRECTLY

Costs associated with landslides cover both specific and indirect damage to public and private property. The term "direct costs" refers to the costs of replacement, repair, reconstruction, and maintenance of property or installations as a result of direct landslide damage or degradation (Schuster et al., 1986, 1996). Many other costs associated with landslides are indirect; examples include the following:

- Lowered property prices of regions at risk of landslides.
- Tax income loss on devalued assets as a consequence of landslides.
- Industrial, farming, and woodland production losses, as well as income losses from tourism, as a consequence of harm to property or buildings or disruptions to transportation networks.
- Productivity loss in humans or domestic animals as a result of illness, disease, or psychological distress.
- Costs associated with programs designed to deter or reduce landslide behavior.

3.6 LANDSLIDE-ASSOCIATED PROBLEMS AND THEIR MITIGATION & MANAGEMENT MEASURES

Slope instability accompanied by sudden mass migration of soils or rocks over a continuous surface of shear is known as landslides (Varnes, 1996). When forces influencing volatility outnumber forces supporting slope stability, landslides occur (Conforth, 2005). Floods (Larsen and Simon, 1993; Stern, 1995), earthquakes (Garwood et al., 1979; Restrepo and Álvarez, 2006), or land use by humans and land practices like road building and urbanization (Larsen and Simon, 1993; Stern, 1995; Pande et al., 2021; Orimoloye et al., 2022) could all cause them (Walker and Del Moral, 2003; Sidle and Ochiai, 2006) Soil erosion are typical soil erosion mechanisms that give great examples of the complex interplay between disruption and succession. Since soil stability varies greatly both spatially and temporally, landslide recovery using biological methods is challenging on landslide surfaces. Stabilization of key colonization and natural land cover (Pande et al., 2018, 2022), amendments of natural resources, dispersal facilitation to alleviate oligarchy bottlenecks, focus on genetically trivial organisms, and encouragement of contact with the surrounding environment will all help to facilitate the regeneration of self-sustaining populations on landslides. Soil bioengineering strategies that facilitate successional processes may help to streamline restoration efforts (Figure 3.3). Eight out of every ten construction-triggered landslides occur in India, which is about 28% of the total, and then in China, amounting around 9% of the total (6%). On the other hand, rainfall contributed 16% of all landslides in India, that predominantly (*ca.* 77%) happened during the monsoon season. India has had the highest proportion of landslides caused by mining, at 12%, followed by Indonesia (11.7%) and China (10%).

3.7 BIOENGINEERING STRATEGY IN LANDSLIDE RISK MITIGATION

The very first concept, *ingenieurbiologie* (Engineering Biology), was coined by Arthur Von Kruedener in 1951 to describe the usage of potential plant species to conduct bioengineering functions in Germany. It was then mistranslated as biological engineering or bioengineering. Additionally, it is referred to as biotechnical engineering within wider fields of research, such as ecological engineering (Bergen et al., 2001), reclamation sciences (Polster, 2008), or literally soil protection (Hudson, 1982). Additionally, to keep the research focused on its environmental intent, soil bioengineering was embraced. The bioengineering application successfully addresses landslides, cliffs, rock fall, and other associated concerns. This application is advanced and an adaptive solution that helps stabilize slopes and sites on hillsides and enhances site efficiency by improved status of soil nutrients. Biotechnology includes the use of live plants and other auxiliary

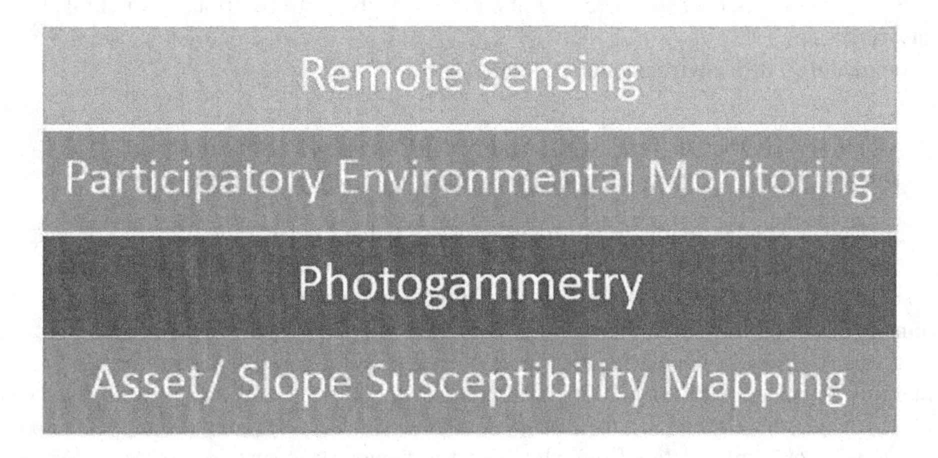

FIGURE 3.3 Strategies of bioengineering for landslide mitigation.

resources to strengthen hills, acts as an advanced strategy to shield the slope from surface stress, reduces the danger of floor tipping, and maximizes surface drainage. For bioengineering applications, it is important to pick species with colonizing behavior, rapid growth, and a thick and extensive root structure with adventitious root, with quick and easy propagation. In order to restore habitats in destroyed/degraded landslides, the implementation of bioengineering is also important. This technology is also an economical option that incorporates local materials and encourages the local public to be interested in management and upkeep. One of the core benefits of bioengineering in protecting slopes is that plants and mechanical frameworks could work together to reinforce each other or complement each other. In some ways, field experiments (White, 1978) found that integrated structural and vegetative protection systems were cheaper than the usage alone of vegetation or structure. Vegetative techniques by themselves are typically much cheaper than earth restoration mechanisms or other natural processes. On the other side, their efficiency in avoiding soil erosion or stopping movement in the slope may also be significantly reduced under harsh conditions. The application for bioengineering blends into the terrain, and thus, it is economically viable, but both geological and traditional earth retention systems are not visually impacted on the landscape. As a result, native plants may be integrated into the architecture. Furthermore, the introduction of indigenous species into their native habitats further improves the soil biodiversity, thereby improving the soil fertility which perhaps helps further in enhancing the vegetation. Consequently, improved vegetation adds more soil organic carbon, thus helping in mitigating the ill effects of climate change. Additionally, mycorrhizal networks that are associated with the plant root systems could potentially contribute the soil strength while adding recalcitrant protein compounds such as glomalin. More importantly, root nexus developed between the trees further strengthens the soil aggregations in a matured ecosystem.

3.8 APPLICATION

The type of erosion and slope collapse mechanism determine the application of soil bioengineering techniques.

Soil bioengineering approaches have great attributes:

1. Initial costs are lower, and long-term maintenance costs are lower than those associated with traditional approaches;
2. Low-cost potential plant species are implemented;
3. Potential impacts such as wildlife habitat, increased water management, and aesthetic appeal;
4. Enhanced strength over time because of plant roots' ability to contribute structural stability; and
5. Compatibility with environmental regulations.

3.9 PARTICIPATION OF THE SOCIAL ENVIRONMENT IN THE APPLICATION AND SUCCESSION OF SOIL BIOENGINEERING

Public knowledge helps in maintaining the progress of habitat regeneration. From a management perspective, understanding the motivations of those who engage in conservation is extremely beneficial because it assists administrators in developing restoration curriculum, encouraging group engagement, and obtaining fund (Stone et al., 2008; Le et al., 2012). Indigenous awareness methods and values have grown in popularity as a basis of expertise for ecosystem planning and resource sustainability (Menzies and Butler, 2006; Armstrong et al., 2007). Traditional as well as local communities' strategies have preserved their ecosystem and given services over several years and offer an exceptional pool of information from which creative responses to current and potential problems may be extracted (Laureano, 2000; Menzies and Butler, 2006; Armstrong et al., 2007).

Soil bioengineering technological tools are prevalently used in Nepal to address slope erosion issues, particularly where superficial landslides are associated with highway construction (Devkota et al., 2006; Mathema and Joshi, 2010).

3.9.1 Seedlings

Seeding is the process of applying seed mixtures of turf, forb, and woody plants to sloped regions. This mechanism results in the formation of a superficial fibrous rooting region in the upper foot of the surface that holds ground soil together and protects them from run-off, storm, and thawing cycles erosion. Seeding is often used in conjunction with other planting methods to address the majority of erosion management problems. Slopes may be seeded manually or by planting seeds in tiny holes along the hill. Broadcast seeding is the most often used methodology in research projects which is considered as quick and effective if the slope soil has been slightly roughened and mulch has been applied to avoid desiccation and wind transport. Hydroseeding brings this application strategy a step further by combining crop, water, manure, and mulch into a single combination. Hydroseeding is advantageous in difficult-to-reach regions and over vast slope faces. Drilling soil holes is a more time-consuming operation; however, it significantly decreases the seed amounts needed. This approach appears to be particularly effective on gentle hills, in narrow fields, and for seed stock of woody plants. Usually, the soil hole is 3 inches in diameter and 4 inches wide. After adding a low-dose fertilizer capsule into the opening later three and a half inches of soil are added above, accompanied by approximately 20 plants. If this is completed, the opening should be filled in accordance with the seed supplier's specifications.

3.9.2 Bare Root Planting

Container and bare root planting entails inserting individual or clusters of rooted plants into sloped gaps. This strategy is usually used for woody plants or plants that will grow to a standardized root coverage over time. To maintain a consistent vegetation system, it is a safe idea to use transplants from surrounding locations. By using rooted plant resources, you may escape the vital germination time associated with seeding or the root growth cycle associated with cuttings. It is a smart choice to use plant groupings or bunch plantings with subtly different rooting and foliage characteristics to ensure rapid stabilization performance. Bunch planting often enables slope-adapted plants to gradually overtake the plant population. Mulching is advised around all plants as a minimum requirement.

3.9.3 Live Stakes

Live stakes are woody plant pieces that have been sliced into fragments and inserted into the hill. Stakes are typically made of hardy species that develop readily and gradually evolve into mature woody shrubs that reinforce the soil. Stakes can be installed in autumn or spring, based on the dormant season of the initial plant. Generally, live stakes are 2–3 feet tall and 12–12 inches in diameter. To ensure proper installation, stakes should have a flat top and a diagonal bottom. This form of stabilization may be used alone or in conjunction with other planting strategies. If used parallel to the hill, stakes should be spaced according to species guidelines. Each row should have stakes arranged in a diamond shape at or around one another. By angling the stake horizontally, you may promote root development over the entire length of the stake below ground. Softly tape the stake through drilled channels that are significantly wider in diameter than the stake. If the top part of the stake is hurt, it can be replaced. Mulching is a safe procedure to follow after installation. With contour wattling, live staking may be used to protect wattles along a hill (Figure 3.4).

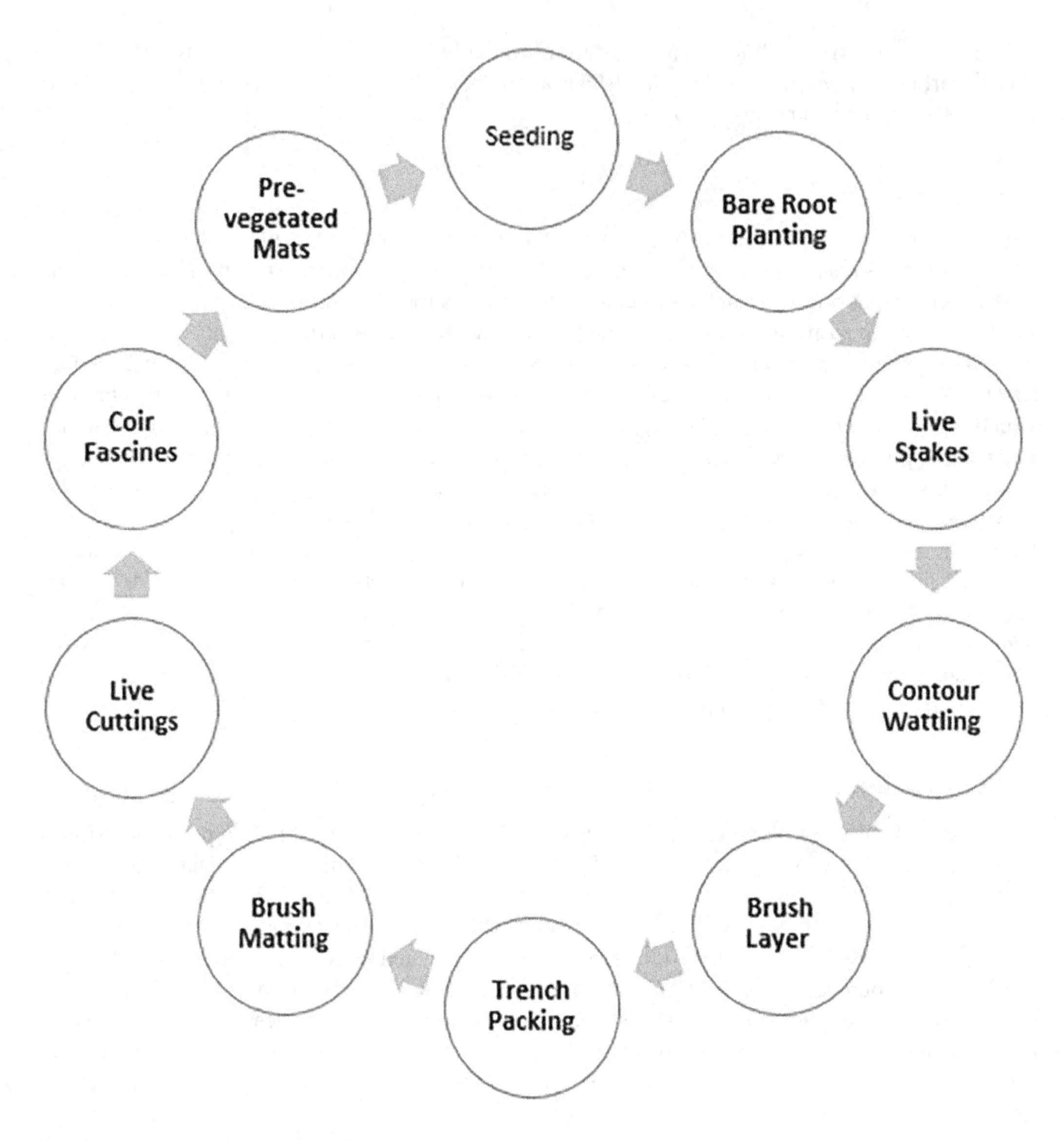

FIGURE 3.4 Installation methods.

3.9.4 Contour Wattling

Contour wattling is a procedure for securing shallow soil structures against landslides. Such a strategy entails coiling or bundling lengths of woody plant fiber (also called live fascines). These wraps, which range in diameter from 8 to 10 inches, were placed consistently over hill contour lines. Cabling around the slope assists in intercepting surface runoff by routing it laterally prior to erosion. Wattles aid in sediment trapping by forming walls that shield downslope areas from material falls or erosion.

3.9.5 Brush Layer

Brush layer planting entails inserting live woody plant material into the slopes layer into contour trenches. This technique is well suited to heavily damaged or eroded cut and fill areas. Layering reinforces the surface, making it more resistant to shallow-seated landslides. Brush layers trap debris as it falls down the hill. Brush layering is detrimental to thick soil formations and must be used sparingly to combat gully erosion. This approach is time consuming and is not conducive to the construction of water channels.

3.9.6 Trench Packing

This technique is being used to delay or disperse the movement of water by putting live plants parallel to the movement in a trench. To mitigate wave effect, potential plants are positioned parallel to the shoreline in trenches. Numerous trenches can be used, each containing a different herb, depending on the distance to water. To dissipate wave energy, a large planting area is typically needed. In upland zones, trench packing slows and spreads water over the soil surface, limiting the likelihood of erosion. Additionally, trench packing may help to monitor shallow seeps, safeguard wetland development and reconstruction, and safeguard unused highways.

3.9.7 Brush Matting

This technique supports streambanks by enveloping them in a mattress-like coating of branches that protects the surface and slows the flow of water. The mat is made up of interwoven, normally dead, branches that are held in place by live stake, rope, twine, or live branches. Live stakes are sometimes cut from dormant willow. Brush matting assists in sedimentation and the growth of vegetation on banks. This technique necessitates a significant quantity of branches.

3.9.8 Live Cuttings

This process may help to anchor resources and raise the planting area on a hill. Live cuttings range in length from 18 inches to 4 feet. Longer cuttings were used for live staking of wattles, whereas small cuttings are used for planting (Tosi, 2007).

3.9.9 Coir Fascines

They are wattles constructed with the help of outer husk of coconut's fibers. Since coir has a higher density, it cannot sink and decays slowly. Coir fascines are often used engineering technique is often used for streambank and wetland preservation projects that provide a natural appearance. Coir fascines are positioned such that their tips are flush with the water's surface. To produce a more realistic appearance, live plants may be incorporated into coir fascines (Tosi, 2007).

3.9.10 Pre-Vegetated Mats

Pre-vegetated mats are living plants that have been grown on a biodegradable movable mat. They are available in a variety of sizes and materials that are easily transported later, assembled in one piece. These mats are usually 4 by 8 feet in size for easy facilitation. Mats are cultivated in nurseries for up to a year or longer in order to have a stable base for plants. Without special packaging, thin mats may be rolled up and shipped (Figure 3.5). Due to the weight of thick sheets, they must be treated using heavy machinery. Mats that have been pre-vegetated are constructed using coir or any other slowly degradable substances and may contain a variety of plant species; hence, these mats are typically used in wetlands or riversides. Usually, the majority of pre-vegetated mats are purchased on a custom basis 1 or 2 years in advance (Tosi, 2007).

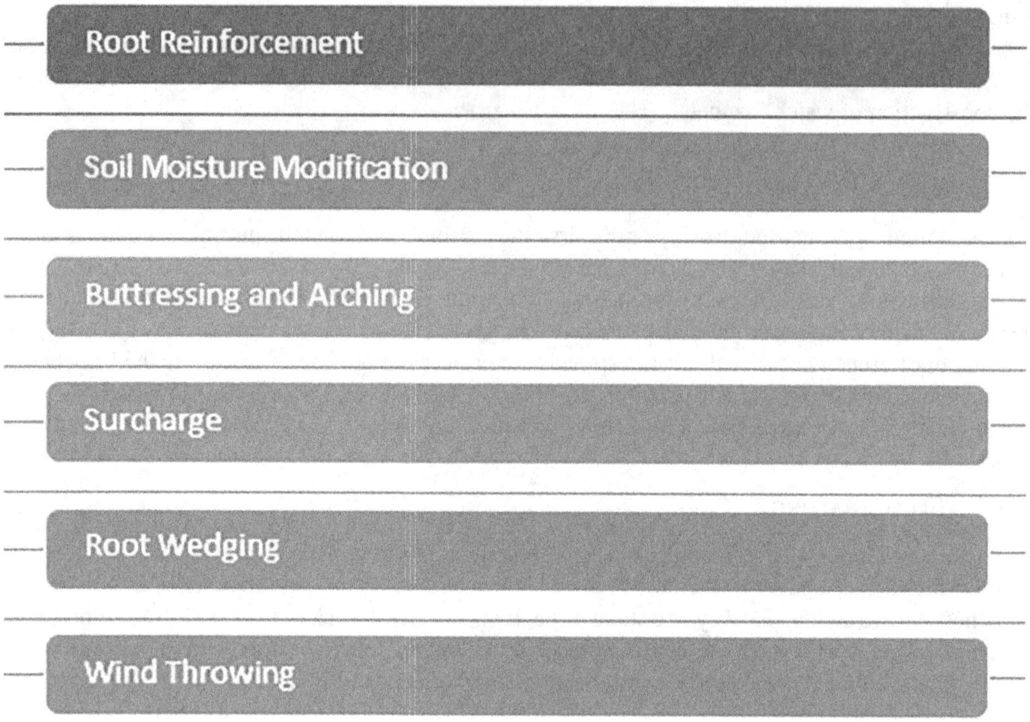

Root Reinforcement

Soil Moisture Modification

Buttressing and Arching

Surcharge

Root Wedging

Wind Throwing

FIGURE 3.5 Indicators influencing slope via vegetation.

3.10 SLOPE STABILITY AND ECOSYSTEM RESTORATION

To enhance slope stabilization, it is important to sustainably manage landslides and sediments' development in the hilly regions or waterways. Trees aid in water infiltration, protect the soil surfaces, strengthen and fertilize the soil, and enhance biological development in the soil. By incorporating plants into ecological regeneration or reconstruction programs, we will help restore habitat systems and functions, as well as general ecological infrastructure. However, foliage has the ability to destabilize slopes as well. For instance, when the high winds occur, taller trees may serve as a trigger, resulting in their uprooting, compromising the mechanical integrity of the slope (Mitchell, 2013). To understand the role of potential plants on the hydrology of slopes, it is necessary to know about soil hydrology, which is a major cause of soil erosion. Although precipitation events are often attributed to their initiation, which alters the water pressures that causes a slope to collapse, suction reduces when moisture infiltrates into soil on a hill, resulting in the loss of intensity and potential collapse. A high water content (or low suction) along with lower observable soil cohesion and increased landslide risk, while a lower water level (or high suction) is combined with stronger apparent cohesion and decreased landslide risk (Fredlund, 1979).

3.11 NATIVE SPECIES IN ECOSYSTEM REGENERATION

Traditional ecosystem conservation methods, as well as modern bioengineering applications, are based on a fundamental principle known as "Selection of Species." The "Species of Choice" can be chosen due to its home field advantage and perhaps demands less care and attention. The chosen plant species must have a deep and comprehensive root system, as well as be hard, quick-growing, and captivating (Sastry et al., 1990). Natural species may be more effective and useful in regeneration and bioengineering systems because it has fewer predators than other species and because disruptions enable non-seeded species to germinate and grow (Munshower, 1993). Global Environment Outlook has planned to include using native species extensively and numerous alternate soil bioengineering techniques (GEO, 2000). The importance of natural vegetation in the protection and stabilization of landslides cannot be overstated. Plant not only covers the surface from rain and wind disruption, but its complex root system also often serves as coherence, connecting loose soil, and prevents soil erosion.

3.12 BARRIER TO ECOLOGICAL RESTORATION OF LANDSLIDES

The first impediment to ecological regeneration and landslide bioengineering is persistent soil erosion (Walker and Shiels, 2008). Effective measures to strengthen slopes, such as blocking reservoirs or installing retaining walls like gabions (Chou et al., 2007), are needed until ecological regeneration and bioengineering can proceed because the site has reached an unsustainable stage that unaided recovery is impossible (Whisenant, 1999). These measures would be effective if the initial disruption is properly alleviated. Extreme compression or reduced surface variability of the soil can even restrict possible trajectories of succession (Walker and Del Moral, 2003) or change competitive balances among organisms (Walker and Shiels, 2008). Several efforts at slope restoration fails, and apparently, long-term obstacles are buried or undercut by new erosion. Natural covering, such as mulching or netting, nonetheless work rather than resurfacing slopes and may mitigate landslides, but they may prevent plant colonization, particularly if there is just a tenuous connection to the surface of the ground. Similarly, fertilizers can encourage the colonization and growth of stabilizing plants, especially if the plants develop quickly, have many lateral roots, and uniformly cover the soil. Sudden development without proper land cover may have negative consequences by raising degradation by runoff channeling or soil removal through raindrops hitting bare soil under plants >0.3 m above the surface (Morgan, 2007). Sowing natural plants into eroded soil may be less expensive (however, it needs more effort) than physical attempts, and the effects may be more robust to episodic disruptions, so residual soil or seed banks are less disrupted. Furthermore, the speeding up of natural recovery cycles would almost definitely facilitate succession over time. The effectiveness of regeneration and engineering activities is often susceptible to ongoing degradation, which can kill plants in their infant stages as well as reset plant community systems (Walker and Shiels, 2008).

3.13 INDIGENOUS PEOPLE AND THEIR SUPPORT

Numerous studies exist on the importance of indigenous knowledge for natural resource management in India, but indigenous knowledge and practices are not well reported.

For the most part, such information tends to exist for two reasons:

1. It has a practical purpose in the cultures involved, and
2. It has a clear and complex dissemination mechanism across practice and oral tradition.

Not all societies have an equal level of awareness in catastrophe prevention. It's been discovered that such experience is more prevalent in homogenous and tribal societies than in migrant communities. Communities with a deep sense of unity and cohesion possessed a greater level of understanding about disaster prevention. The more self-sufficient and endogenous a culture is, the more often it has a rich store of indigenous information. Traditional knowledge is under pressure throughout all communities, owing in part to the consequences of modernity (Nakamura et al., 2007).

Indigenous people traditionally reside in historically, culturally, geographically, and systemically distinct native lands; indigenous or territorial self-identification; and seek to maintain their ethnic, geographic, and organizational distinctness (UNDP, 2000). Traditional culture is described as knowledge that originates in a culture, has informal means of distribution, has been acquired over many centuries, and is adaptable, which is ingrained in a community's lifestyle as a form of survival (Shaw and Baumwoll, 2008). Tribal communities in Bangladesh's Chittagong Hill Tracts deal with landslides as well as forest management, constructing houses according to the slopy regions (Ahmed, 2017). To fight pasture loss and desertification caused by climate change, Mongolia's nomadic people use cultural knowledge and activities such as seasonal relocation, taboos, and customs, as well as preserve differences in livestock for grazing (Youlin et al., 2011).

3.14 INDIGENOUS KNOWLEDGE IN ECOSYSTEM RESTORATION

Native people provide a variety of traditional expertise on both ecosystem conservation and catastrophe risk mitigation. Indigenous societies, in particular, have accumulated complex cultural structures and social beliefs relevant to natural resource protection and catastrophe risk mitigation that have been handed down over centuries. Many indigenous groups, for example, track environmental indices in order to forecast hazards in advance. Plant growth and flowering cycles, animal activity, and bird nesting height were used to forecast floods, drought, and some common threats, and early alerts were given to the population. However, because of climate change and variability, conventional forecasting metrics and forecasts are becoming highly inaccurate. Locals will need to change their findings and predictions if required, as well as integrate new information and technologies, to ensure that effective coping strategies are utilized. Consequently, as bioengineering activities are implemented alongside traditional experience, landslides and other natural disasters could be reduced (Yin and Li, 2001; Zhao, 2005). To secure vital assets such as transportation roads and buildings, it might be appropriate in certain cases to mix ecosystem-based methods with designed infrastructure investments (e.g., gullies, groynes). The significance of indigenous people interested in ecosystem restoration is well recognized. Knowing the motivation behind people's decisions to invest in restoration is extremely helpful for planners, as it can aid them in planning restoration curriculums, encouraging group engagement, and making funding determinations (Stone et al., 2008; Evans et al., 2008; Le et al., 2012). Modern environmental awareness activities and values are gaining acceptance as a basis of expertise for habitat architecture and resource sustainability (Menzies and Butler, 2006; Armstrong et al., 2007). Traditional and local people preserve their ecosystem, give services for several years, and also offer an exceptional pool of information, which serves as a source and novel response to current and potential problems (Laureano, 2000; Menzies and Butler, 2006; Armstrong et al., 2007). Many developed countries provide an integrated solution to slope stabilization that incorporates bioengineering steps with a focus on adequate vegetation cover, such as trees and grasses. Furthermore, the regional sustainable environmental training course designed for Asia in partnership with national partner organizations recommends a variety of natural and constructed infrastructure to mitigate natural hazards (Hussain Shah, 2008). CBNRM (Community-based natural resource management) typically relies on local and indigenous people experience for the collective and nuanced knowledge, and gene pool preserved and established by people who have long histories of experiences with the naturalistic world. These rich, local, and conventional knowledge structures often usually employ integrated ecosystem-based management approaches, especially in the management of landslides and other environmental problems (Xu et al., 2007).

3.15 SELECTION OF POTENTIAL INDIGENOUS PLANTS BASED ON THEIR SOIL REINFORCEMENT AND ANCHORING CAPABILITIES

Shrubs and trees have deeper roots than most vegetation and may infiltrate possible soil surfaces, anchoring the earth. Additionally, small roots connect the soil surrounding a tree to a radius at least 1.5 times the diameter of the canopy. As a result, the efficacy of forests in defending slopes is dependent on their rooting depth in relation to possible failure planes, as well as their density and distribution of roots. The reinforcing characteristics of roots are often affected by root branches, the elasticity of root, and root–soil cohesion (Gray and Sotir, 1996). Since roots are known to be extremely malleable, their use is limited to the finest roots. Conversely, from a technical standpoint, a significant difference exists among delicate and thin roots, which act similarly to cord components in terms of flexural strength, and structural roots, which act identically to pillars in terms of parallel shear stress (Fournier et al., 2006; Reubens et al., 2007). Due to their flexural rigidity, it is incredibly challenging to quantify structural roots near the slip surface using a simple geometrical norm (Wu, 2007). In forests, roots play an important role in anchoring the tree to the soil, and their mechanical associations with the soil are analyzed extensively utilizing finite element models (FEM). These models incorporate root architecture, and structural roots are treated as discrete beam components (Dupuy et al., 2005a, b; Dupuy et al., 2007; Fourcaud et al., 2008). On the lower slopes of the Siwalik Hills, the Gangetic floodplain, and the Dun Valleys, the mixture of *S. robusta* and grassland is common. Root depths and spread are significant for slope stability because the deeper the tree roots penetrate, the more sensitive they can travel through and connect (Cammeraat et al., 2005; Oh et al., 2010). Numerous plant species identified by the indigenous people are now considered by soil scientists and engineers, restoring slopes in various affected areas. These species possess a variety of characteristics that enable them to rapidly stabilize an unstable or failed slope. The pioneering poplar (*Populus sp.*) and willow (*Salix sp.*) are the most common tree/shrub species since they reproduce from plant cuttings, or "live poles", when implanted in moist soil (Wilkinson, 1999; McIvor et al., 2014). Willow trees, in particular, are utilized in riparian areas for a variety of purposes, including streambank defense and nitrogen and sediment control (Kuzovkina and Volk, 2009). Seedlings and saplings that have taproot/fibrous root systems grow at the surface soil, and similarly, cuttings (those without a root system) may be buried up to 2.0 m in the soil. Each slope is thus strengthened by these poles (Rey, 2009). Both poplar and willow poles rapidly grow large lateral root structures capable of interlocking enough with neighboring trees on hillslopes and riverbanks (McIvor et al., 2009; Douglas et al., 2010).

Poplar, vetiver grass, and willow species could be used as relatively new species in a vegetation succession; moreover, these plants were suggested by eco-engineers for hills stabilization. While above-mentioned species have been shown to be extremely effective at strengthening slopes, there is a variety of risks associated with a single species implementation for slope stabilization and safety. These threats involve widespread destruction or diminished "efficiency" as a result of pest and disease incursions, as well as a restricted capacity to respond to environmental changes. Planting of mono-specific invasive plants is being established, especially if the species is exotic to the area where it is planted. Similarly, such organisms can halt succession processes and minimize native plant species colonization, for example, by establishing dense thickets, collecting accessible resources, and evading predators (Walker et al., 2010). Additionally, the disadvantage associated with using only early successional plants, also in their natural habitats, is that they can be transient (Stokes et al., 2014). Tropical plants such as *Tectonagrandis* and *Coffeaarabica* have root systems that extend to a depth of up to 4 m. Conversely, depending on the species and the climatic and soil environments, root biomass and root reinforcement decrease rapidly with depth (Stokes et al., 2009). Moreover, based on the variety, forest vegetation may greatly improve soil intensity at depths greater than one meter (Bischetti et al., 2009; Phillips and Marden, 2004) (Tables 3.1 and 3.2).

3.16 RECOMMENDATIONS

Although civil engineering systems like dams, barriers, retaining basins, and some other constructed terraforming and irrigation alteration are examples of possible remedies and are extremely beneficial for treating soil erosion, these methods generate a significant amount of carbon dioxide, are expensive and often risky to install, interrupt local and regional ecosystem systems, need some continuous maintenance, and ultimately require repair or replacement. In comparison, sustainable

TABLE 3.1
Controlling Soil Erosion and Stabilizing Slopes

Engineering Functions	Hydrological Effects of Vegetation	Erosion Mitigation by Vegetation (Rey and Others, 2004)	Successful Defense against Erosive Agents (Gyssels and Poesen, 2003)
Catchment	Interception	Active defense	Rain drop interception (Woo et al., 1997)
Armoring	Evaporation	Passive defense	Improved water penetration in soil (Cerda, 1998)
Reinforcement	Preservation		Thermal control (Roviera et al., 1999)
Anchoring	Leaf drip		Root systems' soil fixation
Supporting	Lake forming		Improved water penetration in soil
Drainage	Water uptake		Rain drop interception

TABLE 3.2
Popular Plant Species Used in Soil Bioengineering

S. No.	Binomial Name	Common Name	Family	Life Forms
1	*Amaranthusblitoider*	Matweed	Amarantaceae	Tree
2	*Ipomaeapes-caprae*	Goat's foot	Convolvulaceae	Plant
3	*Ammanniabaccifera*	Monarch red stem	Lythraceae	Grass
4	*Lagasceamollis*	Silk leaf	Asteraceae	Tree
5	*Sennaalata*	Candle brush	Fabaceae	Plant
6	*Mesophaerumsuaveolens*	Pignut	Laminaceae	Tree
7	*Chromolaenaodorata*	Christmas bush	Apocynaceae	Tree
8	*Calotropisgigantae*	Milk weed	Martyniacae	Plant
9	*Martyniaannua*	Cat's claw	Fabaceae	Plant
10	*Crotalaria pallidaAiton*	Rattle box	Euphorbiaceae	Plant
11	*Chrsopogenaciclatus*	Golden false beardgrass	Poaceae	Grass
12	*Guilandinabonduc L.*	Fever nut	Fabaceae	Tree
13	*Jatrophagossypiifolia L.*	Bellyache bush	Euphorbiaceae	Plant
14	*Ficusbenghalensis L.*	Banyan	Moraceae	Tree
15	*Tamarixericoides*	Athechavuku	Tamariaceae	Shurb
16	*Persucaria*	Knotweed	Polygonaceae	Plant
17	*Lantana camera L.*	Wild sage	Verbenaceae	Shrub
18	*Echinopsechinatus*	Globe thistle	Asteraceae	Herb
19	*Vachellianilotica (L.)*	Gum Arabic tree	Fabaceae	Tree
20	*Cappariszeylanica L.*	Ceylon caper	Capparaceae	Shrub
21	*Vallarissolanacea*	Bread flower	Apocynaceae	Shrub
22	*Ludwigiaperennis L.*	Water primrose	Onagraceae	Tree
23	*Saccharumspontaneum L.*	Wild sugarcane	Poaceae	Grass
24	*Alangiumsalvifolium (L.f.)*	Sage leaved	Cornaceae	Tree
25	*Cocculushirsutus (L.)*	Broom creeper	Menispermaceae	Tree
26	*Tephrosiapurpurea (L.)*	Wasteland weed	Fabaceae	Plant
27	*Acacia nilogtica (L.)*	Gum Arabic tree	Mimosaceae	Tree
28	*Achyranthesaspera L.*	Devil's horsewhip	Amaranthaceae	Plant
29	*Justiciaadhatoda L.*	Malabar nut	Acanthaceae	Tree

(*Continued*)

TABLE 3.2 (*Continued*)
Popular Plant Species Used in Soil Bioengineering

S. No.	Binomial Name	Common Name	Family	Life Forms
30	*Albizialebbeck (L.)*	Women's tongue tree	Mimosaceae	Tree
31	*Bambusa bamboos (L.) voss*	Bamboo	Poaceae	Tree
32	*Cocosnucifera L.*	Coconut	Arecaceaea	Tree
33	*Kyllinga trips* Var. Roxb.	White water sedge	Cyperacae	Plant
34	*Cuscutareflexa Var. Roxb.*	Giant Dodder	Cuscutaceae	Plant
35	*Borassusflabellifer L.*	Toddy palm	Arecaceae	Tree
36	*Ricinuscommunis L.*	Castor oil plant	Euphorbiaceae	Plant
37	*Cassia alata L.*	Craw plant	Caesalpiniaceae	Tree/plant
38	*Vitexnegundo L.*	Horseshoe vitex	Vitaceae	Tree
39	*Syzygiumcumuni (L.)*	Jamun	Myrtaceae	Tree
40	*Zizphusoenoplea (L.)*	Wiled jujube	Rhamnaceae	Shrub
41	*Amaranthusviridis L.*	Pigweed	Amaranthaceae	Herb
42	*Urenalobata L.*	Pink burr	Malvaceae	Plant
43	*Buteamonosperma (L.)*	Flame of forest	Fabaceae	Plant
44	*Ranunculus sceleratus L.*	Butter cup	Ranunculaceae	Plant
45	*Verbascumchinense (L.)*	Kutki	Scorphulariaceae	Plant
46	*Cyanthilliumcinereum (L.)*	Little iron weed	Asteraceae	Plant
47	*Linderniarotundifolia L.*	Round leaf Lindernia	Linderniaceae	Plant
48	*Mecardoniaprocumbens (Mill)*	Baby jump up	Plantaginaceae	Plant
49	*Centauriumpulchellum*	Slender centaury	Gentinaceae	Herb
50	*Boerhaviadiffusa L.*	Four o' clock plant	Nyctaginaceae	Plant
51	*Rumexdentatus L.*	Toothed Dock	Polygonaceae	Herb
52	*Argemone Mexicana L.*	Flowering Thistle	Papveraceae	Plant
53	*Centellaasiatica (L.)*	Pennywort	Apiaceae	Plant
54	*Cyperusrotundus L.*	Coco grass	Cyperaceae	Grass
55	*Hygrophilaauriculata (Schumact.) Heine*	Kokilaksha	Acanthaceae	Plant
56	*Scopariadulcis L.*	Sweet broom	Plantaginaceae	Shrub
57	*Solanumsisymbriifolium Lam*	Sticky night shade	Solanaceae	Plant
58	*Terminaliaarjuna (Roxb.ex DC.)*	Arjun tree	Combretaceae	Plant
59	*Andrographisechioids*	False water willow	Acanthaceae	Tree
60	*Phalaris minor Retz*	Canary grass	Poaceae	Plant
61	*Indigoferalinnaei Ali*	Birdsville indigo	Fabaceae	Plant
62	*Justiciagandarussaburm f.*	Willow-leaved justicia	Acanthaceae	Shrub
63	*Ludwigiaadscendens (L.)*	Water primrose	Onagraceae	Plant
64	*Polygonumplebeium R.Br.*	Small knotweed	Polygonaceae	
65	*Pontederiavaginalis Burm f.*	Oval-leaved pondweed	Pontederiaceae	Plant
66	*Acmellapaniculata*	Panicled spot flower	Asteraceae	Herb
67	*Afrohybanthusenneaspermus*	Spade flower	Violaceae	Plant
68	*Wahlenbergiamarginata*	Southern rockball	Campanulaceae	Herb
69	*Matricariadiscoidea*	Pineapple weed	Asteraceae	Plant
70	*Cardaminepratensis*	Cuckoo flower	Brassicaceae	Plant
71	*Cosmos sulphureus*	Sulphur cosmos	Asteraceae	Herb
72	*Eclipta prostrate*	False Daisy	Asteraceae	Plant
73	*Polypogonmonospeliensis (L.)*	Rabbit foot grass	Poaceae	Grass
74	*Dalbergiasisso Roxb. Ex DC.*	Rose wood	Fabaceae	Tree

interventions provide a lower environmental impact to encourage ecological functions (Walker and Shiels, 2008) and provide a larger variety of ecosystem resources. Additionally, natural techniques are more robust to current threats such as hurricanes and earthquakes. While much remains unknown about how an ecological solution reacts to abiotic and biotic perturbations, interacts with physical environments, and meets the needs of indigenous communities and habitats, even partial use of ecological tools in combination with conventional bioengineering methods may have immediate advantages that eco-engineers should be informed.

3.17 CONCLUSION

The emerging demand for cost-effective methods for restoring ecosystems and environmental resources has resulted in an expansion in the use of novel nature-based techniques. The last several decades have seen technological advancements focusing primarily on technical skills and more recently on the ecological factors that influence their performance in various applications. Latest global economic and ecological developments are paving the way for ecological regeneration involving a broader range of stakeholders. Ecological conservation efforts will serve as the foundation for potential environmentally integrated communities by combining ecological expertise, technological application, and social engagement. The southern continents are committed to developing soil bioengineering initiatives. Using shared information is a necessary cog in the globalized machine in which we now find ourselves as key partners/associates. This gives us no alternative but to continue adjusting "alien" technology to our unique environmental circumstances. The suggested interventions and initiatives need to be developed and studied, with the possibility of improvements and modifications along the way, in order to demonstrate their efficacy in the targeted area and those in need. This suggests that the existing research is merely a starting point for creating more novel interventions as our understanding of associated hydrologic characteristics, local ecosystems, and social realities grows. Perhaps, "further study is essential" would continue being a conventional conclusion in subsequent studies for a long time, but the challenge is to continuously collect more precise basic evidence in order to develop more successful socio-ecological conservation proposals. This can only be accomplished by increased participation, group engagement, increased investment, and information sharing. Slope stabilization plant species must be screened for their potential to develop and thrive in the target setting, which is characterized by its unique physical, chemical, ecological, and biological attributes. Following the identification of organisms based on their suitability for the climate, specific mechanisms can be used to search for plant traits that are particularly useful for stabilizing slopes or fighting erosion. Combinations of species should be welcomed since, in the majority of situations, slope sustainability can only be achieved by the development of successional systems that need minimal interference and have a long-term solution for regeneration and conservation. Although the regeneration of native habitats and provision of a diverse range of habitat resources can be beneficial in certain cases, traditional land use on slopes also relies on the longevity of one or two organisms rather than natural vegetation succession for slope stabilization. The relationship between slope hydrology and vegetation forms, as well as the effect of vegetation and soil fauna on soil structure, physical, chemical, and ecological processes, needs substantial evidence by conducting multiple studies based in varied localities with soils that are diverse in nature. A deeper understanding of the services rendered by vegetation on slopes, in addition to its stabilizing properties, is needed. More precise modeling experiments in space and time would provide valuable opportunities for civil and geotechnical engineering societies, which are still hesitant to use soft engineering systems and related vegetation. Through joint programs, coordination, training, and schooling, knowledge of soil bio- and eco-engineering techniques must be greatly increased.

REFERENCES

Ahmed, B. (2017). Community Vulnerability to Landslides in Bangladesh (Ph.D. Thesis). University College London (UCL), UK.

Armstrong, M., Kimmerer, R.W., and Vergun, J. (2007). Education and research opportunities for traditional ecological knowledge. *Frontiers in Ecology and the Environment*, 5(4):w12–w14.

Aronson, J., Florest, C., LeFloc'h, E., Ovalle, C., and Pontanier, R. (1993). Restoration and rehabilitation of degraded ecosystems in arid 1 and semiarid regions. 1. A view from the South. *Restoration Ecology*, 1:8–17.

Bergen, S.D., Bolton, S.M., and Fridley, J.L. (2001). Design principles for ecological engineering. *Ecological Engineering*, 18:201–210.

Bischetti, G.B., Chiaradia, E.A. et al. (2009). Root cohesion of forest species in the Italian Alps. *Plant and Soil*, 324:71–89.

Cammeraat, E., van Beek, R. et al. (2005). Vegetation succession and its consequences for slope stability in SE Spain. *Plant and Soil*, 278(1):135–147.

Cerda, A. (1998). The influence of aspect and vegetation on seasonal changes in erosion under rainfall simulation on a clay soil in Spain. *Canadian Journal of Soil Science*, 8(2):321–330.

Chou, W.C, Lin, WT, and Lin, C.Y. (2007). Application of fuzzy set theory and PROMETHEE technique to evaluate suitable 4. Eco-technology method: a case study in Shihmen Reservoir Watershed, Taiwan. *Ecological Engineering*, 31:269–280.

Conforth, D.H. (2005). *Landslides in Practice: Investigations, Analysis and Remedial/Preventive Options in Soils*. Wiley, Hoboken.

Devkota, B.D., Paudel, P., Omura, H., Kubota, T., and Morita, K. (2006). Uses of vegetative measures for erosion mitigation in Mid Hill areas of Nepal. *Kyushu Journal of Forest Research*, 59:265–268.

Douglas, G.B., McIvor, I.R., Potter, J.F., and Foote, L.G. (2010). Root distribution of poplar at varying densities on pastoral hill country. *Plant and Soil*, 333:147–161.

Dupuy, L., Fourcaud, T., and Stokes, A. (2005a). A numerical investigation into the influence of soil type and root architecture on tree anchorage. *Plant and Soil*, 278:119–134.

Dupuy, L., Fourcaud, T., and Stokes, A. (2005b). A numerical investigation into factors affecting the anchorage of roots in tension. *The European Journal of Soil Science*, 56:319–327.

Dupuy, L., Fourcaud, T., Lac, P., and Stokes, A. (2007). A generic 3D finite element model of tree anchorage integrating soil mechanics and real root system architecture. *American Journal of Botany*, 94:1506–1514.

Evans, S.M., Gebbels, S., and Stockill, J.M. (2008). Our shared responsibility': participation in ecological projects as a means of empowering communities to contribute to coastal management processes. *Marine Pollution Bulletin*, 57:3–7.

Fourcaud, T., Zhang, X., Stokes, A., Lambers, H., and Körner, C. (2008). Plant growth modelling and applications: the increasing importance of plant architecture in growth models. *Annals of Botany*, 101:1053–1063.

Fournier, M., Stokes, A., Coutand, C., Fourcaud, T., and Moulia, B. (2006). Tree biomechanics and growth strategies in the context of forest functional ecology. In: Herrel, A., Speck, T., and Rowe, N. (eds.) *Biomechanics: A Mechanical Approach to the Ecology of Animals and Plants*. CRC Press, LCC, USA, pp. 1–34.

Fredlund, D.G. (1979). Second Canadian geotechnical colloquium: appropriate concepts and technology for unsaturated soils. *The Canadian Geotechnical Journal*, 16:121–139.

Garwood, N.C., Janos, D.P., and Brokaw, N. (1979). Earthquake-caused landslides: a major disturbance to tropical forests. *Science*, 12(205):997–999.

GEO. (2000). Technical guidelines on landscape treatment and bioengineering for man-made slopes and retaining walls. Geotechnical Engineering Office, Government of Hong Kong SAR, Hong Kong.

Gray, D.H. and Sotir, R.B. (1996). *Biotechnical and Soil Bioengineering Slope Stabilization: A Practical Guide for Erosion Control*. John Wiley & Sons, New York.

Gyssels, G. and Poesen, J. (2003). The importance of plant root characteristics in controlling concentrated flow erosion rates. *Earth Surface Processes and Landforms*, 28(4):371–384.

Hudson, N.W. (1982). *Soil Conservation* (2nd ed.). Batsford, London.

Hungr, O., Evans, S.G., Bovis, M., and Hutchinson, J.N. (2001). Review of the classification of landslides of the flow type. *Environmental and Engineering Geoscience*, 7:221–238.

Hussain Shah, B. (2008). Field manual on slope stabilization. UNDP Pakistan and ERRA. http://preventionweb.net/english/58.professional/publications/v.php?id=13232

Kuzovkina, Y.A. and Volk, T.A. (2009). The characterization of willow (Salix L.) varieties for use in ecological engineering applications: co-ordination of structure, function and autecology. *Ecological Engineering*, 35:1178–1189.

Larsen, M.C. and Simon, A. (1993). A rainfall intensity-duration threshold for landslides in a humid-tropical environment, Puerto Rico. *Geografiska Annaler*, 75A:13–23.

Laureano, P. (2000). Ancient water catchment techniques for proper management of mediterranean ecosystems. *Water Science and Technology: Water Supply*, 7:237–244.

Le, H.D., Smith, C., Herbohn, J., and Harrison, S. (2012). More than just trees: assessing reforestation success in tropical developing countries. *The Journal of Rural Studies*, 28(1):5–19.

Mathema, P. and Joshi, J. (2010). Assessment of small-scale landslide treatment in Nepal. *Banko Janakari*, 20:3–8.

McIvor, I.R., Douglas, G.B., and Benavides, R. (2009). Coarse root growth of Veronese poplar trees varies with position on an erodible slope in New Zealand. *Agroforestry Systems*, 76:251–264.

McIvor, I., Sloan, S., and Pigem, L.R. (2014). Genetic and environmental influences on root development in cuttings of selected Salix and Populus clones—a greenhouse experiment. *Plant and Soil*, 377:25–42.

Menzies, C.R. and Butler, C. (2006). Understanding ecological knowledge. In: Menzies, C. (ed.) *Traditional Ecological Knowledge and Natural Resource Management*. University of Nebraska Press, Lincoln, pp. 1–20.

Mitchell, S.J. (2013). Wind as a natural disturbance agent in forests: a synthesis. *Forestry*, 86:147–157.

Morgan, R.P.C. (2007). Vegetative-based technologies for erosion control. In: Stokes, A., Spanos, I., and Norris, J.E. (eds.) *Eco- and Ground 20. Bio-Engineering: The Use of Vegetation to Improve Slope Stability*. Springer, New York, pp. 265–272.

Munshower, F.F. (1993). *Practical Handbook of Disturbed Land Revegetation*. Lewis Publishers, London, UK.

Nakamura, H., Nghiem, Q.M, and Iwasa, N. (2007). Reinforcement of tree roots in slope stability: a case study from the Ozawa slope in Iwate Prefecture, Japan. In: Stokes, A., Spanos, I., Norris, J.E., and Cammeraat, L.H. (eds.) *Eco- and Ground Bio-Engineering: The Use of Vegetation to Improve Slope Stability. Developments in Plant and Soil Sciences*. Springer, Dordrecht, vol. 103, pp. 81–90.

Orimoloye, I.R., Olusola, A.O., Belle, J.A. et al. (2022). Drought disaster monitoring and land use dynamics: identification of drought drivers using regression-based algorithms. *Natural Hazards*, 112:1085–1106. https://doi.org/10.1007/s11069-022-05219-9

Oh, H.J., Lee, S., and Soedradjat, G. (2010). Quantitative landslide susceptibility mapping at Pemalang area, Indonesia. *Environmental Earth Sciences*, 60(6):1317–1328.

Pande, C.B., Moharir, K.N., Khadri, S.F.R. et al. (2018). Study of land use classification in an arid region using multispectral satellite images. *Applied Water Science*, 8:123. https://doi.org/10.1007/s13201-018-0764-0

Pande, C.B., Moharir, K.N., Singh, S.K., Varade, A.M., Elbeltagie, A., Khadri, S.F.R., and Choudhari, P. (2021). Estimation of crop and forest biomass resources in a semi-arid region using satellite data and GIS. *Journal of the Saudi Society of Agricultural Sciences*, 20(5):302–311.

Pande, C.B., Kadam, S.A., Jayaraman, R., Gorantiwar, S., and Shinde, M. (2022). Prediction of soil chemical properties using multispectral satellite images and wavelet transforms methods. *Journal of the Saudi Society of Agricultural Sciences*, 21(1):21–28.

Phillips, C. and Marden, M. (2004). Reforestation schemes to manage regional landslide risk. *Landslide Hazard and Risk*. John Wiley and Sons, New York, pp. 517–547.

Polster, D.F. (2008). Soil bioengineering for land restoration and slope stabilization. Course material for training professional and technical staff. Polster Environmental Services, Duncan, British Columbia.

Restrepo, C. and Álvarez, N. (2006). Landslides and their contribution to land cover in the mountain of Mexico and Central America. *Biotropica*. 38:446–457.

Reubens, B., Poesen, J., Danjon, F., Geudens, G., and Muys, B. (2007). The role of fine and coarse roots in shallow slope stability and soil erosion control with a focus on root system architecture: a review. *Trees-Structure and Function*, 21:385–402.

Rey, F. (2009). A strategy for fine sediment retention with bioengineering works in eroded marly catchments in a mountainous Mediterranean climate (Southern Alps, France). *Land Degradation and Development*, 20:210–216.

Rey, F., Ballais, J.L., Marre, A., and Roviera, G. (2004). Role de la vegetation dans la protection contrel'e´rosionhydrique de surface. *Comptes Rendus Geoscience*, 336(11):991–998.

Roviera, G., Robert, Y., Coubat, M., and Nedjaı,' R. (1999). Erosion etstadesbiorhexistasiquesdans les ravines du Saignon (Alpes de Haute Provence): essai de mod' elisationstatistique des vitessesd'e'rosionsurmarnes. *Etudes de Géographie Historique*, 28:109–115.

Sastry, T.C.S. and Kavathekar, K.Y. (1990). (Ed.). Plants for Reclamation of Wastelands. Publication and Information Directorate, 25. Council of Scientific & Industrial Research, Hillside Road, New Delhi, India.

Schuster, R.L. and Fleming, R.W. (1986). Economic losses and fatalities from landslides. *The Bulletin of Engineering Geology and the Environment*, 23:11–28.

Schuster, R.L., Nieto, A.S., O'Rourke, T.D., Crespo, E., and Plaza-Nieto, G. (1996). Mass wasting triggered by the 5 March 1987 Ecuador earthquakes. *Engineering Geology*, 42(1):1–23.

Shaw, R.N. and Baumwoll, J. (2008). *Indigenous Knowledge for Disaster Risk Reduction: Good Practices and Lessons Learned from Experiences in the Asia-Pacific Region*. United Nations International Strategy for Disaster Reduction (UNISDR), Bangkok, Thailand.

Sidle, R.C. and Ochiai, H. (2006). Landslide processes, predictions and land use, American Geographical Union. *Water Resources Management*, 30(18):312.

Stern, M. (1995). Vegetative recovery on earthquake triggered landslide sites in the Ecuadorian Andes. In: Churchill, S.P., Balslev, H., Forero, E., and Luteyn, J.L. (eds.) *Biodiversity and Conservation of Neotropical Montane Forests*. The New York Botanical Garden, Bronx, pp. 207–220.

Stokes, A. and Atger, C. (2009). Desirable plant root traits for protecting natural and engineered slopes against landslides. *Plant and Soil*, 324(1):1–30.

Stokes, A., Douglas, G.B., and Fourcaud, T. (2014). Ecological mitigation of hillslope instability: ten key issues facing researchers and practitioners. *Plant and Soil*, 377:1–23. https://doi.org/10.1007/s11104-014-2044-6

Stone, K., Bhat, M., Bhatta, R., and Mathews, A. (2008). Factors influencing community participation in mangroves restoration: a contingent valuation analysis. *Ocean and Coastal Management*, 51:476–484.

Tosi, M. (2007). Root tensile strength relationships and their slope stability implications of three shrub species in the Northern Apennines (Italy). *Geomorphology*, 87:268–283.

United Nations Development Programme (UNDP). (2000). UNDP and Indigenous Peoples: A Practice Note on Engagement. New York, USA.

Varnes, D.J. (1996). Slope movement, types and processes: landslides, analysis and control. In Schuser, R.L. and Krizek, R.S. (eds.) Special Report 176. United States National Academy of Sciences Transportation Research Board, Washington, DC, pp. 11–33.

Vaughn, K.J., Porensky, L.M., Wilkerson, M.L., Balachowski, J., Peffer, E., and Riginos, Y.T.P. (2010). Restoration ecology. *Nature Education Knowledge*, 3(10):66.

Vidya, U.K., Pande, B.C., Jayaraman, R.A.A., Atre, S.D., Gorantiwar, S.A., and Gavit, K.B. (2021). Surface water dynamics analysis based on sentinel imagery and Google Earth Engine Platform: a case study of Jayakwadi dam. *Sustainable Water Resources Management*, 7:44. https://doi.org/10.1007/s40899-021-00527-7

Walker, L.R. and Del Moral, R. (2003). *Primary Succession and Ecosystem Rehabilitation*. Cambridge University Press, Cambridge, pp. 442.

Walker, L.R. and Shiels, A.B. (2008). Post-disturbance erosion impacts carbon fluxes and plant succession on recent tropical landslides. *Plant and Soil*, 313:205–216.

Walker, L.R., Walker, J., and del Moral, R. (2007). Forging a new alliance between succession and restoration. In Walker, L.R., Walker, J., and Hobbs, R.J. (eds.) *Linking Restoration and Ecological Succession*. Springer, New York, pp. 1–18.

Walker, L.R., Landau, F.H., Velázquez, E., Shiels, A.B., and Sparrow, A. (2010). Early successional woody plants facilitate and ferns inhibit forest development on Puerto Rican landslides. *Journal of Ecology*, 98:625–635.

Whisenant, S.G. (1999). *Repairing Damaged Wildlands: A Process Orientated Landscape-Scale Approach*. Cambridge University Press, Cambridge, pp. 324.

White, C.A. (1978). Best management practices for the control of erosion and sedimentation due to urbanization of the Lake 44. Tahoe Region of California, Proceedings, International Symposium on Urban Hydrology, Runoff, University of Kentucky, Lexington, Kentucky, pp. 233–245.

Wilkinson, A.G. (1999). Poplars and willows for soil erosion control in New Zealand. *Biomass Bioenergy*, 16:263–274.

Woo, M.K., Fang, G., and DiCenzo, P.D. (1997). The role of vegetation in the retardation of rill erosion. *Catena*, 29(2):145–159.

Wu, T.H. (2007). Root reinforcement: analyses and experiments. In Norris, J.E., Stokes, A., Mickovski, S.B., Cammeraat, E., van Beek, L.P.H., Nicoll, B., and Achim, A. (eds.) *Slope Stability and Erosion Control: Ecotechnological Solutions*. Springer, Dordrecht, pp. 21–30.

Xu, J. and Melick, D.R. (2007). Rethinking the effectiveness of public protected areas in southwestern China. *Conservation Biology*, 21(2):318–328.

Yin, H. and Li, C. (2001). Human impact on floods and flood disasters on the Yangtze River. *Geomorphology*, 41:105–109.

Youlin, Y., Jin, L.S., Squires, V., Kyung-soo, K., and Hye-min, P. (2011). Combating Desertification & Land Degradation: Proven Practices from Asia and the PacificThe Korea Forest Service (KFS) and United Nations Convention to Combat Desertification (UNCCD), Daejeon City, Republic of Korea.

Zhao, B. (2005). Estimation of ecological service values of wetlands in Shanghai, China. *Chinese Geographical Science*, 15:151–156.

4 Sustainable Ecosystem Development and Landscaping for Urban and Peri-Urban Areas

R. Gobinath, G.P. Ganapathy, E. Gayathiri, Mehmet Serkan Kırgız, Nihan Naiboğlu, André Gustavo de Sousa Galdino, and Jamal Khatib

4.1 INTRODUCTION

Urbanization is increasing exponentially in the world and is drastically altering the habitats of species and is undoubtedly known to have a significant influence on ecosystems. Biodiversity restoration is vital to urban development and management, which is often a significant symbol for the degree of greening. Challenges including the lack of urban green space and plant species have been barriers to the development of environmentally sustainable communities. Interestingly, nature coexistence and modernization, and also the coordination of economic development and ecology are the priorities that citizens are pursuing (Liang Yao-qin, 2008). Globally, research mostly on concerns of populations in urban environments has emphasized the key role of urban greening, not just for socioeconomic and aesthetic benefits, but also for environmental purposes. In urban contexts, issues such as land degradation, river floods, and loss of biodiversity are recurrent; thus ecological planning and architecture approaches have also been implemented in the last decade. The implementation of bioengineering and environmental concepts on an urban scale is, in particular, a fascinating novelty. Almost all human practices, such as recovery, overgrazing, extraction of resources, and non-systematic land use are decreasing plant diversity, impacting the natural ecosystem on which people depend (Czech et al., 2000; Jeffrey et al., 2001). Both ecologists (Palmer et al., 2004) and urban engineers (Sukopp et al., 1995) had the need to further connect ecology and the architecture realm and imply an interdisciplinary approach. This chapter will help us all to analyze and explore how the vegetation aspect of urban patch forms may enhance ecological function through design, plant ecologists, and architects can become interested. In general, landscape architects have so far underutilized the implementation of economic and environmental regeneration principles. To follow the goal of environmental protection, environmental designers, regardless of the size or nature of the initiative, need to explore and harness their native regional flora and habitats. Recognizing the role of design in the urban ecosystem makes it possible for plant ecologists to see new urban vegetation as an instrument for enhancing the environmental products and services it supports and benefits in the communities.

Urban studies are being more understood because of the complexity of urban settings. When more than half of the world's inhabitants reside in cities, it is interesting that two-thirds of the world's population is expected to be urban by 2050 (UNDESA, 2014), which could show a detrimental effect on Flora and Fauna biodiversity. Types of habitats suitable for most plants and animals have been destroyed by the construction of buildings, new roads, and other public amenity structures, leaving the urban environment fragile and insecure. Through the growth of cities, suburbs and infrastructure

DOI: 10.1201/9781003303237-4

funding, urban sprawl modifies native and traditional agricultural environments (McDonnell and Pickett, 1990; Faeth et al., 2005). Urban greening plays an important part in changing the climate. It is therefore crucial to take appropriate steps to create healthy green space environments and to shape habitat networks by incorporating point, line, soil, vertical, inlaid and ringed green space patterns. Land usage transition, fragmentation of habitats and disruptions that are intrinsically high in urban environments are projected to impact human dynamics, species richness and composition, and also community-responsive and non-adaptive dynamics (Rivkin et al., 2018). When it comes to human achievement, the cities have been scrutinized throughout history and continue to do so now. For instance, the introduction of contemporary science into the cities has given birth to novel concepts related to urban development across the globe. Landscape science is one of the newest environmental science fields that provide innovative solutions for citizens and the world centered on philosophy and science. Analysis of the city from a landscape point of view involves a coherent theoretical structure, and this essay aims to accomplish a small part of this goal. In landscape literature, one of the difficulties faced by specialists, and in particular theorists in this area, is the absence of an absolute stabilization of the role of language and concepts, which has contributed to the development of different interpretations between individuals. These variations are seen in a vast number of experiments, as well as in prototypes, which have often contributed to major inconsistencies. One of these inconsistencies applies to the word urban landscape. In recent decades, the idea of an urban environment has been commonly utilized in theoretical foundations, conversations, and specialized texts such as urban design, landscape architecture, urban engineering, geography, geology, and so on. On the one side, considering the large nature of the idea of landscape and, in particular, its spread to the theoretical foundations of the various sciences, and on the other hand, the presence of common fields of study in the areas alluded to above, it is important to clarify and formulate theoretical frameworks for urban landscape. Of course, this argument is significant; the prominence of the idea of urban landscape in the science, technical, and decision-making culture suggests its necessity and relevance, rather than the exact meaning of the word. This chapter by means of a descriptive-analytical approach, aims to analyze the ideas of theorists and to define theoretical facets of the environment, as well as to propose ideas on the urban landscape structure and its goals. Moreover, this chapter will highlight several proposals and innovative concepts that will be appropriate for the protection of plant diversity and the sustainable utilization of ecological changes during urbanization in order to preserve biodiversity wealth. In view of the theoretical zonal vegetation guidance, one can affirm that indigenous plants are the bedrock of green spaces because of their excellent ability to adapt, as they have evolved their metabolisms, biological and ecological characteristics that have adapted climate-wise to the local environment through a long evolutionary process. In addition, in a city with a significant shortage of water, more emphasis should be put on anti-drought, drought-resistant, and water-saving plants. In the meanwhile, the greening of plant communities' intermediate layers can be increased. Together, a multi-structured society of trees, shrubs, and herbs will encourage natural diversity and the survival of biodiversity. Therefore, the optimization of architectural systems and acceptable configurations within scarce urban land resources is desirable. In particular, underlying shifts in the dynamics of green space and spatial trends have been critical for extracting natural, social, and economic benefits. This chapter highlights the information available on sustainable ecosystem development and landscaping mitigation strategies available by sustainable and eco-friendly development of urban and peri-urban plant species development schemes. Technological advances that reflect the socioeconomic characteristics of cities and urban complexes have to be developed.

4.2 LANDSCAPES

Landscape describes how the key picture of people from the landscape applies to some terms, like nature, elegance, region, city, and garden. "The outcomes indicate a positive connotation of landscape and its high relevance for individuals" (Hokema, 2015). The idea of landscape was established in the 15th century in Europe and started with the Renaissance and modernity, simultaneously

(Berque, 2013). That is traditionally the product of the difference between the field of science and the world of phenomena in modern times (Berque, 1995). Indeed, the first activity in the emergence of the landscape is the Cogito Cartesian, which is regarded as the foundation of the ontology of modernity and suggests an infinite modern subject (Table 4.1). At this moment, industrialized humans are seeking to individualize the landscape and establish a landscape of existence by fracturing the unity between man and nature (Berque, 1995; Simmel, 2007). The landscape is, in reality, the perfect sample of the dialectic between the world and man, nature and culture, object and subject formed by modern absolute human real purpose (Alehashemi et al., 2017).

TABLE 4.1
List of Plant That Controls Soil Erosion

S.No	Binomial Name	Common Name	Family	Life Forms
1	*Amaranthus blitoider*	Matweed	Amarantaceae	Tree
2	*Ipomaea pes-caprae*	Goat's foot	Convolvulaceae	Plant
3	*Ammannia baccifera*	Monarch red stem	Lythraceae	Grass
4	*Lagascea mollis*	Silk leaf	Asteraceae	Tree
5	*Senna alata*	Candle brush	Fabaceae	Plant
6	*Mesophaerum suaveolens*	Pignut	Laminaceae	Tree
7	*Chromolaena odorata*	Christmas bush	Apocynaceae	Tree
8	*Calotropis gigantae*	Milk weed	Martyniacae	Plant
9	*Martynia annua*	Cat's claw	Fabaceae	Plant
10	*Crotalaria pallida Aiton*	Rattle box	Euphorbiaceae	Plant
11	*Chrsopogen aciclatus*	Golden false beardgrass	Poaceae	Grass
12	*Guilandina bonduc L.*	Fever nut	Fabaceae	Tree
13	*Jatropha gossypiifolia L.*	Bellyache bush	Euphorbiaceae	Plant
14	*Ficus benghalensis L.*	Banyan	Moraceae	Tree
15	*Tamarix ericoides*	Athechavuku	Tamariaceae	Shurb
16	*Persucaria*	Knotweed	Polygonaceae	Plant
17	*Lantana camera L.*	Wild sage	Verbenaceae	Shrub
18	*Echinops echinatus*	Globe thistle	Asteraceae	Herb
19	*Vachellia nilotica (L.)*	Gum Arabic tree	Fabaceae	Tree
20	*Capparis zeylanica L.*	Ceylon caper	Capparaceae	Shrub
21	*Vallaris solanacea*	Bread flower	Apocynaceae	Shrub
22	*Ludwigia perennis L.*	Water primrose	Onagraceae	Tree
23	*Saccharum spontaneum L.*	Wild sugarcane	Poaceae	Grass
24	*Alangium salvifolium (L.f.)*	Sage leaved	Cornaceae	Tree
25	*Cocculus hirsutus (L.)*	Broom creeper	Menispermaceae	Tree
26	*Tephrosia purpurea (L.)*	Wasteland weed	Fabaceae	Plant
27	*Acacia nilogtica (L.)*	Gum Arabic tree	Mimosaceae	Tree
28	*Achyranthes aspera L.*	Devil's horsewhip	Amaranthaceae	Plant
29	*Justicia adhatoda L.*	Malabar nut	Acanthaceae	Tree
30	*Albizia lebbeck (L.)*	Women's tounge tree	Mimosaceae	Tree
31	*Bambusa bamboos (L.) voss*	Bamboo	Poaceae	Tree
32	*Cocos nucifera L.*	Coconut	Arecaceaea	Tree
33	*Kyllinga trips Var. Roxb.*	White water sedge	Cyperacae	Plant
34	*Cuscuta reflexa Var. Roxb.*	Giant Dodder	Cuscutaceae	Plant
35	*Borassus flabellifer L.*	Toddy palm	Arecaceae	Tree
36	*Ricinus communis L.*	Castor oil plant	Euphorbiaceae	Plant
37	*Cassia alata L.*	Craw craw plant	Caesalpiniaceae	Tree/ plant

(Continued)

TABLE 4.1 (*Continued*)
List of Plant That Controls Soil Erosion

S.No	Binomial Name	Common Name	Family	Life Forms
38	*Vitex negundo L.*	Horse shoe vitex	Vitaceae	Tree
39	*Syzygium cumuni (L.)*	Jamun	Myrtaceae	Tree
40	*Zizphus oenoplea (L.)*	Wile jujube	Rhamnaceae	Shrub
41	*Amaranthus viridis L.*	Pigweed	Amaranthaceae	Herb
42	*Urena lobata L.*	Pink burr	Malvaceae	Plant
43	*Butea monosperma (L.)*	Flame of forest	Fabaceae	Plant
44	*Ranunculus sceleratus L.*	Butter cup	Ranunculaceae	Plant
45	*Verbascum chinense (L.)*	Kutki	Scorphulariaceae	Plant
46	*Cyanthillium cinereum (L.)*	Little iron weed	Asteraceae	Plant
47	*Lindernia rotundifolia L.*	Round leaf Lindernia	Linderniaceae	Plant
48	*Mecardonia procumbens (Mill)*	Baby jump up	Plantaginaceae	Plant
49	Centaurium pulchellum	Slender centaury	Gentinaceae	Herb
50	*Boerhavia diffusa L.*	Four'o clock plant	Nyctaginaceae	Plant
51	*Rumex dentatus L.*	Toothed Dock	Polygonaceae	Herb
52	*Argemone Mexicana L.*	Flowering Thistle	Papveraceae	Plant
53	*Centella asiatica (L.)*	Pennywort	Apiaceae	Plant
54	*Cyperus rotundus L.*	Coco grass	Cyperaceae	Grass
55	*Hygrophila auriculata (Schumact.) Heine*	Kokilaksha	Acanthaceae	Plant
56	*Scoparia dulcis L.*	Sweet broom	Plantaginaceae	Shrub
57	*Solanum sisymbriifolium Lam*	Stickey night shade	Solanaceae	Plant
58	*Terminalia arjuna (Roxb.ex DC.)*	Arjun tree	Combretaceae	Plant
59	*Andrographis echioids*	False water willow	Acanthaceae	Tree
60	*Phalaris minor Retz*	Canary grass	Poaceae	Plant
61	*Indigofera linnaei Ali*	Birdsville indigo	Fabaceae	Plant
62	*Justicia gandarussa burm f.*	Willow leaved justicia	Acanthaceae	Shrub
63	*Ludwigia adscendens (L.)*	Water primrose	Onagraceae	Plant
64	*Polygonum plebeium R.Br.*	Small knotweed	Polygonaceae	
65	*Pontederia vaginalis Burm f.*	Oval leaved pondweed	Pontederiaceae	Plant
66	*Acmella paniculata*	Panicled spot flower	Asteraceae	Herb
67	*Afrohybanthus enneaspermus*	Spade flower	Violaceae	Plant
68	*Wahlenbergia marginata*	Southern rockball	Campanulaceae	Herb
69	*Matricaria discoidea*	Pineapple weed	Asteraceae	Plant
70	*Cardamine pratensis*	Cuckoo flower	Brassicaceae	Plant
71	*Cosmos sulphureus*	Sulphur cosmos	Asteraceae	Herb
72	*Eclipta prostrate*	False Daisy	Asteraceae	Plant
73	*Polypogonmon ospeliensis (L.)*	Rabbit foot grass	Poaceae	Grass
74	*Dalbergia sisso Roxb.Ex DC.*	Rosewood	Fabaceae	Tree

The definition of the environment can also be said not only to be restricted to a joint project between numerous disciplines such as geography, design, sociology, and ecology. Since the landscape does not only have a physical reality, it also has other elements, such as psychological, emotional, and spiritual. Therefore, several scholars aim to obtain a holistic perspective on the landscape by proposing breaching distinctions between various fields and introducing integrative methods, from human and natural sciences to arts (Arnaiz-Schmitz et al., 2018; De Groot et al., 2010; Huu et al., 2018; Kaplan, 2009; Klug, 2012). Landscapes are beneficial for people as living environments, for their pleasure and well-being and also for the affective relationships of people to specific locations (Frick and Buchecker, 2008; Hermes et al., 2018; Manzo and Devine-Wright,

2013; Rewitzer et al., 2017; Ridding et al., 2018; Wartmann and Purves, 2018). Land use management programs that have been established rely primarily on environmental measures (Ståhl et al., 2011), with indicators of interpretation frequently tending to lag behind that of physical landscape metrics. The landscape character evaluations, which identify and characterize landscape character, are a notable exception (Fairclough et al., 2018a).

4.2.1 Urban Landscape

More than 50% of the world's population actually resides in urban areas, and there are over 20 metropolitan cities with populations of more than 10 million inhabitants (UN, 2014; Pickett et al., 2011). As people begin to move to urban areas, it is fundamental to properly protect and maintain the area's natural resources. The landscapes play a crucial role in the mutual conceptions of sustainability research, where they (the ecosystems) describe the dynamic relationships between environmental and human processes (Wu, 2013). Wu takes sustainability and defines it as the long-term capability of the landscape to provide all the resources that maintain and improve our overall wealth/health. Urban landscaping has long been an integral aspect of the dynamic urban environment, but the value inevitably enhances the entry of much of the world into towns and cities. It offers a modern scientific structure for interpreting practical works, based on the analysis of the relationship among natural environment, spatial dynamics, and environmental behavior (Meng, 2011). Natural variations had also been one of the most impactful fields of applied ecology study. Many urbans are being induced by uncertainty and unreasonable cycles in nature. Useful urban landscape models of order also perform a key role in sustainable urban planning. Fortunately, existing urban landscape patterns have been destroyed in several areas (Figure 4.1). Studies on urban landscape dynamics, complex laws, and rational protection are thus becoming highly important (Zhang, 2009). The Comprehensive System Architecture makes it possible to take a systematic look at the various levels of the community that are interconnected but also tries to unravel the facets that help form the character of the city. Urban heritage is perceived to be an important solution to urban protection in which the character of the community can be maintained, whilst its morphological, cultural, social, ecological, and economic characteristics are shown in the design phase (Kırmızı and Karaman, 2021).

4.2.2 Peri-Urban Landscape

Regions of transformation from rural to urban landscape are known as peri-urban and are clustered between the outer boundaries of urban and suburban populations and rural areas. Peri-urban areas were always overlooked in local and urban development. These were viewed as a place for urban sprawl and the site of regional and trans-regional services and infrastructure. Peri-urban areas perform diverse territorial and environmental responsibilities in terms of economic, environmental, and social functions. However, both urban and peri-urban populations benefit from the many agro-ecosystem services offered by agricultural property (Tacoli, 1998). Land cover change in peri-urban areas is inevitable due to population growth and economic activities. Unplanned land cover change leads to environmental degradation and ecosystem services loss. There are substantial geographical divisions between residential and rural regions. Research in the peri-urban ecosystem has also gradually expanded over the last decade as investigators attempt to understand the essence of the environment that connects the urban and rural landscape (Vita Zlender, 2021). The use of wastewater in peri-urban agriculture demonstrates another method in which water and agricultural products migrate from the rural to the urban. In an age of increasing urbanization and climate change, wastewater created by cities is vital to peri-urban livelihoods (Qvistrom and Saltzman, 2006). The two key factors have to be analyzed thoroughly. First, as this sort of landscape has only recently been recognized, there is no universal agreement regarding the space to which it refers. The word has gained prominence (Lambert and Rosencrantz, 2011; Simon et al., 2006) and is thus

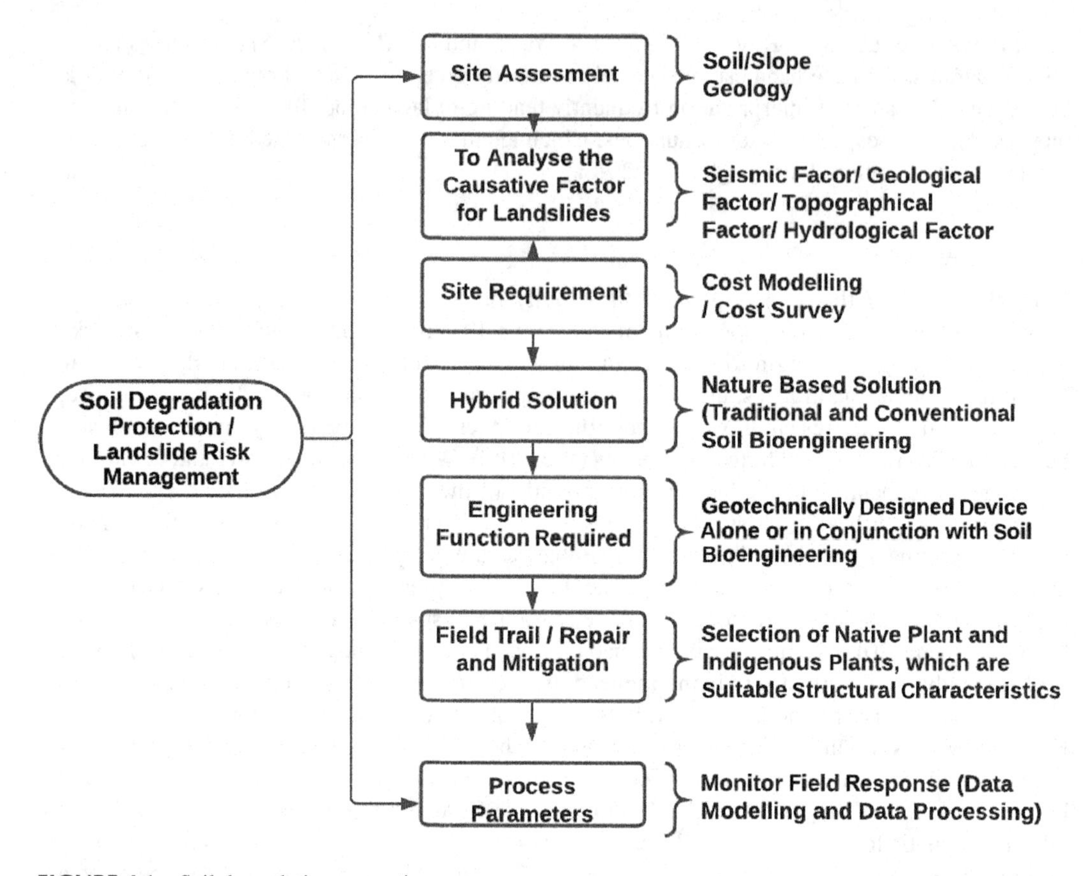

FIGURE 4.1 Soil degradation protection.

used in this document. The government operators/sectors are the backbone and act as a key factor in maintaining the stabilization of the peri-urban environment by overseeing, allowing us to manage reforms in land use and emerging technologies at public entities (La Rosa et al., 2018; Tosics and Ravetz, 2011), which can be done by processes such as urban development plans.

4.3 CHALLENGES

India has the fastest pace of transition among the BRIC nations in the urban population, according to the United Nations, and will stay over 2% annually over the next three decades (United Nations, 2014). Based on existing population models, by 2050, an estimated 854 million individuals will reside in Indian cities. Hence, there is a need for developing urban and peri-urban landscaping in order to cater to the change in population and population demographics. The implementation of these following conceptual methods will provide additional standard protocols:

i. How and where to facilitate the protection of current ecological roles and facilities
ii. How emerging technology can be used to further turn and transform land and water sources.

All such principles demonstrate the need to incorporate structural and non-structural interventions to reduce natural calamities risks (Mascarenhas and Miguez, 2002). It requires a shift in current trends in land use reform and urban growth to enable local municipalities to implement strategies for the proper management of urban/peri-urban floodplains. Without careful production of knowledge, cities will be overwhelmed with environmental challenges. Soil bioengineering and civil engineering development have been a driving force to face the challenges.

4.3.1 Role of Horticulture

In the field of phenotypic selection, eco-evolutionary input has also been studied in urban environments. In addition, the implementation of ecological regeneration concepts and techniques provides real alternatives to the host of environmental design problems faced in both the wild and urban world, and this is why bioengineering has recently started to be implemented in the cities as well. In urban environments, in particular, horticultural species could encourage rapid revegetation in ecologically harsh environments, contributing to problematic invasive species in both the immediate project environment and more importantly, the ecosystems around it (Simmons et al., 2007). In addition to this, horticulture plant species not only enhance the aesthetic value and biodiversity of the urban vicinities but also act as major sinks for air pollutants, alongside their usual carbon sequestration abilities (Safeena et al., 2021). Furthermore, there is growing evidence that certain plant species such as Chinese Brake Fern (*Pteris vittata*) have bioremediation capabilities in water and water contamination due to heavy metals (Bertin et al., 2022). More importantly, horticulture plants further promote the infiltration of rainwater, thereby preventing urban areas from natural calamities such as floods.

4.3.2 Impact of Urbanization on Plant Life in Riparian Areas

As a draw to human development, the riparian areas are seriously impacted by urbanization. Next, urbanization within riparian zones expands the region of impervious soil, which limits the infiltration of runoff and contributes to the destruction of initial riparian habitat (White, 2006). So, identifying how well urbanization impacts vegetation in riparian areas is essential, because these habitat communities offer a range of vital ecological resources and preserve a high degree of biodiversity. A few river basins have not improved owing to urban/suburban development, and a few small towns and settlements remain spread around these rivers. The contribution of species diversity to urbanization between rivers is still incompletely understood, which could impede the introduction of appropriate Eco-engineering management practices. Significant data can be collected for the maintenance of these regions, while comparing the variety of plant organisms and the composition of sites at varying stages of urbanization. So there's a need of an hour, to analyze plant species within the same land-use type but under different degrees of urbanization. It is therefore necessary to allow distinctions between artificial (located inside urban areas) and semi-natural (located outside urban areas) ecosystems of two rivers with differing degrees of urbanization and growth. It is quite important for the researchers to examine the urbanization and environmental variables that are affecting the richness of plant species diversity and the composition of species in artificial and semi-natural environments along the gradient of urbanization. It is expected that the impact of urbanization influences on vegetation which differ with the type of habitat and the dominant factor of the species will therefore vary with the type of habitat.

4.3.3 Issues in Urban Ecology

Not much research has been done on the significance of the plants that are present in the city for the provision of environmental resources and for its sustainability. Changing the diversity, distribution, and population structure of flora may have practical implications when these individual species play important roles (see, e.g., Holling, 1973; Chapin et al., 1998; Rosenfeld, 2002; Norberg, 2004).

4.3.4 Urban Ecology from a Landscape Perspective

To understand and maintain a population, the structures, roles, and geography (e.g., communities and cities) must be studied and linked to their social, cultural, and geographic origins (Grimm and Redman, 2004). The interaction between ecological characteristics (e.g., the existence of various functional groups) and the ecosystem services people would enjoy in urban centers is not well understood. Since urban planning is driven by human values, there is a need to increase awareness about the ecological criteria for ecosystem services that improve human well-being. Many services at Eco-engineering rely in particular on the existence of particular species in various natural vegetation, and of these species, the existence is in turn being determined by several influences, socio-economical, as well as ecological. As a consequence, the various vegetation would provide diverse climatic roles to the ecosystem and hence provide different environmental services. To ensure that the environmental resources are neither disrupted nor people access to their needs, at least in some sections of the region, ecological succession phases and different kinds of sites among different urban contexts should be prepared and controlled for (Flores et al., 1998, see also Nyström and Folke, 2001; Bengtsson et al., 2003). There are a lot of ecosystem resources that have been defined as fundamental or essential, but most of those are extremely arbitrary and likely to alter.

4.3.5 Parks and Public Green Space

Parks, community gardens, and countless other types of private and public green spaces make up our local and global green infrastructure framework collectively. Green infrastructure (GI) has been defined as an integrated green infrastructure network that maintains the principles and functions of natural environments and provides associated benefits to human populations (Benedict and McMahon, 2006). While some argue that Western civilizations prefer "open savannah environments with few trees," studies show that parks include between 30% and 40% of the world's trees. Most visitors tend to flock to well-kept open spaces (Shanahan et al., 2015a) (Coombes et al., 2010). In contrast, two Maryland studies indicated that low-income blacks had equal access to parks (Abercrombie et al., 2008; Boone et al., 2009), but the scale of the parks differed with larger parks in predominantly white block populations (Boone et al., 2009). Because of the possible health advantages of green areas, a second phase has been sparked by concern about whether parks and green space are equally accessible in poor and marginalized environments (Taylor et al., 2007) and an expansion of the agenda for urban environmental justice (Anguelovski, 2015; Jennings et al., 2012). The quality of neighborhood facilities such as parks and greenways will also be impacted by the location of residents and the level of sources of pollution (Payne-Sturges and Gee, 2006).

4.3.6 Multiple Benefits of Vegetated Land Cover to Health and Well-being

Humanity's very existence depends on being among plants. Biologists call this subconscious desire to be among vegetation a name: biophilia. By using combinations of plants and art (and even fragrances, sound and light effects), we believe that we will be able to improve our health. Studying green spaces' dual benefits for both nature and humans of robust habitats is a dynamic undertaking that will involve wider mechanisms that capture how vegetation enhances biological integrity and contributes to socio-ecological benefits (Bull et al., 2016; Cumming, 2011). Urban vegetation may offer a wide range of threats and hazards to health and well-being. Trees with a high leaf area index, for instance, may have a better potential to reduce air pollutants, but plants should have sufficient biogenic hydrocarbon emissions to reduce ozone formation (Taha et al., 1997). While less green space exposure could be correlated with a higher likelihood of morbidity and mortality for certain health conditions (Donovan et al., 2013; Kuo, 2015), the findings have nevertheless been mixed. For example, when analyzing the effect of green spaces on physical exercise, findings have been contradictory (Hartig et al., 2014), reduced body weight (James et al., 2015), and the incidence of levels of local crime (Locke et al., 2017; Wolf and Robbins, 2015). However, numerous forms of urban vegetative cover (e.g., herbaceous, shrub ground, and forest) may often provide essential and beneficial health-related factors such as physical activity and body mass index relationships (Tsai et al., 2015). Urban green spaces decrease heat, balance emissions, and minimize stormwater runoff. So they immediately enhance health by providing physical, social, and psychological renewal for city inhabitants. So knowing how these bonuses function is critical. Previously, accessibility, quality, amenities, attractiveness, and security were considered important features of urban green areas that influence utilization. But the relationship is more complex. Green space's utility for exercise or social activities is more likely to match the observed benefits than its character (Jennings et al., 2016a) and public policy is starting to follow suit (National Park Service, 2011; UK Health Department, 2010). Larson et al. (2016a), for instance, examined the role of public parks in cities across the USA in multiple realms of subjective well-being (social, financial, community, and physical). They noticed that a favorable correlation with health was demonstrated by variables such as park quantity, consistency, and access. Recently, Kuo (2015) described several possible mechanisms (21) to explain how nature impacts human well-being and wellness as a promising key mechanism in the partnership between nature and health.

4.3.7 Plant Resources for Building the Urban Environments

Research has shown that large trees can absorb significant amounts of carbon dioxide, particulate matter, and other pollutants from the atmosphere each year and release oxygen through photosynthesis. Intelligent landscape management can reduce water, air, and soil pollution (Isaifan, 2020). Parks and urban green spaces impact people's health by providing them with an inexpensive setting for recreation and relaxation. Landscape methods strive to develop tools and ideas for allocating and managing land in order to accomplish social, economic, and environmental goals in locations where agriculture, mining, and other profitable land use conflict with environmental and biodiversity goals (Rumi Naito, 2022) Plants species provide food, fiber, building material, fuel, and other intangible benefits to humans. They have been associated with reduced stress, increased pain tolerance, and improved mental functioning in people. They contribute to cleaner, healthier air, thus improving our well-being and comfort. The understanding of the role of trees, in particular, in promoting both human and ecological health is increasing. The benefits occur with scenes of nature, individual plants indoors, gardens outdoors, parks, and forests. The usage of native plants for soil erosion management in urban environments is growing increasingly in global popularity. The collection of well-adapted native species and other technical knowledge also must be acquired in order to advance technological propagation methods, cultural establishment, and management requirements. Thus, there is minimal research accessible on native plants for urban environments from agencies and private consultants. Native plants include a broad variety of uses, low expense, aesthetic appeal, and the intrinsic capacity to enrich an environment. Plants have modified themselves morphologically to absorb the pollutants and provide us clean environment. Some plants also develop biochemical defense mechanisms which enable them to detoxify the harmful chemicals. Trees with profuse branching, large size, and hairy leaves help in trapping the dust (Gheorghe, 2011).

As native plants need less nutrients (Artificial fertilizers that leach into the groundwater) than hybrid plants and non-native plants, they have the ability to enhance water quality (Shivega, 2017). In addition, wetlands can be preserved, stream banks strengthened, border areas provided, erosion managed, air purification supported, landscapes can be enhanced, wildlife habitat increased, and pollutes minimized in the runoff of rainwater. Native plant species and countryside grasslands contribute to the protection of biodiversity. It is important to provide biodiversity corridors within the cities to encourage plant and animal movement after experiencing global warming. Urban environments need to be permeable to wildlife; they need to keep as well as nurture human and even non-human species living there. In addition, the presence of water sources will also promote biodiversity.

Few organizations or private experts offer details about the usage of native plants as a means of care for issues with natural resources. That marketplace niche may be filled by the Urban Plant Materials Platform. The National Regulator for Compulsory Standards (NRCS) and Natural Resources Conservation Service (USDA) evaluated native plants at public sites throughout Asia and hundreds of native varieties of native trees, shrubs, perennials, and grasses were included in the plant list. The framework for native plants is an excellent showcase to educate the public on how to use native plants in urban environments. The benefits of native biodiversity can be reaped by urban ecosystems. Their usage as an alternative to the conventional lawn as a water-conserving solution reduces the need for complex and expensive watering devices. In addition, maintenance and chemical applications-related expenses are significantly decreased.

4.3.8 Some of the Main Problems Confronting Urban Plants Include

Many landscapers, planners, and horticulturists are responsible for greening urban neighborhoods.

Many of these issues have a significant effect on the range of native plant species used effectively in urban plantings (BIO Intelligence Service, 2014).

- The building materials reflect the sun, which is harmful to the plant.
- Salt overflow, salt mist, and salinated soils.
- Flooding occurs when there is an excessive amount of water present, as well as during the dry season.
- Soil compaction is one of the big issues that impact land usage.
- Heavy metal-rich asphalt and other components in cities wash through soils, hampering or stopping plants from receiving nutrients.
- Urine with excess urea and salt will kill trees.

Bioengineers aim to choose varieties of exotic, evergreen, and tropical plants to improve the diversity and productivity of local urban flora (Vogt, 2017).

4.3.9 Exotic Species in Urban/Peri-Urban Lanscaping

Exotic (i.e., alien) plant species are gradually seen as unsustainable design features of urban environments in many areas of the world (Hitchmough, 2011). Several studies indicate that exotic vegetation plays an important compensatory function for biodiversity in urban areas. It is proposed that in urban environments where non-native plants still live so that both natives and non-natives can be cultivated. Great consideration must be taken in the conservation of green spaces by the state. The daytime solar heat and nighttime lighting would shape the urban heat island influence. If trees are sited inconveniently, it may contribute to changes in subsidence in the surrounding environment due to tree roots and drier summers. Tree roots are also responsible for clay soil instability.

4.3.10 Role of Evergreen Plant in Urban/Peri-urban Landscaping

Generally, and worldwide, an evergreen plant is a plant that has leaves throughout the year long. For instance, the reason that during a year a plant has leaves does not mean it never loses them. The re-growth of the subsequent leaves, though, is normally too swift that without leaves, the plant is not seen. While the term "evergreen" is often used to refer to pine, fir, spruce, and several other woody trees, it is also likely to refer to a variety of diverse evergreen plant taxa that come in a variety of sizes, shapes, colors, and textures. Their uniqueness and attention-grabbing qualities have made them renowned. Because of their prominent and noticeable appearances throughout the year, many landscape designers and amateur consumers also choose evergreen species for landscape designs or plantations (Yalcinalp, 2010).

4.4 CONCLUSION

The ability to manage the urban/peri-urban using soil bioengineering involves not only the use of a wider range of resources to envision the environmental services connected with a given landscaping but also a deeper awareness of how this knowledge may be useful for the management of cultural environments by administrators and residents. We have seen that taking constructive steps to enhance landscaping regulation services will offer important benefits to the protection of established/developed heritage and the well-being of urban citizens. According to studies conducted, climate change puts major stress on urban environments, thus mitigating the impacts would be inevitable." These studies show the main role played by eco-engineers in greening cities on sociological, economic, and aesthetic grounds (Jeremy, 2015). Issues including land degradation, floods of waterways, and loss of biodiversity in urban areas are extensive. Policy and research now stress the importance of soil bio-engineers role in developing urban, peri-urban social-ecological structures. Our results indicate that by presenting quantifiable evidence that can help management policies for urban and peri-urban environments; the implementation of an environmental strategy can well offer a basis for informing land use decisions. The merits of retention or transfer depend on a variety of variables that are subject to change, and urban/peri-urban land allocation decisions need to be carefully evaluated in light of existing socio-economic factors that have a strong effect on land use change processes, especially when permanent losses are involved. In such situations, the provision of more efficient solutions to housing needs in growing cities with a view to minimizing urban flooding may entail the conversion of chosen urban growth areas at higher density rates to compensate for the protection of strategic natural resources for the management of the Ecosystem. Setting green fields in an urban background undoubtedly brings challenges to urban farming. Using indigenous or well-established plant communities in growing urban plantings that are well-suited, resilient, and aesthetically superior could result in a long-term sustainable ecosystem that functions and yields various ecosystem services.

REFERENCES

Alehashemi, A., Mansouri, S.A., and Barati, N. (2017). Urban infrastructures and the necessity of changing their definition and planning Landscape infrastructure; a new concept for urban infrastructures in 21st century. *The Monthly Scientific Journal of Bagh- E Nazar*, 13(43), 5–18.

Anguelovski, I. (2015). From toxic sites to parks as (green) LULUs? New challenges of inequity, privilege, gentrification, and exclusion for urban environmental justice. *Journal of Planning Literature*, 0885412215610491.

Arnaiz-Schmitz, C., Schmitz, M., Herrero-Jáuregui, C., Gutiérrez-Angonese, J., Pineda, F., and Montes, C. (2018). Identifying socio-ecological networks in rural-urban gradients: diagnosis of a changing cultural landscape. *Science of the Total Environment*, 612, 625–635.

Benedict, M.A. and McMahon, E.T. (2006). *Green Infrastructure*. Linking landscapes and communities, Washington-Covelo-London.

Bengtsson, J., Angelstam, P., Elmqvist, T., Emanuelsson, U., Folke, C., Ihse, M., Moberg, F., and Nyström, M. (2003). Reserves, resilience and dynamic landscapes. *AMBIO: A Journal of the Human Environment*, 32, 389–396.

Berque, A. (1995). Les raisons du paysage: de la Chine antique aux environnements de synthèse. Fernand Hazan, France.

Berque, A. (2013). *Thinking Through Landscape*. Routledge, New York.

Bertin, P.N., Crognale, S., Plewniak, F. et al. (2022). Water and soil contaminated by arsenic: the use of microorganisms and plants in bioremediation. *Environmental Science and Pollution Research*, 29, 9462–9489. https://doi.org/10.1007/s11356-021-17817-4

BIO Intelligence Service. (2014). Soil and water in a changing environment, Final Report prepared for European Commission (DG ENV), with support from HydroLogic.

Bull, J., Jobstvogt, N., Böhnke-Henrichs, A., Mascarenhas, A., Sitas, N., Baulcomb, C., Lambini, C., Rawlins, M., Baral, H., and Zähringer, J. (2016). Strengths, weaknesses, opportunities and threats: a SWOT analysis of the ecosystem services framework. *Ecosystem Services*, 17, 99–111.

Chapin, F.S., Sala, O.E., Burke, I.C., Grime, J.P., Hooper, D.U., Lauenroth, W.K., Lombard, A., Mooney, H.A., Mosier, A.R., Naeem, S., Pacala, S.W., Roy, J., Steffen, W.L., and Tilman, D. (1998). Ecosystem consequences of changing biodiversity—experimental evidence and a research agenda for the future. *Bioscience*, 48, 45–52.

Coombes, E., Jones, A., and Hillsdon, M. (2010). The relationship of physical activity and overweight to objectivity measured green space accessibility and use. *Social Science & Medicine*, 70(6), 816–822.

Cumming, G.S. (2011). Spatial resilience: integrating landscape ecology, resilience, and sustainability. *Landscape Ecology*, 26(7), 899–909.

Czech, B., Krausman, P.R., and Devers, P.K. (2000). Economic associations among causes of species endangerment in the United States. *Journal of Biological Sciences*, 50, 593–601.

De Groot, R.S., Alkemade, R., Braat, L., Hein, L., and Willemen, L. (2010). Challenges in integrating the concept of ecosystem services and values in landscape planning, management and decision making. *Ecological Complexity*, 7(3), 260–272.

Donovan, G.H., Butry, D.T., Michael, Y.L., Prestemon, J.P., Liebhold, A.M., Gatziolis, D., and Mao, M.Y. (2013). The relationship between trees and human health: evidence from the spread of the emerald ash borer. *The American Journal of Preventive Medicine*, 44(2), 139–145. https://doi.org/10.1016/j.amepre.2012.09.066

Faeth, S.H., Marussich, W.A., Shochat, E., and Warren, P.S. (2005). Trophic dynamics in urban communities. *Bioscience*, 55, 399–407. https://doi.org/10.1641/0006–3568(2005)055[0399:TDIUC]2.0.CO;2

Fairclough, G., Sarlöv Herlin, I., and Swanwick, C. (2018). Landscape character approaches in global, disciplinary and policy context: an introduction. In *Routledge Handbook of Landscape Character Assessment* (pp. 3–20). Routledge, New York.

Flores, A., Pickett, S.T.A., Zipperer, W.C., Pouyat, R.V., and Pirani, R. (1998). Adopting a modern ecological view of the metropolitan landscape: the case of a greenspace system for the New York City region. *Landscape and Urban Planning*, 39, 295–308.

Frick, J. and Buchecker, M. (2008). Gesellschaftliche Ansprüche an den Lebens-und Erholungsraum: eine praxisorientierte Synthese der Erkenntnisse aus zwei Forschungsprogrammen. In M. Buchecker, J. Frick, and S. Tobias (Eds.), Gesellschaftliche Ansprüche an den Lebens-und Erholungsraum: eine praxisorientierte Synthese der Erkenntnisse aus zwei Forschungsprogrammen. Eidg. Forschungsanstalt für Wald, Schnee und Landschaft WSL.

Gheorghe, I.F. and Ion, B. (2011). The effects of air pollutants on vegetation and the role of vegetation in reducing atmospheric pollution. T*he Impact of Air Pollution on Health, Economy, Environment and Agricultural Sources*. IntechOpen, London. https://doi.org/10.5772/17660

Grimm, N.B. and Redman, C.L. (2004). Approaches to the study of urban ecosystems: the case of Central Arizona—Phoenix. *Urban Ecosystems*, 7, 199–213.

Hartig, T., Mitchell, R., de Vries, S., and Frumkin, H. (2014). Nature and health. *Annual Review of Public Health*, 35(1), 207–228. https://doi.org/10.1146/annurev-publhealth-032013–182443

Hermes, J., Van Berkel, D., Burkhard, B., Plieninger, T., Fagerholm, N., Von Haaren, C., and Albert, C. (2018). Assessment and valuation of recreational ecosystem services of landscapes. *Ecosystem Services*, 31, 289–295. https://doi.org/10.1016/j.ecoser.2018.04.011

Hitchmough, J. (2011). Exotic plants and plantings in the sustainable, designed urban landscape. *Landscape and Urban Planning*, 100(4), 380–382. https://doi.org/10.1016/j.landurbplan.2011.02.017

Hokema, D. (2015). Landscape is everywhere, the construction of landscape by US-American laypersons. In Bruns, D., Kühne, O., Schönwald, A., and Theile, S. (Eds.) *Landscape Culture - Culturing Landscapes, The Differentiated Construction of Landscapes*. Springer, Wiesbaden, pp. 69–80.

Holling, C.S. (1973). Resilience and stability of ecological systems. *Annual Review of Ecology and Systematics*, 4, 1–23.

Huu, L.H., Ballatore, T.J., Irvine, K.N., Nguyen, T.H.D., Truong, T.C.T., and Yoshihisa, S. (2018). Socio-geographic indicators to evaluate landscape cultural ecosystem services: a case of mekong delta, vietnam. *Ecosystem Services*, 31, 527–542.

Isaifan, R.J. and Baldauf, R.W. (2020). Estimating economic and environmental benefits of urban trees in desert regions. *Urban Forestry & Urban Greening*, N/A, https://doi.org/10.3389/fevo.2020.00016

James, P., Banay, R.F., Hart, J.E., and Laden, F. (2015). A review of the health benefits of greenness. *Current Epidemiology Reports*, 2(2), 131–142.

Jeffrey, D.K., Alissa, M., and Ralph, J.A. (2001). Integrating urbanization into landscape level ecological assessments. *Ecosystem*, 4, 3–18.

Jennings, V., Johnson Gaither, C., and Gragg, R. (2012). Promoting environmental justice through urban green space access: a synopsis. *Environmental Justice*, 5(1), 1–7.

Jennings, V., Larson, C., and Larson, L. (2016). Ecosystem services and preventive medicine: a natural connection. *The American Journal of Preventive Medicine*, 50(5), 642–645. https://doi.org/10.1016/j.amepre.2015.11.001

Jeremy, G.C., Cavan, G., Connelly, A., Guy, S., Handley, J., and Kazmierczak, A. (2015). Climate change and the city: building capacity for urban adaptation. *Progress in Planning*, 95, 1–66. https://doi.org/10.1016/j.progress.2013.08.001

Juliane, V., Gillner, S., Hofmann, M., Tharang, A., Dettmann, S., Gerstenberg, T., Schmidt, C., Gebauer, H., Riet, K., Berger, U., and Roloff, A. (2017). Citree: a database supporting tree selection for urban areas in temperate climate. *Landscape and Urban Planning*, 157, 14–25. https://doi.org/10.1016/j.landurbplan.2016.06.005

Kaplan, A. (2009). Landscape architecture's commitment to landscape concept: a missing link?. *Journal of Landscape Architecture*, 4(1), 56–65.

Kırmızı, O. and Karaman, A. (2021). A participatory planning model in the context of Historic Urban Landscape: the case of Kyrenia's Historic Port Area. *Land Use Policy*, 102, 105130. https://doi.org/10.1016/j.landusepol.2020.105130

Klug, H. (2012). An integrated holistic transdisciplinary landscape planning concept after the Leitbild approach. *Ecological Indicators*, 23, 616–626.

Kuo, M. (2015). How might contact with nature promote human health? Promising mechanisms and a possible central pathway. *Frontiers in Psychology*, 6, 12.

La, D., Rosa, D., Geneletti, M., Spyra, C., Albert, C., Fürst, A.H., Perera, U., Peterson, G.M., and Pastur, L.I. (Eds.). (2018). Sustainable planning for peri-urban landscapes. *Ecosystem Services from Forest Landscapes*. Springer, Chem, pp. 89–126.

Lambert, A. and Rosencrantz, E. (2011). The (mis) Measurement of Periurbanization. *Metropolitics*. https://www.metropolitiques.eu/The-mis-measurement-of.html

Larson, L., Jennings, V., and Cloutier, S.A. (2016). Public parks and wellbeing in urban areas of the United States. *PLoS One*, 11(4), e0153211.

Locke, D.H., Han, S.H., Kondo, M.C., Murphy-Dunning, C., and Cox, M. (2017). Did community greening reduce crime? Evidence from New Haven, CT, 1996–2007. *Landscaping Urban Plan*, 161, 72–79.

Manzo L.C. and Devine-Wright, P. (2013). *Place Attachment: Advances in Theory, Methods and Applications*. Taylor & Francis, Routledge. https://doi.org/10.4324/9780203757765

Mascarenhas, F. and Miguez, M. (2002). Urban flood control through a mathematical cell model. *Water International*, 27(2), 208–218. http://doi.org/10.1080/02508060208686994

McDonnell, M.J. and Pickett, S.T.A. (1990). Ecosystem structure and function along urban-rural gradients: an unexploited opportunity for ecology. *Ecology*, 71, 1232–1237. https://doi.org/10.2307/1938259

Meng, C., Yang Y., and Hu H. (2011). A GIS-based urban landscape study of Lanzhou city, China. 19th International Conference on Geoinformatics (2011). https://doi.org/10.1109/geoinformatics.2011.5981091

National Park Service Health and Wellness Executive Steering Committee. (2011). Healthy Parks Healthy People US Strategic Action Plan. Retrieved from: https://www.nps.gov/public_health/hp/hphp/press/1012–955-WASO.pdf

Norberg, J. (2004). Biodiversity and ecosystem functioning: a complex adaptive systems approach. *Limnology and Oceanography*, 49, 1269–1277.

Nyström, M. and Folke, C. (2001). Spatial resilience of coral reefs. *Ecosystems*, 4, 406–417.

Palmer, M., Bernhardt, E., et al., (2004). Ecology for a crowded planet. *Science*, 304, 1251–1252.

Payne-Sturges, D. and Gee, G.C. (2006). National environmental health measures for minority and lowincome populations: tracking social disparities in environmental health. *Environmental Research*, 102(2), 154–171.

Pickett, S., Cadenasso, M., Grove, J.M., Boone, C.G., Groffman, P.M., Irwin, E., Kaushal, S.S., Marshall, V., McGrath, B.P., and Nilon, C.H. (2011). Urban ecological systems: scientific foundations and a decade of progress. *The Journal of Environmental Management*, 92(3), 331–362.

Qvistrom, M. and Saltzman, K. (2006). Exploring landscape dynamics at the edge of the city: spatial plans and everyday places at the inner urban fringe of Malm¨o, Sweden. *Landscape Research*, 31(1), 21–41. https://doi.org/10.1080/01426390500448534

Ramos, I.L., Ferreiro, M., Colaço, M.C., and Santos, S. (2013). Peri-urban landscapes in metropolitan areas: using transdisciplinary research to move towards an improved conceptual and geographical understanding. AESOP-AC S P Joint Congress, Dublin.

Rewitzer, S., Huber, R., Gret-Regamey A., and Barkmann, J. (2017). Economic valuation of cultural ecosystem service changes to a landscape in the Swiss Alps. *Ecosystem Services*, 26, 197–208. https://doi.org/10.1016/j.ecoser.2017.06.014

Ridding, L.E., Redhead, J.W., Oliver, T.H., Schmucki, R., McGinlay, J., Graves, A.R., Morris, J., Bradbury, R.B., King, H., and Bullock, J.M. (2018). The importance of landscape characteristics for the delivery of cultural ecosystem services. *Journal of Environmental Management*, 206, 1145–1154. https://doi.org/10.1016/j.jenvman.2017.11.066

Rivkin, L.R., Santangelo, J.S., Alberti, M., Aronson, M.F.J., de Keyzer, C.W., and Diamond, S.E. (2018). A roadmap for urban evolutionary ecology. *Evolutionary Applications*, 18, 1–15. https://doi.org/10.1111/eva.12734

Rosenfeld, J.S. (2002). Functional redundancy in ecology and conservation. *Oikos*, 98, 156–162.

Rumi, N., Jiaying, Z., and Kai, M.A. (2022). Chan, An integrative framework for transformative social change: a case in global wildlife trade. *Sustainability Science*, 17, 171–189. https://doi.org/10.1007/s11625-021-01081-z

Safeena, S.A., Shilpa Shree, K.G., Kumar, P.N., Saha, T.N., and Prasad, K.V. (2021). Studies for determination of air pollution tolerance index of ornamental plant species grown in the vertical landscape system. *Biological Forum – An International Journal*, 13(1), 388–399.

Shanahan, D., Lin, B., Gaston, K., Bush, R., and Fuller, R. (2015). What is the role of trees and remnant vegetation in attracting people to urban parks? *Landscape Ecology*, 30(1), 153–165.

Shivega, G.W. and Aldrich-Wolfe, L. (2017). Native plants fare better against an introduced competitor with native microbes and lower nitrogen availability [published online ahead of print, 2017 Jan 24]. *AoB Plants*, 9(1), plx004. https://doi.org/10.1093/aobpla/plx004

Simmel, G. (2007). The philosophy of landscape. *Theory, Culture & Society*, 24(7–8), 20–29.

Simon, D., McGregor, D., and Thompson, D. (2006). Contemporary perspectives on the peri- urban zones of cities in developing areas in. In: McGregor, D.F.M., Simon, D., and Thompson, D.A. (Eds.). *The Peri-Urban Interface: Approaches to Sustainable Natural and Human Resource Use*. Earthscan, London, pp. 1–17.

Stahl, G. Allard, A., Esseen, P.A. Glimskär, A.A. Ringvall, J., Svensson, S., Sundquist, P., Christensen, Å.G., Torell, M., Högström, K., Lagerqvist, L., Marklund, B.N., and Inghe, O. (2011). National Inventory of Landscapes in Sweden (NILS)—scope, design, and experiences from establishing a multiscale biodiversity monitoring system. *Environmental Monitoring and Assessment*, 173(1–4), 579–595. https://doi.org/10.1007/s10661-010-1406–7

Sukopp, H., Numata, M., and Huber, A. (1995). *Urban Ecology as the Basis of Urban Planning*. SPB Academic Publishing, The Hague.

Tacoli, C. (1998). Rural-urban interactions: a guide to the literature. *Environment and Urbanization*, 10, 147–166.

Taha, H., Douglas, S., and Haney, J. (1997). Mesoscale meteorological and air quality impacts of increased urban albedo and vegetation. *Energy and Buildings*, 25(2), 169–177.

Taylor, W.C., Floyd, M.F., Whitt-Glover, M., and Brooks, J. (2007). Environmental justice: a framework for collaboration between the public health and parks and recreation fields to study disparities in physical activity. *Journal of Physical Acitivity and Health*, 4(1), S50–S63.

Tosics, I., Ravetz, J., Piorr, A., Ravetz, J., and Tosics, I. (Eds.). (2011). Synthesis Report, Peri-Urbanisation in Europe. Towards European Policies to Sustain Urban-Rural Futures (2011), pp. 80–87

Tsai, W.-L., Floyd, M.F., Leung, Y.-F., McHale, M.R., amd Reich, B.J. (2015). Urban vegetative cover fragmentation in the U.S. *The American Journal of Preventive Medicine*, 50, 509–517. https://doi.org/10.1016/j.amepre.2015.09.022

UN DESA–United Nations, Department of Economic and Social Affairs, Population Division. (2014). World Urbanization Prospects: The 2014 Revision. United Nations, New York.

United Kingdom Department of Health. (2010). Healthy lives, healthy people: our strategy for public health in England (Vol. 7985): The Stationery Office.

United Nations. (2014). World urbanization prospects: the 2014 revision, highlights. Department of Economic and Social Affairs. Population Division, United Nations.

United Nations, Department of Economic and Social Affairs, Population Division. (2014). World Urbanization Prospects: The 2014 Revision, Highlights (ST/ESA/SER.A/352).

Vita, Z. (2021). Characterisation of peri-urban landscape based on the views and attitudes of different actors. *Land Use Policy* 101, 105181. https://doi.org/10.1016/j.landusepol.2020.105181 R

Wartmann F.M. and Purves, R.S. (2018). Investigating sense of place as a cultural ecosystem service in different landscapes through the lens of language. *Landscape and Urban Planning*, 175, 169–183. https://doi.org/10.1016/j.landurbplan.2018.03.021

White, M.D. and Greer, K.A. (2006). The effects of watershed urbanization on the stream hydrology and riparian vegetation of Los Peñasquitos Creek, California. *Landscape and Urban Planning*, 74, 125–138.

Wolf, K.L. and Robbins, A.S. (2015). Metro nature, environmental health, and economic value. *Environmental Health Perspectives*, 123(5), 390–398.

Wu, J. (2013). Landscape sustainability science: ecosystem services and human well-being in changing landscapes. *Landscape Ecology*, 28(6), 999–1023. https://doi.org/10.1007/s10980-013-9894–9

Yalcinalp, E., Var, M., and Pulatkan, M. (2010). Evergreen plants in urban parks and their importance regarding landscape architecture, a sample of Trabzon City. *Acta Horticulturae*, 881, 277–285. https://doi.org/10.17660/actahortic.2010.881.37

Yao-qin, L., Jing-wen, L., Jing, L., and Sanna, V. (2008). Impact of urbanization on plant diversity: a case study in built-up areas of Beijing. *Forestry Studies in China*, 10(3), 179–188. https://doi.org/10.1007/s11632-008-0036-4

Zhang, M. (2009). Landscape pattern dynamics and environmental impacts based on RS image. Second International Conference on Environmental and Computer Science. https://doi.org/10.1109/icecs.2009.28

5 Climate Change Impact of Fluoride Contamination on Human Health in Dry Zone of Sri Lanka

M. D. K. L. Gunathilaka, V. P. I. S. Wijeratne, Lasantha Manawadu, and Chaitanya. B. Pande

5.1 INTRODUCTION

Global climate change has and will have diverse impacts all over the world. Sri Lanka is also victimized due to the intensified impacts of climate change. These impacts are heavily varied from region to region where the climate and topography are diverse. Among the dreadful outcomes of climate change, rising seawater levels, eroded beaches, shifting temperatures, increased weather anomalies, dying forests, and natural disasters are more common on the island where some regions are highly vulnerable. The most emphasized is the dry zone of the country. In the last two decades of the 20th century, Sri Lanka experienced many extreme weather events (Deheragoda and Karunanayake, 2003). Several previous studies have documented significant trends in rainfall and temperature in the country (Herath and Rathnayake, 2005; Karunathilaka et al., 2017). Murphy (2011) mentioned climate change increases in temperature in the country which is faster than the rate of global warming. The average temperature over 1900–1917 and 2000–2017 showed Sri Lanka had experienced warming of around 0.8°C over the 20th century. A general trend in the decreasing rainfall in the latter part of the 20th century is documented. At the same time, the number of consecutive dry days has increased (ADB, 2020).

Based on the 1961–1990 average annual rainfall data (Figure 5.1), the rainfall predicted for 2050 showed to increase by 14% and 5% for A2 and B2 scenarios. Colombo and Galle are two locations in the wet climate zone that are predicted to increase average annual rainfall by 32% (A2) and 24% (B2) respectively. Among the dry climate zones, the average annual rainfall is also predicted to increase particularly in Jaffna, Mannar, Puttalam and Hambanthota. Some dry zone areas including Anuradhapura, Batticaloa and Trincomalee are predicted to decrease the average rainfall. Especially the northeast monsoon from December to February is predicted to decrease by 34% (A2) and 26% (B2) (De Silva, 2006). The highest increase in temperature is from Anuradhapura while the lowest is from Nuwara Eliya. During the northeast monsoon period, the overall increase in mean annual air temperature (Figure 5.2) across the island is predicted to increase by 1.6°C (A2) and 1.3°C (B2). The highest increase in temperature compared to the baseline (1961–1990) is predicted in Jaffna and Kurunagala by 1.8°C (A2) and 1.4°C (B2) (De Silva, 2006). A noticeable decline in rainfall, in the dry zone, combined with an increase in temperature, evapotranspiration and soil moisture deficit (Figure 5.3) will have a serious impact on the water resources, agroecology and biodiversity of the dry zone. Hence the dry zone inevitably becomes familiar with the impacts of climate change.

Sri Lanka has been experiencing an astounding coastal erosion rate of 0.30–0.35 m a year. This continuing erosion adversely impacts almost 55% of the coastline around the island (Baba, 2010).

DOI: 10.1201/9781003303237-5

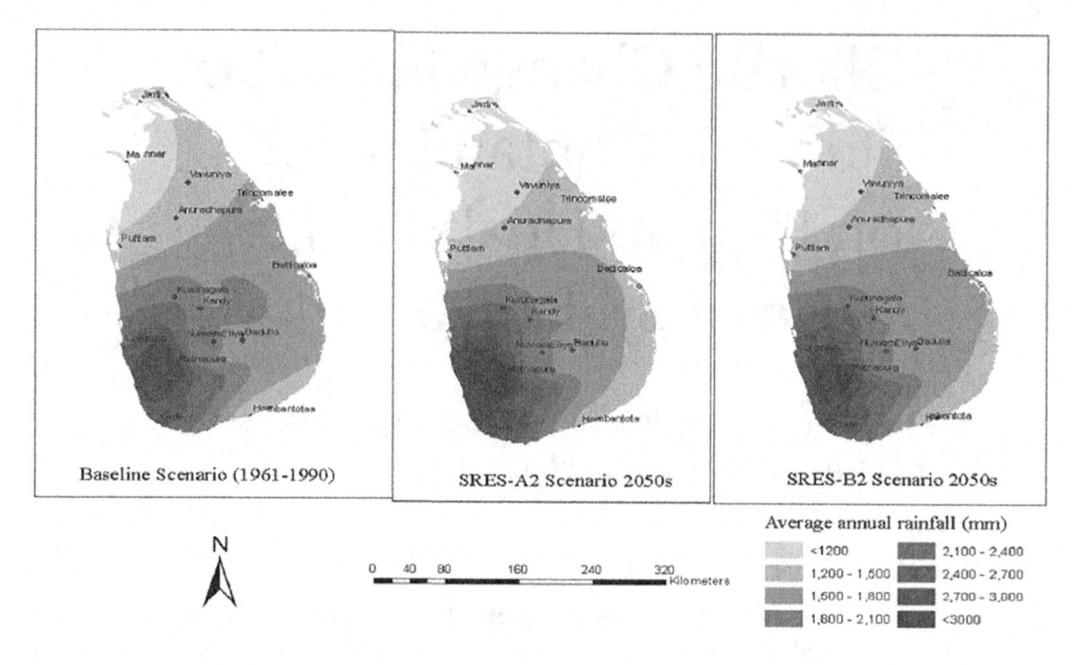

FIGURE 5.1 Spatial variation in average annual rainfall.

Source: De Silva (2006).

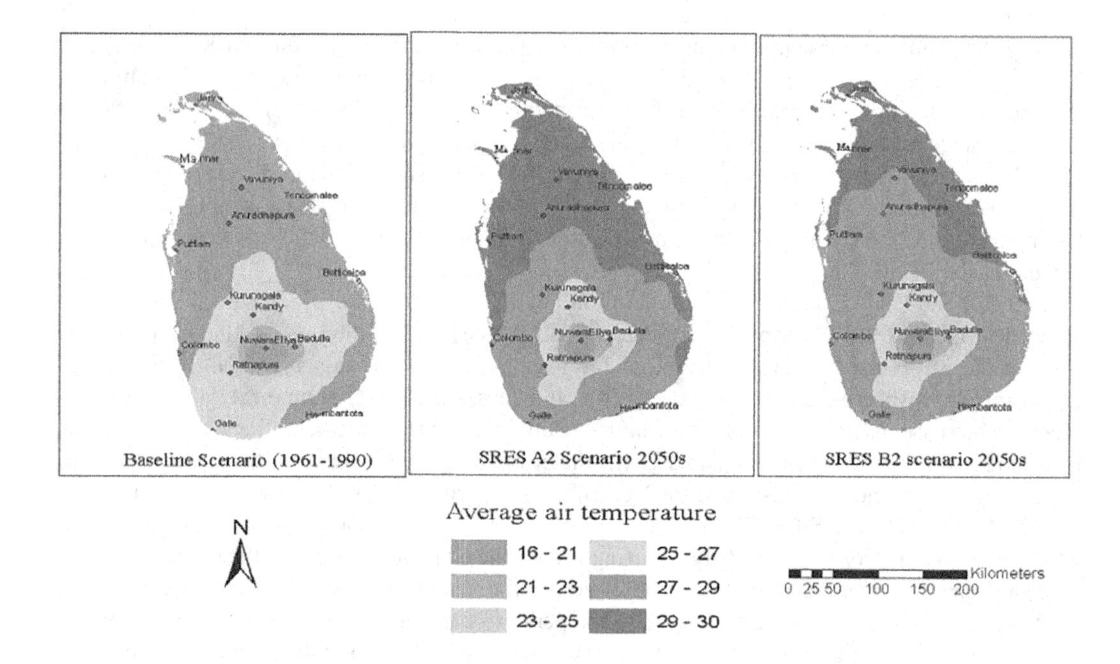

FIGURE 5.2 Spatial variation in average air temperature.

Source: De Silva (2006).

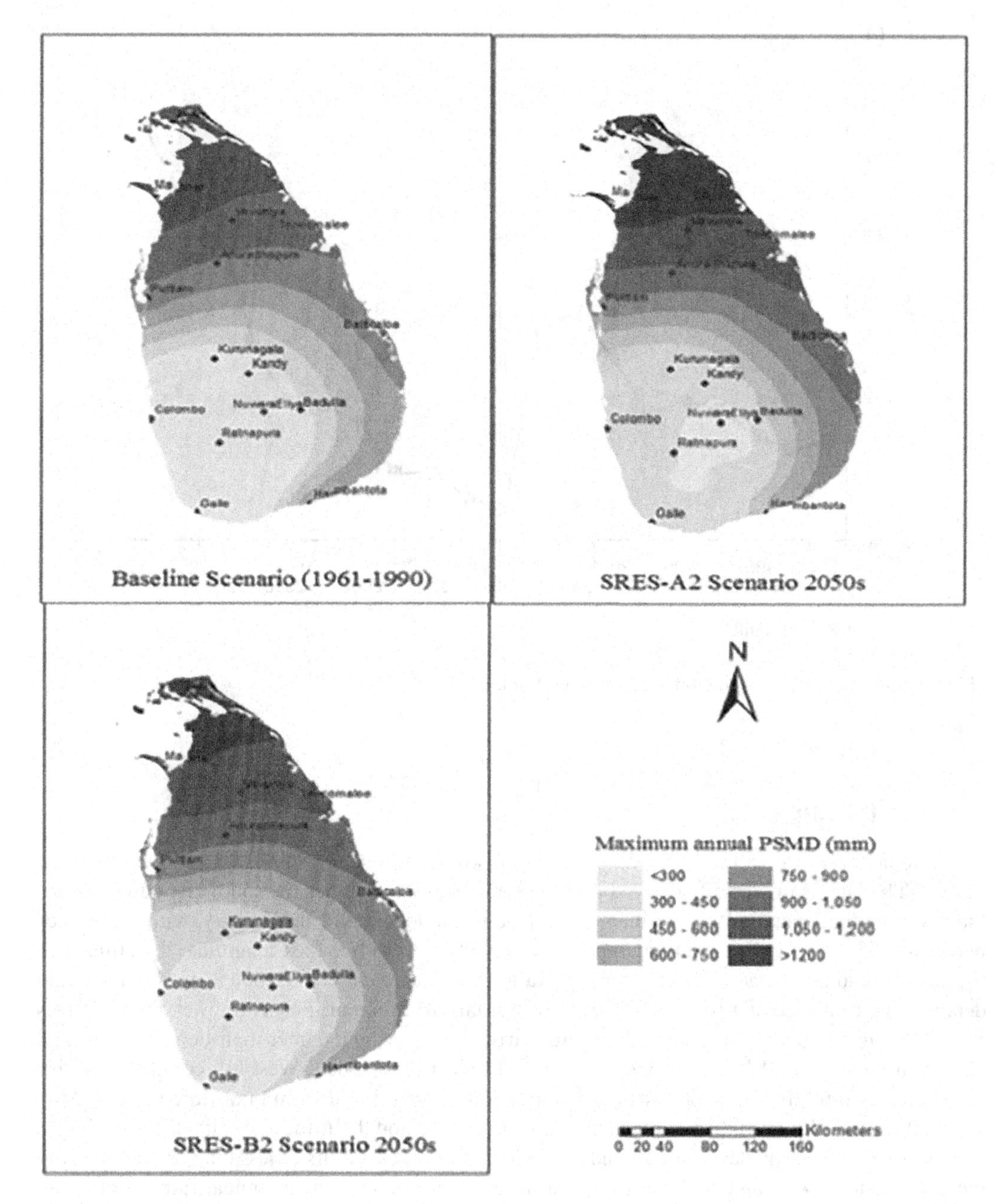

FIGURE 5.3 Spatial variation in maximum annual potential soil moisture deficit.

Source: De Silva (2006).

This will further intensify the saltwater intrusion in low-lying areas like Kalpitiya. There is plenty of literature that highlights climate change in Sri Lanka. Baba (2010) described and brought numerous statistics and information on the current scenario of climate change in the country's prospects.

Not only considering the current climate change scenarios, the climate change impact vulnerability in Sri Lanka is higher and ranked 100th out of 181 countries in the 2017 ND-GAIN Index (Figure 5.4).

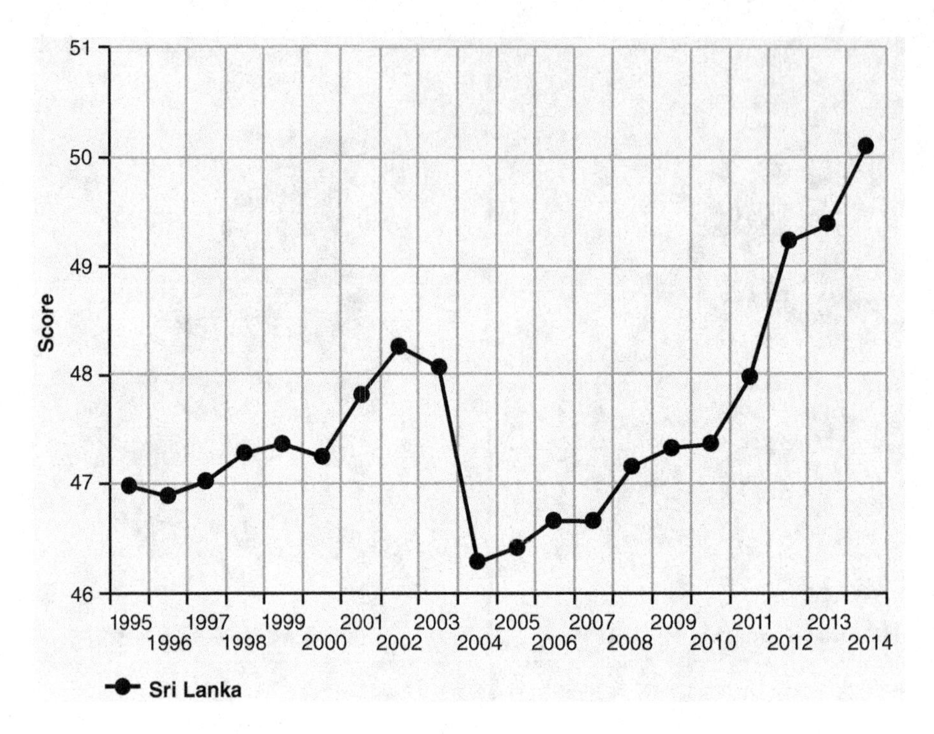

FIGURE 5.4 Climate change vulnerability in Sri Lanka.

Source: ADB (2020).

5.2 STUDY AREA

The total land area of Sri Lanka is 65,200 km^2; 430 km from north to south and 224 km from east to west. The land is universally divided into the wet zone that occupies the southwest quadrant and the dry zone that occupies the north and east. Due to the location of the country, with the tropics between 5°55′ to 9°51′ North latitude and between 79°42′ to 81°53′ East longitude, the climate of the island could be characterized as tropical. Rainfall in the country; monsoonal, convectional and depressional rains account for a major share of rainfall. Multiple origins as the wet zone receives from 2,000 to 5,500 mm mean annual rainfall, particularly during the inter-monsoonal seasons and also during the South-West monsoon that brings heavy rain to South-West hill country. The dry zone receives rainfall from October to December and lesser rainfall from mid-March to mid-May for most areas. The rainfall pattern in the dry zone is bimodal. In the long dry season May to August, there are strong desiccating winds and mean monthly rainfalls of less than 50 mm. During the extreme northwest and southwest, the driest territories receive mean annual rainfall of about 100 mm. Regional differences observed in air temperature over the country are mainly due to altitude rather than latitude. The mean annual temperature at sea level is 20°C with bit variation. In the lowlands, up to an altitude of 100–150 m, the mean annual temperature varies between 26.5°C and 28.5°C, with an annual temperature of 27.5°C. The mean annual temperature at Nuwara Eliya at 1,800 m sea level is 15.9°C. During the long dry season, the temperature rises from 30°C to 32°C.

The island is divided into three major zones based on the relief pattern i.e., the lower coastal country, the intermediate zone and the hill zone. The central part of the southern half of the island is mountainous with heights of more than 2.5 km. The low country of the wet zone is hilly and dissected by many perennial rivers and streams. The core regions of the central highlands contain many complex topographical features such as ridges, peaks, plateaus, basins, valleys and escarpments. The high country of the wet zone rises in steps to plateau at about 450 m and 750 m, with higher ranges rising to 1,500–1,800 m, and peaks like Piduruthalagala (2,524 m), Sri Pada (2,238 m) and Knuckles (1,863 m) (Dent and Goonewardene, 1993). At elevations up to about 1,000 m natural vegetation is lowland evergreen and semi-evergreen rainforest, giving way above 1,000 m to the montane rainforest, but only fragments survive. Most of the land has been cleared for tea plantation at high to lower elevations, rubber at moderate elevations, terraced rice fields and vegetable gardens. The dry zone peneplains are undulating with isolated hills (monadnocks) that occurred due to denudation. This plain is drained by a few large and many seasonal streams. Agriculture is dependent on water storage in larger to small tanks and irrigation systems and canals drive through croplands. The topographical features strongly affect the spatial patterns of winds, seasonal rainfall, temperature, relative humidity and other climatic elements, particularly during the monsoon season.

5.3 CLIMATE CHANGE AND FLUORIDE CONTAMINATION IN GROUNDWATER

Climate change is a scientific fact, even though it is a debatable topic among world powers. Along with climate change, water scarcity has become a recurrent problem in many developing countries. Water scarcity may affect the quality and quantity of groundwater and in turn alters the chemical and physical composition of groundwater. Similar to the dilemma in climate change, fluoride in drinking water is a critical but not an emphasized problem in the world (Figure 5.5). Despite having nearly 200 million people vulnerable to dental and skeletal fluorosis it is not discussed concerning climate change and its impacts on groundwater quality where most people rely on for drinking purposes. Most of them are from tropical countries like China, Sri Lanka, India, Argentina, Mexico and tropical Africa. These countries are well-identified for having a very high incidence rate of dental and/or skeletal fluorosis caused due to excessive use of fluoride-containing drinking water (Edmunds and Smedley, 2005).

The high evaporation rates increase the concentrations of many dissolved ions including F^-, $Ca2^+$, Na^+, Mg^{2+} and PO^{43-} in the groundwater which would almost certainly have a detrimental effect on the health of the people living in areas with dry climates, when such water is consumed in large quantities (Dissanyake and Chandrajith, 2019). Soil–water interaction and the residence time of fluoride are significant factors that manage the fluoride concentration in water (Dissanyake

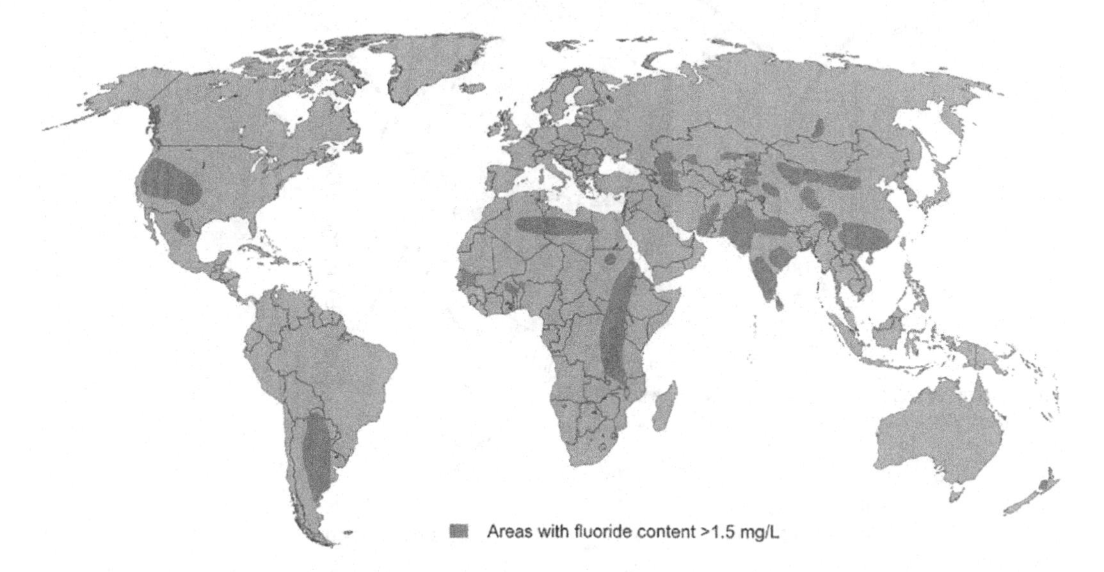

FIGURE 5.5 Groundwater fluoride exceeding WHO guidelines for drinking water of 1.5 mg/L.

Source: Dissanyake and Chandrajith (2019).

and Chandrajith, 2019) (Figure 5.6). Generally, the concentration of fluoride in groundwater is controlled by the solubility of $CaF_{2,}$ whereas the dissolution of Ca^{2+} in groundwater is controlled by the solubility of calcite and fluorite (Dissanyake and Chandrajith, 2019). Considering the dissolution of calcite and fluorite they are inversely proportional to each other. The activities of calcium, fluoride and carbonate are independent in groundwater (Kim et al., 2012). It is known apart from residence time, soil conditions, evapotranspiration and precipitation influence the fluoride concentrations in groundwater whereas all these factors are affected by climate change when ambient temperature rises (Watts et al., 2017). For example, the dry zone soil in Sri Lanka tends to accumulate salts in the soil due to the effect of drought and evaporation which then enrich the groundwater with many cations (Dissanyake and Chandrajith, 2019).

Geologically, Precambrian high-grade metamorphic rocks are dominated in Sri Lanka can be divided into three major lithotectonic units namely, the Highland complex, the Vijayan complex and the Wanni complex (Cooray, 1994). Most Precambrian rocks consist of fluoride-bearing minerals. Minerals like fluorite tourmaline and topaz found in many locations in the dry zone contribute to the general geochemical cycle of fluorine in the physical environment. Being a tropical humid country humid climatic conditions intensify the rock weathering contributing to fluorides readily into the solution (Chandrajith et al., 2012). A technically important feature to note is that the rocks and minerals in both wet and dry zones are similar and thus the fluoride variation is mainly a climatic feature and not a geologic one (Chandrajith et al., 2012).

Among the elements that are geochemically fractionated are those categorized as essential and toxic. Living beings often suffer from diseases caused by either excess or deficiency of such elements. Their distribution in soil, plants and water is controlled by geochemical parameters (Dissanayake and Chandrajith, 2007). Even fluoride is an essential element but if it is digested excessively causes ill health. Since Anuradhapura and Polonnaruwa in the dry zone tend to have higher concentrations

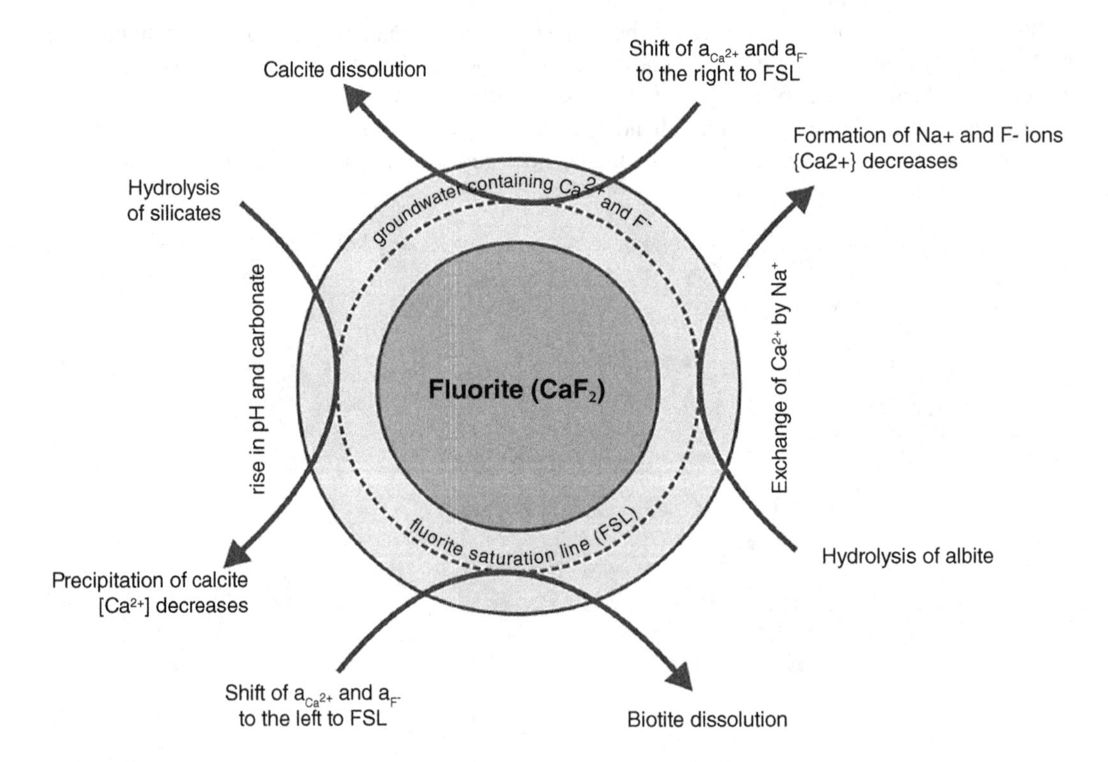

FIGURE 5.6 Interaction dynamics of fluoride in groundwater.

Source: Dissanyake and Chandrajith (2019).

of fluoride in groundwater these terrains could be geochemically identified as 'geochemical provinces' (Dissanayake and Chandrajith, 2007). Accordingly, four geochemical provinces can be delineated which is apparent with a heavy influence of climate on the nature of the chemical composition of the geochemical provinces. Typically primary rocks influence the chemical composition of the geological materials in the land. However, the climate changes the final composition of the soils and water on account of either leaching or evaporation. Hence, the dry zone in Sri Lanka is chemically markedly different from the wet zone and this resulted in some diseases.

Higher groundwater fluoride exists at a relatively higher pH and lowers electric conductivity. Lower electric conductivity values indicate lesser dissolution of carbonate minerals in the water, in other words, the presence of meta-igneous rocks which consist of fewer carbonate minerals. Higher groundwater fluoride is available under oxidizing conditions with lower chemical oxygen demand (COD) levels. The lower COD values reflect the less dissolved organic carbon in water indicating that water rich in fluoride is less contaminated by organic matter (Drever, 1982). Higher fluoride exists under lower COD values with higher DO levels. Also, higher fluoride in water does not comply with nutrient concentrations. So, the evidence is higher fluoride is not due to artificial contaminations such as fertilizer. Hence fluoride may leach into groundwater due to the weathering of fluoride-rich minerals in the basement rock (Dissanayake and Weerasooriya, 1986).

The rain mainly received from northeastern monsoons, the groundwater recharge process is initiated in the dry zone. This is the local process of groundwater recharging in the region. Typically, the northeastern monsoon period is shorter and extends from December to February and brings rather a low amount of precipitation compared to southwestern monsoons. The percolation through soils during this short wet period takes place slowly (Chandrapala and Wimalasuriya, 2003). Thus water saturated with fluoride can percolate down to the deep groundwater via fracture zones (Jayawardana et al., 2010). Finally, the wells received this fluoride mixed water, particularly during the dry period and hence oxidation-reduction potential values are maintained within the range of the groundwater stability field (Jayawardana et al., 2010). Rains with a lower pH fluoride concentration of water further rise due to the dissolution of water-soluble metal-fluoride complexes in soils due to the closely linked nature of pH with oxidation-reduction potential value (Latha et al., 1998; Zhu et al., 2006). Exclusively, the acid rains were reported in the dry zone of Sri Lanka during November and December and from April to May (Jayawardana et al., 2010). In this respect, fluoride contamination in groundwater in the dry zone of Sri Lanka is evident as a consequence of climate change-related processes interrelated with the geochemistry of the country results in medical geology.

5.4 GROUNDWATER RESOURCES AND CLIMATE CHANGE IN SRI LANKA

Groundwater resources as a renewable resource possess 13% of total renewable water resources in Sri Lanka (Table 5.1). Apart from it, 60% of water consumed is obtained from groundwater as well (Table 5.2). Groundwater resources were broadly used in the Jaffna Peninsula and dry zone. It was in the 1960 and 1970 that the scientific research on groundwater in Sri Lanka was first initiated as a consequence of the rapidly increased rural farmer population and intensive agriculture practices (Wijesinghe and Kodithuwakku, 1990). After 1978, the attention paid to groundwater resources continued to increase. By the end of 1980, there were more than 12,000 boreholes within the country (Ranasinghe and Wijesekara, 1990; Senarath, 1990). Now dry zone minor irrigation schemes consist of groundwater extraction wells (Gunawardena and Pabasara, 2016). Mainly there are six types of groundwater aquifers in Sri Lanka. They are shallow karstic pioneer in the Jaffna peninsula, coastal sand, deep confined, lateritic, alluvial and shallow regolith aquifers in the hard rock regions (Panabokke and Perera, 2005) (Figure 5.7). Shallow karstic aquifers get recharged via the infiltration of rainfall. During the northeastern monsoons, these aquifers get fully recharged and with less rainfall water level drops rapidly. Deep aquifers need further adequate understanding and are least utilized (Gunawardena and Pabasara, 2016). There are seven identified and more than 60 m deep with relatively high discharge rates. Coastal aquifers are of three types, including;

TABLE 5.1
Water Resource in Sri Lanka

Internal Renewable Water Resources	53 km²/year
	47 as a % of rainfall
Surface water entering the country	0 km²/year
Groundwater as a % of total renewable water resources	13
Total renewable water resources	53 km²/year
Total renewable water resource per capita	2,549 m³/inhabitant/year

Source: Lacombe et al. (2019).

TABLE 5.2
Water Uses in Sri Lanka

Total Water Use as a % of Total Renewable Water Resources	25
Total water use	12.9 km³/year
Agricultural water use	11.3 km³/year
Industrial water use	0.8 km³/year
Municipal water use	0.8 km³/year
Surface water use	5.1 km³/year
Groundwater use	7.8 km³/year
Groundwater used as a % of total water use	60
Groundwater used as a % of total renewable groundwater	15
Agricultural water use as a % of total water use	87
Agricultural land (103 km²)	27.4
Irrigated area (103 km²)	5.7
Irrigated area as a % of agricultural land	21.0

Source: Lacombe et al. (2019).

(1) shallow aquifers on coastal spits and bars i.e., Kalpitiya, Mannar Island, (2) shallow aquifers in raised beaches i.e., Nilaveli, Kalkudah, and (3) moderately deep aquifers on old red and yellow sands of prior beach plains i.e., Katunayake, Chilaw. The supply of groundwater in those shallow aquifers is limited but highly used in intensive agriculture as well as for the tourism industry. The significant feature of coastal sand aquifers is that they are recharged during the wet period and contract in the dry season (Panabokke and Perera, 2005). Accordingly, saline and brackish boundaries are fluctuating and contaminated due to salt intrusion i.e., Kalpitiya. Alluvial aquifers have a reliable volume of groundwater and can be utilized throughout the year. The water level of these aquifers is not reduced even in extreme weather conditions. Despite low yield and transmissivity, shallow regolith aquifers have adequate groundwater capacity. Agro-well farming heavily depends on these shallow regolith aquifers. Laterite aquifers have considerable water-holding capacity which is dependent on the depth of the laterite (Cabook) formation. These aquifers get recharged quickly with the first rains (Gunawardena and Pabasara, 2016).

There are various groundwater issues in Sri Lanka. The major and the most discussed issue is water pollution. It is mainly due to intensive agriculture dependent on chemical fertilizers and pesticides to meet the rising demand. Groundwater resources are heavily contaminated with nitrate, chlorides, sulphate, and heavy metals (Gunawardena and Pabasara, 2016). The nitrate concentration over exceeds while cadmium (Cd) level exceeded the World Health Organization limits for drinking

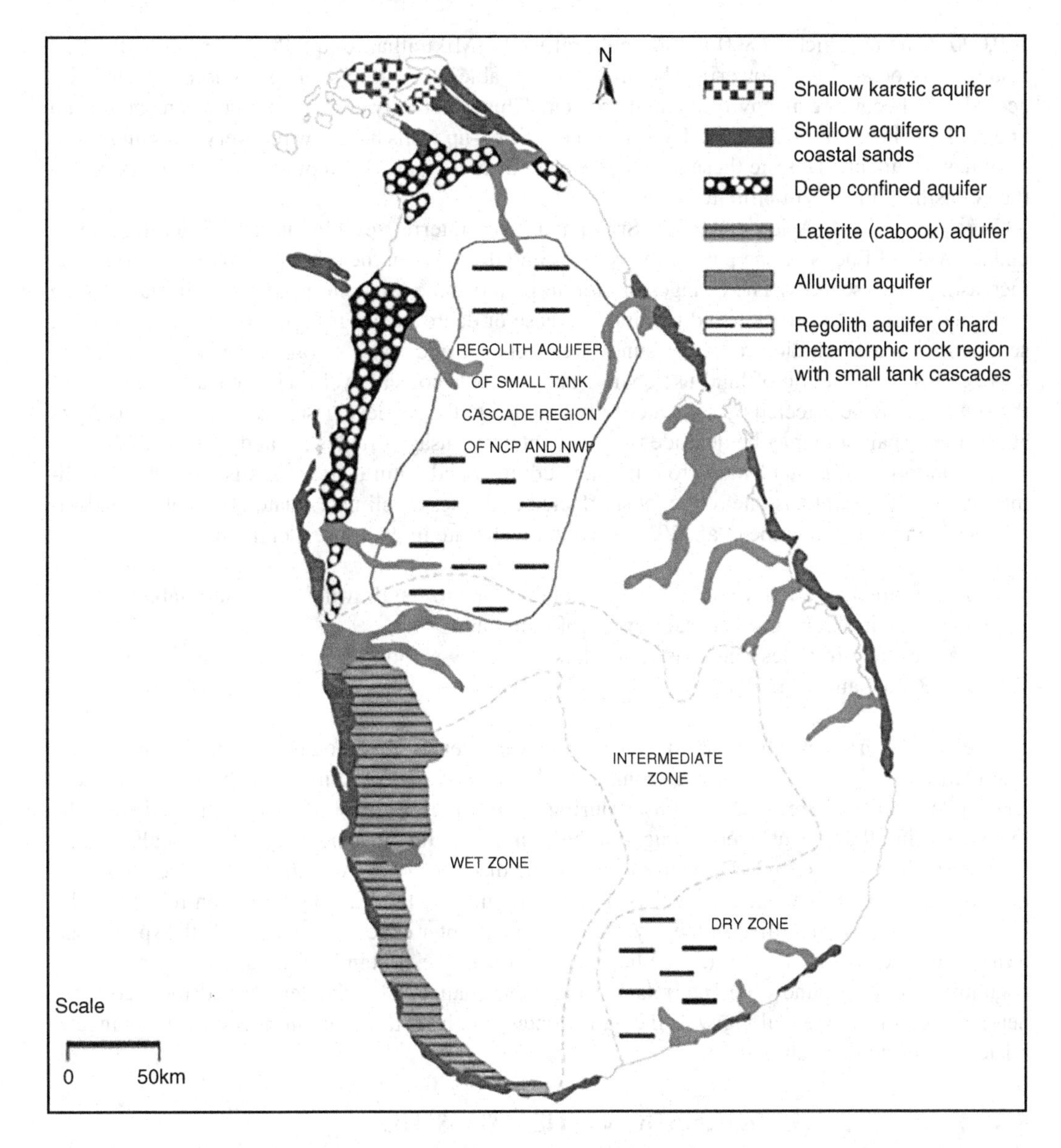

FIGURE 5.7 Groundwater resources in Sri Lanka.

Source: Panabokke and Perera (2005).

water standards (Mikunthan et al., 2013). The causes of groundwater pollution are extensive use of nitrogen-containing fertilizers and pesticides, shallow groundwater table and high population density while coastal areas are vulnerable due to over-extraction of groundwater. For example, the Kalpitiya area was found as the most polluted groundwater by nitrate approximately 39.8^+ to 88 mgl^{-1} as NO^{-3} (Liyanage et al., 2000). During the rainy season, the fertilizer is washed down to the groundwater table and nitrate is washed away with the runoff and infiltrated toward the water table. Groundwater in the area was identified as not suitable for domestic consumption at least during the dry season (Matara et al., 2014). Further, excessive extraction of groundwater has led to a decrease in the quality and quantity of groundwater. Literary pieces of evidence provide appropriate examples to portray the increase of agro-wells in large quantity. During 1992–2020, 18,000 agro-wells received subsidies for construction (Jayakodi, 2006). Jaffna peninsula alone dug over

100,000 wells of which 17,860 are agricultural wells (Mikunthan et al., 2013). Heavy extraction is always associated with lowering the groundwater table, drying up sources, later abandoned and accordingly becoming an environmental problem. This is not only confined to the Jaffna peninsula and could be identified within the dry zone. Those dried-up wells have low recovery rates and water becomes unsuitable. Despite these issues, the groundwater resources in dry zones are at risk due to the intensified impacts of climate change.

Like many South Asian countries, Sri Lanka faces interrelated climate risks. The frequency and intensity of floods in one part of the country and droughts in the other part are associated with increasing water scarcity. The changes in rainfall patterns and land cover changes especially in the dry zone intensify the erosion and possibly increase or decrease groundwater recharge. Gradually, sedimentation and siltation reduce groundwater recharge and water storage capacities as a consequence of the interaction of land-use change and the hydrological cycle. The coastal dry zones of the country may be affected by saltwater intrusion due to the sea level rise. Overall water quality is at risk once again and may be degraded through intensive usage (Tables 5.1 and 5.2).

The pollution of groundwater from human activities and natural processes is worsened by climate change. Sri Lanka is one of the most affected countries by climate-related risks of groundwater contamination (Lacombe et al., 2019). Two main climate impacts are identified.

- Aquifer recharge reduction in dry zone areas leads to the drawdown of water tables, followed by horizontal or vertical transfer of pollutants.
- Saline water intrudes into coastal aquifers as the dry-season flow decreases in river deltas i.e., Kalu Ganga, Mi Oya

Aquifers in the dry zone are contaminated with elevated levels of pathogens, nitrate, chlorine, fluoride heavy metals and sulphates (Lacombe et al., 2019). Coastal aquifers in the dry zone areas cover 1,250 km^2 and are intensively used during the dry period. Freshwater is recharged from the monsoon rainfall and is utilized throughout the year. Regarding groundwater-related health issues, Chronic Kidney Disease (CKD) is prominent in the dry zone of the country and is frequently and heavily discussed in academia as well as in the community. Fluoride contamination-related health issues could also be identified in the dry zone. So this chapter expects to discuss (1) the spatial pattern of fluoride contamination, (2) the health risk of fluoride contamination, and (3) how fluoride contamination in groundwater interrelates to climate change. The chapter applied for a comprehensive literature review along with thematic content analysis to highlight how climate change is related to fluoride contamination.

5.5 FLUORIDE CONTAMINATION & HEALTH ISSUES IN DRY ZONE SRI LANKA

Groundwater is the main drinking water source in most dry zone areas (Dissanayake, 2005). Considering the weather statistics, the average annual rainfall in the dry zone is <1,750 mm with monthly rainfall varying in a bimodal distribution indicating temporal variations in the water supply (Wickramasinghe et al., 2021). Groundwater is typically considered to be less abundant in the hard-crystalline terrain due to its extremely low porosity (Wright, 1992). Besides, the subsurface channel systems generated by weathering along joints, faults, shear zones and fractures settled endodontically in the earth's upper crust, supports the occurrence and movement of groundwater (Jayasuriya and Jayasena, 2019).

Many residents suffer from dental fluorosis, skeleton fluorosis, cholera, typhoid, hepatitis, and dysentery due to a lack of clean drinking water (Baba, 2010; Herath et al., 2017). Fluorine is a chemical element that is the lightest halogen and available at standard conditions as a highly toxic, pale yellow diatom gas. It is one of the most reactive elements. Fluorine containing mineral fluorspar or fluorite was first coined by the German physician and mineralogist Georgius Agricola

(Britannica, 2021). Fluoride either refers to the fluorine ion or to a compound that contains the element fluorine (Figure 5.8). It is an essential element for human beings. Fluoride causes the mineralization processes of bones and teeth, cementing crystals to provide more strength and hardness (Singh et al., 2001). Chloride is particularly available in rocks, soil, and water as well as in biological chains in living organisms. It has higher electronegativity and reactivity fluoride occurs naturally in water due to the weathering of rocks that contain fluoride which minerals such as hornblende, biotite, appetite, and fluorite (Totsche et al., 2000). Fluoride is readily available in water with lower Ca and Na (Larsen and Widdowson, 1971). Fluoride also leaches into water (Figure 5.8) from anthropogenic sources including phosphate fertilizers and electronic waste materials (Gilpin and Johnson, 1980; Arnesen and Krogstad, 1998). Further, Latha et al. (1998) and Zhu et al. (1980) cited by Jayawardena et al. (2010) explain fluoride as a major component in acidic soils, iron hydroxides serve as a significant sink for fluoride in soil resulting into the enhancement of fluoride concentration in water under acidic conditions. Fluorine occurs naturally as fluoride in an extensive range of pH and under positive oxidation-reduction potential as fluorine is not highly redox-sensitive (Jayawardana et al., 2010). It also can be available as Hydrogen Fluoride under strongly acidic conditions (Takeno, 2005).

The optimum content of fluoride essential for the growth of bones and formation of dental enamel must be contained in drinking water less than 1.5 mg/L (WHO, 2011) (Table 5.3). While the Sri Lankan standard is less than 1.0 mg/L (SLSI, 2013) due to the amount of water intake is larger in Sri Lanka than in temperate countries as Sri Lanka is located in a tropical climate (Herath et al., 2017). The highest concentration of fluoride, 7.0 mg/L, was identified in the Anuradhapura district, followed by 6.8 mg/L in the Moneragala district (Herath et al., 2017). A recent study shows the samples from Anuradhapura and Moneragala exceeding the average fluoride concentration of 1.0 mg/L (Herath et al., 2017). High values in the standard deviation represent the high variation of fluoride concentrations of each well, even among wells within the same district (Table 5.4). The fluoride concentration of well water in Sri Lanka ranged from 0 to 7.0 mg/L. Eighty percent of wells contained a fluoride concentration of 0–1.0 mg/L, 10.1% of wells contained 1.0–1.5 mg/L of fluoride, and 9.9% contained a fluoride concentration of more than 1.5 mg/L (Herath et al., 2017). Table 5.4 demonstrates that the dry zone is more responsible for high concentrations of fluoride than the wet zone.

Several studies also paid attention to spatial distribution and movement patterns of fluoride and sources of fluoride (Dissanayake and Weerasooriya, 1986; Dissanayake, 1991, 1993;

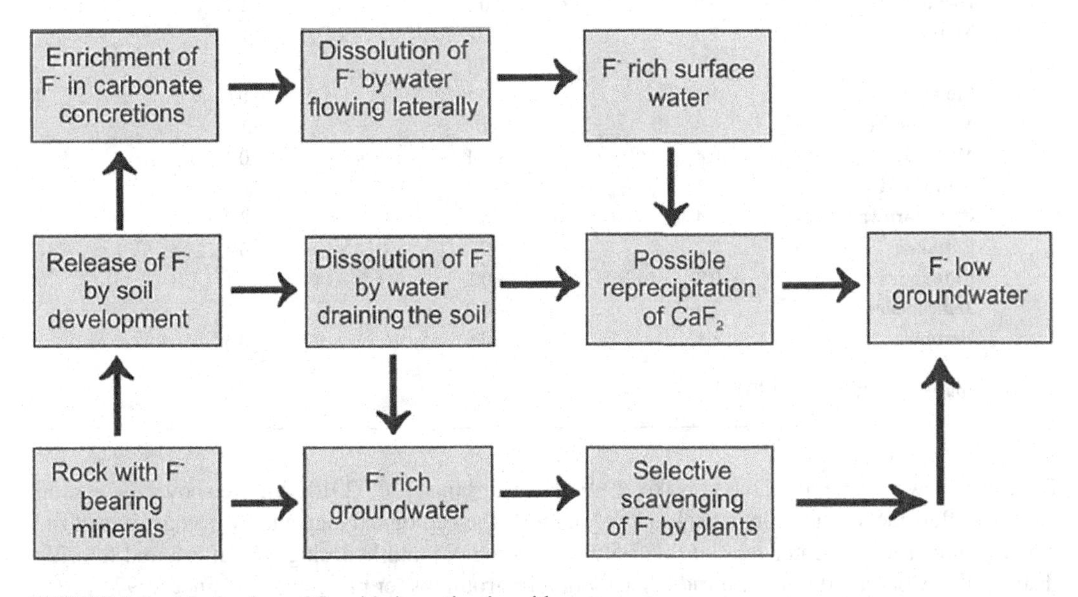

FIGURE 5.8 Mechanism of fluoride ingestion in arid areas.

Source: Dissanyake and Chandrajith (2019).

TABLE 5.3
WHO Guidelines for Fluoride in Drinking Water (1971)

Concentration of Fluoride (mg/L)	Impact on Health
0.0–0.5	Limited growth and fertility, dental caries
0.5–1.5	Promotes dental health, prevents tooth decay
1.5–4.0	Dental fluorosis (mottling of teeth)
4.0–10.0	Dental fluorosis, skeletal fluorosis (pain in back and neck bones)
>10.0	Crippling fluorosis

Source: Chandrajith et al. (2012).

TABLE 5.4
Fluoride Concentrations of Water Samples for each District in Sri Lanka

District	Maximum Value (mg/L)	Average Value (mg/L)	Standard Deviation (md/L)
Ampara	5.8	0.8	1
Anuradhapura	7.0	1.1	0.9
Badulla	2.3	0.6	0.5
Batticloa	0.7	0.2	0.2
Colombo	0.1	0.0	0.0
Galle	5.8	0.1	0.6
Gampaha	0.1	0.0	0.0
Hambanthota	2.3	0.6	0.4
Jaffna	1.3	0.2	0.2
Kalutara	0.7	0.1	0.1
Kandy	0.4	0.2	0.1
Kegalle	0.3	0.1	0.1
Kilinochchi	1.3	0.2	0.3
Kurunegala	5.0	0.7	1.2
Matale	2.6	0.4	0.6
Matara	0.6	0.1	0.1
Mannar	1.2	0.4	0.4
Moneragala	6.8	1.4	1.3
Mullaitivu	0.4	0.1	0.1
Nuwara Eliya	0.1	0.04	0.03
Polonnaruwa	3.4	0.8	0.7
Puttalam	2.2	0.3	0.4
Ratnapura	0.2	0.1	0.1
Trincomalee	2.4	0.8	0.7
Vavuniya	3.1	0.8	0.6

Source: Herath et al. (2017).

Dharmagunawardhane and Dissanayake, 1993; Jayawardana et al., 2010). Figure shows the spatial pattern of fluoride distribution in Sri Lanka along with the geological map and hydrology map of the country indicating the inter-relationship between geological and hydrological pattern and fluoride distribution. Compared to the fluoride distribution in groundwater has expanded since 1997.

Along with high fluoride concentration dry zone regolith aquifers are well popular for higher fluoride contaminations and associated endemic diseases and chronic renal failure (Chandrajith et al., 2006). Almost 90% of school children less than 15 years old suffer from dental fluorosis in the Anuradhapura district belonging to the dry zone of Sri Lanka (Herath et al., 2017). The kidney except for the pineal gland is highly exposed to high concentrations of fluoride than all other soft tissues in the human body. Already damaged kidney accumulates more fluoride. One particular feature is those who suffer from chronic kidney disease in tropical countries tend to develop skeletal fluorosis (Dissanyake and Chandrajith, 2019). There is another risk of fluoride contamination identified by Lantz et al. (1987). A case of fluoride intoxication; potomania of Vichy water, highly mineralized water containing 8.5 mg/L of fluoride (Lantz et al., 1987). According to the observation made by Lantz et al. (1987) a relationship between osteosclerosis and end-stage renal failure determined that the long duration of high fluoride intake and the absence of other causes of renal insufficiency suggested a causal relationship between fluoride intoxication and renal failure. Accordingly, it is visible that Anuradhapura also reported a higher concentration of fluoride in groundwater (Herath et al., 2017) which may further become a risk if it exceeds 8.5 mg/L of fluoride. However, literary pieces of evidence demonstrate that 1.6% out of 407 wells in Thambutthegama had fluoride levels ranging from 4 to 10 mg/L (Chandrajith et al., 2012). Therefore, Anuradhapura can be considered as a hotspot of fluoride-contaminated groundwater. Polonnaruwa is also a significant area prevalent in fluoride contamination. Kavudulla, Galamuna, Lankapura, Medirigiriya, Hingurakgoda and Thamankaduwa are other areas in Polonnaruwa found to have contaminated drinking water. Also, Wellawaya, Thanamlwila and Buttala regions in Moneragala district and Hambanthota district have higher fluoride levels (Chandrajith et al., 2012).

Not only higher concentrations and lower concentrations of fluoride also create health issues such as exposure over a long period causes kidney failure (Dissanayake and Chandrajith, 2007). This is further described by Harinarayan et al. (2006) where chronic fluoride intoxication is associated with renal tubular dysfunction concentrating on the possibility that fluoride toxicity may be responsible for both bone and kidney disease. The glomerular filtration rate is very sensitive to fluoride exposure (Ando et al., 2001). At low dosages, fluoride interacts with cellular systems including oxidative stress and modulation of intracellular redox homeostasis and some others emphasizing the toxicity of fluoride to human beings (Cittanova et al., 1996). Moreover, the literature shows that further experiments had been carried out to identify the health risks of intake of fluoride-contaminated water.

"Some subcellular determinants of fluoride cytotoxicity" and to determine whether sub-toxic fluoride exposure affects tubular cell vulnerability to superimposed adenosine triphosphate (ATP) depletion and nephrotoxic attack (Zager and Iwata, 1997). They concluded that (1) fluoride induces dose-dependent cytotoxicity in cultured human proximal tubular cells, (2) this occurs via Ca2+ and phospholipase A2- (PLA2) dependent mechanisms, (3) partial cytosolic PLA2 depletion subsequently results and (4) sub-toxic fluoride exposure can acutely increase cell resistance to further attack (Dissanyake and Chandrajith, 2019).

In all these aspects, biological consequences of fluoride exposure on mammalian cells vary such as homeostasis, metabolism, oxidative stress, cell respiration, cell migration, proliferation, gene expression, signaling, apoptosis and exocytosis endocytosis recycling.

5.6 CONCLUSION

The science dealing with natural geological factors and the health of living beings; man and animals is defined as medical geology pronounces the influence of ordinary environmental factors on the spatial distribution of health issues. Such health issues are prominent in the tropical zone where countries experience the humid climate and geogenic factors while most countries are less developed and struggle to cope with various socio-economical as well as environmental issues. The Sri Lanka scenario is the same. The groundwater is the main source of drinking water in the dry zone of Sri Lanka. Several studies carried out for drinking water quality in dry zones have identified various problems including fluoride contamination which is a result of climate change. The chemical composition of the rocks, soils, minerals and precipitation, air temperature evapotranspiration and many other factors regulate the fluoride contamination in groundwater in the dry zone of Sri Lanka. Covering two-thirds of the landmass in the country's dry zone receives lower rainfall which in turn interrelates with various factors to result in fluoride contamination in groundwater. Nearly 80% of groundwater consumers in the dry zone suffer from chronic fluoride poisoning mainly leading to dental fluorosis and some cases of skeletal fluorosis. To conclude proper consolidation of groundwater-related stakeholders formulate is sustainable and equity basis groundwater policy is recommended to control the risk of fluoride contamination. Summing up, the movement of groundwater has a significant effect on fluoride distribution and concentration heavily controlled by precipitation received as rainfall and fracture density of the basement rock in the dry zone of Sri Lanka.

REFERENCES

ADB. 2020. *Climate Risk Country Profile: Sri Lanka.* Washington: Asian Development Bank (ADB).

Ando, M., Tadano, M., Yamamoto, S., Tamura, K., Asanuma, S., Watanabe, T., Kondo, T., Sakurai, S., Ji, R., Liang, C., Chen, X., Hong, Z. and Cao, S. 2001. Health effects of fluoride pollution caused by coal burning. *Science of the Total Environment*, 271: 107–116.

Arnesen, A. and Krogstad, T. 1998. Sorption and desorption of fluoride in soil polluted from the almuminum smelter at Ardal in western Norway. *Water Air Soil Pollution*, 103: 357–373.

Baba, N. 2010. Sinking the pearl of the Indian Ocean: climate change in Sri Lanka. *Global Majority*, 1(1): 1–13.

Britannica. 2021. Retrieved 05 07, 2021, from https://www.britannica.com/science/fluorine

Chandrajith, R., Wijewardana, G., Dissanayake, C. and Abeygunasekara, A. 2006. Biomineralogy of human urinary calculi (kidney stones) from some geographic regions of Sri Lanka. Environmental Geochemistry and Health, 28: 393–399.

Chandrajith, R., Padmasiri, J., Dissanayake, C. and Prematilaka, K. 2012. Spatial distribution of fluoride in groundwater of Sri Lanka. *Journal of the National Science Foundation of Sri Lanka*, 40 (4): 303–309.

Chandrapala, L. and Wimalasuriya, M. 2003. Satellite measurements supplemeted with meteorological data to operationally estimate evaporation in Sri Lanka. *Agricultural Water Management*, 58: 89–107.

Cittanova, M. L., Lelongt, B., Verpont, M. C., Ggeniteau-Legendre, M., Wahbe, F., Prie, D., Coriat, P. and Ronco, P. M. 1996. Fluoride ion toxicity in human kidney collecting duct cells. *Anesthesiology*, 84(2): 428–435.

Cooray, P. 1994. The precambrian of Sri Lanka: a historical review. *Precambrian Research*, 66: 3–18.

Dent, D. L. and Goonewardene, L. K. P. A. 1993. Resource assessment and land use planning in Sri Lanka: a case study, In: Environmental Planning Issues. The Environmental Planning Group (Vol. 4, pp. 1–139). London: The International Institute for Environment and Development.

Dharmagunawardhane, H. and Dissanayake, C. 1993. Fluoride problems in Sri Lanka. *Environmental Management and Health*, 4: 9–16.

Dissanayake, C. 1991. The fluoride problem in the groundwater of Sri Lanka-enviornmnetal management and health. *International Journal of Environmental Studies*, 19: 195–203.

Dissanayake, C. 1993. The hydrogeochemical atlas of Sri Lanka: applications in environmental studies. In: W. Erdelen, C. Preu, and M. B. Ishwaran (Eds.), *Ecology and Landscape Management in Sri Lanka* (pp. 491–504). Germany: Margraf Scientific Books.

Dissanayake, C. 2005. Water quality in the dry zone of Sri Lanka-some interesting health aspects. *Journal of the National Science Foundation*, 33(3): 161–168.

Dissanayake, C. and Chandrajith. R. 2007. Medical geology in tropical countries with special reference to Sri Lanka. *Environmental Geochemistry and Health*, 29: 155–162.

Dissanyake, C. and Chandrajith, R. 2019. Fluoride and hardness in groundwater of tropical regions - review of recent evidence indicating tissue calcification and calcium phosphate nanoparticle formation in kidney tubules. *Ceylon journal of Science*, 48(3): 197–207.

Dissanayake, C. and Weerasooriya, S.1986. Fluorine as an indicator of mineralization hydrogeochemistry of a Precambrian mineralized belt in Sri Lanka. *Chemical Geology*, 28: 257–270.

Edmunds, M. and Smedley, P. 2005. Fluoride in natural waters-occurance, control and health aspects (Selinus., Alloway., Centemo., Finkelman., Fuge., Lindh., et al., Eds.). *Essentials of Medical Geology*, 5: 301–329.

Gilpin, L. and Johnson, A. 1980. Florine in agricultural soils of southeastern Pennsylvania. *Soil Science Society of America Journal*, 24: 255–258.

Harinarayan, C., Kochupillai, N., Madhu, S., Gupta, N. and Meunier, P. 2006. Fluorotoxic metabolic bone disease: an osteo-renal syndrome caused by excess flouride ingestion in the tropics. *Bone*, 39(4): 907–914.

Herath, S. and Rathnayake, U. 2005. Changing rainfall and its impact on landslides in Sri Lanka. *Journal of Mountain Science*, 2(3): 2018–224.

Jayakodi, A. N. 2006. Large diameter shallow Agro-wells – a national asset or a burden for the nation? *Journal of Agricultural Sciences*, 2(1): 1–10.

Herath, H. M., Ayala S., Kubota, K., Kawakami, T., Nagasawa, S., Motoyama, A., Weragoda, S. K., Chaminda G. G. T. and Yatigammana, S. K. 2017. Potential risk of drinking water to human health in Sri Lanka. *Environmental Forensics*, 18(3): 241–250.

Jayasuriya, C. and Jayasena, H. 2019. Refined GIS mapping to reinvestigate groundwater mining ptential surrounding the manmade reserviours and tributories in the Deduru Oya basin, Sri Lanka. *International Journal of Economic and Environmental Geology*, 10(4): 7–13.

Jayawardana, D., Pitawala, H. and Ishiga, H. 2010. Groundwater quality in difference climatic zones of Sri Lanka: focus on the occurance of Fluoride. *International Journal of Environmental Science and Development*, 1(3): 244–250.

Karunathilaka, K., Dabare, H. and Nandalal, K. 2017. Changes in rainfall in Sri Lanka during 1966–2015. *Journal of the Institution of Engineers*, 50(2): 39–48.

Kehelpannala, K., 1997. Deformation of a high-grade gondwana fragment. *Gondwana Research*, 1: 47–68.

Kim, S., Kim, K., Ko, K., Kim, Y. and Lee, K. 2012. Co-contamination of arsenic and fluoride in the groundwater of unconsolidated aquifers under reducing environmnets. *Chemosphere*, 87(8): 851–856.

Lantz, O., Jouvin, M., Vernejoul, M. and Druet, P. 1987. Fluoride-induced chronic renal failure. *American Journal of Kidney Diseases*, 10(2): 136–139.

Larsen, S. and Widdowson, A. E. 1971. Soil fluorine. *The Journal of Soil Science*, 197(122): 210–221.

Liyanage, C., Thabrew, M. and Kuruppuarachchi, D. 2000. Nitrate pollution in ground water of Kalpitiya: an evaluation of the content of Nitrate in the water and food items cultivated in the area. *Journal of National Science Foundation*, 28(2): 101–122.

Singh, P., Barjatiya, M., Dhing, S., Bhatnagar, R., Kothari, S. and Dhar, V. 2001. Evidence suggesting that high intake of fluoride provokes nephrolithiasis in tribal population. *Urological Research*, 29: 238–244.

SLSI. 2013. *SLS Standard 614*. Colombo: Sri Lanka Institue of Standard Instititue.

Takeno, N. 2005. *Atlas of Eh-pH Diagrams*. Japan: Geological Society.

Totsche, K. U., Wilcke, W., Korbus, M., Kobza, J. and Zech, W. 2000. Evaluation of fluoride induced metal mobilization in soil columns. *Journal of Environmental Quality*, 29(2): 454–459.

Watts, N., Adger, W. N., Ayeb-Karlsson, S., Bai, Y., Byass, S., Campbell-lendrum, D. D., et al. (2017). The Lancet countdown: tracking progress on health and climate change. *The Lancet*, 389(10074): 1151–1164.

WHO. 2011. *Guidelines for Drinking Water Quality* (4 ed.). Geneva: World Health Organization.

Wickramasinghe, B., Jayasena, H., Perera, K. and Rajapakse, R. 2021. Assessment of hydrogeological scenario in a cross-section from Anamaduwa to Kalpitiya in Northwest Sri Lanka. *Ceylon Journal of Science*, 50(1): 83–96.

Wright, E. 1992. The hydrogeology of crystalline basement aquifers in Africa. In E. Wright, & W. Burgess (Eds.), *Hydrogeology of Crystalline Basement Aquifers in Africa* (pp. 1–27). London: Geological Society Special Publication.

Zager, R. and Iwata, M. 1997. Inorganic fluoride. Divergent effects on human proximal tubular cell viability. *The American Journal of Pathology*, 150(2): 735.

Zhu, M., Xie, M. and Jiang, X. 2006. Interaction of fluoride with hydroxyaluminum-montmorillonite complexes and implications for flouride-contaminated acidic soils. *Applied Geochemistry*, 21: 675–683.

6 Application of Geospatial Technology in Catchment Modeling Using SCS-CN Method for Estimating the Direct Runoff on Barakar River Basin, Jharkhand

Arishmita Ghosh, Kalyan Kumar Bhar, Abhay M. Varade, and Sandipan Das

6.1 INTRODUCTION

Flooding induced by precipitation events over rivers is a primary concern worldwide (Gupta et al., 2019). The rivers, primarily rain-fed, can turn monstrous many a time due to excessive downpours, thus causing havoc throughout its catchment area (Larabi et al., 2018; Ranatunga et al., 2017; Shadeed and Almasri, 2010). Therefore, delineating the watershed or catchment basin through catchment modeling in such a situation becomes the need of the hour for mounting various evacuation strategies in case of flood-related exigencies. This is required not merely after flood evacuation strategies but also as requisites for prior actions against such flash flood events. Remote Sensing and Geographic Information System technology are widely applied in developing flood forecasting models, damage assessment, estimating runoff within the watershed, and determining stormwater drainage (Kannan et al., 2008; Meshram et al., 2017; Singh et al., 2017; Rana et al., 2020; Al-Ghobari and Dewidar, 2021; Meraj et al., 2021).

Damodar has a total catchment area of 25,820 km^2 and is often referred to as the sorrow of Bengal as it causes torrential floods during the monsoon period (Pandey et al., 2012; Ghosh and Guchhait, 2016; Bera and Singh, 2021). It has been partially tamed by constructing dams, including Konar, Panchet, Tilaiya, and Maithon (Ghosh et al., 2014). Barakar, the major tributary of the Damodar River, has two dams. Barakar, along with other tributaries, carries a considerable volume of monsoon water to the lower Damodar basin, flooding over the plains of West Bengal. The rainfall-runoff extent of the Barakar catchment can be estimated utilizing one of the most widely accepted methods of runoff estimation in the watershed: the SCS-CN method used in this instance to assist in the estimation of the runoff volume within the Barakar catchment based on the amount of rainfall in the area. The hydrologists widely accepted the technique for their hydrological studies (Rizeei et al., 2018; Al-Ghobari et al., 2020). The SCS-CN method was developed by the NRCS and USDA and is extensively discussed in the National Engineering Handbook (NEH-4), Section of Hydrology as a widely used method of runoff depth estimation (USDA, 1972; Singh et al., 2015).

DOI: 10.1201/9781003303237-6

6.2 STUDY AREA

The study is the Barakar River basin, including the Maithon and Tilaiya Dam catchments. It has a latitudinal extent of 24°00′N–24°30′N and a longitudinal extent of 85°00′E–87°00′E. In Jharkhand, it originates from the Padma village in the Padma block of the Barhi subdivision of the Hazaribagh district. Then it flows with a flux over the other districts of Jharkhand, namely Koderma, Giridih, Dhanbad, and Jamtara. The catchment area of the Barakar extends over the districts of Chatra, Hazaribagh, Koderma, Giridih, Dhanbad, Jamtara, and Deoghar in Jharkhand and a very small portion of the catchment stretches into the Bardhaman district of West Bengal. Among the four major dams built over the Damodar River, two had been made over the Barakar River, namely the Tilaiya Dam and the Maithon Dam.

Tilaiya is the upstream dam of the Barakar River situated in the Hazaribagh district of Jharkhand, having a catchment area of 984 km^2 and Maithon is the downstream dam situated in the Dhanbad district of Jharkhand having a catchment of 6,391.74 km^2. Tilaiya Dam is 366 m long and 30.18 m high from the river bed level and has a hydroelectric power station on its left bank. Maithon Dam is composed of two earthen and concrete structures spanning 4,860 m, and the concrete dam is 43.89 m high above the river bed level. On the left bank of the river is the hydel power station over the Maithon Dam, like the Tilaiya power station, but it is built underground, which is a first of its kind in India. Among the 15 small and medium-sized tributaries of the Barakar, Barsoti and Usri are the two major tributaries. They are distinct apart because the Barsoti joins the Barakar from the south and the Usri flows in from the north into the Barakar. The Barakar, after flowing for such a long extent, enters the Damodar in the Dishergarh alongside Asansol in the Paschim Bardhaman district of West Bengal.

6.3 MATERIALS AND METHODS

6.3.1 Data Sources

The analyses in the project had been done through the two most important geospatial techniques – Remote Sensing and GIS. The base map for the study area was taken from the IGIS map portal, and then the study area was delineated from it. SRTM DEM data-1 Arc-Second (30 m resolution) obtained from USGS Earth Explorer was applied to delineate the catchment area. The Landsat 8 OLI/TIRS data-30 m resolution was collected from USGS Earth Explorer for preparing the LULC classification for both the years 2010 and 2017. To prepare the soil texture map and the HSG map, the digital soil map shapefile was collected from the Digital Soil Map of the World (DSMW) portal of the Food and Agriculture Organization (FAO). Daily rainfall data in millimeters of 2010 and 2017 was collected from the India-WRIS website.

6.4 SCS-CN METHOD

The rainfall-runoff estimation was mainly done with the help of the most well-known method – the SCS CN method (Shi et al., 2020; Verma et al., 2020; Xiaojun et al., 2021). The USDA initially introduced the SCS-CN model for the United States' conditions (Bartlett et al., 2016). The model is extensively used for computing direct runoff (Satheesh kumar et al., 2017; Bal et al., 2021; Soulis, 2021). In doing so, it mainly uses conceptual or empirical equations of rainfall and watershed coefficients as inputs. Watershed coefficients are based on the Curve Number (CN), which is primarily based on the land cover soil complex. SCS assumes that the proportion of actual retention to potential maximum retention equals the ratio of actual runoff to that of maximum likely runoff, where rainfall less initial abstraction equals maximum likely runoff. The empirical relationship is shown below

$$\frac{F}{S} = \frac{Q}{P - Ia} \tag{6.1}$$

Where,

F = actual retention (mm)
S = Maximum retention potential (mm)
Q = Amount of runoff accumulated (mm)
P = accumulated rainfall depth (mm)
Ia = initial abstraction (mm)

When the runoff starts, all additional rainfall becomes a part of either the runoff or actual retention; hence the actual retention is the difference between the rainfall minus initial abstraction and runoff. Therefore,

$$F = P - Ia - Q \tag{6.2}$$

Combining Equations (6.1) and (6.2), we get,

$$Q = \frac{(P - Ia)2}{(P - Ia + S)} \tag{6.3}$$

In order to eliminate the need for estimating Ia and S, the SCS performed a regression analysis using the recorded rainfall and runoff data from smaller drainage basins and obtained the following relationship:

$$Ia = 0.2S \tag{6.4}$$

Combining Equations (6.3) and (6.4), we get,

$$Q = \frac{(P - 0.2S)2}{P + 0.8S} \text{ for } P > 0.2S \tag{6.5}$$

6.4.1 Estimation of S

The parameter S, which is the potential maximum retention in the equation, depends on the soil-vegetation-land use (SVL) complex and Antecedent Moisture Condition (AMC) characteristics in a watershed. There is a lower and upper limit of S for each SVL complex. To make this parameter more linear, S has been converted to the Curve Number (CN). Hence the equation is:

$$S = \frac{25{,}400}{\text{CN}} - 254 \tag{6.6}$$

6.4.2 Determination of CN

The CN value can be determined through two conditions.

6.4.2.1 Hydrological Soil Type

The soil is categorized into four hydrological categories based on the capacity of the water infiltration rate. They are

Group A: High capacity (>7.62 mm) consists of deep, well to excessively well-drained sands or gravels. Specific - loam-clay texture soils.

Group B: Medium capacity (3.81–7.62 mm). This characterizes the loamy, sandy texture soils;

Group C: Low capacity (1.27–3.81 mm). Specific - loam-clay texture soils.
Group D: Very low capacity (0–1.27 mm). Specific - loam-clay and clay-loamy texture.

6.4.2.2 Antecedent Moisture Condition (AMC)

The water content of the soil at a specific moment is referred to as the antecedent moisture condition (AMC). The volume and rate of infiltration flow are affected by the antecedent moisture condition. The SCS developed three classes of AMC and named them I, II, and III, mainly depending on the previous five days' rainfall. The three AMC classes correlate to the soil characteristics listed below.

AMC I: Soils are dry but not to the wilting point; satisfactory cultivation has taken place. Soil conditions are dry. Precipitation ranges from 12.7 mm in the dry season to 35.6 mm in the wet season.

AMC II: Normal soil moisture conditions or average conditions. The dry season precipitation ranges from 12.7 to 28 mm, while the wet season precipitation ranges from 35.6 to 53.4 mm.

AMC III: Heavy rainfall, or light rainfall, and low temperatures occurred within the last 5 days. Saturated soil conditions. Specific to the precipitations >28 mm, in the dry season, respective >53.4 mm, in the wet season.

6.4.3 Selection of CN

The standard CN values are mainly assigned to each land use soil group combination. The Antecedent Moisture Condition (AMC II), also known as the average or normal soil moisture condition, is used to assign these CN values. The state of vegetation growth is mostly determined by the hydrologic situation. Based on this the curve number which is assigned is the CN II stating the normal or average condition (Table 6.1).

From these CN II values, the formula for calculating weighted CN values for each land use soil group combination is:

$$\text{WCN} = \frac{(\text{Area} * \text{CN})}{\text{Total Area}} \tag{6.7}$$

Then from the WCN values of the total watershed, the corresponding CN I and CN II values are estimated with the formulae:

$$\text{CN I} = \frac{4.2 * \text{CN II}}{10 - (0.058 * \text{CN II})} \tag{6.8}$$

$$\text{CN III} = \frac{23 * \text{CN II}}{10 - (0.13 * \text{CN II})} \tag{6.9}$$

Vandersypen et al. (1972) developed the relationship between initial abstraction and retention of water by the catchment for Indian conditions as $Ia_{=}0.3S$.

Therefore, the

$$Q = \frac{(P - 0.3S)2}{P + 0.7S} \text{ for } P > 0.3S \tag{6.10}$$

In this study, we have used the formula mentioned above to estimate the runoff depth in our catchment.

TABLE 6.1
Runoff Curve Numbers based on AMC-IIi for Hydrological Soil Cover Complexes (SCS, 1986)

Description of Land Use	Hydrologic Soil Group			
	A	*B*	*C*	*D*
Paved parking lots, roofs, driveways	98	98	98	98
Streets and Roads				
Paved with curbs and storm sewers	98	98	98	98
Gravel	76	85	89	91
Dirt	72	82	87	89
Cultivated (Agricultural Crop) Land				
Without conservation treatment (no terraces)	72	81	88	91
With conservation treatment (terraces, contours)	62	71	78	81
Pasture or Range Land				
Poor (<50% ground cover or heavily grazed)	68	79	86	89
Good (50%–75% ground cover; not heavily grazed)	39	61	74	80
Meadow (grass, no grazing, mowed for hay)	30	58	71	78
Brush (good, >75% ground cover)	30	48	65	73
Woods and Forests				
Poor (small trees/brush destroyed by over-grazing or burning)	45	66	77	83
Fair (grazing but not burned; some brush)	36	60	73	79
Good (no grazing; brush covers ground)	30	55	70	77
Open Spaces (Lawns, Parks, Golf Courses, Cemeteries, etc.):				
Fair (grass covers 50%–75% of the area)	49	69	79	84
Good (grass covers >75% of the area)	39	61	74	80
Commercial and Business Districts (85% impervious)	89	92	94	95
Industrial Districts (72% impervious)	81	88	91	93
Residential Areas				
1/8 Acre lots, about 65% impervious	77	85	90	92
1/4 Acre lots, about 38% impervious	61	75	83	87
1/2 Acre lots, about 25% impervious	54	70	80	85
1 Acre lots, about 20% impervious	51	68	79	83

6.5 METHODOLOGY

The spatial and non-spatial data were gathered from a variety of sources and then processed to produce the end product, direct runoff. The SRTM DEM data was used to prepare the watershed polygon, the soil map was used for the HSG map and the LANDSAT 8 data was used to prepare the LULC map. Further, the rainfall data collected over that region was used to estimate the direct runoff (Figure 6.1).

6.6 RESULTS AND DISCUSSION

6.6.1 Watershed Delineation

The two pour-points were taken at the Maithon Dam and the other at Tilaiya Dam to delineate two different watersheds, though one is within the other. But for getting two individual watersheds, one as the catchment area of the Tilaiya Dam and another one as the catchment area of the Maithon Dam, both were delineated (Figure 6.2).

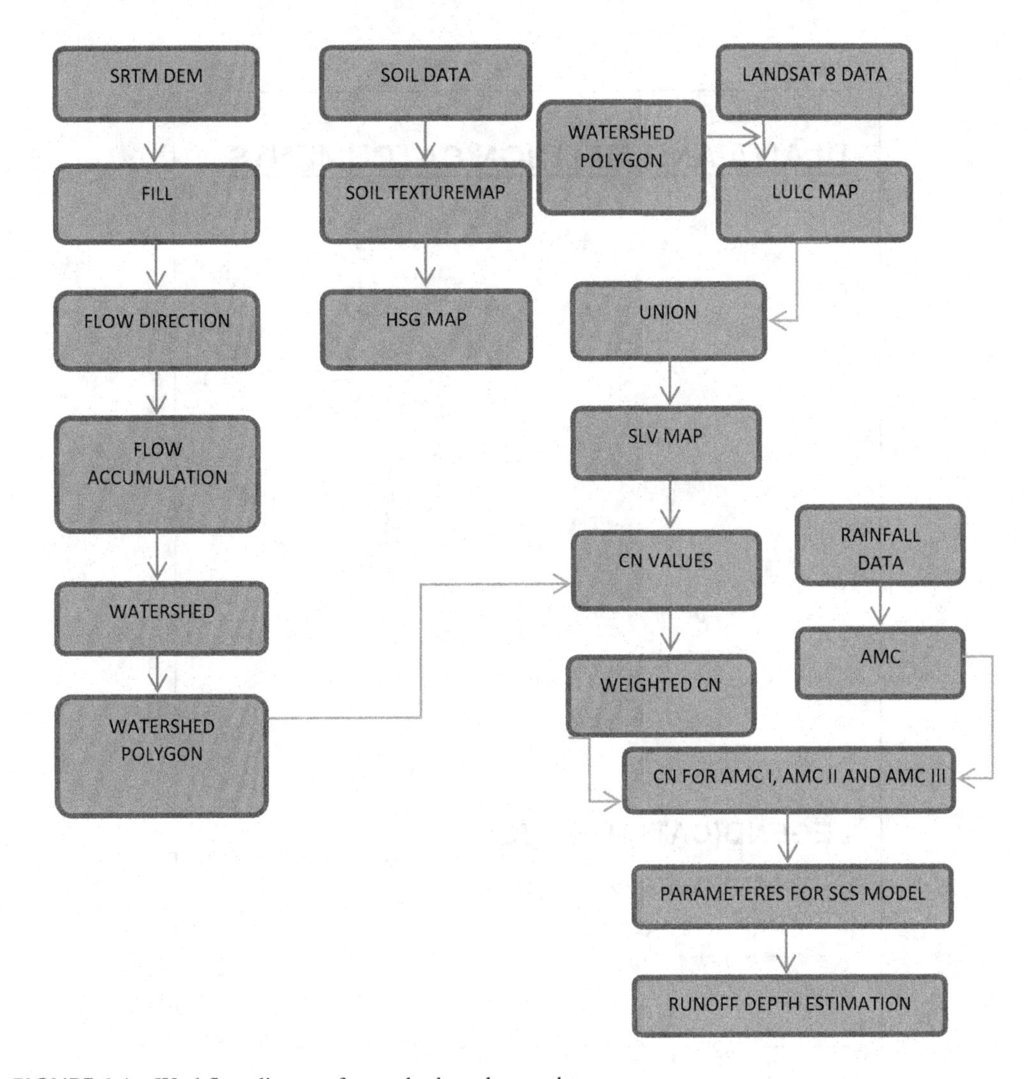

FIGURE 6.1 Workflow diagram for methods and procedures.

6.6.2 Landuse-Landcover Map

The catchment area was extracted from the Landsat 8 satellite data using the watershed boundary, and then the land-use landcover map was prepared using the supervised image classification method. The catchment was classified into seven different classes, including agricultural land (54.43%), scattered vegetation (13.35%), open forest (11.04%), settlements (8.06%), barren land (5.80%), dense forest (4.41%), and water bodies (2.88%). The overall classification accuracy was 80.65%, and the Kappa accuracy was 0.7743. The agricultural land (54.43%) occupies the majority of the area of the Maithon catchment (Figure 6.3). In the case of the Tilaiya catchment, the maximum area is covered by agricultural land (38.89%) followed by open forest (14.75%), settlements (13.33%), dense forest (12.77%), scattered vegetation (12.43%), waterbodies (4.78%), and barren land (3.01%) (Figure 6.4).

6.6.3 Soil Map

The ESRI soil shape file was used to prepare the soil texture map from the given soil codes. The Maithon catchment is mainly covered by sandy clay loam followed by loam. Sandy loam covers

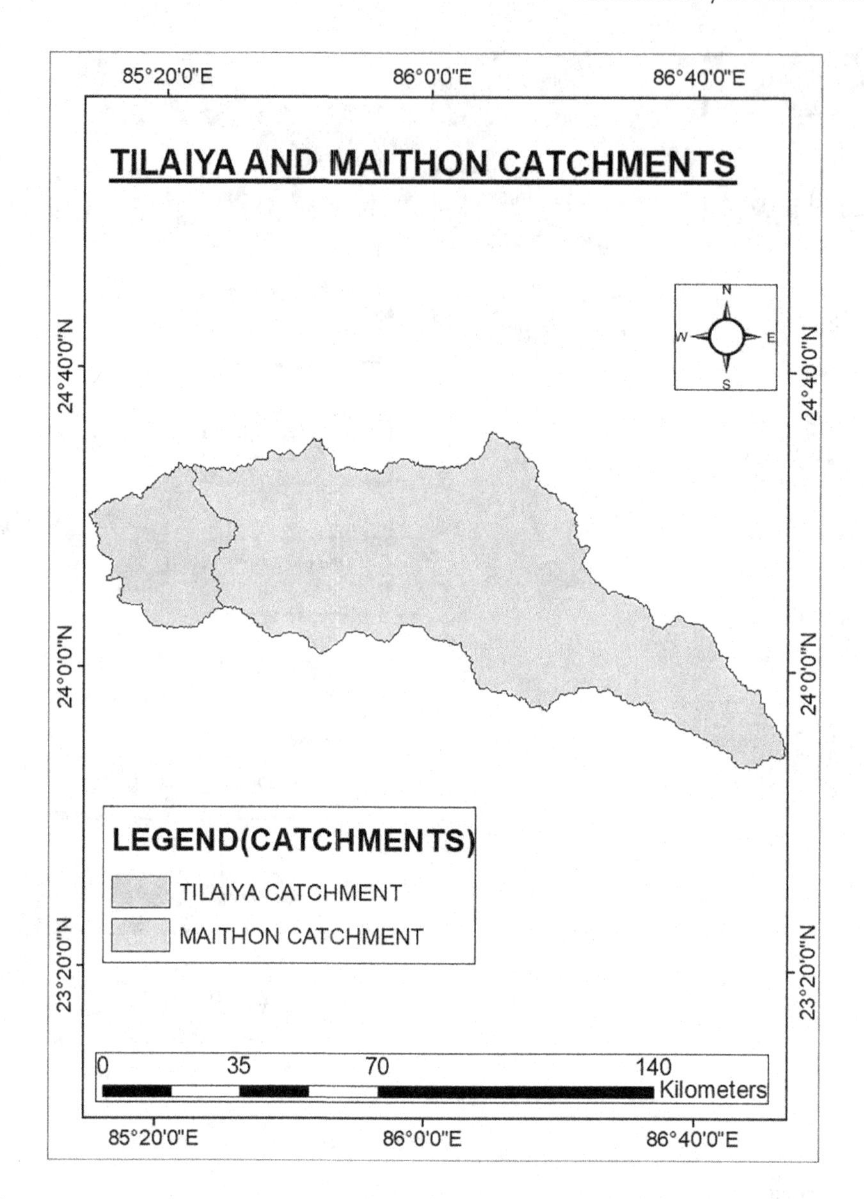

FIGURE 6.2 Map showing Tilaiya and Maithon catchments.

a very small part of the catchment area. Next, the Hydrological Soil Group (HSG) map was prepared from the soil texture map concerning the water infiltration capacity of the soil types, which displayed the properties of the Hydrological Soil Group types. Sandy clay loam which is highest in content over the catchment area, shows the property of low infiltration and low water transmission rates. Therefore, it is categorized into Hydrological soil group C (HSG-C). The second soil texture type was Loam which displays the property of being highly drained with both high infiltration and water transmission rates, hence categorized under the Hydrological soil group A (HSG-A) (Figure 6.5).

Sandy loam covered the least area in the Maithon catchment as a whole, which showed the property of being moderately drained with moderately fine to moderately coarse-textured soils having moderate water transmission rates. Therefore, it is categorized under the Hydrological soil group type B (HSG-B) (Figure 6.5). In the case of the Tilaiya catchment, the soil texture group

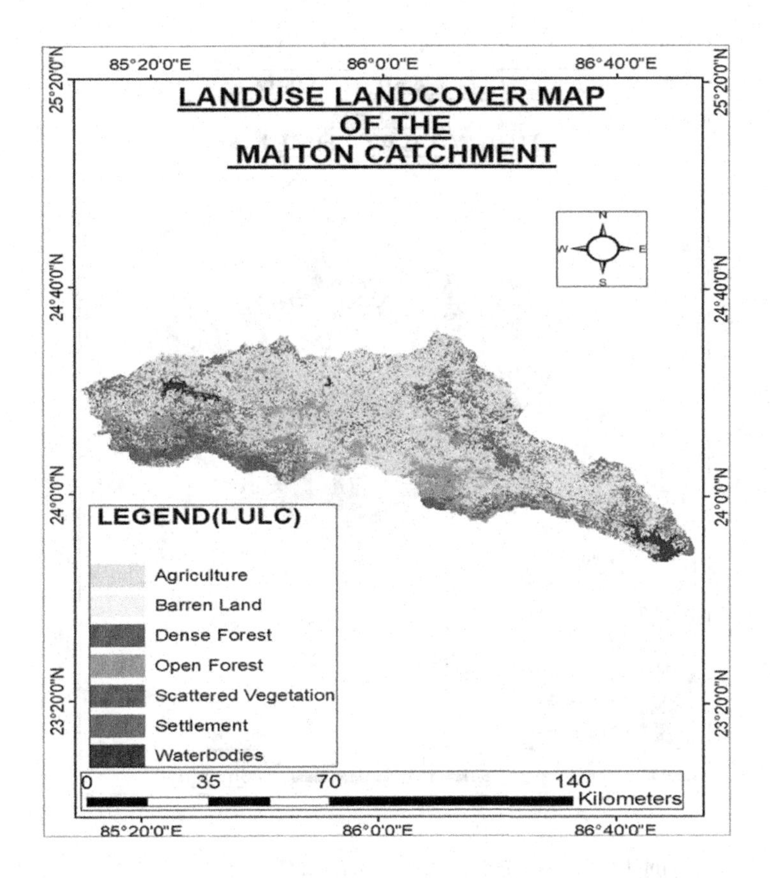

FIGURE 6.3 Map showing LULC area under Maithon catchment.

which falls in that catchment is only sandy loam and sandy clay loam. The sandy loam covers the maximum part of the Tilaiya catchment and the sandy clay loam covers only small parts of the Tilaiya catchment. The Tilaiya catchment thus has two Hydrological soil group types, Group A and Group B (HSG-A and HSG-B) categorized likewise in the case of the Maithon catchment (Figure 6.6).

6.6.4 Soil Landuse Vegetation Complex (SLV) Map

After preparing the land-use land-cover map and the hydrological soil group map, both thematic maps were combined to form the Soil Land Use Vegetation complex (SLV) map. This was created by performing the union operation of the HSG map over the LULC map. The result of this was an SLV map showing the soil infiltration rates depending on the land-use land-cover type of the catchment (Figure 6.7).

6.6.5 Curve Number (CN) and Weighted Curve Number (WCN) Values

From the Soil Land use Vegetation complex (SLV) map, the CN and WCN values were estimated for both Tilaiya and Maithon catchment, depending on the HSG soil type and the LULC classes falling under each soil type. Firstly, the area was calculated for each LULC class and the CN values were estimated according to Table 6.1 where the CN values are given considering the AMC II condition and the HSG soil group under which the particular LULC class has fallen. This method was

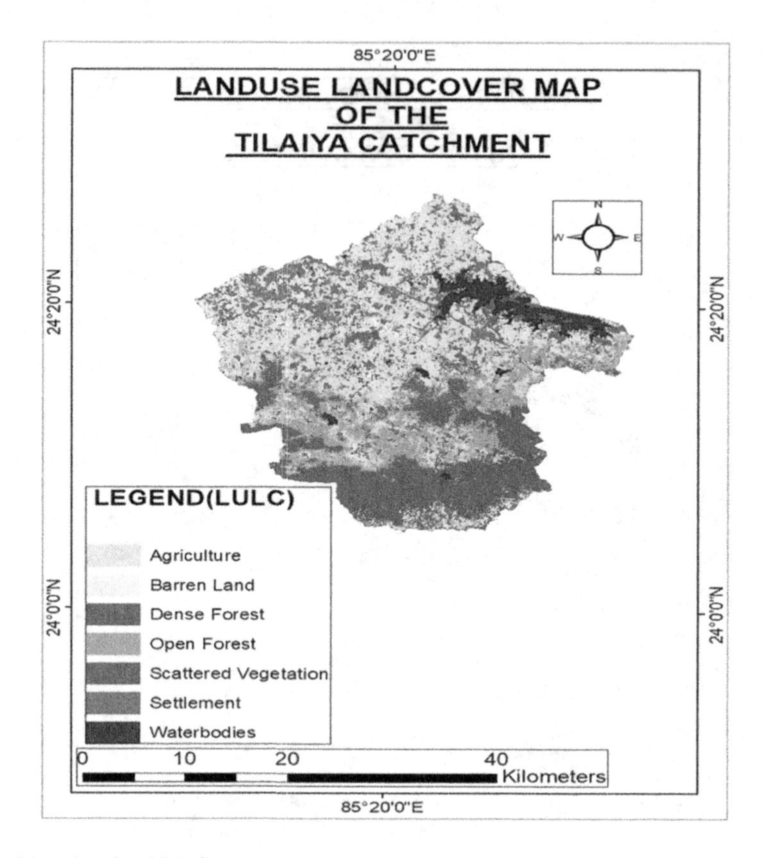

FIGURE 6.4 Map showing LULC area under Tilaiya catchment.

followed for both the Maithon and Tilaiya catchments, of which Maithon has three HSG groups i.e., A, B, and C, and Tilaiya has two HSG groups A and C (Tables 6.2 and 6.3). The resultant CN and WCN values of the Maithon catchment are as follows:

The resultant CN and WCN values of the Tilaiya catchment are as follows:

6.7 ANTECEDENT MOISTURE CONDITIONS, CN I AND CN III VALUES

Depending on the rainfall data and the previously calculated CN-II values, antecedent moisture condition was estimated for both Maithon and Tilaiya catchments. The rainfall data for the years 2010 and 2017 have been used in the research study to show the runoff depth and preceding condition of both the Maithon and Tilaiya catchment for both a drought year and a flood year, respectively. After calculating the AMC based on the rainfall data, the CN-I and CN-III values were calculated from the resultant AMC and CN-II values.

6.8 POTENTIAL MAXIMUM RETENTION (S) AND RUNOFF DEPTH (Q)

The potential maximum retention (S) and the runoff depth (Q) were calculated, based on which valid runoff depth for both the catchments was obtained (Tables 6.4 and 6.5).

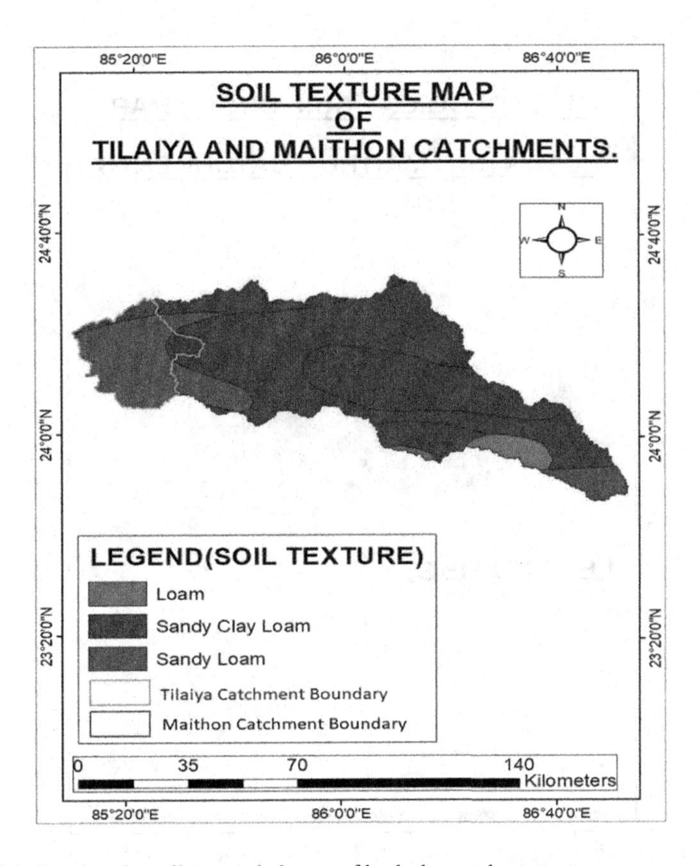

FIGURE 6.5 Map showing the soil textural classes of both the catchments.

From the above results and charts, it is evident that in the year 2017, the runoff, even after the maximum potential retention, was almost half of the storage capacity of the dams, causing floods in that year in both catchments (Table 6.4, Figure 6.8). In contrast to that, in 2010, the runoff was very low in the Maithon catchment, and almost excessively low to no runoff occurred in the Tilaiya Catchment, thus leading to drought in that year (Table 6.5, Figure 6.9).

6.9 CONCLUSION

This study has shown the rainfall-runoff depths in the Maithon and Tilaiya catchments over the Barakar River Basin, Jharkhand, for 2 years, 2010 and 2017. This purpose was to get a clear picturization of the Barakar River basin and its dams during the heavy and low rainfall years, respectively. This is because Barakar and the other minor tributaries of the Damodar cause destruction mainly in the rainy seasons in the lower reaches of the Damodar River basin. The research study was carried out to know exactly how much runoff occurs in the catchment concerning the amount of rainfall during the rainy season. In contrast, what happens to the catchments is little or almost no rainfall. The SCS-CN method, one of the most important physical hydrological models, was applied to calculate the region's rainfall-runoff depth. The Tilaiya and Maithon catchments have

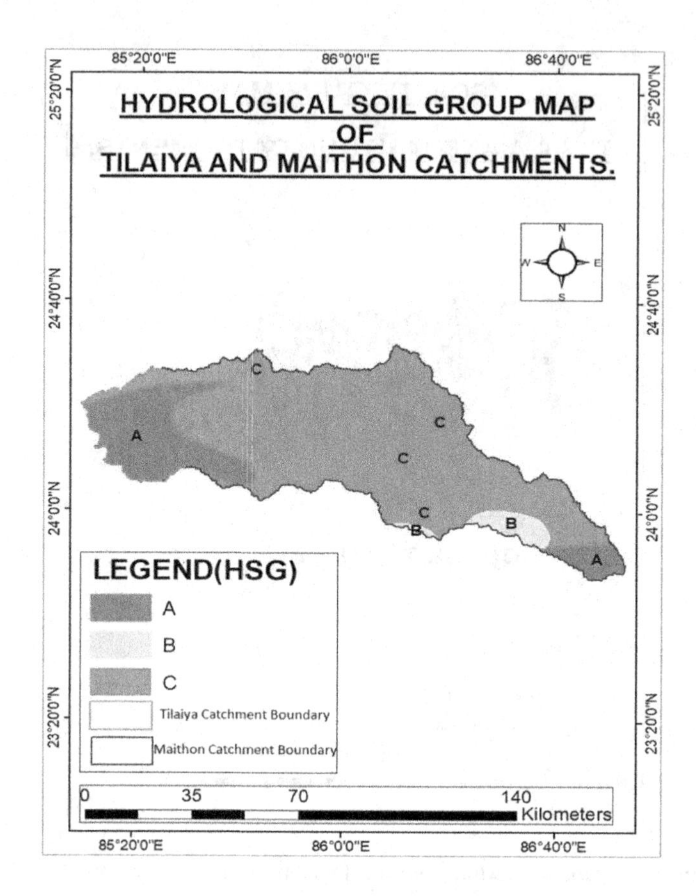

FIGURE 6.6 Map showing the hydrological soil groups for both the catchments.

been delineated, and the recent flood and drought years have been estimated depending on the runoff depth of the region. This will lead to a clear view of the catchment during the flood year and how much water gets accumulated in both the catchments and the dams. We could also get a clear image of the catchments during the least rainfall year, i.e., drought conditions. Therefore, it is necessary to release the stored waters from the dams during the flood and drought seasons to protect the dam from breaking during the floods and provide the required water to the agricultural fields and the surrounding areas during the drought year, respectively. Thus releasing the stored waters from the dams during the highest rainfall year leads to extreme flood and havoc and even loss of life and property in the catchments and the lower reaches of the Damodar River basin because the river flows downwards, toward Bardhamman district and enters West Bengal. In such circumstances taking precursory measures by keeping a check over the daily rainfall-runoff condition and acting immediately, responsibly, and necessarily becomes the need of the hour to combat such situations and prevent the loss of life and property.

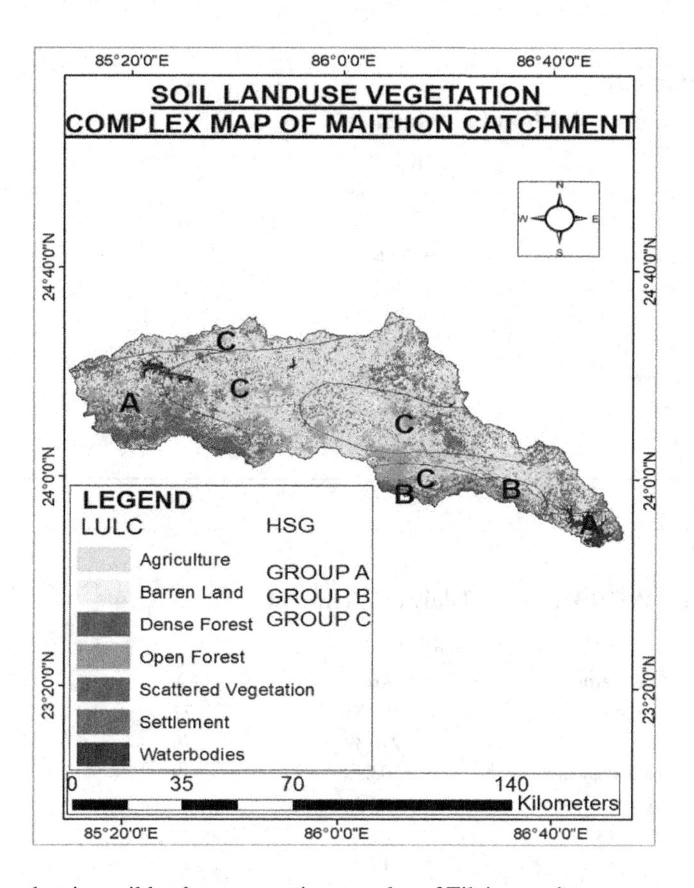

FIGURE 6.7 Map showing soil land use vegetation complex of Tilaiya catchment.

TABLE 6.2
CN and WCN values of Maithon Catchment

Soil Type A

ID	Classname	Area	CN	WCN = Area*CN/Total
1	Settlement	142.27	77	8.40
2	Agriculture	455.97	76	26.59
3	Scattered vegetation	184.41	49	6.93
4	Barren land	27.23	77	1.60
5	Water bodies	107.61	98	8.09
6	Open forest	162.14	36	4.47
7	Dense forest	223.54	30	5.14
	Total	1,303.21		61.25

Soil Type B

ID	Class Name	Area	CN	WCN = Area*CN/Total
1	Settlement	7.4183	86	2.91
2	Agriculture	65.97	85	25.59
3	Scattered vegetation	78.23	68	24.28
4	Barren land	2.27	86	0.89
5	Water bodies	5.03	98	2.25
6	Open forest	43.84	60	12.00
7	Dense forest	16.29	55	4.09
	Total	219.09		72.03

(Continued)

TABLE 6.2 (*Continued*)
CN and WCN values of Maithon Catchment

		Soil Type C		
ID	Class Name	Area	CN	WCN = Area*CN/Total
1	Settlement	365.83	91	6.82
2	Agriculture	2,963.33	90	54.66
3	Scattered vegetation	592.24	79	9.58
4	Barren land	342.20	91	6.38
5	Water bodies	72.27	98	1.45
6	Open forest	500.86	73	7.49
7	Dense forest	42.34	70	0.60
	Total	4,879.08		87.00

TABLE 6.3
Estimated CN and WCN Values of Tilaiya Catchment

		Soil Type A		
ID	Classname	Area	CN	WCN = Area*CN/Total
1	Settlement	100.83	77	9.81
2	Agriculture	290.89	76	27.95
3	Scattered vegetation	103.33	49	6.40
4	Barren land	23.48	77	2.28
5	Water bodies	30.73	98	3.80
6	Open forest	119.03	36	5.41
7	Dense forest	122.52	30	4.64
	Total	790.84		60.33
		Soil Type C		
ID	**Class Name**	**Area**	**CN**	**WCN = Area*CN/Total**
1	Settlement	26.96	91	14.56
2	Agriculture	82.22	90	43.93
3	Scattered vegetation	15.99	79	7.50
4	Barren land	5.46	91	2.95
5	Water bodies	15.23	98	8.86
6	Open forest	22.56	73	9.77
7	Dense forest	0.01	70	0.004
	Total	168.44		87.59

TABLE 6.4
Rainfall & Runoff for Tilaiya & Maithon Catchments for the Year 2017

		Year-2017	
	Rainfall (mm)	Runoff (mm)	Runoff (cubic millimetres)
Tilaiyacatchment	2,088.16	173.24	1,66,189,100.4
Maithoncatchment	6,470.51	1,024.71	65,59,631,339

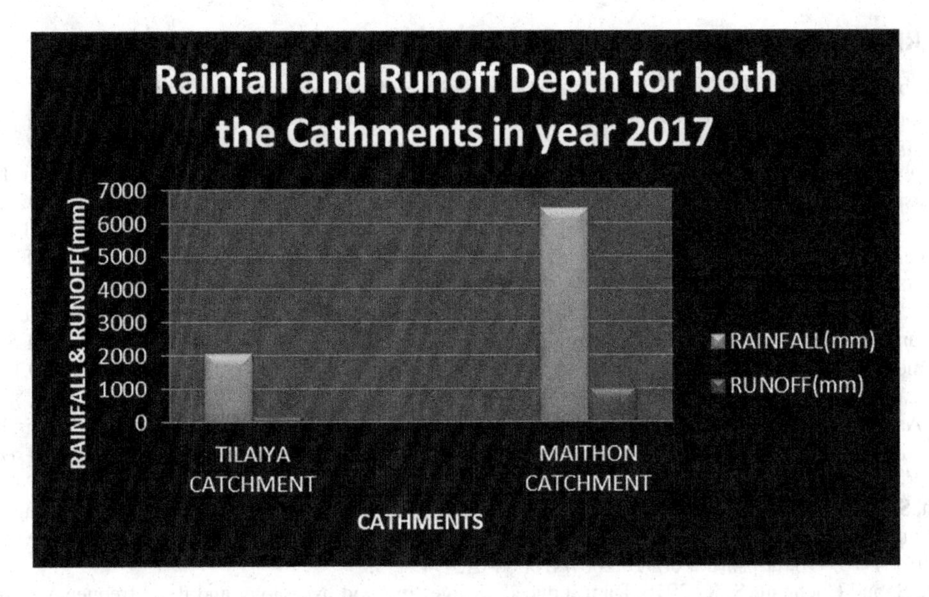

FIGURE 6.8 Rainfall-runoff depth for both the catchments in the year 2017.

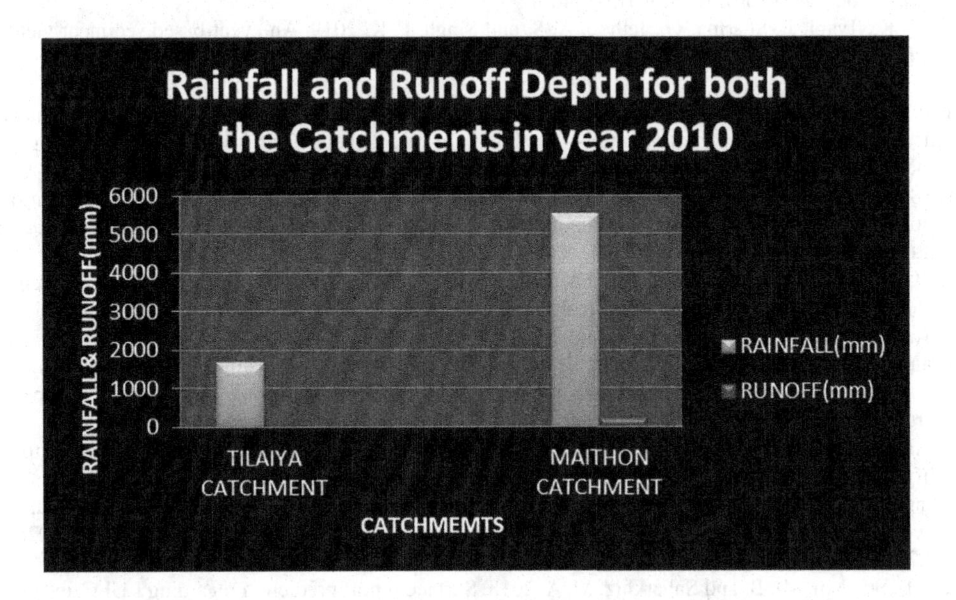

FIGURE 6.9 Rainfall-runoff depth for both the catchments in the year 2010.

TABLE 6.5
Rainfall & Runoff for Tilaiya & Maithon Catchments for the Year 2010

	Year-2010		
	Rainfall (mm)	**Runoff (mm)**	**Runoff (cubic millimetres)**
Tilaiya catchment	1,669.74	3.16	3,039,592.27
Maithon catchment	5,522.29	225.76	14,45,220,126

REFERENCES

Al-Ghobari, H. and Dewidar, A. Z. 2021. Integrating GIS-based MCDA techniques and the SCS-CN method for identifying potential zones for rainwater harvesting in a semi-arid area. *Water*, 13(5): 704. https://doi.org/10.3390/w13050704.

Al-Ghobari, H., Dewidar, A. and Alataway, A. 2020. Estimation of surface water runoff for a semi-arid area using RS and GIS-based SCS-CN method. *Water*, 12(7): 1924. https://doi.org/10.3390/w12071924.

Bartlett, M. S., Parolari, A. J., McDonnell, J. J. and Porporato, A. 2016. Beyond the SCSCN method: a theoretical framework for spatially lumped rainfall-runoff response. *Water Resources Research*, 52: 4608–4627. https://doi.org/10.1002/2015WR018439.

Bal, M., Dandpat, A. K. and Naik, B. 2021. Hydrological modeling with respect to impact of land-use and land-cover change on the runoff dynamics in Budhabalanga river basing using ArcGIS and SWAT model. *Remote Sensing Applications Society and Environment*, 23: 100527. https://doi.org/10.1016/j.rsase.2021.100527.

Bera, A. and Singh, S. K. 2021. Comparative assessment of livelihood vulnerability of climate induced migrants: a micro level study on Sagar Island, India. *Sustainability Agri Food Environmental Research*, 9(2): 1–15. https://doi.org/10.7770/safer-V9N2-art2324.

Ghosh, S. and Guchhait, S. K. 2014. Hydrogeomorphic variability due to dam constructions and emerging problems: a case study of Damodar River, West Bengal, India. *Environment, Development and Sustainability*, 16(3): 769–96. https://doi.org/10.1007/s10668-013-9494-5.

Ghosh, S. and Guchhait, S. K. 2016. Dam-induced changes in flood hydrology and flood frequency of tropical river: a study in Damodar River of West Bengal, India. *Arabian Journal of Geosciences*, 9(2): 1–26. https://doi.org/10.1007/s12517-015-2046-6.

Gupta, S. K., Tyagi, J., Sharma, G., Jethoo, A. S. and Singh, P. K. 2019. An event-based sediment yield and runoff modeling using soil moisture balance/budgeting (SMB) method. *Water Resources Management*, 33(11): 3721–3741.

Kannan, N., Santhi, C., Williams, J. R. and Arnold, J. G. 2008. Development of a continuous soil moisture accounting procedure for curve number methodology and its behavior withdifferent evapotranspiration methods. *Hydrological Processes*, 22: 2114–2121.

Larabi, S., St-Hilaire, A., Chebana, F. and Latraverse, M. 2018. Multi-criteria process-based calibration using functional data analysis to improve hydrological model realism. *Water Resources Management*, 32: 195–211. https://doi.org/10.1007/s11269-017-1803-6.

Meraj, G., Singh, S. K., Kanga, S. and Islam, M. N. 2021. Modeling on comparison of ecosystem services concepts, tools, methods and their ecological-economic implications: a review. *Modeling Earth Systems and Environment*, 8: 15–34. https://doi.org/10.1007/s40808-021-01131-6.

Meshram, S. G., Sharma, S. K. and Tignath, S. 2017. Application of remote sensing and geographical information system for generation of runoff curve number. *Applied Water Science*, 7: 1773–1779. https://doi.org/10.1007/s13201-015-0350-7.

Rana, V.K., Maruthi, T. and Suryanarayana, V. 2020. GIS-based multi criteria decision making method to identify potential runoff storage zones within watershed. *Annals of GIS*, 26(2): 149–168.

Ranatunga, T. S., Tong, T. Y. and Yang, Y. J. 2017. An approach to measure parameter sensitivity in watershed hydrological modelling. *Hydrological Sciences Journal*, 62(1): 76–92. https://doi.org/10.1080/02626667.2016.1174335.

Rizeei, H., Pradhan, M., B. and Saharkhiz, M. A. 2018. Surface runoff prediction regarding LULC and climate dynamics using coupled LTM, optimized ARIMA, and GIS-based SCS-CN models in tropical region. *Arabian Journal of Geosciences*, 11: 53. https://doi.org/10.1007/s12517-018-3397-6.

Satheeshkumar, S., Venkateswaran, S. and Kannan, R. 2017. Rainfall–runoff estimation using SCS–CN and GIS approach in the Pappiredipatti watershed of the Vaniyar sub basin, South India. *Modeling Earth Systems* and *Environment*, 3: 24. https://doi.org/10.1007/s40808-017-0301-4.

Singh, P. K., Mishra, S. K., Berndtsson, R., Jain, M. K. and Pandey, R. P. 2015. Development of a modified SMA based MSCS-CN model for runoff estimation. *Water Resources Management*, 29: 4111–4127. https://doi.org/10.1007/s11269-015-1048-1.

Singh, A. K., Jasrotia, A. S., Taloor, A. K., Kotlia, B. S., Kumar, V., Roy, S., Ray, P. K. C., Singh, K. K., Singh, A. K. and Sharma A. K. 2017. Estimation of quantitative measures of total water storage variation from GRACE and GLDAS-NOAH satellites using geospatial technology. *Quaternary International*, 444 (Part A): 191–200.

Soulis, K. X. 2021. Soil conservation service curve number (SCS-CN) method: current applications, remaining challenges, and future perspectives. *Water*, 13(2): 192. https://doi.org/10.3390/w13020192.

Shi, W. and Wang, N. 2020. An improved SCS-CN method incorporating slope, soil moisture, and storm duration factors for runoff prediction. *Water*, 12(5): 1335. https://doi.org/10.3390/w12051335.

Shadeed, S. and Almasri, M. 2010. Application of GIS-based SCS-CN method in West Bank catchments, Palestine. *Water Science and Engineering*, 3(1): 1–13. https://doi.org/10.3882/j.issn.1674-2370.2010.01.001.

USDA, 1972. *Soil Conservation Service, National Engineering Handbook*. Hydrology Section 4. Chapters 4–10. Washington, DC: USDA.

Vandersypen, D. R., Bali, J. S. and Yadav, Y. P. 1972. *Handbook of Hydrology*. New Delhi: Soil Conservation Division, Ministry of Agriculture, Government of India.

Verma, S., Singh, P. K., Mishra, S. K., Singh, V. P., Singh, V. and Singh, A. 2020. Activation soil moisture accounting (ASMA) for runoff estimation using soil conservation service curve number (SCS-CN) method. *Journal of Hydrology*, 589: 125114. https://doi.org/10.1016/j.jhydrol.2020.125114.

Xiaojun, G., Peng, C., Xingchang, C., Li, Y., Ju, Z. and Yuqing, S. 2021. Spatial uncertainty of rainfall and its impact on hydrological hazard forecasting in a small semiarid mountainous watershed. *Journal of Hydrology*, 595: 126049. https://doi.org/10.1016/j.jhydrol.2021.126049.

7 Hydro-Geospatial Investigation to Propose Water Conservation Sites for Water Management in Limestone Terrain

Kamal Kishor Sahu, Amit P. Multaniya, and Manish Kumar Sinha

7.1 INTRODUCTION

The primary fresh water supply found below the surface of the earth is groundwater, which is also a vital resource for various human activities such as agriculture, household use, and supply for the industrial and manufacturing sectors etc. (CGWB, 2008; Ghazavi et al., 2018; Kaliraj, 2015). More than 85% of India's rural household water needs, 50% of its urban water needs, and more than 50% of its irrigation needs are met by groundwater, which is rapidly running out in many places as a result of uncontrolled withdrawal in various sectors. With rapid development and stupendous increase of population in the recent past, the areas have been reduced day by day to natural infiltration, hence the expansion for natural recharge of the groundwater is declining. In contrast to natural recharge; the artificial recharge innovative technique is the use of water to replenish the aquifer artificially (Bhattacharya, 2010; Mukherjee et al., 2017; Senthilkumar et al., 2019). The process by which groundwater is growing at a faster rate than it would under natural replenishment conditions is known as artificial recharge. Therefore, any artificial facility that recharges an aquifer with water may be deemed an artificial facility (CGWB, 2014). The rate of recharge is one of the variables used to evaluate groundwater resources that is the most challenging to accurately determine (Mukherjee et al., 2011). Moreover, the sustainability of groundwater, which is quickly depleting, is coming under pressure from population growth and overexploitation. In addition to focusing on groundwater potential zones, it is now essential to support sustainable groundwater conservation during the severe water shortage (Mallick et al., 2019; Preeja et al., 2011; Raviraj et al., 2017). *Water efficiency* is a tool of water conservation. That results in more efficient water use and thus reduces water demand (Kumari and Singh, 2016). However locating the potential sites for artificial recharge is very difficult and depends on many factors including rainfall, drainage density, lineament density, slope, soil, landuse/land cover, geology, and geomorphology (Andualem and Demeke, 2019; Gnanachandrasamy et al., 2018; Magesh et al., 2012).

The objective of water conservation efforts is to ensure availability for future generations, the natural replacement should be greater than the rate of withdrawal of fresh water from an ecosystem (Kumari and Singh, 2016). One of the most useful steps in the growth and management of water resources is the construction of water conservation structures and sites. Ponds, canals, bavadis, and reservoirs have been used to store and furnish surface water worldwide since the dawn of civilization, but this practice is neither entirely scientific nor demand-driven

DOI: 10.1201/9781003303237-7

(Hashemi et al., 2012; Rajasekhar et al., 2018). Geospatial Technologies like Remote Sensing (RS) and GIS have been useful in the estimate and management of essential water resources in the upcoming Municipal Corporation of India (Chenini and Ben Mammou, 2010; Chenini et al., 2010; Singh et al., 2013). In recent times, the geospatial technologies and MIF methodology have been helpful in water conservation and identifying artificial recharge structures with relatively accurate results (Rajasekhar et al., 2018; Raviraj et al., 2017). Geospatial tool is also competent to demarcate the high-potential sites for artificial recharge in unapproachable areas by traditional methods (Solomon and Quiel, 2006). The selection of potential locations for artificial recharge is influenced by a number of variables, including Lithology, Lineament, Geomorphology, Drainage, Rainfall, Land use/Land cover, Slope, Soil, and TWI were integrated into the spatial domain of GIS (Raviraj et al., 2017).

Birgaon Municipal Corporation Region is one among the emerging population and is surrounded by Raipur and Durg urban areas of Chhattisgarh, India and comes under a critical category in terms of groundwater (CGWB, 2013; Sinha et al., 2021; Vaibhav Prakashrao Deshpande, 2021). The Birgaon Municipal Corporation has a total of 40 wards, with a total population of 1.25 Lakh (nagarnigambirgaon.com). Birgaon Municipality has also maintained the water supply and sewerage system of 21,284 houses. 14 MLD water capacity plant has been set up in the municipal area for supplies of water to 25 while in the remaining 15 wards, water has been supplied through tankers throughout the year. Due to the unavailability of drinking water sources in the surrounding area of Birgaon, water supply in all wards has been a major issue for the Birgaon Municipal Corporation (Sinha et al., 2019b).

The southwest monsoon contributes to the water flow in the stream, and only during the monsoon do these streams retain water. However, due to inadequate recharge structures across the stream, water is lost as runoff, so the people who live in this area are completely dependent on an alternative source, namely groundwater, to meet all of their domestic needs. As a result, the municipal area's current groundwater development condition is classified as a highly exploited area, and it is essential to support the groundwater recharge structure in order to improve the groundwater scenario of the area.

7.2 STUDY AREA

7.2.1 Location

Birgaon Municipal Corporation Region lies approximately 21°20′65″–21°15′33″ north latitude and 81°36′25″ –81°38′45″ east longitude, as the study area's location map is shown in Figure 7.1. Birgaon Municipal Corporation Region is at an altitude of 298.15 m and covers an area of approximately 35 km^2 with a total Population census of 2011–1.25 Lakh. The Municipal area is bounded by geographical features; the Chhokra Nala in the east and north, the Kharun in the west and the Mumbai-Kolkata Railway line in the south.

7.2.2 Climate

The climate of the study area is tropical wet and dry. Temperatures remain moderate throughout the year, except from March to June, which can be extremely hot. Temperatures in Birgaon can hit 48°C from April to June during the summer. A pleasant break from the heat is provided by the monsoon season, which lasts from late June to October. Rainfall in Birgaon is typically 1,292 millimetres (50.9 in). When it comes to low temps and low humidity, winter lasts from November to January. In the summer, it is between 44°C and 47°C, and in the winter, it is between 10°C and 25°C. However, extremes in temperature can be observed with scales falling to less than 10°C–48°C (Sinha et al., 2022).

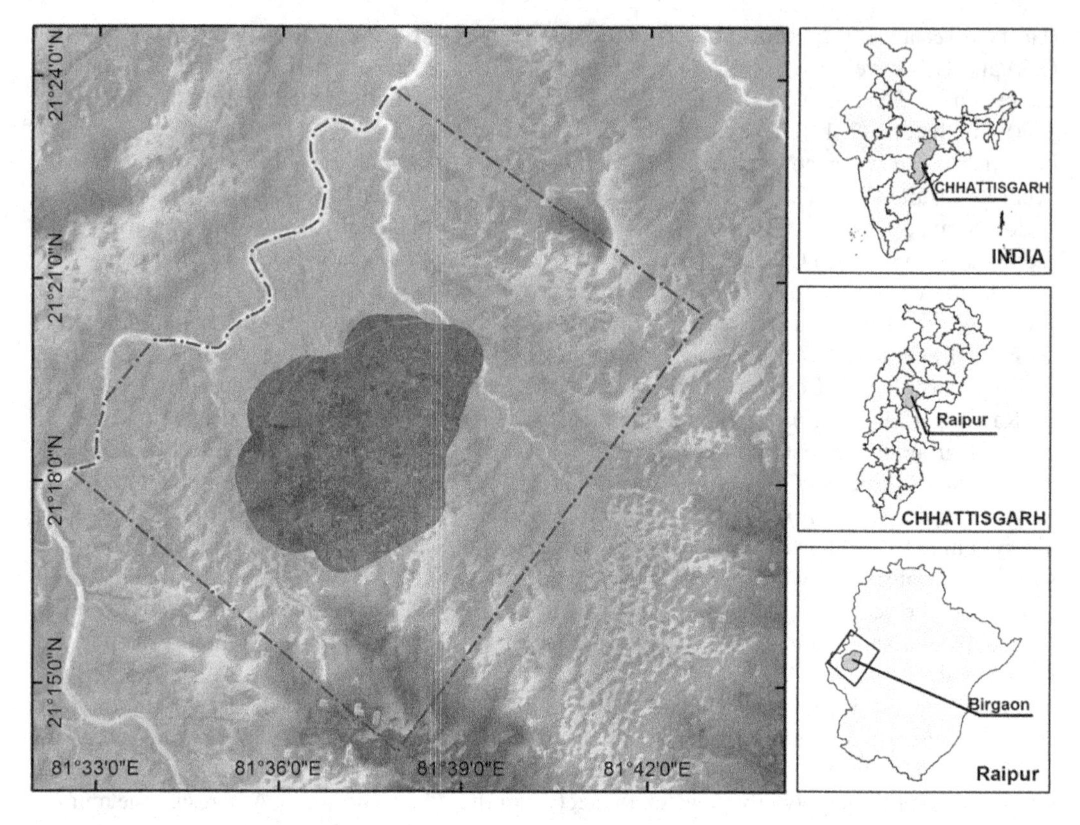

FIGURE 7.1 Location map of the study area.

7.2.3 Geology

Ferruginous glauconitic arenite and Limestone and dolomite are the major rock-types in the area after Alluvium-sand/silt dominant. The study area may be divided into two geological formations such as major rocks of Chandi Formation, Raipur Group, and alluvial deposits of Quaternary found on the buried plain area (CGWB, 2012b).

7.2.4 Lineament

The lineament map quantifies the length of linear features in the earth. In the study region, the South-East and North-West are primarily the directions of lineament trends. Near lineament intersection zones, there is a significant potential for groundwater (CGWB, 2012c; CGWB, 2014).

7.2.5 Geomorphology

Geomorphologically the area has matured types of land forms and can be broadly divided into three prominent geomorphic units. These are: (1) Dissected pediplain and (2) Pediment made by Proterozoic shale-limestone dolomite area; (3) Alluvial Plain formed by Kharun-Seonath Alluvium (CGWB, 2011; CGWB, 2012c).

7.2.6 Drainage

The primary features of the surface and subsurface formation are reflected in the drainage pattern. More drainage density would result in a greater runoff, which would affect infiltration (Das et al., 2017; Sinha et al., 2019a; Sinha et al., 2017). In general, the study area drainage pattern is dendritic.

7.2.7 Land Use/Land Cover

Different land use patterns affect the infiltration process and also control permeability. The study area mostly consists of Built-up Industry and urban, Agricultural Land, Wastelands followed by water bodies, and Plantation.

7.2.8 Soils

Soil texture influences the infiltration process. So it is the most important factor to control groundwater potential. In this region, the most common types of soil are clay loamy, clay, sandy clay loam, and sandy loam (Tamgadge and NBSSLUP, 2002).

7.2.9 Slope

The slope of the Municipal Corporation region is gentle. The slope of an area is one of the parameters which determines the groundwater recharge capability. Most part of the study area is under a low degree of slope.

7.3 MATERIALS AND METHODS

The methodology for artificial recharge structure in the present study is classified into four steps: (1) Parameter identification affecting the possibility for groundwater; (2) Rank and weighting are assigned after parameter processing to guarantee uniformity; (3) Reclassification of all the thematic layers with computed score of sub-classes; (4) Combining all influencing variables, classifying the output layer into five categories viz. – very good, good, moderate, low, and poor. (5) Suggestion of different recharge structures for suitable areas. The process used in this research to identify appropriate locations for artificial recharging has been demonstrated in Figure 7.2.

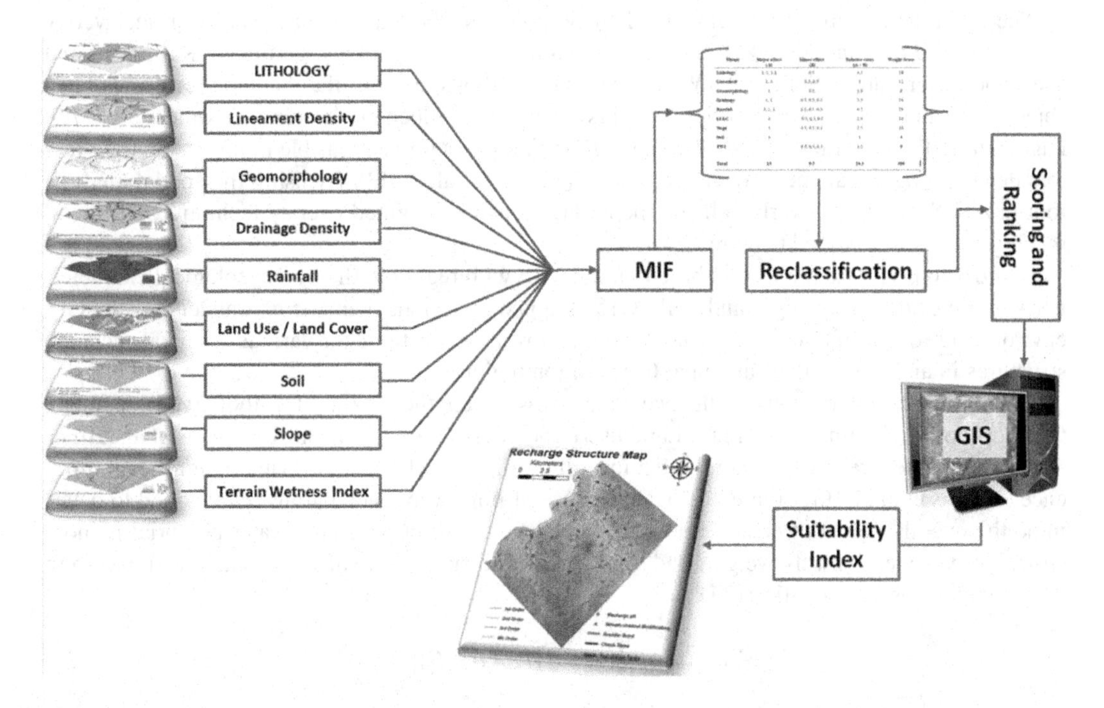

FIGURE 7.2 Workflow for artificial recharge structures.

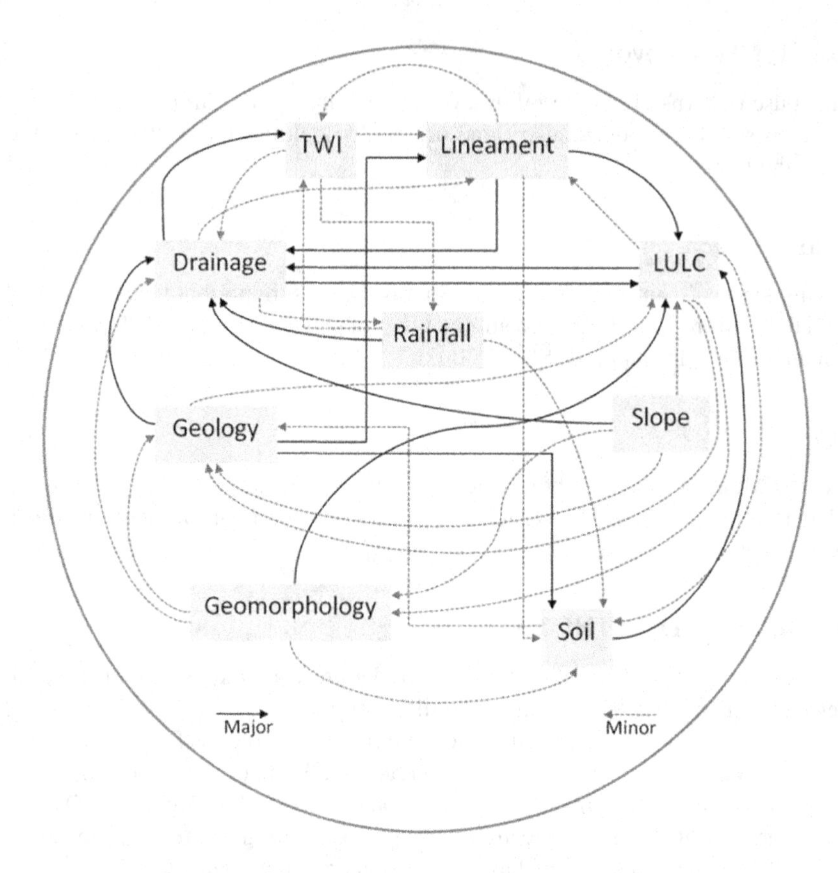

FIGURE 7.3 Interactive influence of factors concerning the artificial recharge structures.

The recharge structure suitability method proposed in the study area is based on an overlay analysis of the nine factors viz. Lithology, Lineament, Geomorphology, Drainage, Rainfall, Land use/Land cover, Slope, Soil, and TWI. Using MIF methods, each of the thematic levels has been categorized and given the appropriate weights for various influencing factors (Etikala et al., 2019; Fagbohun, 2018; Raviraj et al., 2017). The MIF technique is proved feasible to understand the factors determining the artificial recharge zoning (Biswas et al., 2013). By using the spatial analysis tool in a GIS setting to overlap all the theme layers using weighted overlay techniques, artificial recharge zones were found (Figure 7.3).

In order to prioritize the factors based on the MIF with regard to the hydrogeological characteristics of the study area, weight and rank were assigned to various thematic parameters for the GIS environment to identify suitable recharge zones as well as locations for various artificial recharge structures in and around the Municipal Corporation region.

Based on the interactions, influence value was given the credit of Lithology, Lineament, Geomorphology, Drainage, Rainfall, Land use/Land cover, Slope, Soil, and TWI (Das et al., 2017). All the parameters received a score of 0.5 for minor impact and a score of one for significant influence (Biswas et al., 2013) (Table 7.1). The measure of impact for each parameter is added to determine the overall weight of each element. A parameter's impact on groundwater potential is more strongly correlated with its weight, and vice versa. The suggested values for each attribute were determined using the Equation (7.1):

$$\text{MIF Score} = \left[(A+B)/\Sigma(A+B)\right]*100 \tag{7.1}$$

Where the major effect is represented by A and the minor effect is represented by B.

According to the efficacy of each connection, the weights of the different influence factors are assigned. To calculate the artificial recharge potential zones, the impact of each influencing element may be involved. They are also reliant on one another. The rank of each factor was assigned according to Saaty's scale (1–9) of relative importance assessment (Saaty, 2008; Zhang et al., 2009). Further, the ranks were allocated for subclasses with consideration of the review of past researches and field experience during weighted overlay analysis by weight overlay method in a GIS environment.

The topographic, soil, land-use, and hydrogeologic conditions are important factors controlling the suitability of an area for artificial recharge (CGWB, 2012c). The prevalent soil, land use, and slope conditions determine the extent of infiltration. Based on the site selection factors listed in Table 7.2, the appropriate recharge structure has identified locations that are suggested for site-specific recharge structures.

TABLE 7.1
Multi-Influencing Factor (MIF) Matrix Table of Different Thematic Layers

Theme	Major Effect (*A*)	Minor Effect (*B*)	Relative Rates (*A*+*B*)	Weight Score
Lithology	1, 1, 1	0.5	3.5	16
Lineament	1, 1	0.5, 0.5	3	13
Geomorphology	1	0.5, 0.5, 0.5	2.5	11
Drainage	1, 1	0.5, 0.5	3	13
Rainfall	1	0.5, 0.5	2	9
LULC	1	0.5, 0.5, 0.5, 0.5	3	13
Slope	1	0.5, 0.5, 0.5	2.5	11
Soil	1	0.5	1.5	7
TWI		0.5, 0.5, 0.5	1.5	7
Total	12	10.5	22.5	100

TABLE 7.2
Site Selection Criteria for Targeting the Site-Specific Artificial Recharge Structures, Used After (CGWB, 2008; CGWB, 2009; CGWB, 2012a; CGWB, 2012c; Karthick and Chokkalingam, 2018; Raviraj et al., 2017; Senthilkumar et al., 2019)

Type of Recharge Structure	Nature of Topography	Slope Range (in degree)	Site-Specific Landuse	Soil Type	Stream Order	Benefitted Area Cover
Percolation Tank	Plain surface to gentle slope	<3	1. Uncultivable land, 2. Barren land, 3. Long-term fallow land, 4. Wasteland surrounded by cultivable land, 5. Shrub vegetative cover	Coarse alluvium, sandy clay loam, unconsolidated soil with gravels and boulders	2nd and 3rd	1. Cultivable land with sufficient number of wells, 2. Dry land agriculture, 3. Area surrounding the rural settlements
Recharge Pit	Plain surface to gentle slope	<3	1. Uncultivable land, 2. Barren land, 3. Long-term fallow land	Coarse alluvium, sandy clay loam, unconsolidated soil with gravels and boulders	2nd and 3rd	1. Cultivable land with sufficient number of wells, 2. Single crop areas for providing irrigation to a limited area

(*Continued*)

TABLE 7.2 (*Continued*)
Site Selection Criteria for Targeting the Site-Specific Artificial Recharge Structures, Used After (CGWB, 2008; CGWB, 2009; CGWB, 2012a; CGWB, 2012c; Karthick and Chokkalingam, 2018; Raviraj et al., 2017; Senthilkumar et al., 2019)

Type of Recharge Structure	Nature of Topography	Slope Range (in degree)	Site-Specific Landuse	Soil Type	Stream Order	Benefitted Area Cover
Check Dam	Gentle to moderate slope surface	3–5	1. Runoff area surrounded by cultivable land 2. Gentle slope of hilly terrain with elevation of less than 2 m comprises a small order stream flow across	Coarse alluvium, sandy soil mixture of gravels and boulders, sandy clay loam	2nd and 3rd	1. Irrigable land under well irrigation, 2. Upland area of contour cultivable land or graded bunding
Stream channel modification	Gentle to moderate slope surface	3–5	1. Wasteland surrounded by cultivable land 2. Barren land, 3. Long-term fallow land	Coarse alluvium, sandy clay loam,	2nd and 3rd	1. Irrigable land under well Irrigation 2. Surface storage dams exist upstream of the recharge sites
Boulder bund	Moderate to steep slope surface	>5	1. Shrub vegetative cover 2. Uncultivable land	Sandy soil, sandy clay loam	1st and 3rd	1. Upland area of contour cultivable land or graded bunding

The following aspects were considered while fixing the locations of the recharge structures:

- Habitation location
- Recharge water adequacy
- Aquifer hydrogeological properties
- Condition of slope/terrain

7.4 RESULTS AND DISCUSSION

Birgaon is one of the important Municipal Corporations in Raipur district Chhattisgarh of central India. This is a new municipal corporation of Chhattisgarh's capital and the fast-developing industrial activities are taking place in the region. This region has witnessed a high growth rate in population, along with a rapid increase in industrial infrastructure and residential colonies. The demand for fresh water in the newly formed Municipal Corporation is rapidly increasing with the huge amount of constant development projects in the region. However, the region experiences periodic water stress conditions due to seasonal precipitation patterns and scarcity of surface water resources. Therefore, management of available water resources is critical, to fulfill water requirements in the area. However, surplus use of groundwater should be carried out mutually with artificial recharging in regulation to uphold the long-term sustainability of water resources. Artificial recharge Replenishment of groundwater by aquifers in the hot and dry regions of the area is essential, as the intensity of normal rainfall is grossly inadequate to produce any moisture surplus under normal infiltration conditions. This Study signifies an important step in addressing a critical gap in the identification of artificial recharge zones and to enhance the competence of recharge structures. In this study, a geospatial approach was used to delineate potential artificial recharge sites in Birgaon Municipal Region due to their ability to deal with large volumes of spatial information. Identification of suitable sites for artificial recharge was conducted through MIF for various factors in the GIS environment such as Lithology, Lineament density, Geomorphology, Drainage density, Rainfall, Land use/Land cover, Slope, Soil, and TWI were reclassified and assigned ranks and weights to integrate them to obtain the map of artificial recharge zones. All the factors were analyzed against each other in an Interactive influence comparison matrix. The rank provided from 1 to 9 to sub-classes of thematic layer according to their infiltration and percolation behavior. One is assigned to poor infiltration and percolation behavior of the layers and nine is for excellent respectively (Table 7.2). The output of the adopted procedure and the final thematic map are discussed here in this section.

7.5 LITHOLOGY

It is well-established fact that geological set-up of an area plays a vital role in the distribution and occurrence of groundwater. The lithology of the area is prepared by using the District Resource Map. The study area is predominantly with Limestone and Dolomite followed by Ferruginous glauconitic arenite. Alluvium sand and silt occupy mostly in a small area near to the Kharun River. All three rock types were assigned the rank based on their groundwater recharge potential (Table 7.2). The area distribution of the rock types in the study area (Figure 7.4a) shows that about 58% of the study area is covered by Limestone and Dolomite followed by 41% Ferruginous glauconitic arenite, very little area (01%) is covered by Alluvium sand and silt.

7.6 LINEAMENT DENSITY

Lineaments are linear or curvilinear structures on the earth's surface, which describe the weaker zone of bed rocks and the area is considered a secondary aquifer in hard rock regions. It increased secondary porosity and permeability and it is also good indicators of the groundwater recharge zone and helps to suggest the structures. Lineaments are digitized from CartoSat-1 satellite imagery. The lineament trends in the study area are predominantly along South-East and North-West. Groundwater potential is high near lineament intersection zones. The lineament layer was used for computing the lineament density map in terms of the length of feature per unit area (km/km^2) from Equation (7.2).

$$L_d = \Sigma_{i=1}{}^{n}(L_i)/A \tag{7.2}$$

L_d is the lineament density, L_i -the sum of the length of all lineaments; A is the area in km^2.

Lineament density is divided into five classes based on density value for recharge potential; maximum value assigned for higher density and least value for lower density (Table 7.2). The areal distribution of lineaments and their density classes are shown in Figure 7.4b.

7.7 GEOMORPHOLOGY

The geomorphology of an area is one of the most important features in evaluating the suitable recharge area and structures. It is delineated from CartoSat-1 satellite imagery. The geomorphology of the study area is dominated by Pediment (59%) followed by Pediplains (34%) and alluvial plains (3%). Total Surface water resources are spread over 4% area. The groundwater recharge potential is represented by distributions of the geomorphic classes shown in Figure 7.4c. Pediments and Pediplains exhibit flat surfaces with shallow weathered material covered and the assigned rank for suitable recharge is poor to moderate in this region but we suggest and assigned moderate rank near to fracture lineaments. Alluvium plain was assigned a high rank because it has a predominant role in groundwater movement and storage (Table 7.2).

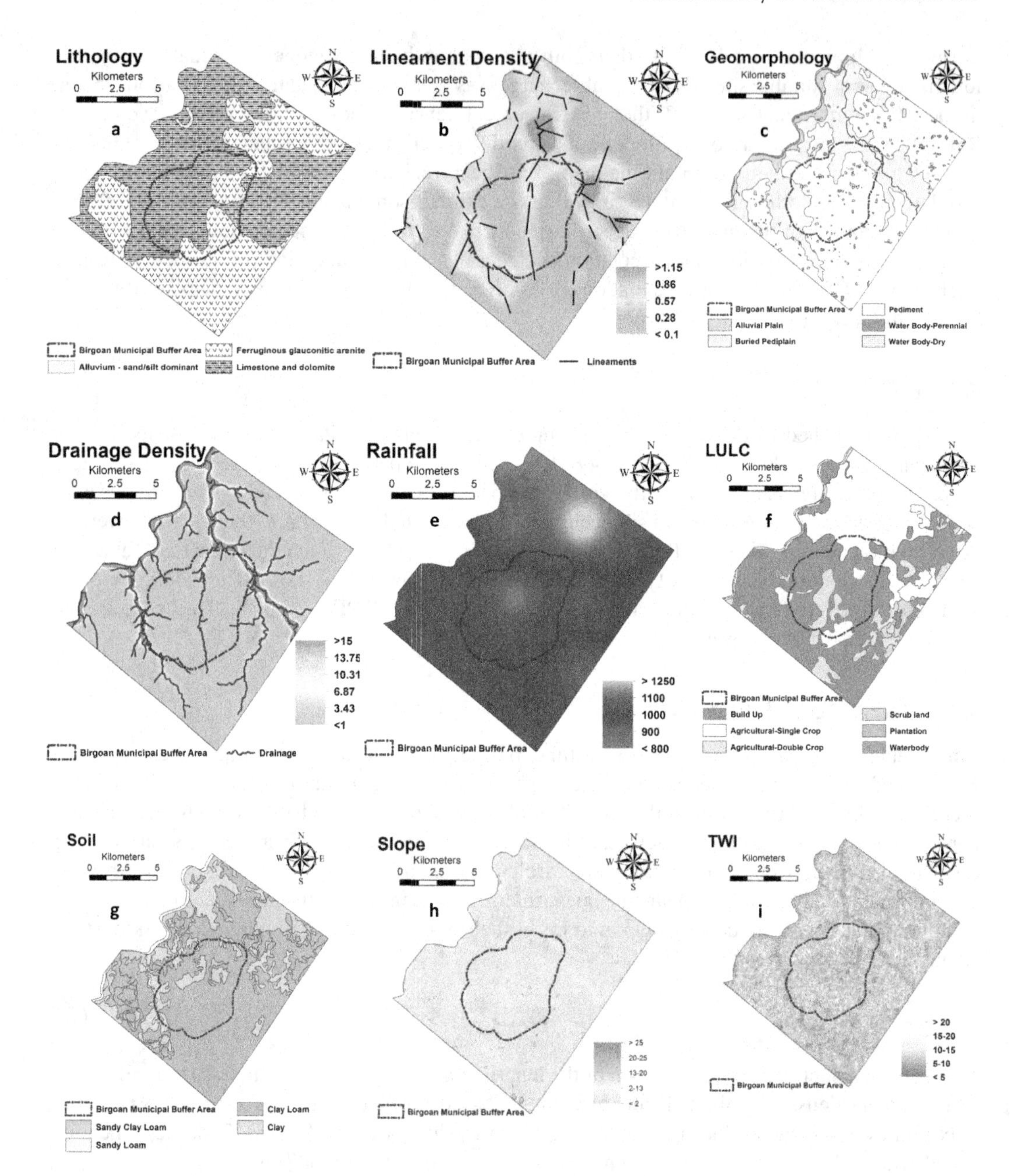

FIGURE 7.4 (a–i) Aquifer system in and around Birgaon Municipal Region – (a) Lithology; (b) Lineament density; (c) Geomorphology; (d) Drainage density; (e) Rainfall; (f) LULC; (g) Soil; (h) Slope; (i) TWI.

7.8 DRAINAGE DENSITY

Drainage density plays a vital role to the identification of hydrogeological features because drainage network and density are controlled in a basic way by the underlying lithology, vegetation type, infiltration rate, slope angle, and the capacity of soils to absorb rainfall. For the drainage density layer, the drainage network for the study area was delineated from Survey of India Toposheets and updated by screen digitization of CartoSat-1 satellite imagery. The drainage layer was used for computing the drainage density map in terms of the length of channels per unit area (km/km^2) by Equation (7.3).

$$D_d = \Sigma_{i=1}^{n}(D_i)/A \quad (7.3)$$

D_d is the drainage density, D_i -the sum of the length of all streams; A is the area in km^2.

The major characteristics of the surface as well as subsurface formation reflected by drainage pattern. The more the drainage density, the higher would be the runoff. The suitability of artificial recharge zones is inversely related to drainage density because of its relation with surface runoff and permeability. The higher rank is assigned to the suitability of the recharge structure area. Also, the lower rank is assigned to no drainage area (Table 7.2). Figure 7.4d illustrates the drainage condition of the study area.

7.9 RAINFALL

The annual rainfall varies from 800 to 1,250 mm (Figure 7.4e). The major part of the study area is receiving good rainfall by the south-east monsoon period (July–September). The normal onset and withdrawal of the monsoon in the study area is in the first week of July and October, respectively. High infiltration occurred in high rainfall areas and low infiltration in low rainfall areas. The higher rank is assigned to high rainfall and vice versa.

7.10 LAND USE/LAND COVER (LULC)

Land use/land cover is an important indicator of the extent of groundwater demand and supply as well as in the selection of artificial recharge sites for groundwater development. The major land use/land cover classes in the study area are built-up (55%), agriculture-single and double crop (40%), scrubland (2%), plantations (1%), and water bodies (2%) (Figure 7.4f). Built-up is assigned a low rank, Agriculture and scrub areas are assigned a moderate rank and plantation is assigned a high rank for recharge zonation study.

7.11 SLOPE

In the study area, the slope varies from 0% to 25% with an almost flat plain. Flat plains make more favorable conditions to hold more rainfall and facilitate recharge as compared to elevated areas with higher slopes having high runoff and low infiltration rates. On the basis of slope, the study area was classified into moderate to high rank. A major area of study comes under high rank (Figure 7.4g).

7.12 SOIL

The type of Soil and its texture is the preliminary role of infiltration and transmission of surface water into an aquifer system. Surface runoff and recharge condition of an area depends on soil condition. Soil data prepared by Soil Map published by the Indian Council of Agricultural Research, National Bureau of Soil Survey (NBSS) and Land Use Planning (LUP) is exploited in this study. On the basis of soil texture, the study area is classified into low to moderate rank. Most of the soil types of the study ranked as moderate class.

7.13 TERRAIN WETNESS INDEX (TWI)

The goal of the model is to predict the distribution of groundwater recharge and discharge areas in the landscape, the most important factor is topography. The topographic wetness index combines local upslope contributing area and slope, and is commonly used to quantify topographic control on hydrological processes. In this methods for computing relative index differ primarily in the way the upslope contributing area is calculated. TWI maps were generated from CartoDEM. The TWI as a secondary topographic index has been widely adopted to illustrate the impact of morphology conditions on the location and size of saturated zones of surface runoff generation. TWI is calculated from Equation (7.4):

$$TWI = \text{In}(A_s / \tan\beta) \tag{7.4}$$

Where A_s is the cumulative upslope area draining through a point (per unit contour length) and $\tan\beta$ is the slope angle at the point.

After integration of all the thematic maps with their weightage the raster output indicates the artificial recharge zones. Earlier, ranks from 1 to 9 were assigned for individual classes of geology, geomorphology, slope, soil, land use/land cover, drainage density, lineament density, and TWI layers based on the influence on water recharge hence the final output map was obtained with 5 classes viz. – Poor, Low, Moderate, Good and Very good and illustrated in Figure 7.5.

TABLE 7.3
Rank Assigned to Classes of each Theme based on Their Influence to Artificial Recharge Structures

Theme	Weight Score	Class	Rank
Lithology	16	Alluvium-sand/silt dominant	8
		Ferruginous glauconitic arenite	6
		Limestone and dolomite	4
Lineament	13	>1.15	8
		0.86	7
		0.57	6
		0.28	4
		<0.10	2
Geomorphology	11	Alluvial plain	8
		Buried pediplain	6
		Pediment	5
		Water body-dry	8
		Water body-perennial	9
Drainage	13	>17.50	2
		12.00	4
		8.50	6
		4.20	7
		<1.00	1
Rainfall	9	>1,250	9
		1,100	8
		1,000	7
		900	6
		<800	5
LULC	13	Built up	2
		Agriculture-single crop	6
		Agriculture-double crop	7
		Scrub land	7
		Plantation	8
		Waterbody	9

(Continued)

TABLE 7.3 (*Continued*)
Rank Assigned to Classes of each Theme based on Their Influence to Artificial Recharge Structures

Theme	Weight Score	Class	Rank
Slope	11	>25	3
		20–25	6
		13–20	7
		2–13	8
		<2	9
Soil	7	Sandy clay loam	6
		Sandy loam	7
		Clay loam	5
		Clay	4
TWI	7	>20	8
		15–20	7
		10–15	6
		5–10	4
		<5	2

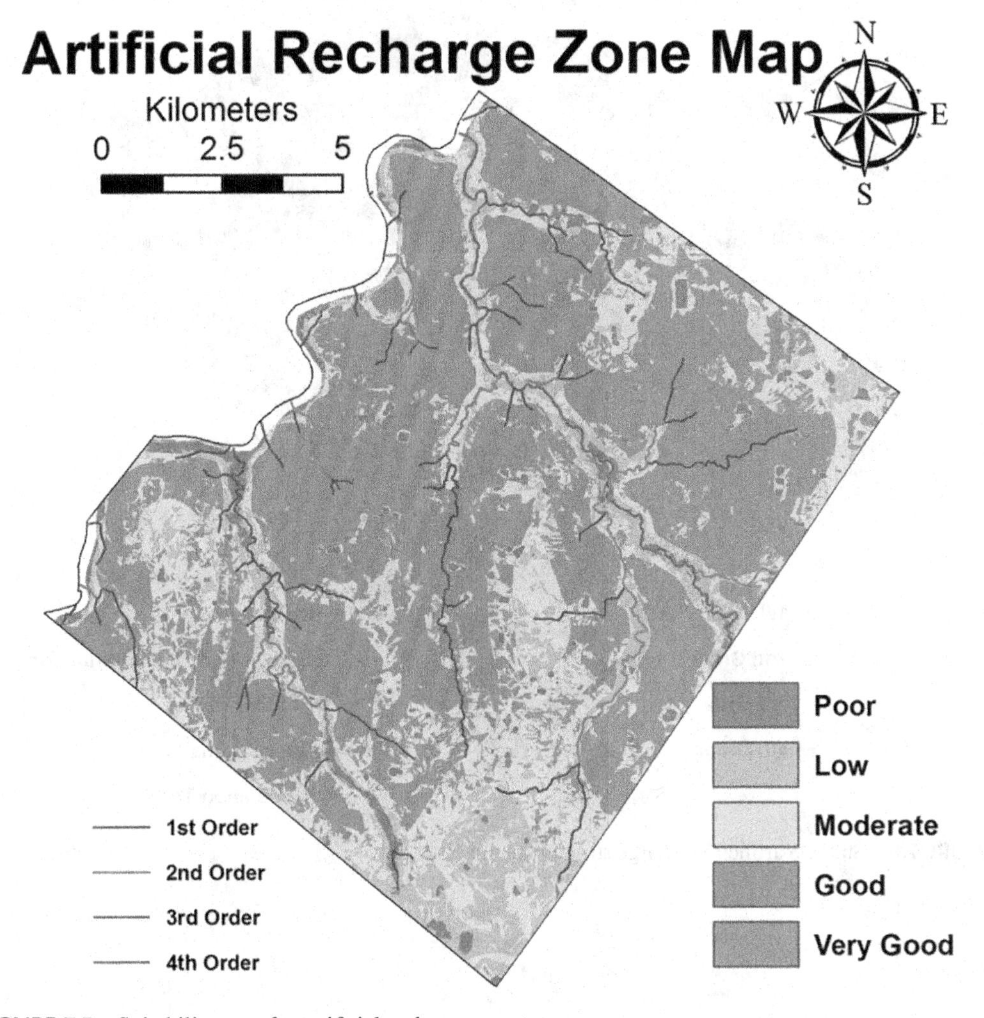

FIGURE 7.5 Suitability map for artificial recharge structures.

7.14 VALIDATION

The potential groundwater recharge zone for the study area is shown in Figure 7.5. It is clearly observed that high recharge potential zones are located in the north to northeast part of the study area. A cross-validation study has been carried out in this part to ensure the recharge potential of groundwater as per field data published by CGWB (CGWB, 2014; Mukherjee et al., 2011) and other geohydrological research done in this zone (Indhulekha et al., 2019; Mondal et al., 2021). The previous report suggested that the well yield capacity in alluvial and pediplains aquifers is about 150 LPM and 100 LPM respectively. The identified recharge zone is also lying in this category in the form of alluvial and buried pediplain (Figure 7.4c). The groundwater fluctuation level in this site ranges from 3 to 10 mbgl (Figure 7.6).

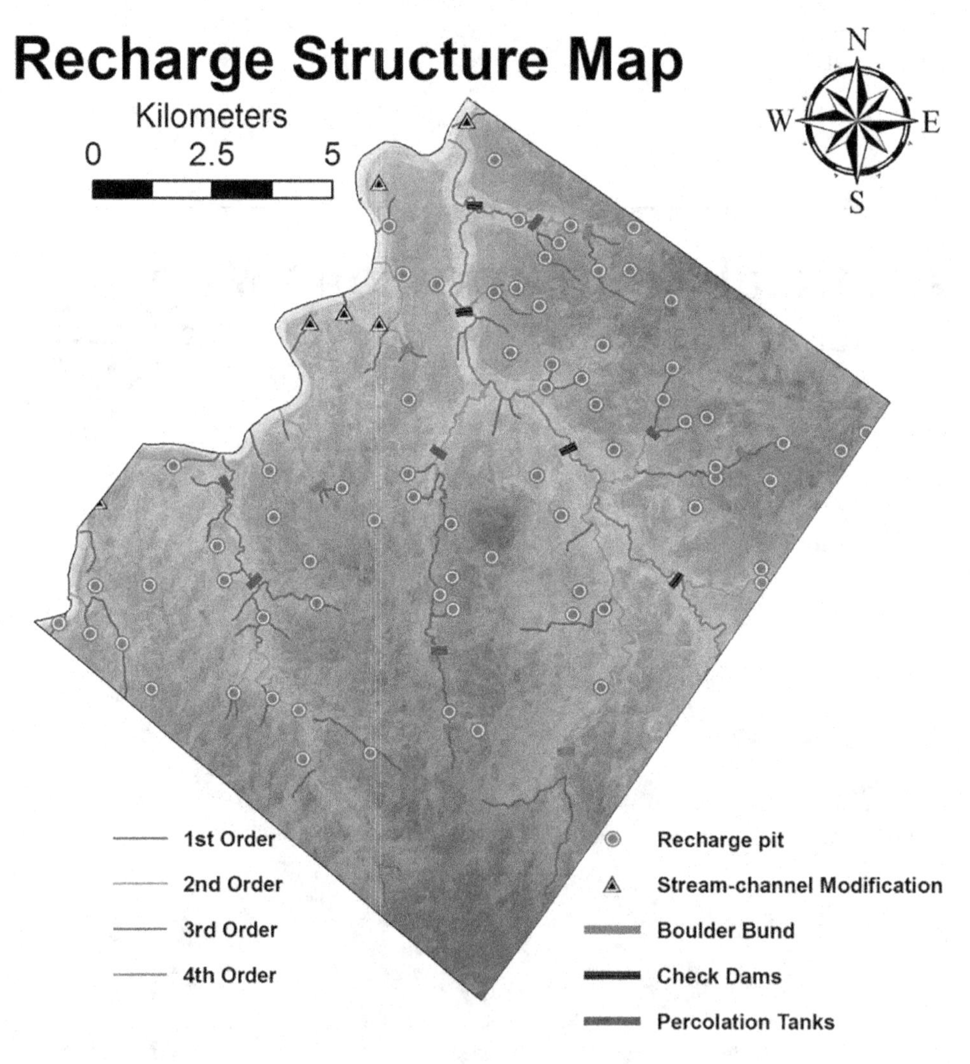

FIGURE 7.6 Suitable artificial recharge structures map.

7.15 CONCLUSION

For planning and implementation of any successful artificial recharge structure with proper scientific investigations for the suitability of the area for recharge in terms of climate, topography, soil, and land use characteristics and hydrogeologic set-up are important for evolving a realistic plan for artificial recharge. Thus, the artificial groundwater recharge zones delineation in the present study using integrated geospatial techniques and the MIF method are reliable and can serve as guidelines for planning future artificial recharge projects for sustainable groundwater utilization. Moreover, this method can be widely used as an important tool for solving groundwater problems of different topography with inherent boundaries of data scarcity and multi-criteria analysis. Birgaon Municipal region was divided into five zones, viz. – "poor", "low", "moderate", "good", and "very good" based on the analysis that influence the groundwater recharge. About 04% and 13% of the study area falls under "poor" and "low" groundwater recharge zones, 24% area under "moderate" zone, 38% under "good" zone, and 21% of the total area falls under the "very good" zone. Results indicate that the north to north-eastern portion of the study area is suitable for an artificial water recharge zone due to its good infiltration capacity by secondary structure and fall under very good and good zones. A relatively very good amount of groundwater depth in the region suggests greater availability of dynamic underground storage during the non-monsoon (dry) period and hence is favorable for artificial groundwater recharge. In addition, the ""moderately suitable" zones have average groundwater depths during the non-monsoon period, advising the availability of underground storage for recharge. The poor availability of groundwater in the region gives an indication of further deteriorating conditions in future. The present situation suggests that further installation of individual bore wells which consume time, energy, and cost should be prevented and appreciate group water supply scheme in this region.

Geospatial technology gives a synoptic view for the planner to identify the problem and accordingly mitigate the local area issues. The study suggested that GIS is a powerful tool for a detail understanding of problems. It's also providing a simple way to integrate different ground information in a single platform. This will also reduce time, money and manpower. The role of GIS and remote sensing for groundwater-potential zone identification, management, and conservation has been fully realized only since the last decade.

7.15.1 Proposed Artificial Recharge Structure

The site suitability analysis assists in the execution of suitable sites for the recharge structures. Based on the artificial recharge zone map, Drainage density and stream order, a map signifying the sites for different recharge structures has been prepared (Figure 7.6). The high gradient area with elevated surface acts as a high runoff zone. The runoff washes away precious topsoil from the upland. The primary objective of this study is to preserve or enhance groundwater resources in various parts of Municipal which include the conservation or disposal of excess rainwater. In general, the sites for recharge structures are identified upstream of the problematic area. They are located mainly on 1st to 3rd-order streams and at the most up to the initial stages of 4th order stream. The suggested recharge structures are given below.

7.15.1.1 Recharge Pits

Recharge pits are structures that conquer the complexity of artificial recharge of phreatic aquifers from surface-water sources. They are made either by constructing an embankment across a water course or by excavating a pit or a combination of both. These structures are suggested in single-crop areas for providing irrigation to low rainfall areas during critical periods.

7.15.1.2 Stream-Channel Modification

Stream channel modification methods are used in alluvial areas, but can also be gainfully used in hard rock areas where thin river alluvium overlies good phreatic aquifers or the rocks are extensively weathered or fractured in and around the stream channel. Artificial recharge through stream channel modifications is more effective if surface storage dams exist upstream of the recharge sites for controlled release of waters.

7.15.1.3 Bolder Bunds

These bunds are low-cost small bunds across 1st to 3rd lower order streams flowing through the plains and valleys where acquisition of land for inundation of large areas is not possible. They are suitable at the hill slopes where catchments are small and stream courses have been intensified by erosion.

7.15.1.4 Check Dams

Check dams are storage structures executed across 2nd to 3rd-order streams and constructed for checking the stream runoff during rainy sessions to store water for specific use beyond the monsoon period. They are also helpful to recharge the groundwater reservoir located in the near vicinity.

7.15.1.5 Percolation Tanks

Percolation tanks are artificially created surface water body which helps to increase groundwater recharge and prevent excessive runoff. These are generally constructed across streams and bigger gullies for submerging a highly permeable land area so that the surface runoff is made to percolate and recharge the groundwater storage. Usually, a percolation tank should not retain water beyond February in the Indian perspective. Suitable conditions for percolation tanks which improve their utilities are presented in lower-order stream channels (2nd, 3rd and 4th) having sufficient weathered zone/loose material/fractures.

There are a number of problems associated with the use of artificial recharge methods. These include disadvantages related to aspects such as effectiveness (e.g., not all of the added water may be recoverable), contamination risks due to injection of poor quality water, siltation, and a lack of awareness about the implications of the recharge method. Therefore, scientific consideration for the selection of an appropriate site for artificial recharge in a specific area is required.

ACKNOWLEDGEMENTS

The authors are thankful to Mr. Mudit Kumar Singh, Director General, CCoST for encouraging and providing support to carry out this Study. The authors would also like to thank the staff of CGSAC group who was involved in this Study for their time and guidance, National Bureau of Soil Survey and Land Use Planning (NBSS–LUP), Nagpur, Geological Survey of India (GSI), Raipur and Central Ground Water Board (CGWB) for their useful data related to the study.

REFERENCES

Andualem, T.G., Demeke, G.G. 2019. Groundwater potential assessment using GIS and remote sensing: a case study of Guna tana landscape, upper blue Nile Basin, Ethiopia. *Journal of Hydrology: Regional Studies,* 24, 1–24.

Bhattacharya, A. 2010. Artificial ground water recharge with a special reference to India. *International Journal of Research and Reviews in Applied Sciences*, 4, 1–4.

Biswas, A., Jana, A. and Mandal, A. 2013. Application of remote sensing, gis and mif technique for elucidation of groundwater potential zones from a part of Orissa coastal tract. Eastern India. *Research Journal of Recent Sciences*, 2, 42–49.

CGWB. 2008. Ground water brochure of Raipur district, Chattisgarh. In: Verma JR, editors, Central Ground Water Board, Government of India. Central Ground Water Board NCCR, Raipur, Chhattisgarh, India, pp. 1–17.

CGWB. 2009. Dynamic ground water resources of Chhattisgarh. In: Aswal NK, editors, Central Ground Water Board, Government of India. Central Ground Water Board NCCR, Raipur, Chhattisgarh, India, pp. 1–267.

CGWB. 2011. Ground Water Scenario in Major Cities of India. In: Mukherjee R, Sahoo M, Naik KC, editors, Central Ground water Board, Ministry of Jal Shakti, Govt. of India.

CGWB. 2012a. Aquifer systems of Chhattisgarh. In: Dhiman SC, Verma JR, Sahoo BK, Naik KC, Kumar S, Sinha SK, et al., editors. Central Ground Water Board, Government of India. Central Ground Water Board NCCR, Raipur, Chhattisgarh, India, pp. 1–73.

CGWB. 2012b. Ground water brochure of Raipur district, Chattisgarh. In: Sahoo M, editor. Central Ground Water Board, Government of India. Central Ground Water Board NCCR, Raipur, Chhattisgarh, India

CGWB. 2012c. Ground water exploration in Chhattisgarh State. In: Verma JR, Naik KC, editors. Central Ground Water Board, Government of India. Central Ground Water Board NCCR, Raipur, Chhattisgarh, India, pp. 1–238.

CGWB. 2013. Pollution – raipur urban agglomerate. In: Naik KC, editor. Central Ground Water Board, Government of India. Central Ground Water Board NCCR, Raipur, Chhattisgarh, India, pp. 1–39.

CGWB. 2014. State hydrogeological report Chhattisgarh. In: Naik KC, Verma JR, Sonkusare MM, editors. Central Ground Water Board, Government of India. Central Ground Water Board NCCR, Raipur, Chhattisgarh, India, pp. 1–263.

Chenini, I., Ben Mammou, A. 2010. Groundwater recharge study in arid region: An approach using GIS techniques and numerical modeling. *Computers & Geosciences,* 36, 801–817.

Chenini, I., Mammou, A.B., El May, M. 2010. Groundwater recharge zone mapping using GIS-Based multi-criteria analysis: a case study in central Tunisia Maknassy Basin. *Water Resources Management,* 24, 921–939.

Etikala, B., Golla, V., Li, P., Renati, S. 2019. Deciphering groundwater potential zones using MIF technique and GIS: a study from Tirupati area, Chittoor District, Andhra Pradesh, India. *Hydro Research,* 1, 1–7.

Fagbohun, B. 2018. Integrating GIS and multi-influencing factor technique for delineation of potential groundwater recharge zones in parts of Ilesha schist belt, southwestern Nigeria. *Environmental Earth Sciences,* 77, 85.

Ghazavi, R., Babaei, S., Erfanian, M. 2018. Recharge wells site selection for artificial groundwater recharge in an urban area using fuzzy logic technique. *Water Resources Management,* 32, 3821–3834.

Gnanachandrasamy, G., Zhou, Y., Bagyaraj, M., Venkatramanan, S., Ramkumar, T., Wang, S. 2018. Remote sensing and GIS based groundwater potential zone mapping in Ariyalur District, Tamil Nadu. *Journal of the Geological Society of India,* 92, 484–490.

Hashemi, H., Berndtsson, R., Kompanizare, M., Persson, M. 2012. Natural vs. artificial groundwater recharge, quantification through inverse modeling. *Hydrology and Earth System Sciences Discussions,* 9, 1–9.

Indhulekha, K., Mondal, C.K., Jhariya, D. C. 2019. Groundwater prospect mapping using remote sensing, GIS and resistivity survey techniques in Chhokra Nala Raipur district, Chhattisgarh, India. *Journal of Water Supply: Research and Technology-Aqua,* 68, 595–606.

Kaliraj, S. 2015. Evaluation of multiple environmental factors for site-specific groundwater recharge structures in the Vaigai River upper basin, Tamil Nadu, India, using GIS-based weighted overlay analysis. *Environmental Earth Sciences,* 74(5), 4355–4380.

Karthick, P., Chokkalingam, L. 2018. Identification of groundwater recharge sites and suitable recharge structures for Thuraiyur taluk using Geospatial technology. *Indian Journal of Geo-Marine Sciences,* 47, 2117–2125.

Kumari, M., Singh, D. 2016. Water conservation: strategies and solutions. *International Journal of Advance Research and Review,* 1, 75–79.

Magesh, N. S., Chandrasekar, N., Soundranayagam, J. P. 2012. Delineation of groundwater potential zones in Theni district, Tamil Nadu, using remote sensing, GIS and MIF techniques. *Geoscience Frontiers,* 3, 189–196.

Mallick, J., Khan, R. A., Ahmed, M., Alqadhi, S. D., Alsubih, M., Falqi, I., et al. 2019. Modeling groundwater potential zone in a semi-arid region of aseer using fuzzy-AHP and geoinformation techniques. *Water,* 11, 2656.

Mondal, K. C., Jhariya, D. C., Mandal, H. S. 2021. Geoelectric imaging to assess aquifer conditions in Raipur City, Chhattisgarh, India, using schlumberger method. *Journal of the Geological Society of India,* 97, 943–950.

Mukherjee, A., Gupta, A., Ray, R. K., Tewari, D. 2017. Aquifer response to recharge–discharge phenomenon: inference from well hydrographs for genetic classification. *Applied Water Science,* 7, 801–812.

Mukherjee, R., Sahoo, M., Naik, K. C. 2011. Raipur City, Chhattisgarh. In: Groundwater Scenario in Major Cities of India. Central Ground Water Board NCCR, Raipur, Chhattisgarh, India, pp. 188–195.

Preeja, K. R., Joseph, S, Thomas, J., Vijith, H. 2011. Identification of groundwater potential zones of a tropical river basin Kerala, India using remote sensing and GIS techniques. *Journal of the Indian Society of Remote Sensing,* 39, 83–94.

Rajasekhar, M., Raju, S. G, Raju, S. R., Basha, I. U. 2018. Data on artificial recharge sites identified by geospatial tools in semi-arid region of Anantapur District, Andhra Pradesh, India. *Data in Brief*, 19, 462–474.

Raviraj, A., Kuruppath, N., Kannan, B. 2017. Identification of potential groundwater recharge zones using remote sensing and geographical information system in Amaravathy basin. *Journal of Remote Sensing & GIS*, 6, 2.

Saaty, T. L. 2008. Decision making with the analytic hierarchy process. *International Journal of Services Sciences*, 1, 83–98.

Senthilkumar, M., Gnanasundar, D., Arumugam, R. 2019. Identifying groundwater recharge zones using remote sensing & GIS techniques in Amaravathi aquifer system, Tamil Nadu, South India. *Sustainable Environment Research*, 29, 1–29.

Singh, A., Panda, S. N., Kumar, K. S., Sharma, C. S. 2013. Artificial groundwater recharge zones mapping using remote sensing and GIS: a case study in Indian Punjab. *Environmental Management*, 52, 61–71.

Sinha, M. K., Baghel, T., Baier, K., Verma, M. K., Jha, R., Azzam, R. 2019a. Impact of Urbanization on Surface Runoff Characteristics at Catchment Scale. Springer, Singapore, pp. 31–42.

Sinha, M. K., Baier, K., Azzam, R., Baghel, T., Verma, M. K. 2019b. Semi-distributed Modelling of Stormwater Drains Using Integrated Hydrodynamic EPA-SWM Model. Springer, Singapore, pp. 557–567.

Sinha, M. K., Baier, K., Azzam, R., Verma, M. K., Kumar, S. 2022. Impacts of climate variability on urban rainfall extremes using statistical analysis of climatic variables for change detection and trend analysis. In: Kumar P, Nigam GK, Sinha MK, Singh A, editors. Water Resources Management and Sustainability. Springer, Singapore, pp. 333–387.

Sinha, M. K., Rajput, P., Baier, K., Azzam, R. 2021. GIS-based assessment of urban groundwater pollution potential using water quality indices. Groundwater Resources Development and Planning in the Semi-Arid Region. Springer, Singapore, pp. 293–313.

Sinha, M. K., Rajput, P., Verma, M. K., Baier, K., Jha, R., Azzam, R. 2017. Response of urban lakes due to informal urban settlements: a case study in Raipur city. In: Pandey VK, Lakpade R, Arora S, Dave AK, Sinha J, editors. 1st Asian Conference on Water and Land Management for Food and Livelihood Security, IGKV Raipur and India Soil Conservation Society of India New Delhi, India, pp. 1–30.

Solomon, S., Quiel, F. 2006. Groundwater study using remote sensing and geographic information systems (GIS) in the central highlands of Eritrea. *Hydrogeology Journal,* 14, 1029–1041.

Tamgadge, D. B. 2002. NBSSLUP. National Bureau of Soil Survey and Land Use Planning- Soils.

Vaibhav Prakashrao Deshpande, M. K. S. A. S. 2021. Identification of critical ground water potential zones using AHP and geospatial techniques. *Design Engineering*, 13, 1774–1786.

Zhang, Z., Liu, X., Yang, S. 2009. A note on the 1–9 scale and index scale in AHP. In: Shi Y, Wang S, Peng Y, Li J, Zeng Y, editors. *Cutting-Edge Research Topics on Multiple Criteria Decision Making.* Springer, Berlin, Heidelberg, pp. 630–634.

8 Rainfall Spatiotemporal Variability and Trends in the Semi-Arid Ecological Zone of Nigeria

Saheed Adekunle Raji, Shakirudeen Odunuga, and Mayowa Fasona

8.1 INTRODUCTION

The availability and utilization of climate knowledge are undoubtedly of vast importance to human survival on Earth. This is because several human activities are tied to it: from agriculture, health, industrial development, and economy, to the reduction of disaster risks, climate information is central to socioeconomic development. In fact, the subject of climate and climate-related issues has consistently been within the top five agenda items at international gatherings over the last century due to its connection to human survival and security (Fasona et al., 2007; Jellason et al., 2020). Rainfall, as an element of climate and a key component of global precipitation, remains crucial within the realm of climate dynamics. Any significant change in the amount, distribution, and spread of rainfall in an area will have a profound impact on human activities as well as other rain-associated natural systems and processes (Abiodun et al., 2013). This is why it is important to acquire credible data on rainfall and extract useful information to safeguard humanity as a whole. Assessing the pattern of changes in rainfall is even more important due to its multidimensional nature, particularly in terms of spatiotemporal trends, variability, anomalies, and multiscale heterogeneity (Oguntunde et al., 2011).

Various research has been conducted worldwide in an attempt to narrow gaps in the spatiotemporal assessment of rainfall. However, the global perspective of these studies renders their outcomes less feasible for regional or local resource management purposes due to obvious spatial differences, data scale, and resolution, as well as the accuracy of possible impacts for decision-making. Therefore, regional rainfall assessment becomes critical not only for capturing their peculiarities but also for proper horizontal data assessment, sustainable resource management, climate-related hazard and risk reduction purposes, among others (Animashaun et al., 2020). Regional rainfall assessment will reduce large-scale uncertainty that is inevitable from global-scale observations and present a representative rainfall image for future planning. In the tropics, especially in Africa, regional basin heterogeneity exists, such that rainfall distribution within the tropical semi-arid region is vastly different from the coastal swamp forest areas (Incoom et al., 2020; Ayanlade et al., 2018; Ekpoh and Nsa, 2011). Okpara et al. (2013) stated that from 1969 to the end of the first decade of the 21st century, Africa has experienced a drastic reduction in mean annual rainfall with marked regional differences. The drought of the 1970s to the early 1990s contributed as much as 39%, with roughly 62% occurring within the Sudan and Sahelian ecological zones (Nicholson and Grist, 2001; IPCC, 2007).

The report of the Intergovernmental Panel on Climate Change (IPCC) stated that rainfall variability tends to increase, especially within the global drylands (IPCC, 2007, 2014). Within the

DOI: 10.1201/9781003303237-8

context of Africa, the IPCC asserted that while some African countries might experience an increase in rainfall over time, the extent of rainfall variability within the semi-arid and Sahelian ecosystems would intensify. Numerous studies have investigated the validity of these claims based on rainfall trends across different spatiotemporal analyses (monthly, interannually, annual, decades, 30-year periods, half a century, or a century) using different methodologies that were based on diverse data structures, which could be point or grid-based. The accompanying results of these studies showed that different countries and regions exhibit different rainfall trends, in varying directions and magnitudes (Incoom et al., 2020; Jamshadali et al., 2021; Kumar et al., 2021; Sahoo and Yadav, 2021; Elbeltagi et al., 2022a). First, Jamshadali et al. (2021) conducted a study to explore the spatial variability of South Asian summer monsoon rainfall using APHRODITE rainfall products covering 1951–2015. The outcome revealed that the western coastal belts of India and central Indian regions experienced an increase in extreme rain events along with clearly marked, albeit increasing, rainfall variabilities. Kumar et al. (2021) investigated rainfall trends, variability, and dynamics in the Punjab region of India from 1981 to 2020 using data acquired from Climate Hazards Group InfraRed Precipitation with Station data (CHIRPS) (Elbeltagi et al., 2022b). The results showed that yearly rainfall and monsoon rainfall had increased by 5.67% and 32.18 mm, respectively, based on the World Meteorological Organization (WMO) standard. Incoom et al. (2020) studied the frequency of water scarcity in the savannah region of Ghana using rainfall data from nine climatic stations, analyzing trends with the Mann-Kendall test while anomalies were based on the standard precipitation index. Although the observed trends were not significant at the annual phase, periodic drought periods were detected, such that the years 1983 and 2015 recorded the most profound drought spells, and increasing rainfall variabilities indicated intensifying instances of aridity within the northern fringes of Ghana.

In Nigeria, recent literature is filled with quantitative assessments of rainfall trends across different temporal analyses, focusing on the 20th century (e.g., Ayanlade et al., 2021; Animashaun et al., 2020; Shiru et al., 2019; Ayanlade et al., 2018; Akinsanola and Ogunjobi, 2017; Abiodun et al., 2013; Oguntunde et al., 2011; Olaniran, 2002). Oguntunde et al. (2011) evaluated changes in spatial and temporal rainfall patterns in Nigeria over a century (from 1901 to 2000) using data acquired from the Global Gridded Climatology (CRU TS 2.1), which covers synoptic stations nationwide. The results of the study showed a strong latitudinal influence on rainfall distribution, with roughly 90% of the Nigerian landscape exhibiting negative trends, and only 22% of these changes being statistically significant at a 0.05 level of significance. Akinsanola and Ogunjobi (2017) conducted a four-decade study (1974–2013) on spatiotemporal trends of rainfall on yearly and interannual scales using data acquired from 23 in-situ meteorological stations in Nigeria. The study employed autocorrelation, non-autocorrelated assessments (modified Mann-Kendall test), and other tests of magnitude and homogeneity. The outcome of the study revealed an increasing trend in annual rainfall, while at the interannual level, significant pre-wet and post-wet trends were detected, with an increase in wet season rainfall ranging between 13.9% and 15.4%. Most recently, Ayanlade et al. (2021) examined spatiotemporal changes in precipitation and their impact on vegetation dynamics within the six ecological zones in Nigeria using spaceborne and gridded rainfall data correlated with the normalized difference vegetation index (NDVI) analysis. The study revealed a significant seasonal relationship between rainfall and vegetation changes. Rainfall variabilities within the dry season in the Sahel and Sudan ecological zones significantly influenced ecological functionalities. The functional productivity of ecological zones directly correlates with the character and direction of rainfall, especially in Nigeria, where roughly 70% of the populace engages in small-scale rain-fed agriculture, and any shift in rainfall affects livelihoods (Elbeltagi et al., 2022c; Elbeltagi et al., 2021; Animashaun et al., 2020; Oguntunde et al., 2011).

Existing studies within the semi-arid ecological zone of northern Nigeria on the trends and variability of rainfall are very limited. Studies such as Usman et al. (2018), which analyzed rainfall trends across northern Nigeria, confined the study temporally to a 35-year period using only 10 synoptic stations. Ekpoh and Nsa (2011) conducted a monotonic trend assessment within the north-western part of Nigeria using two decades of rainfall data. Also, Adejuwon and Dada (2021) conducted a drought-based standard period index assessment of the north with very limited consideration of the importance of rainfall trends and variability. Meanwhile, Jellason et al. (2020) stated that the resilience of dryland smallholder farmers, especially in the north-western part of Nigeria, is often tested by a cloudy and blurry understanding of annual and interannual possibilities in rainfall behavior, apart from the unpredictable social and other environmental challenges. While suggesting that more innovative long-term rainfall studies are essential to building science-to-farm relationships, there is a need for a better understanding of the indicators of climate change in rainfall patterns and trends.

This study, therefore, is an attempt to address the identified research problem and fill the observed gap within the Sokoto-Rima Basin of the north-western axis of Nigeria. The overarching aim of the study was to examine the link between rainfall trends and variabilities in the past climate (1951–2017) and the potential future climate (2018–2050). To achieve this, the following research questions were formulated: (1) What is the nature of rainfall climatology in the Sokoto-Rima Basin between the past and the future climate? (2) What are the trends in rainfall across monthly, annual, interannual, and decadal timescales for each climatic period in the Sokoto-Rima Basin? and (3) What are the implications of these trends for agriculture and water ecosystem services? It is anticipated that the findings of this study will assist decision-makers in the Sokoto-Rima Basin and the entire Sudano-Sahelian ecological zone of Nigeria in implementing effective adaptation and mitigation strategies to address potential changes in rainfall.

8.2 STUDY AREA

The Sokoto-Rima Basin (SRB) (Figure 8.1) is an expansive semi-arid belt in West Africa that spans across the administrative boundaries of three countries: Nigeria, Niger, and the Benin Republic. This study focuses on the Nigerian section, which is located between latitudes 8°11′–13°53′N of the Equator and longitudes 3°26′–9°24′E of the Greenwich Meridian, covering a total area of approximately 925,796.5 km^2 (roughly one-tenth of Nigeria's land area). The geographical location of the SRB indicates its direct exposure to the tropical environmental conditions of the Sahelian ecosystem in the West African subregion. As a result, its overall climatology is governed by the West African monsoon (Vizy and Cook, 2018). This implies that precipitation exhibits a distinct seasonality. Historically, annual precipitation rarely exceeds 1,200 mm, and there is a northward reduction in rainfall (Raji et al., 2021). Also, Abiodun et al. (2013) affirmed that precipitation within this region is influenced by mesoscale processes, which exert significant control over its pattern and distribution, accounting for up to 75% of the variability. Daily temperatures, which play a crucial role in shaping the ecological response to climate variations, also exhibit seasonality, with an annual average of about 30°C (Oguntunde et al., 2011; Vizy and Cook, 2018).

The annual rainfall hydrograph in the SRB typically exhibits a unimodal pattern, with the peak occurring in August. Similar to other ecological basins in West Africa, the SRB experiences two distinct seasons. The rainy (wet) season usually begins in May or June, depending on the prevailing atmospheric conditions, and lasts until September or early October. In contrast, the dry season spans from November to March and is characterized by minimal or no rainfall, accompanied by dusty and cold winds known as the Harmattan season, which blows from the northwest. This phenomenon is specifically influenced by the annual movement of the intertropical convergence zone (ITCZ) and its associated weather patterns. Due to the concentration of rainfall within a 3–5-month period, there is significant surface runoff, leading to high stream discharge, particularly in riverine communities such as Sokoto, Gusau, Argungu, and others. This favorable hydrological condition supports extensive agricultural cultivation and animal husbandry in the SRB.

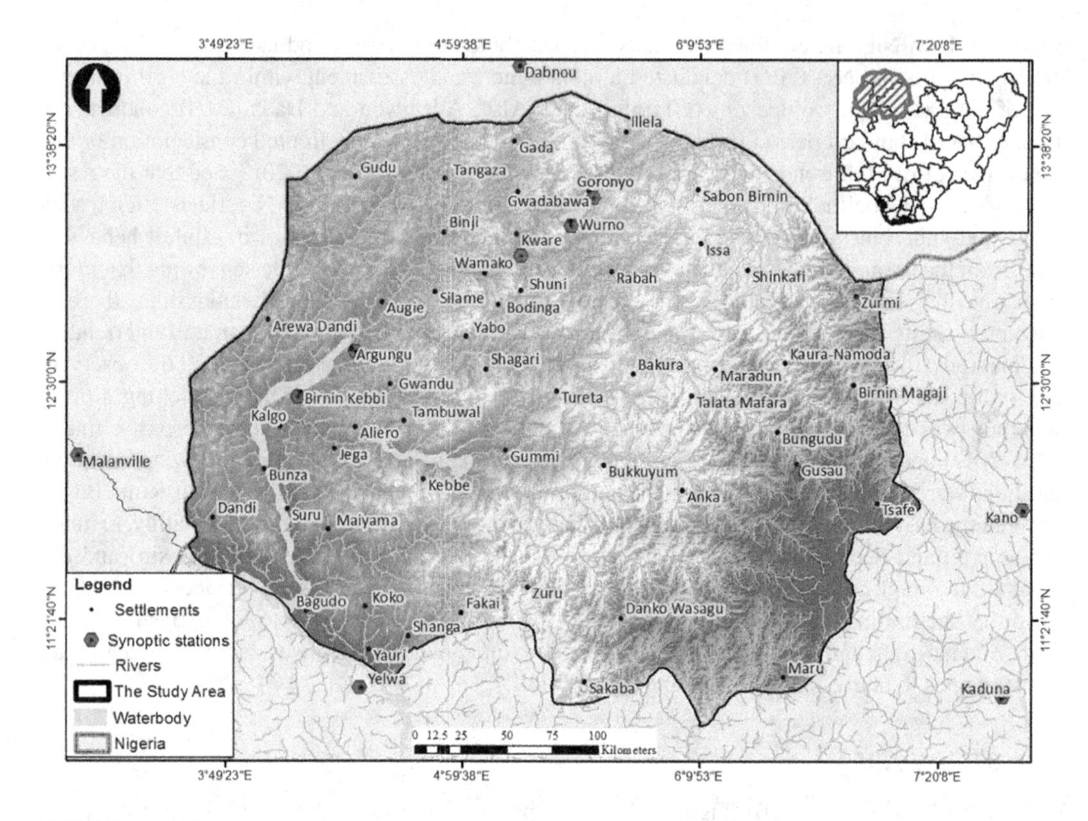

FIGURE 8.1 The study area in the context of Nigeria with the physical properties.

8.3 MATERIALS AND METHODS

8.3.1 Data Sources

Rainfall data were obtained from the archives of the Nigeria Meteorological Agency (NIMET) and supplemented with data from Princeton University's station-based data available through the Princeton Climate Analytics (PCA) for Africa portal at ***https://platform.princetonclimate.com/PCA_Platform/***. The NIMET data were collected from synoptic stations located in nine cities: Sokoto, Yelwa, Birnin Kebbi, Argungu, Gusau, Goronyo, Wurno, Kano, and Kaduna. Data for neighboring countries, Benin Republic (Malanville), and Niger Republic (Dabnou), were obtained from the PCA portal. However, it should be noted that some synoptic stations had incomplete or missing values in the NIMET data. Specifically, there was no data recorded for Birnin Kebbi from 1963 to 1964, no data for Argungu in 1963, and none for Gusau in 1965. The study period was chosen as 1951–2017, spanning 67 years, which aligns with the duration of a climatic period according to Aguado and Burt (2013) and Ojo et al. (2001). This data period provides a sufficient timeframe to observe any variability and changes in climatic patterns. To address the observed data gaps, statistical regression was used with data from highly correlated synoptic stations, and the derived trend line equations were utilized to fill the missing data. The climatic data analysis considered the period from 1951 to 2017, which was used to define the climate.

Conversely, the future climate was defined to cover the period 2018–2050. It is worth noting that 2018 was chosen as a continuation of the last year of the past climate period. The downscaled data from the MRI-CGCM3 model was obtained. This data was generated by the Meteorological Research Institute of Japan as part of the World Climate Research Programme's (WCRP)

Coupled Model Intercomparison Project Phase 5 (CMIP5). The data was based on Representative Concentration Pathway (RCP) 4.5, which indicates a moderate increase in the likelihood of extreme climatic events while still being suitable for ecosystem service modeling (Hao and Yu, 2018). The data was downloaded from the Lawrence Livermore National Laboratory, a platform developed in collaboration with the Earth System Grid Federation (ESGF), through the portal at ***https://esgf-node.llnl.gov/search/cmip5/***. The selection of this data was based on its empirical suitability for accurately representing the monthly precipitation and temperature scenarios in Nigeria, including the observable drivers of climatic events such as the mesoscale factor (Abiodun et al., 2013; Shiru et al., 2019).

8.3.2 Trend and Variability Detection

Mann-Kendall's test and standardized anomaly index were used for trend and variability analysis at monthly, interannual, and annual levels. Mann-Kendall's test is a non-parametric measure of monotonic trend and is less sensitive to outliers. It basically tests the hypothesis of no trend against the alternative hypothesis of a changing trend. This is expressed in Equations (8.1) and (8.2).

$$S = \sum_{i-1}^{n-1} \sum_{j+i+1}^{n} \sin\left(x_j - x_k\right) \tag{8.1}$$

$$\sin\left(x_j - x_k\right) = \begin{cases} +1 \text{ if } \left(x_j - x_k\right) > 1 \\ 0 \text{ if } \left(x_j - x_k\right) = 0 \\ -1 \text{ if } \left(x_j - x_k\right) < -1 \end{cases} \tag{8.2}$$

where: S is the Mann-Kendall's test statistics; x_i and $_j$ are the sequential data values of the time series in the years i, and j (j>i) and n are the length of the time series. Prior to the analysis of Mann-Kendall trends, it is expected that the data series must be devoid of multicollinearity. This is to ensure that any form of correlation within the data is removed or reduced to the minimum such that it does not infiltrate the true trend expected of the original dataset. Hence, the Pearson correlation coefficient was tested to ensure trend-free pre-whitening. The corresponding Sen's slope for the expression of the magnitude of the climate data is expressed in Equation (8.3):

$$\beta = \text{median}\left(\frac{x_j - x_i}{j - i}\right), \quad j < i \tag{8.3}$$

where β is Sen's slope estimate, and β>0 signifies an increasing trend in the climate data series. Conversely, the data series presents a decreasing trend during the specific time period.

The standardized anomaly index was computed in order to determine how "normal" or "usual" a particular climatic year can be compared to the long-term situation. This is mathematically expressed in Equation (8.4);

$$Z = (x - \mu) / \sigma \tag{8.4}$$

where: Z is the standardized anomaly index, x is the annual climatic parameter value, μ is the long-term mean annual climatic parameter value over a period of observation, and σ is the long-term standard deviation over the period of observation. And, the interpretation of the standardized anomaly index was based on the Hänsel et al. (2016) classification presented in Table 8.1.

TABLE 8.1
Anomaly Index Classification Used in the Study

S/No.	Anomaly Index	Index Description
1	≥ 2.00	Extremely wet
2	1.50 to 1.99	Very wet
3	1.00 to 1.49	Moderately wet
4	0.50 to 0.99	Slightly wet
5	−0.49 to 0.49	Near normal
6	−0.99 to −0.50	Slightly dry
7	−1.49 to −1.00	Moderately dry
8	−1.99 to −1.50	Very dry
9	≤ 2.00	Extremely dry

Source: Hänsel et al. (2016).

8.3.3 Spatial Interpolation

The inverse distance weighted (IDW) method was employed to generate climate maps. IDW is a spatial interpolation technique that predicts cell values in a raster format using a limited number of sample space (Pratap et al., 2021). The technique used the hypothesis of the first law of geography, which states that closer events are more auto-correlated than those farther apart. Hence, its prediction technique is based on the fact that each known point has a local influence that shrinks with increasing space, thereby giving greater weights to points closest to the prediction location based on the distance decay effect. de Smith et al. (2018) mathematically expressed IDW in Equation (8.5):

$$Z_j = k_j \sum_{i=1}^{n} \left(\frac{1}{d_{ij}^{\propto}}\right) z_i \qquad (8.5)$$

where: z_j is the spatially predicted value, d_{ij} is the spatial interval between the known and the predicted value, z_i is the known value and k_j is an adjustment to ensure that the weights add up to unity (1).

8.4 RESULTS AND DISCUSSION

The result and description are given in different heads as follows.

8.5 RAINFALL CHARACTERISTICS

8.5.1 Statistical Overview of Rainfall Climatology

Rainfall climatology represents the overall precipitation patterns of a particular area. Table 8.2 presents a summary of long-term temporal statistics for each of the selected synoptic stations within the SRB during the past climate period. Rainfall exhibits significant variation across these stations, with Dabnou (97.9%) and Goronyo (83.6%) having the highest coefficients of variation (CV). The rainfall variability in Dabnou is reflected in its wide range of rainfall values, spanning from 70.6 to 1,363.4 mm/year (mean = 394.5 ± 386.2 mm/year). Similarly, Goronyo also displays notable variability, with rainfall values ranging from 138.6 to 976.6 mm/year (mean = 409.6 ± 342.6 mm/year). On the other hand, Kaduna exhibits the lowest CV at 33.5%, and its rainfall values range from 833.4 to 1,792 mm/year (mean = 1,171.7 ± 392.6 mm/year). For the entire SRB, rainfall values range from 226.3 to 1,206.8 mm/year (mean = 754.2 ± 168.7 mm/year) with a CV of 22.4%. Looking toward the future (2018–2050), it is anticipated that rainfall variability will decrease, as indicated in Table 8.3. Dabnou is expected to have a reduced CV of 20.9%, while the rainfall

TABLE 8.2
Annual Rainfall Statistics of Selected Synoptic Stations of the Sokoto-Rima Basin (1951–2017)

Synoptic Station	Minimum (mm/year)	Maximum (mm/year)	Mean (mm/year)	Standard Deviation (mm/year)	CV (%)
Argungu	230.5	1,404.9	637.7	405.2	63.5
Birnin Kebbi	283.5	1,623.4	695.8	455.7	65.5
Goronyo	138.6	976.6	409.6	342.6	83.6
Gusau	468.7	1,908.9	959.3	538.5	55.9
Kano	459.0	1,694.3	720.1	455.6	63.3
Kaduna	833.4	1,792.0	1,171.7	392.6	33.5
Sokoto	173.0	1,175.1	552.8	352.0	63.7
Wurno	283.9	1,082.0	446.8	325.0	72.7
Yelwa	558.7	1,884.6	1,031.4	443.8	43.0
Dabnou	70.6	1,363.4	394.5	386.2	97.9
Malanville	314.1	1,505.9	585.4	384.1	65.6
Sokoto-Rima Basin	226.3	1,206.8	754.2	168.7	22.4

TABLE 8.3
Annual Rainfall Statistics of Selected Synoptic Stations of the Sokoto-Rima Basin (2018–2050)

Synoptic Station	Minimum (mm/year)	Maximum (mm/year)	Mean (mm/year)	Standard Deviation (mm/year)	CV (%)
Argungu	1,009.7	1,618.4	1,264.2	210.0	16.6
Birnin Kebbi	1,101.0	1,745.7	1,369.0	219.1	16.0
Goronyo	770.5	1,381.7	997.0	193.8	19.4
Gusau	1,139.8	1,935.8	1,413.3	234.6	16.6
Kano	1,082.8	2,006.0	1,369.1	250.8	18.3
Kaduna	1,259.9	1,971.4	1,680.0	195.3	11.6
Sokoto	817.6	1,405.2	1,056.1	195.6	18.5
Wurno	734.3	1,367.1	954.0	195.1	20.4
Yelwa	1,353.4	2,082.1	1,641.7	209.1	12.7
Dabnou	658.0	1,183.3	845.5	176.6	20.9
Malanville	1,165.6	1,912.5	1,471.8	222.9	15.1
Sokoto-Rima Basin	1,032.3	1,599.9	1,278.3	194.9	15.2

amount is projected to range from 658 to 1,183.3 mm/year (mean = 845 ± 176.6 mm/year). Similarly, Kaduna will continue to exhibit the lowest CV at 11.6%, with projected rainfall distribution ranging from 1,259.9 to 1,971.4 mm/year (mean = 1,680 ± 195.3 mm/year). Clearly, the future climatology of the SRB (CV of 15.2%, rainfall distribution range of 1,032.3–1,559.9 mm/year with an average of 1,278.3 ± 194.9 mm/year) suggests an increase in rainfall throughout the entire area.

The spatial distribution of annual rainfall statistics is illustrated for the past and future climates, respectively. In the first period, a distinct northward decrease in rainfall was observed, with the highest isohyet value of 1,800 mm observed in southern areas around Yelwa and parts of Gusau. The location-specific isohyet value of 1,650 mm/year around the Talata-Mafara-Maradun-Birnin-Magaji axis in the east and Goronyo-Wurno-Rabah axis in the northern region demonstrated the influence of location on rainfall distribution. The most notable feature of rainfall distribution in the SRB is the 1,500 mm/year isohyet, which traverses the entire north-south axis. In the later period (2018–2050), the north-south delineation is more precisely depicted, with a range of 2,000 mm/year around Yelwa in the south to 1,300 mm/year around Gada at the northernmost end of the SRB, following an increasing latitudinal trend. The spatial pattern of the coefficient of variation for the two periods is shown in (1951–2017) and (2018–2050), displaying an eastward variation in rainfall. Although the extent of variability decreased, the influence of location remains largely unchanged, with exceptions in Goronyo in the northern axis in the future period. The temporal rainfall distribution and averages are represented as cumulative distribution functions (CDF) plotted against the normal distribution in Figures 8.2 (1951–2017) and 8.3 (2018–2050). The CDF helps in setting thresholds for assessing specific rainfall events, and the extent to which these thresholds conform to normalcy within the distribution can be determined using the normal distribution curve. Thus, the CDF shows

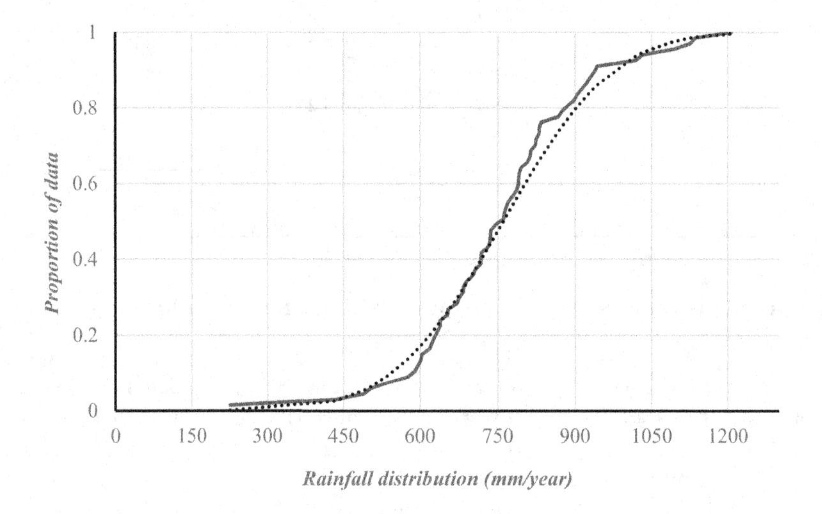

FIGURE 8.2 Rainfall distribution of the Sokoto-Rima Basin plotted as a cumulative distribution function (in blue) compared to a normal distribution curve (in brown) spanning the past climate.

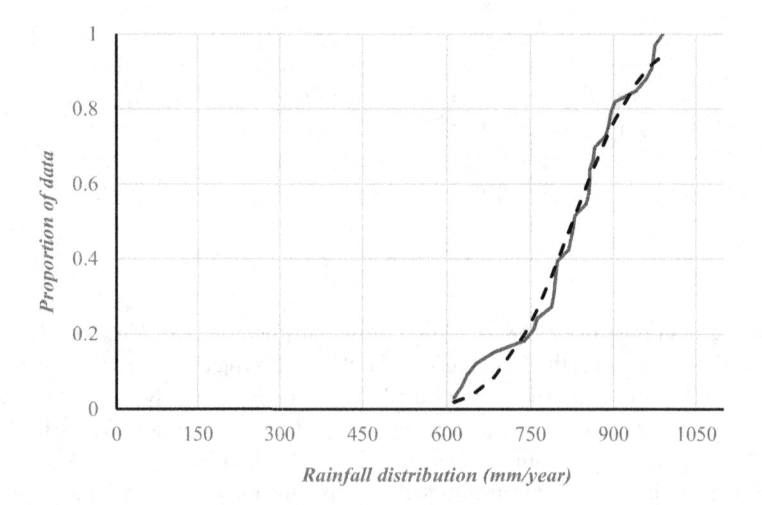

FIGURE 8.3 Rainfall distribution of the Sokoto-Rima Basin plotted as a cumulative distribution function (in blue) compared to a normal distribution curve (in brown) for the future climate).

that approximately 50% of the rainfall is around 750 mm/year for the period 1951–2017, and it is expected to increase to around 850 mm/year in the subsequent years (2018–2050). Most of these rains are expected to be regionally distributed, particularly toward the south and some parts of the eastern axis of the SRB.

8.5.2 Rainfall Variability and Anomaly Patterns

The spatial distribution of rainfall variability for the two periods of study is depicted in Figures 8.4 (past climate) and 8.5 (future climate). In the past climate, a variability range of 0.96–1.99 was recorded, with specific regional delineations: the eastern axis comprising Birnin Kebbi, Kalgo, and Argungu showed low variability values, while clusters of high variability were observed within the northern section covering the Kware-Sokoto-Bodinga-Shuni axis. Generally, a southward decrease in variability was observed. This pattern is expected to persist in the future, as shown in Figure 8.5, with a clearly marked north-south decrease in the extent of rainfall variability. These findings align with previous studies by Akinsanola and Ogunjobi (2017) and Oguntunde et al. (2011). For example, Oguntunde et al. (2011) stated that the geographical distribution of rainfall variability within the short-grass ecological zone of northern Nigeria reflects a progressive decrease in rainfall variability due to possible changes in global climate patterns and the potential influence of the El Niño-Southern Oscillation (ENSO).

With respect to rainfall anomaly, two distinct periods can be identified for the past climate in the SRB, as shown in Figure 8.6: (1) from 1951 to 1969, a well-defined period with 15 "dry" years and three intervening "normal" years (1956, 1961, and 1965); and (2) from 1970 to 2017, a 48-year period characterized by approximately 2/3 normal years and 1/3 "dry" years. However, the decadal

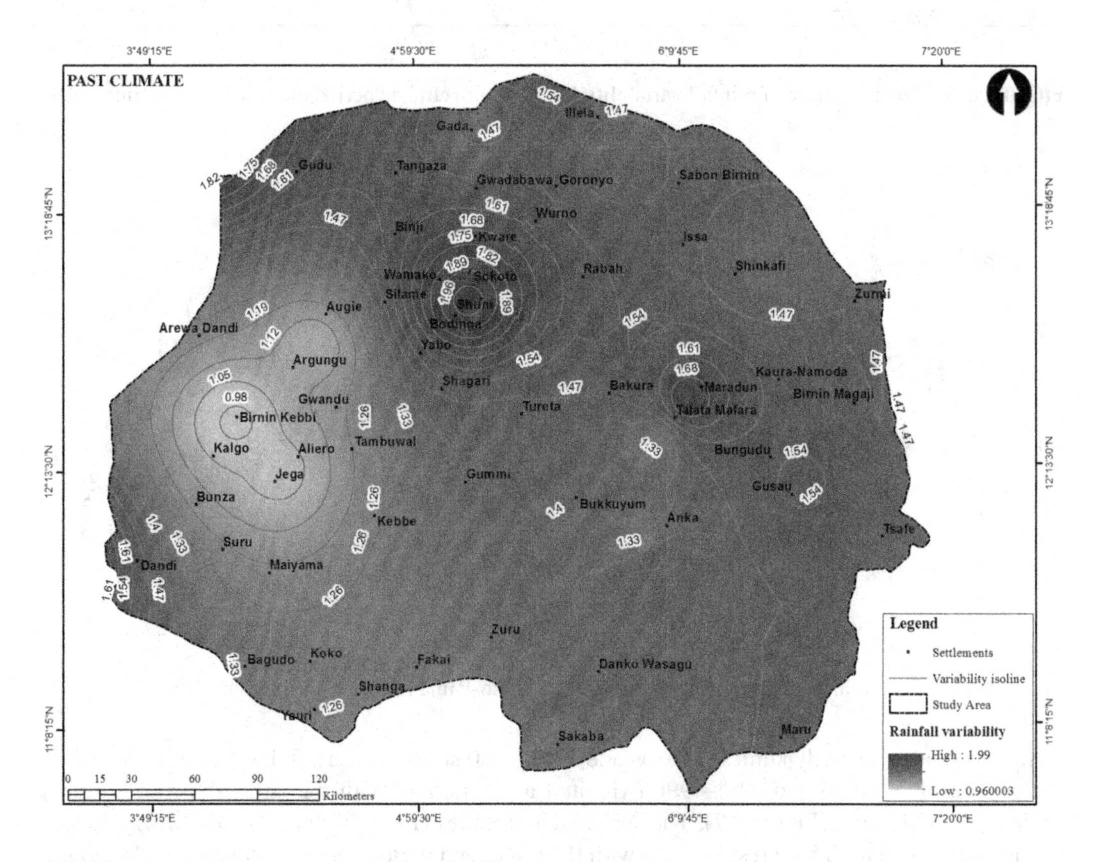

FIGURE 8.4 Spatial pattern of rainfall variability for the past climate over the Sokoto-Rima Basin.

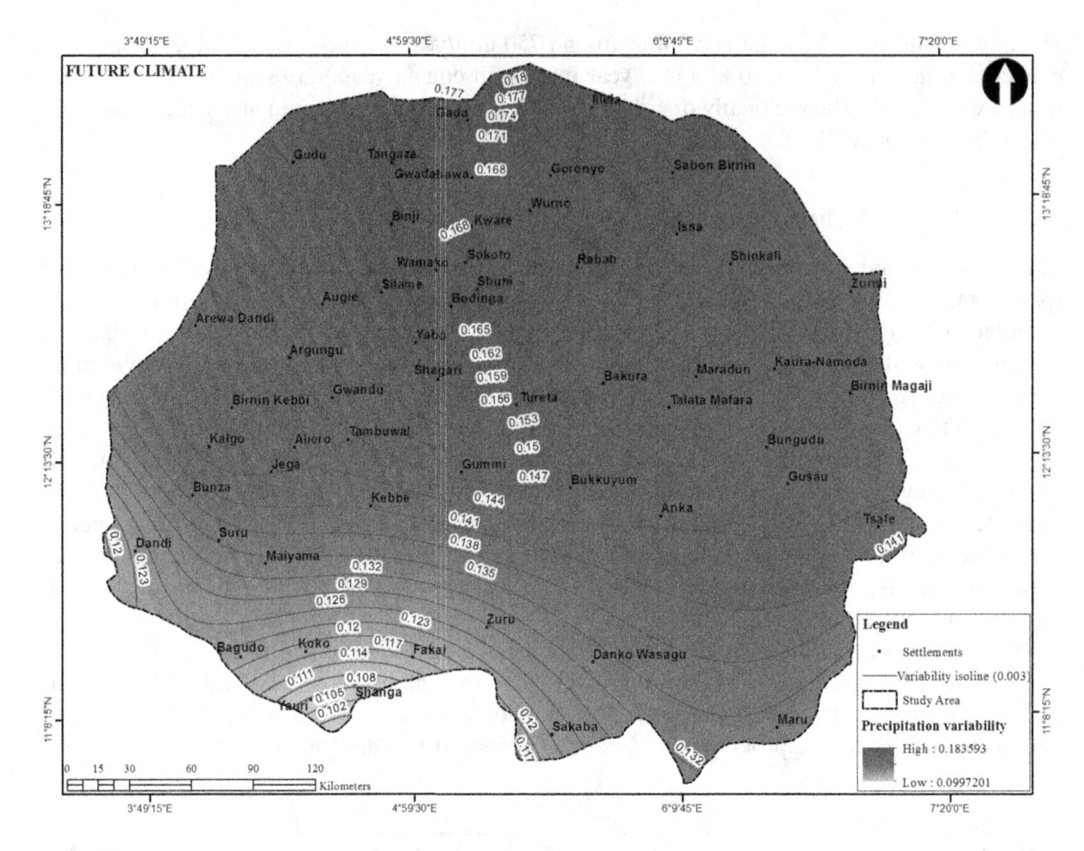

FIGURE 8.5 Spatial pattern of rainfall variability for the future climate period over the Sokoto-Rima Basin.

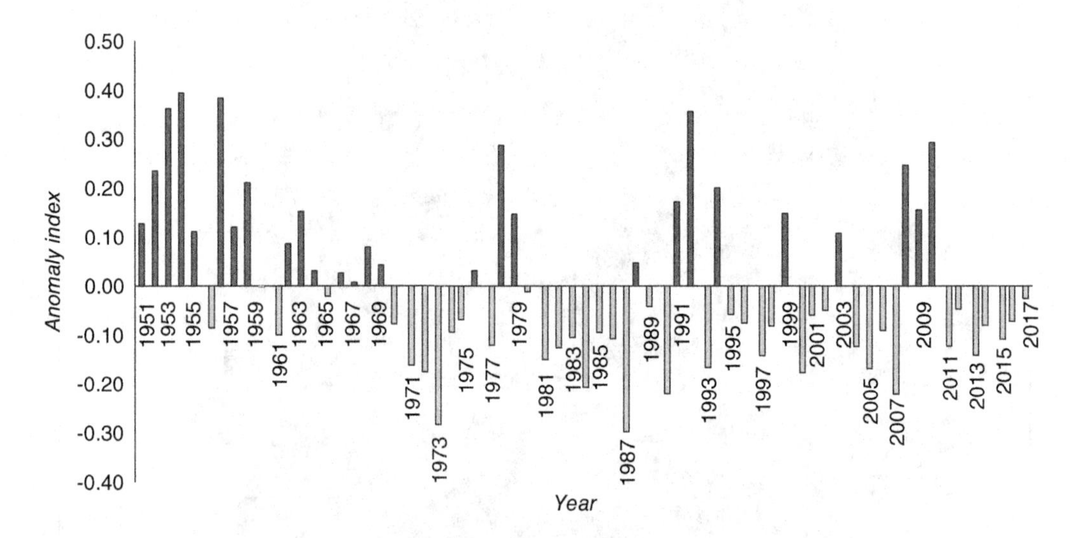

FIGURE 8.6 Temporal rainfall anomaly index of the Sokoto-Rima Basin for the past climate.

analysis revealed some dynamics. The decade 1951–1960 showed a certain level of wetness, while the decades 1971–1980 and 1981–1990 exhibited a dry period, with the latter showing a higher occurrence of drought (Figure 8.7). The subsequent decades (1991–2000 and 2001–2010) returned to "normal" anomaly. These results align with the prolonged periods of low wetness and drought in northern Nigeria reported in the literature by Nicholson and Grist (2001), Oguntunde et al. (2011), and Akinsanola and Ogunjobi (2017). Specifically, Akinsanola and Ogunjobi (2017) suggested that the non-constant general circulation shape, driven by tropical discontinuities and easterlies across West Africa, can be identified as driving factors of rainfall anomalies. Oguntunde et al. (2011) added that the northward shift in rainfall anomaly could be related to latitudinal influence.

The expected level of anomaly in rainfall for the climatic period 2018–2050 indicates a progression from the previous period (i.e., 1951–2017). As depicted in Figure 8.8, the annual anomaly index will generate two distinct rainfall anomaly patterns. The first period (2018–2033) is expected to be dominated by 12 "dry" years and 6 "wet" years, while the second period (2034–2050) is anticipated to be somewhat balanced with 8 "dry" years and 7 "wet" years. This suggests that the future climate is anticipated to experience roughly two decades of wetness compared to 13 years of potential dry spells. The decadal rainfall anomaly analysis (Figure 8.9) indicates that the decade 2018–2027 is expected to be dry, while the subsequent decades (2028–2037 and 2038–2047) are anticipated to be wet.

The spatial pattern of the rainfall anomaly index is depicted in Figure 8.10 for the past climate and Figure 8.11 for the future climate. During the past climate, the index ranged from –1.28 to

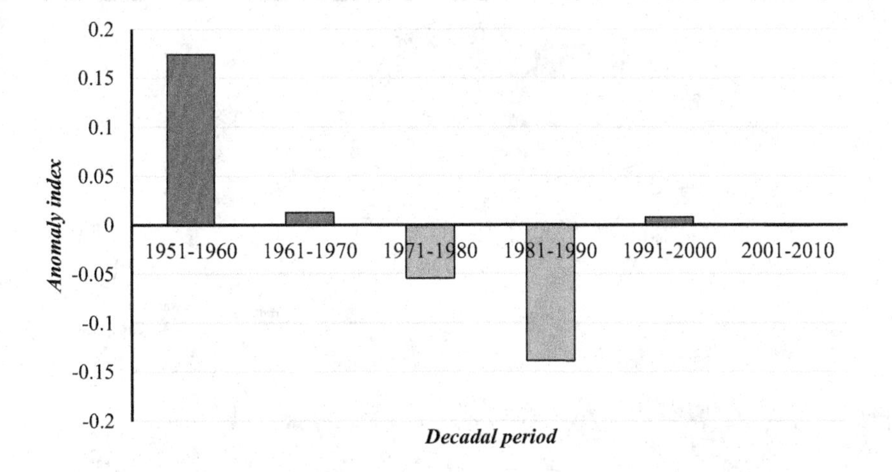

FIGURE 8.7 Decadal rainfall anomaly index of the Sokoto-Rima Basin for the past climate.

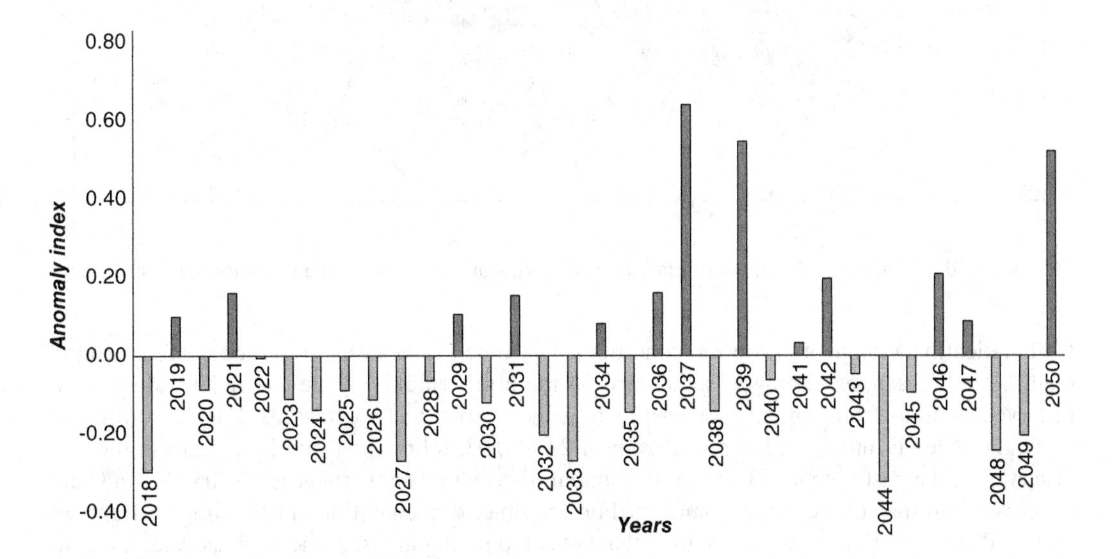

FIGURE 8.8 Temporal rainfall anomaly index of the Sokoto-Rima Basin for the future climate.

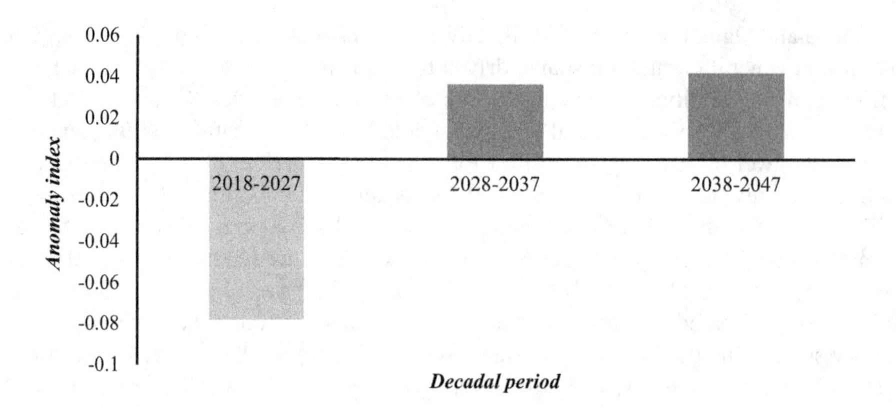

FIGURE 8.9 Decadal rainfall anomaly index of the Sokoto-Rima Basin for the future climate.

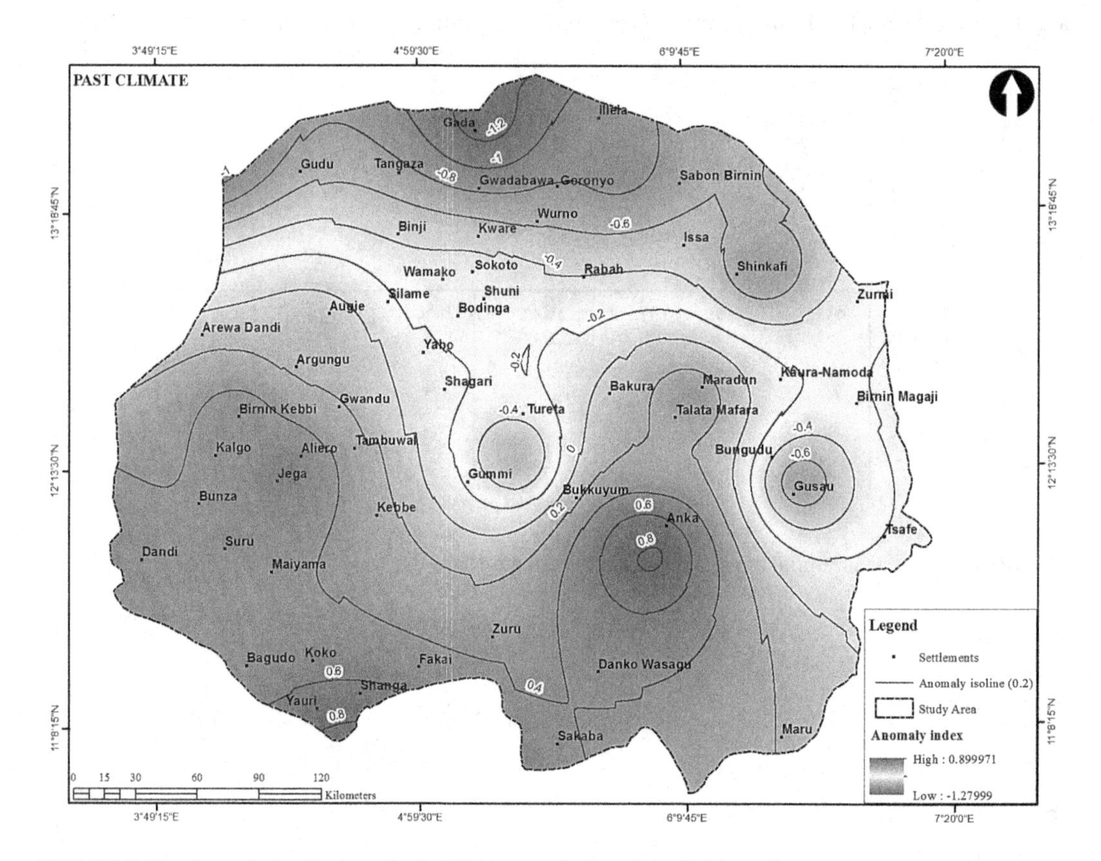

FIGURE 8.10 Spatial distribution of rainfall anomaly index of the Sokoto-Rima Basin for the past climate.

0.90, indicating a range of dry to wet conditions. Dry index values were observed in areas around Gummi and Gusau, extending to some parts of the northern axis, while wet index values were observed in areas from Anka in the east to Jega in the west and southwards. For the future climate, the index is anticipated to range from –1.24 to 0.93, following a similar pattern to the one observed in the past climate. However, the spatial distribution suggests possible dryness in some areas with normal to wetness anomalies. For example, dry conditions in Gummi will spread eastwards to Bukkuyum, extending to Anka and surrounding areas. Areas such as Augie, Gudu, Gada, and Illeila will maintain their dry conditions throughout 2050. Sokoto, Shuni, Bodinga,

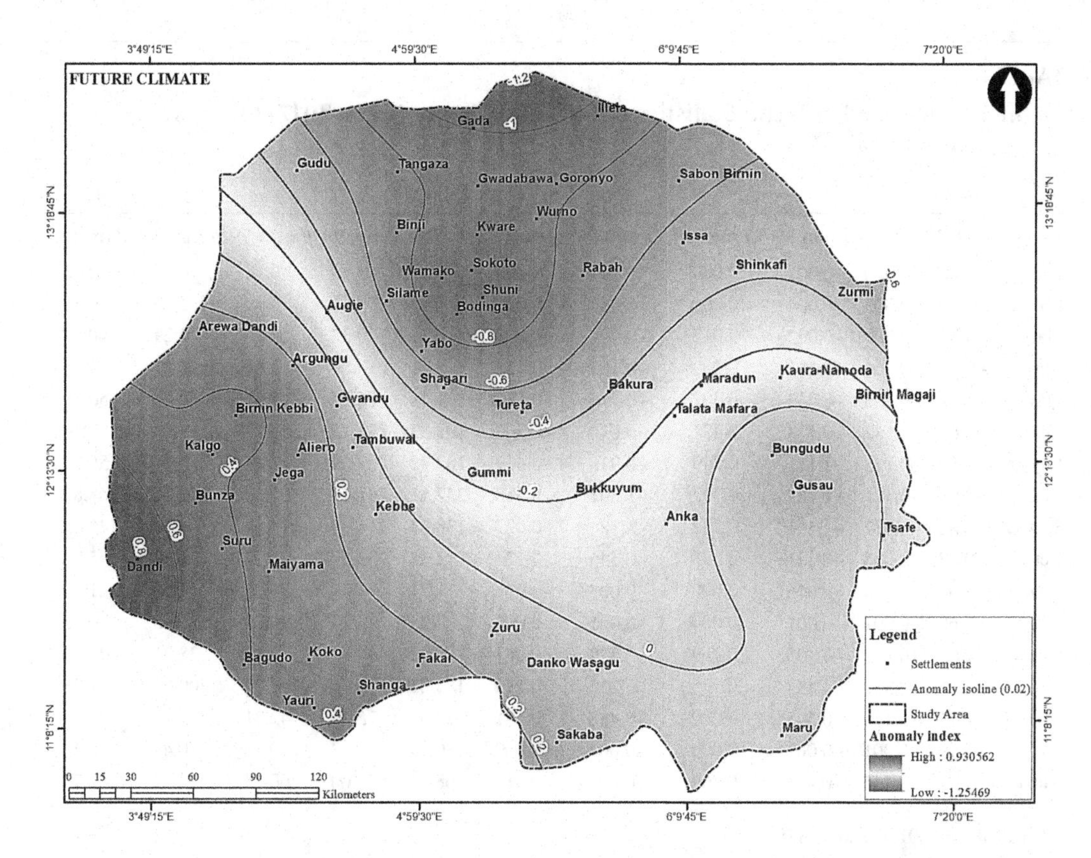

FIGURE 8.11 Spatial distribution of rainfall anomaly index of the Sokoto-Rima Basin for the future climate.

Wamako, Yabo, and Shagari will experience increased dryness compared to normal rainfall patterns during the future climate. Akinsanola and Ogunjobi (2014) established that communities within the SRB have experienced fluctuating rainfall patterns, and this trend is expected to continue in the future as the influence of global warming and climate change affects the climate system of tropical West Africa.

8.6 ANNUAL, INTERANNUAL, AND MONTHLY TRENDS

The results of the non-parametric Mann-Kendall test used to investigate significant trends at the annual, interannual, and monthly periods for the past climate are presented in Table 8.4. During the past climate, a decreasing monotonic trend at a significance level of 0.05 was observed at the annual level. This indicates that the null hypothesis is rejected for the entire past climate. However, there is a significant variation in monthly trend patterns. Specifically, 10 months of the year showed a decreasing trend, with January ($Z=-2.6$, $\alpha=0.01$), February ($Z=-2.6$, $\alpha=0.01$), July ($Z=-2.6$, $\alpha=0.1$), November, and December ($Z=-2.6$, $\alpha=0.01$) having the highest and equal magnitude of Z-scores. On the other hand, increasing rainfall trends were observed in March ($Z=0.80$, $\alpha>0.1$) and October ($Z=0.69$, $\alpha>0.1$). In summary, a predominantly decreasing monotonic rainfall trend was observed within the months of the past climate in the SRB. Additionally, rainfall patterns within the first and last quarters of the year had a significant influence on the past climate in the SRB. Interannually, the dry season, pre-wet season, wet season, and post-wet season showed M-K statistics of –2.99, –1.02, –1.35, and –1.13 at $\alpha=0.001$, and $\alpha>0.1$, respectively. This indicates a significant decreasing trend in rainfall amount within the SRB. The Sen's slope analysis across the respective periods and seasons of assessment confirms the decreasing trends.

TABLE 8.4
Rainfall Mann-Kendall Trend Statistics of the Past Climate (1951–2017) Over the Sokoto-Rima Basin

Time	M-K	Sen's Slope Estimate									
Series	Test Z	Q	Q min 99	Q max 99	Q min 95	Q max 95	B	β min 99	β max 99	β min 95	β max 95
Jan	−2.60**	0.000	−0.001	0.000	−0.001	0.000	0.17	0.18	0.15	0.18	0.16
Feb	−2.60**	−0.002	−0.004	0.000	−0.003	−0.001	0.75	0.81	0.67	0.80	0.70
Mar	0.80	0.000	−0.002	0.012	−0.001	0.008	0.25	0.32	−0.10	0.29	−0.04
Apr	−0.04	−0.004	−0.241	0.209	−0.167	0.152	8.71	19.88	2.89	16.41	4.74
May	−1.11	−0.266	−0.847	0.307	−0.718	0.181	61.09	76.27	45.57	73.78	50.10
June	−0.63	−0.142	−0.773	0.471	−0.650	0.306	103.33	122.13	80.78	117.93	87.36
July	−0.54	−0.178	−1.111	0.749	−0.915	0.489	182.72	223.69	147.08	214.47	156.05
Aug	−0.94	−0.421	−1.709	0.717	−1.391	0.388	237.51	281.71	201.48	270.26	208.81
Sept	+−1.82	−0.537	−1.401	0.252	−1.184	0.027	138.27	174.96	116.54	164.75	126.63
Oct	0.69	0.068	−0.281	0.573	−0.145	0.424	21.32	33.57	−2.16	31.63	5.99
Nov	−2.60**	−0.003	−0.006	0.000	−0.005	−0.001	1.21	1.31	1.08	1.28	1.12
Dec	−2.60**	−0.001	−0.001	0.000	−0.001	0.000	0.23	0.25	0.21	0.25	0.21
MAM	−1.02	−0.122	−0.405	0.159	−0.329	0.087	30.42	40.84	19.49	38.25	23.27
JJA	−1.35	−0.327	−0.883	0.325	−0.730	0.170	183.85	206.99	157.20	200.03	161.38
SON	−1.13	−0.172	−0.595	0.220	−0.488	0.131	57.65	75.05	42.21	72.13	45.17
DJF	−2.99**	−0.001	−0.002	0.000	−0.002	0.000	0.39	0.42	0.35	0.41	0.36
Annual	−1.76+	−0.165	−0.420	0.066	−0.363	0.017	68.37	76.00	60.24	74.44	62.73

***Trend at $\alpha=0.001$ level of significance.
**Trend at $\alpha=0.01$ level of significance.
*Trend at $\alpha=0.05$; level of significance.
+Trend at $\alpha=0.1$ level of significance.
The blank cell shows the significance level is greater than 0.1.

The spatial distribution of the M-K trend for the past climate over the SRB is shown in Figure 8.12, revealing that rainfall trends in the SRB are not only regional but also location-specific. The figure also illustrates a gradual shift from a decreasing monotonic trend to an increasing trend. Gusau (Z=−0.3), Birnin Kebbi (Z=−0.3), Dandi (Z=−0.4), Sokoto (Z=−0.2), and Gada (Z=−0.2) all exhibited decreasing rainfall trends. On the other hand, Shinkafi, DankoWasagu, Bukkuyum, Tureta, and Gummi all showed increasing rainfall trends ranging from 0.1 to 0.2, significant at $\alpha>0.01$. Overall, this indicates a shift in rainfall trend from the north and from the east to the central areas, highlighting the influence of orography on rainfall distribution over the SRB (Akinsanola and Ogunjobi, 2014).

Also, Table 8.5 shows the trend of the future climate's rainfall across the annual, interannual, and seasonal periods. During this phase, an increasing trend will be observed for 9 months, except for December, January, and February, with an M-K test of −2.6 at $\alpha=0.001$ level of significance. This is very different from the trend observed during the past climate, which showed a decreasing trend of the same magnitude. This suggests that the trend of rainfall within these months will continue to have dry spells. Similarly, an increasing monotonic trend will start in March (Z=2.21, $\alpha=0.05$) and peak in September (Z=5.53, $\alpha=0.001$) each year. The seasonal trend pattern will exhibit a dry season, pre-wet, wet, and post-wet seasons, returning M-K statistics of −2.07 at $\alpha=0.05$, 4.23, 5.39, and 4.98 at $\alpha=0.001$, respectively, indicating a shift from a decreasing monotonic trend to an increasing monotonic trend. This suggests that the climate model predicts an increase in rainfall in the Sokoto-Rima Basin in the mid-future. In terms of spatial extent, there will be a progressive shift from the decreasing monotonic rainfall trends observed in the past climate to a progressively wetter future, as displayed in Figure 8.13. The northward shift in the magnitude of the trend will also be observed, as areas with the least monotonic trends (Bagudo, Koko, Shanga, and Yauri) are located in the south, while areas in the north (Gada, Ileilla) have the highest increasing monotonic trends.

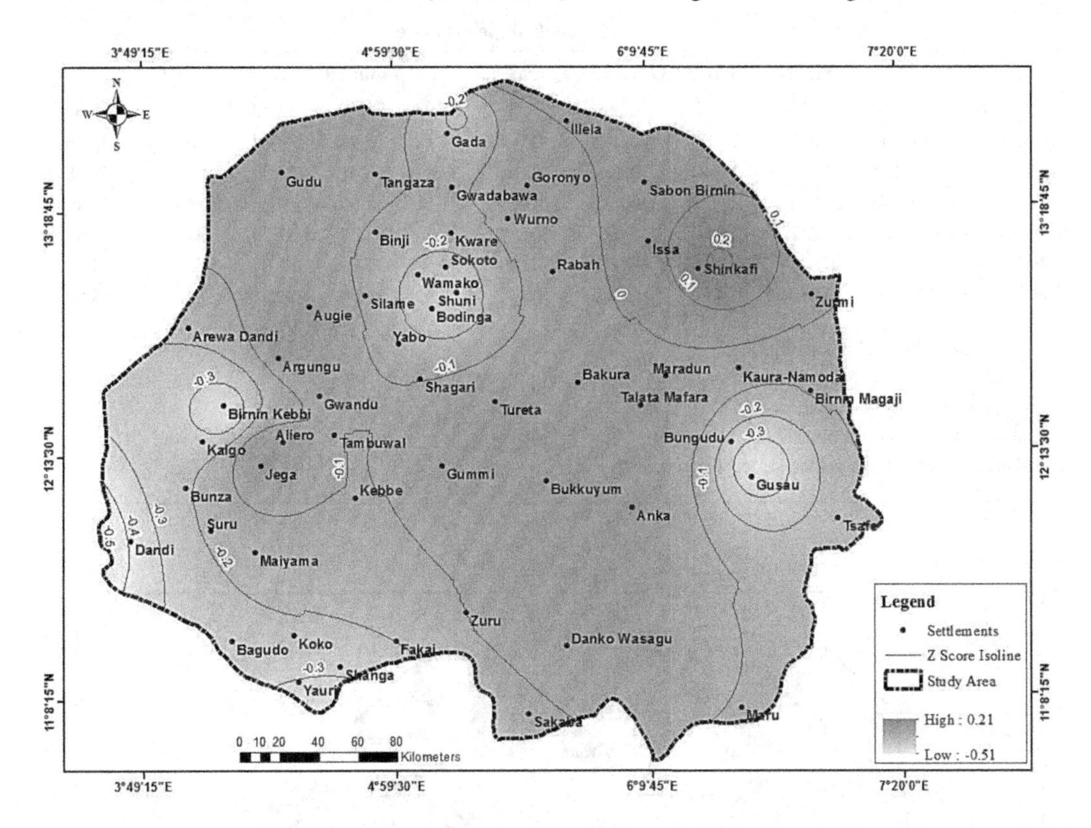

FIGURE 8.12 Spatial distribution of M-K trends (Z-score) for the past climate over the Sokoto-Rima Basin.

TABLE 8.5
Rainfall Mann-Kendall Trend Statistics of the Future Climate (2018–2050) Over the Sokoto-Rima Basin

Time	M-K	Sen's slope estimate									
Series	Test Z	Q	Q min 99	Q max 99	Q min 95	Q max 95	β	β min 99	β max 99	β min 95	β max 95
Jan	−2.08*	−0.021	−0.049	0.005	−0.043	−0.001	33.78	34.31	32.93	34.23	33.15
Feb	−0.76	−0.007	−0.039	0.018	−0.029	0.013	36.22	36.98	35.24	36.62	35.35
Mar	2.21*	0.017	−0.003	0.041	0.001	0.035	38.13	38.79	37.45	38.65	37.63
Apr	4.63***	0.029	0.014	0.043	0.018	0.040	38.93	39.32	38.32	39.25	38.42
May	3.87***	0.038	0.014	0.059	0.019	0.054	36.89	37.53	36.17	37.46	36.51
Jun	3.90***	0.039	0.012	0.063	0.019	0.057	33.76	34.58	33.04	34.36	33.21
Jul	5.15***	0.051	0.024	0.079	0.031	0.070	30.22	30.99	29.48	30.73	29.87
Aug	5.45***	0.044	0.025	0.068	0.030	0.063	28.99	29.52	28.35	29.38	28.46
Sept	5.53***	0.048	0.029	0.068	0.032	0.064	30.85	31.34	30.34	31.24	30.44
Oct	4.87***	0.041	0.022	0.054	0.027	0.050	34.37	35.00	33.82	34.77	34.00

(Continued)

TABLE 8.5 (*Continued*)
Rainfall Mann-Kendall Trend Statistics of the Future Climate (2018–2050) Over the Sokoto-Rima Basin

Time	M-K	Sen's slope estimate									
Series	Test *Z*	*Q*	*Q* min 99	*Q* max 99	*Q* min 95	*Q* max 95	β	β min 99	β max 99	β min 95	β max 95
Nov	0.78+	0.005	-0.013	0.023	–0.009	0.019	35.93	36.70	35.53	36.50	35.60
Dec	–1.00	–0.012	–0.041	0.013	–0.032	0.008	34.12	35.48	33.24	34.98	33.56
MAM	4.23***	0.028	0.012	0.044	0.016	0.040	37.87	38.41	37.32	38.24	37.48
JJA	5.93***	0.043	0.024	0.065	0.028	0.061	31.11	31.60	30.36	31.47	30.49
SON	4.98***	0.031	0.017	0.045	0.021	0.041	33.74	34.08	33.29	33.93	33.41
DJF	–2.07*	–0.015	–0.037	0.004	–0.032	–0.001	34.59	35.28	34.07	35.03	34.22
Annual	4.51***	0.022	0.010	0.036	0.012	0.032	34.32	34.70	33.81	34.60	33.96

***Trend at $\alpha = 0.001$ level of significance.
**Trend at $\alpha = 0.01$ level of significance.
*Trend at $\alpha = 0.05$; level of significance.
+Trend at $\alpha = 0.1$ level of significance.
The blank cell shows the significance level is greater than 0.1.

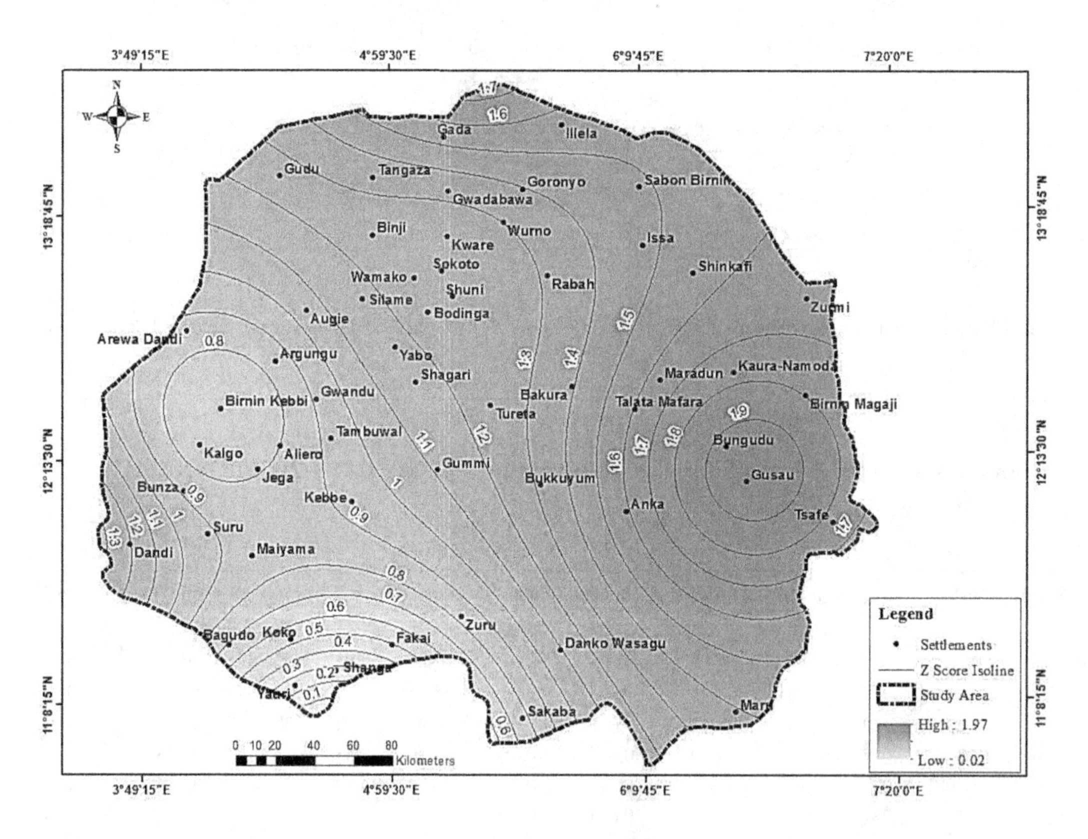

FIGURE 8.13 Spatial distribution of M-K trends (Z-score) for the future climate over the Sokoto-Rima Basin.

8.7 TRENDS DETECTION BASED ON THE WMO STANDARDIZED PERIOD

The WMO establishes a 30-year reference period for the technical analysis of climatic normals and the potential detection of climate change (WMO, 2018). This period was adopted for this study, resulting in the periods 1951–1980 and 1981–2010 for the past climate, and 2018–2047 for the future climate. The results of the temporal trends at the annual and interannual levels are shown in Tables 8.6 and 8.7 for the past climate, and Table 8.8 for the future climate. The period 1951–1980 exhibited a general decreasing monotonic trend, with significant values obtained for January, February, and August, each with a Z-score of –2.783 at a 0.01 level of significance (Table 8.6). June, a crucial month during the wet season, showed the only increasing monotonic trend. At the interannual level, the dry, pre-wet, wet, and post-wet seasons all displayed significant decreasing monotonic trends, indicating an annual persistence of decreasing rainfall amounts during that period. Spatially, a westward and northward shift in the pattern of monotonic trends was observed (Figure 8.14). These shifts were observed in locations such as Gusau, Tsafe, and Bungudu in the east, and Gada and Illeila in the north.

The period 1981–2010, however, exhibited a completely different rainfall trend. As shown in Table 8.7, a general increasing monotonic trend was observed, with significant levels exceeding 0.1. The annual average returned a positive Z-score of 1.213. However, decreasing monotonic trends were detected for May ($Z=-0.178$, $\alpha>0.1$) and September (–0.303, $\alpha>0.1$), indicating a deviation from the

TABLE 8.6
Rainfall Mann-Kendall Trend Statistics for the Period 1951–1980 Over the Sokoto-Rima Basin

Time	M-K	Sen's slope estimate									
Series	Test Z	Q	Q min 99	Q max 99	Q min 95	Q max 95	β	β min 99	β max 99	β min 95	β max 95
Jan	–2.78**	–0.002	–0.003	0.000	–0.003	–0.001	0.19	0.21	0.17	0.20	0.17
Feb	–2.78**	–0.007	–0.014	–0.001	–0.012	–0.003	0.84	0.94	0.75	0.91	0.78
Mar	–1.31	–0.008	–0.071	0.003	–0.045	0.001	0.27	1.55	0.06	1.03	0.09
Apr	–0.34	–0.073	–1.423	0.723	–1.093	0.462	13.25	37.32	0.65	33.41	1.95
May	–0.36	–0.294	–2.595	2.260	–1.988	1.750	61.61	105.66	31.52	93.85	41.46
Jun	0.25	0.155	–2.041	2.561	–1.367	1.886	100.28	127.88	69.65	120.19	80.15
Jul	–0.99	–1.819	–5.582	2.428	–4.298	1.566	212.69	270.97	154.57	252.53	171.49
Aug	–2.28*	–2.998	–6.620	0.442	–5.473	–0.326	286.80	341.54	221.68	320.95	233.73
Sept	–2.39*	–2.438	–5.636	0.334	–5.144	–0.364	174.52	225.63	128.86	216.77	138.93
Oct	–1.52	–0.865	–2.858	0.379	–2.077	0.157	30.23	60.82	9.61	47.94	12.82
Nov	–2.78**	–0.012	–0.023	–0.002	–0.019	–0.004	1.34	1.50	1.20	1.45	1.25
Dec	–2.78**	–0.002	–0.004	0.000	–0.004	–0.001	0.26	0.29	0.23	0.28	0.24
MAM	–0.82	–0.386	–1.483	0.771	–1.204	0.569	32.60	48.11	17.40	44.77	21.39
JJA	–2.18*	–1.642	–4.207	0.348	–3.521	–0.155	201.30	237.36	176.95	223.39	188.52
SON	–2.21*	–1.310	–2.981	0.167	–2.449	–0.163	73.72	101.52	54.42	91.72	58.82
DJF	–3.14**	–0.004	–0.007	–0.001	–0.006	–0.002	0.43	0.47	0.39	0.46	0.40
Annual	–3.14**	–0.881	–1.617	–0.204	–1.404	–0.342	79.00	90.01	68.85	87.21	70.87

***Trend at $\alpha=0.001$ level of significance.
**Trend at $\alpha=0.01$ level of significance.
*Trend at $\alpha=0.05$; level of significance.
+Trend at $\alpha=0.1$ level of significance.
The blank cell shows the significance level is greater than 0.1.

TABLE 8.7
Rainfall Mann-Kendall Trend Statistics for the Period 1981–2010 Over the Sokoto-Rima Basin

	M-K	Sen's Slope Estimate									
Time Series	Test Z	Q	Q min 99	Q max 99	Q min 95	Q max 95	β	β min 99	β max 99	β min 95	β max 95
Jan	1.57	0.001	0.000	0.002	0.000	0.002	0.12	0.17	0.04	0.15	0.06
Feb	1.57	0.003	−0.002	0.011	−0.001	0.009	0.51	0.75	0.19	0.69	0.26
Mar	0.77	0.002	−0.005	0.041	−0.002	0.030	0.03	0.32	−1.46	0.17	−0.97
Apr	0.57	0.104	−0.389	0.569	−0.224	0.422	0.69	25.27	−18.95	17.02	−12.21
May	−0.18	−0.176	−1.952	1.550	−1.404	1.091	49.35	131.63	−28.75	102.95	−11.50
Jun	0.39	0.355	−2.139	2.901	−1.555	2.150	79.93	184.20	−30.90	160.16	1.97
Jul	0.25	0.227	−2.193	2.448	−1.595	1.851	157.28	271.90	56.36	244.13	83.24
Aug	1.36	2.473	−2.278	5.830	−1.254	4.857	86.01	310.71	−53.68	260.02	−9.89
Sept	−0.30	−0.259	−3.082	2.431	−2.174	1.647	116.71	249.23	11.51	204.74	42.44
Oct	1.75^{+}	0.425	−0.481	2.008	−0.076	1.573	1.33	42.32	−66.41	26.72	−49.26
Nov	1.57	0.005	−0.003	0.017	−0.001	0.015	0.82	1.20	0.31	1.10	0.42
Dec	1.57	0.001	−0.001	0.003	0.000	0.003	0.16	0.23	0.06	0.21	0.08
MAM	−0.79	−0.171	−0.893	0.578	−0.676	0.400	29.49	62.21	−5.88	52.59	1.92
JJA	2.03^{*}	1.460	−0.443	2.586	0.067	2.307	92.14	174.48	43.75	150.22	55.05
SON	0.14	0.076	−1.083	1.290	−0.762	0.945	45.64	94.71	−10.01	82.02	5.29
DJF	1.78^{+}	0.001	−0.001	0.005	0.000	0.004	0.27	0.38	0.15	0.34	0.18
Annual	1.21	0.298	−0.461	0.901	−0.245	0.765	44.27	76.47	19.01	66.55	24.05

***Trend at $\alpha = 0.001$ level of significance.
**Trend at $\alpha = 0.01$ level of significance.
*Trend at $\alpha = 0.05$; level of significance.
$^{+}$Trend at $\alpha = 0.1$ level of significance.
The blank cell shows the significance level is greater than 0.1.

TABLE 8.8
Rainfall Mann-Kendall Trend Statistics for the Period 2018–2047 Over the Sokoto-Rima Basin

	M-K	Sen's Slope Estimate									
Time Series	Test Z	Q	Q min 99	Q max 99	Q min 95	Q max 95	β	β min 99	β max 99	β min 95	β max 95
Jan	1.36	0.001	−0.002	0.030	0.000	0.022	0.022	0.06	−0.14	0.04	−0.07
Feb	0.61	0.018	−0.121	0.121	−0.054	0.075	0.920	3.83	0.07	2.52	0.27
Mar	0.32	0.033	−0.397	0.418	−0.284	0.301	4.563	12.43	−0.19	10.39	0.76
Apr	−0.93	−0.268	−1.268	0.533	−0.887	0.329	20.273	38.95	10.11	32.20	11.24
May	0.86	0.619	−0.949	2.814	−0.548	2.393	50.674	70.78	25.51	67.57	31.42
Jun	0.11	0.169	−2.039	2.173	−1.642	1.593	117.316	151.52	87.44	145.84	97.58
Jul	0	−0.013	−2.203	–	–	1.619	179.103	224.94	141.84	213.89	150.64
Aug	1.96^{*}	2.030	−0.909	4.804	0.000	4.131	195.485	230.71	167.36	218.15	175.39

(Continued)

TABLE 8.8 (*Continued*)

Rainfall Mann-Kendall Trend Statistics for the Period 2018–2047 Over the Sokoto-Rima Basin

Time	M-K	Sen's Slope Estimate									
Series	Test Z	Q	Q min 99	Q max 99	Q min 95	Q max 95	β	β min 99	β max 99	β min 95	β max 95
Sept	2.21*	1.453	−0.248	3.330	0.230	2.896	144.393	173.51	116.40	164.76	122.81
Oct	−1.14	−0.405	−1.710	0.795	−1.453	0.414	25.397	56.62	14.18	50.16	16.28
Nov	−0.93	−0.008	−0.329	0.019	−0.266	0.005	0.685	8.18	0.13	6.69	0.42
Dec	0.11	0.000	−0.002	0.011	−0.001	0.003	0.013	0.05	−0.05	0.03	0.
MAM	0.5	0.111	−0.760	1.090	−0.516	0.870	31.225	40.91	16.96	38.10	20.72
JJA	0.96	0.545	−0.776	2.432	−0.500	1.972	171.715	189.42	138.81	186.77	142.14
SON	1.18	0.372	−0.654	1.363	−0.257	1.130	57.042	77.13	46.47	67.57	48.29
DJF	0	0.000	−0.167	0.166	−0.106	0.123	2.550	5.74	0.29	4.52	1.03
Annual	0.93	0.184	−0.317	0.805	−0.194	0.638	64.616	74.60	57.19	72.95	58.07

***Trend at $\alpha=0.001$ level of significance.
**Trend at $\alpha=0.01$ level of significance.
*Trend at $\alpha=0.05$; level of significance.
+Trend at $\alpha=0.1$ level of significance.
The blank cell shows the significance level is greater than 0.1.

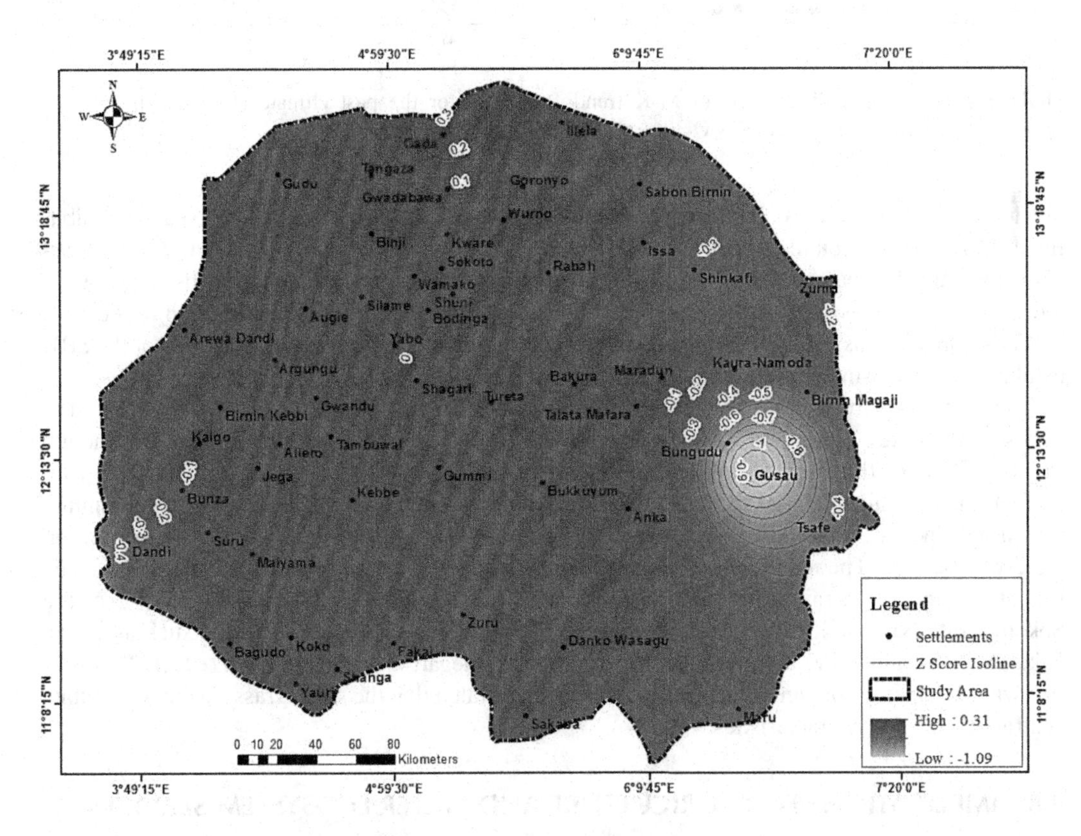

FIGURE 8.14 Spatial distribution of M-K trends (Z-score) for the past climate (1951–1980) over the Sokoto-Rima Basin based on the WMO 30-year data standard.

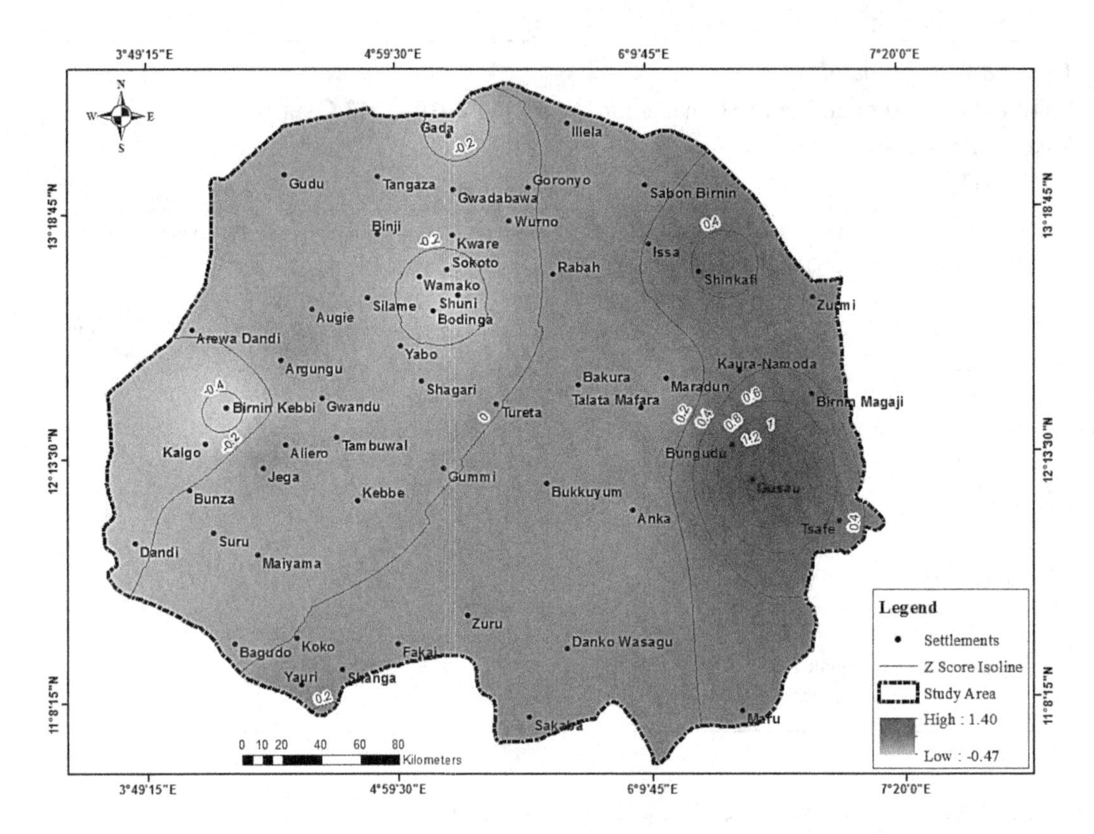

FIGURE 8.15 Spatial distribution of M-K trends (Z-score) for the past climate (1981–2010) over the Sokoto-Rima Basin based on the WMO 30-year data standard.

previous trend. Seasonal trends displayed some fluctuations, with only the pre-wet season exhibiting a decreasing monotonic trend ($Z=-0.785$, $\alpha>0.1$), while the wet ($Z=2.034$, $\alpha=0.05$), post-wet ($Z=0.143$), and dry seasons ($Z=1.784$, $\alpha=0.1$) all exhibited increasing monotonic trends. The magnitude of these trends is supported by the respective Sen's slopes. The spatial distribution of the M-K trend Z-scores for this decade is depicted in Figure 8.15. It shows that approximately 55% of the SRB exhibited an increasing rainfall monotonic trend, particularly in locations in the eastern axis.

The detected increasing monotonic trend for the past climate period (1981–2010) is expected to persist in the future (2018–2047), as suggested by the climate model shown in Table 8.8. Although the magnitude of the trend is lower, it predicts that April ($Z=-0.93$, $\alpha>0.1$), October ($Z=-1.14$, $\alpha>0.1$), and November ($Z=-0.93$, $\alpha>0.1$) will exhibit decreasing monotonic trends. The interannual period will indicate a periodic increasing trend in rainfall, while the magnitude of the dry season trend will be zero. The magnitude of these trends is supported by the respective Sen's slopes. The spatial distribution of rainfall trends (M-K statistics) for this period shows that over 70% of the Sokoto-Rima Basin will experience more rainfall (Figure 8.16). Conversely, areas such as Koko, Fakai, Shanga, and Yauri are more likely to experience a negative rainfall trend at $\alpha<0.1$. This is in line with the findings of Oguntunde et al. (2011), which stated that the short-grass savannah exhibits contrasting rainfall characteristics.

8.8 IMPLICATIONS FOR AGRICULTURE AND WATER ECOSYSTEM SERVICES

In this study, we present an overview of the spatial and temporal variability of rainfall in the SRB. The concise statistical analysis of rainfall climatology demonstrates the influence of latitude on rainfall distribution, with increasing latitude resulting in decreased rainfall amount and

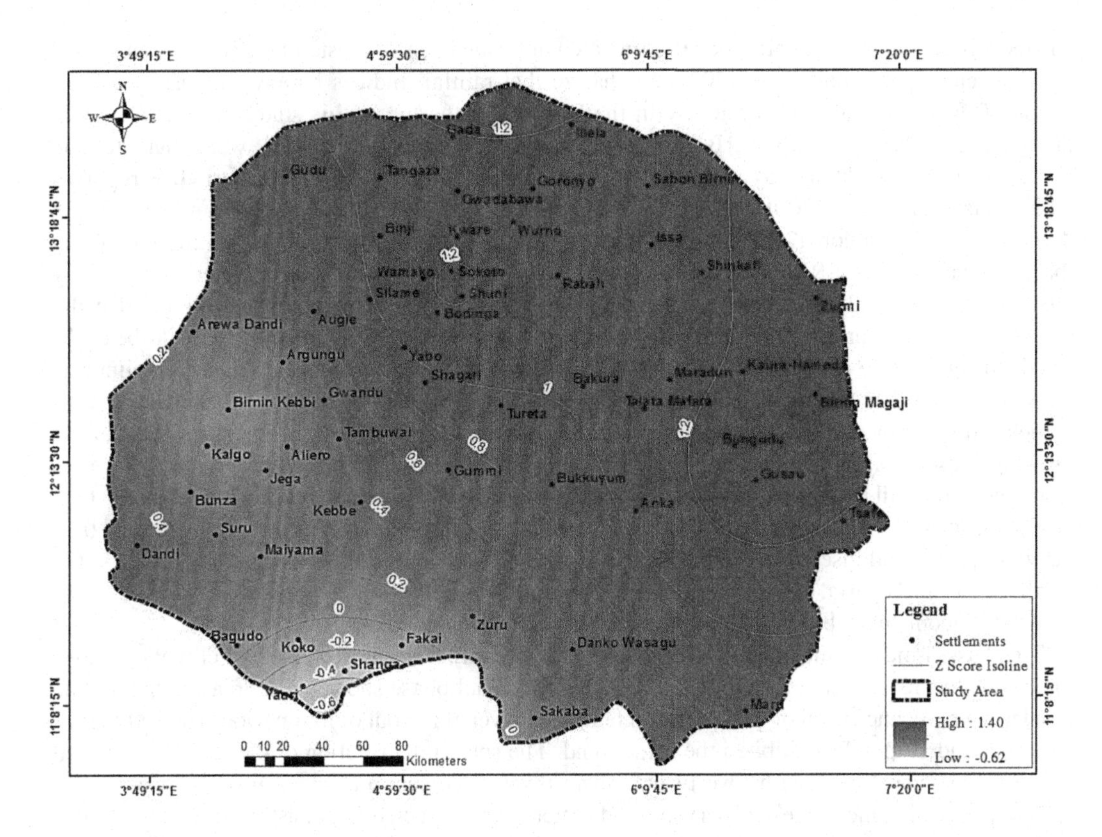

FIGURE 8.16 Spatial distribution of M-K trends (Z-score) for future climate (2018–2047) over the Sokoto-Rima Basin based on the WMO 30-year data standard.

distribution. This is consistent with a study conducted by Oguntunde et al. (2011), which observed a direct latitudinal zonation of rainfall, particularly within the northern zone of Nigeria where the SRB is located. Across the two periods considered in this study, there is a progressive increase in rainfall amount from the past to the future. The projected increase is expected to be 580 mm/year with a mean of ± 148 mm/year. The spatial distribution of these anticipated changes is expected to follow a south-north decreasing pattern, in accordance with the latitudinal influence. Additionally, the coefficient of variation (CV) indicates high variability in rainfall over the SRB across both climatic periods. The spatial pattern of the CV is relatively similar between the two periods, with values ranging from low to high distributed in an eastern-to-northern pattern. These findings pose challenges to the livelihoods of the SRB, which relies on ecosystem services such as agriculture and water resources. Raji et al. (2021) argue that the increasing rainfall amount will lead to intensified floods, emphasizing the need for agricultural practices that consider flood extremes based on local contexts and basin-scale considerations. Prior to this, Akinsanola and Ogunjobi (2014) highlighted the importance of incorporating adaptive and mitigating measures into agricultural practices in rain-sensitive areas of the Sudano-Sahelian landscape in northern Nigeria, where sociocultural activities are directly linked to agriculture.

It is not uncommon to assess the extent of variability and anomalies in rainfall when conducting long-term assessments. In this study, the spatiotemporal assessment of rainfall variability in the SRB revealed a range of 0.96–1.99 during the past climate period, while a decrease is expected in the future with a range of 0.10–0.18. These results are not unique to the SRB alone but are observed in tropical regions worldwide. For example, Jamshadali et al. (2021) investigated rainfall variability in extreme events of the South Asian summer monsoon and found that variability has become less pronounced at the interannual scale, while the contribution of variability to rainfall patterns has increased over time. While global climate change has been suggested as a possible influence on this trend, Jamshadali et al. (2021) emphasize that rainfall indices largely support the observation of decreased rainfall variability in the tropics. Furthermore, this study demonstrates that drought spells observed in the SRB, particularly during the 1971–1980 decade, were characteristic of the Nigerian drought and, to a large extent, the Sahelian drought of the same period, as reported by Nicholson and Grist (2001), Olaniran (2002), Oguntunde et al. (2011), Ekpoh and Nsa (2011), Akinsanola and Ogunjobi (2014), and Animashaun et al. (2020). Animashaun et al. (2020) reported that the drought in the SRB extended southwards to parts of the central area of Nigeria in the early 1980s. The nature and characteristics of drought irregularity were further emphasized by a decadal analysis. On the other hand, it is anticipated that drought occurrence and severity could be replicated during the 2018–2027 period. Drought occurrences and severity can have devastating impacts on water availability for existing agricultural practices in the SRB. Drought is often characterized by delayed onset of rainfall, late start of the rainy season, and instability in the assessment of the growing season, which disrupts the cropping calendar for a specific period. As a result, farmers who rely on rainfall patterns for cultivation and decision-making may face confusion (Okpara et al., 2013). Ultimately, the disrupted growing season and late resumption of farming activities could trigger widespread food insecurity (Oloruntade et al., 2017). Dam operations, which are common in the northern part of Nigeria, also depend on reliable rainfall information. When drought occurs, these activities become unstable, disrupting the established hydrological profile.

Trend assessment at the annual, interannual, and monthly levels during the two climatic periods indicated a decreasing trend. In the past climate, the annual phase showed a decreasing monotonic trend at a significance level of 0.05. At the interannual level, three out of four periods showed a similar trend, and 10 months exhibited the same trend. The spatial distribution of these trends revealed location-specific patterns, with a westward and northward decrease and an eastward increase, albeit with lower trend values. These observed trend patterns are expected to persist in the future climate with location-specific changes in the spatial domain. The dynamics of rainfall trends indicate possible variability and changes in rainfall patterns. Numerous literature sources suggest that trend patterns in the northern semi-arid region of Nigeria could be linked to a series of natural and anthropogenic drivers and forcings, such as the influence of tropical wind systems including the Africa Easterly Jet (AEJ) and the Tropical Easterly Jet (TEJ), the ITCZ, ocean–atmosphere interactions, and anthropogenic factors (Adejuwon and Dada, 2021; Okonkwo et al., 2014; Olaniran, 2002). The Walker circulation, a key component of ocean–atmosphere interaction, including ENSO, has been proposed as a trigger for West African aridity and its associated rainfall trends. It is well-known that aridity resulting from a loss of rainfall has distressing effects on agriculture, not just water resources. Despite the ecological richness of the SRB, the onset of drought leads to a decline in crop yield and livestock production, a reduction in raw materials derived from agricultural products, and disruptions in the overall agricultural food chain. Infestations of insect pests such as locusts and instances of malnutrition may arise, food insecurity triggers an increase in food prices, malnutrition becomes more likely, and the potential for famine emerges. Some of these effects have been documented in parts of the SRB by Adejuwon (2016). The compounded pressures on scarce productive ecosystem services could escalate into resource conflicts, as reported by Apeldoorn (1977), AlaanuloluwaIkhuoso et al. (2020), and Jellason et al. (2020). These findings present a clearly marked rainfall subalternity compared to the tropical and global scenarios.

8.9 CONCLUSION

This study has examined the spatiotemporal variability and trends in rainfall in the Sokoto-Rima Basin during two data periods: 1951 to 2017 and the future projection from 2018 to 2050, resulting in a 100-year study period. Trend, variability, and anomaly tests were conducted across various timescales, including monthly, interannual, annual, and decadal, following the WMO standard. Overall, the study indicates that significant trends observed at each timescale will shift from consistently dry rainfall to lower wet anomalies from the past to future climates. The northward dynamics of variability will remain unchanged, while changes in the spatial distribution of dryness-wetness contexts will be location-specific. These changes have significant implications for the SRB, requiring actions in both research and societal contexts. The dynamics of the spatial distribution of rainfall variability demonstrated that it is 2.5 times stronger than a temporal distribution (from a coefficient of variation of 59%–22.4%). This highlights the importance of considering spatial variability as a crucial component in understanding hydrological processes and dynamics at the basin level. Since the SRB is an agrarian landscape, historical dynamics of cropland and other fundamental land cover types can be utilized to address the gap in ecohydrology. According to Raji et al. (2020), approximately 73% of the arable land in the SRB was dominated by cropland in 2015, and this is expected to increase to 78% by 2050. These land use changes are likely to have a significant impact on the ecohydrology of the watershed. While cropland expansion is expected to contribute to addressing food insecurity in Nigeria through agricultural expansion and increased land productivity, it can also conflict with desertification control measures. It is important to avoid such counterproductive measures while ensuring a thorough understanding of watershed hydrology for sustainability purposes.

The analysis of variability and trends in the rainfall data across the two phases of the study revealed that the SRB has been experiencing periods of drought since the 1970s, 1990s, and early 2000s. These observations indicate the extent of environmental devastation, degradation, and deterioration of ecosystem goods and services in the semi-arid region. Studies conducted by Animashaun (2020), Shiru et al. (2019), Ayanlade et al. (2018), Oguntunde et al. (2011), and Olaniran (2002) have all agreed that long-term variations in the northern axis of Nigeria are influenced by global factors such as the ENSO, the movement of the ITCZ, anomalies in sea surface temperatures (SST), and biogeographic feedback. These factors, along with teleconnections between them, can trigger rainfall anomalies in the SRB and Nigeria as a whole. Therefore, further studies on the interrelationships of these factors in inducing rainfall variability and anomalies within the semi-arid region of West Africa would be beneficial.

ACKNOWLEDGMENTS

The authors appreciate the kind permissions granted by the Princeton Climate Analytics (PCA) of Princeton University for data access. We also salute the efforts of the editors and reviewers for their useful comments.

REFERENCES

Abiodun, B., Lawal, K., Salami, A. and Abatan, A. 2013. Potential influences of global warming on future climate and extreme events in Nigeria. *Regional Environmental Change*, 13, 477–491. https://doi.org/10.1007/s10113-012-0381—7.

Adejuwon, J. O. 2016. Effect of climate variability on school attendance: a case study of Zamfara State in the semi-arid zone of Nigeria. *Weather*, 71(10), 248–254.

Adejuwon, J. O. and Dada, E. 2021. Temporal analysis of drought characteristics in the tropical semi-arid zone of Nigeria. *Scientific African*, 14, e01016. https://doi.org/10.1016/j.sciaf.2021.e01016.

Akinsanola, A. A. and Ogunjobi, K. O. 2014. Analysis of rainfall and temperature variability over Nigeria. *Global Journal of Human-Social Science: B Geography, Geo-Sciences, Environmental Disaster, Management* 14(3), 19.

Akinsanola, A. A. and Ogunjobi, K. O. 2017. Recent homogeneity analysis and long-term spatio-temporal rainfall trends in Nigeria. *Theoretical and Applied Climatology*, 128, 275–289. https://doi.org/10.1007/s00704-015-1701-x.

AlaanuloluwaIkhuoso, O., Adegbeye, M. J., Elghandour, M. M. Y., Mellado, M., Al-Dobaib, S. N. and Salem, A. Z. M. 2020. Climate change and agriculture: the competition for limited resources amidst crop farmers-livestock herding conflict in Nigeria - a review. *Journal of Cleaner Production*, 272, 123104. https://doi.org/10.1016/j.jclepro.2020.123104.

Animashaun, I. M, Oguntunde, P. G., Akinwunmiju, A. S. and Olubajo, B. 2020. Rainfall analysis over the Niger central hydrological area, Nigeria: variability, trend, and change point detection. *Scientific African*, 8, e00419. https://doi.org/10.1016/j.sciaf.2020.e00419.

Apeldoorn, J. V. 1977. Drought in Nigeria. Draft report Vol. 2, Lessons of 1972–1974 disasters, Center for Social and Economic Research, Ahmadu Bello University, Zaria.

Ayanlade, A. J., Nwaezeigwe, O. D., Orimoogunje, O. O. I Olokeogun, O. S. 2021. Rainfall seasonality effects on vegetation greenness in different ecological zones. *Environmental Challenges*, 4, 100144. https://doi.org/10.1016/j.envc.2021.100144.

Ayanlade, A., Radeny, M., Morton, J. F. and Muchaba, T. 2018. Rainfall variability and drought characteristics in two agro-climatic zones: an assessment of climate change challenges in Africa. *Science of the Total Environment*, 630, 728–737. https://doi.org/10.1016/j.scitotenv.2018.02.196.

de Smith, M. J., Goodchild, M. F. and Longley, P. 2018. *Geospatial Analysis: A Comprehensive Guide to Principles, Techniques and Software Tools*. Edinburgh: The Winchelsea Press.

Ekpoh, I. J. and Nsa, E. 2011. Extreme climatic variability in North-western Nigeria: an analysis of rainfall trends and patterns. *Journal of geography and Geology*, 3(1), 51.

Elbeltagi, A., Nasrin, A., Arfan, A., Safwan, M., Ali, M. et al. 2021. Applications of Gaussian process regression for predicting blue water footprint: case study in Ad Daqahliyah, Egypt. *Agricultural Water Management*, 255, 107–152.

Elbeltagi, A., Kumar, N., Chandel, A. et al. 2022a. Modelling the reference crop evapotranspiration in the Beas-Sutlej basin (India): an artificial neural network approach based on different combinations of meteorological data. *Environmental Monitoring and Assessment*, 194, 141. https://doi.org/10.1007/s10661-022-09812-0.

Elbeltagi, A., Attila, N., Safwan, M., Chaitanya, B. P., Manish, K., Shakeel, A. B., József, Z., László, H., János, T., Elza, K., Endre, H. and Csaba, J. 2022b. Combination of limited meteorological data for predicting reference crop evapotranspiration using artificial neural network method. *Agronomy*, 12(2), 516. https://doi.org/10.3390/agronomy12020516.

Elbeltagi, A., Zerouali, B., Bailek, N. et al. 2022c. Optimizing hyper-parameters of deep hybrid learning for rainfall prediction: a case study of a Mediterranean basin. *Arabian Journal of Geosciences*, 15, 933. https://doi.org/10.1007/s12517-022-10098-2.

Fasona, M., Omojola, A., Adeaga, O. and Dabi, D. 2007. Aspects of Climate Change and Resource Conflicts in the Nigeria Savannah. In Report of IPCC/TGICA Expert Meeting on Integrating Analysis of Regional Climate Change and Response Options (pp. 45–55).

Hänsel, S., Schucknecht, A. and Matschullat, J. 2016. The modified rainfall anomaly index (mRAI)—is this an alternative to the standardised precipitation index (SPI) in evaluating 217 future extreme precipitation characteristics? *Theoretical and Applied Climatology*, 123, 827–844. https://doi.org/10.1007/s00704-015-1389-y.

Hao, R. and Yu, D. 2018. Optimization schemes for grassland ecosystem services under climate change. *Ecological Indicators*, 85, 1158–1169.

Incoom, A. B. M., Adjei, K. A. and Odai, S. N. 2020. Rainfall variabilities and droughts in the Savannah zone of Ghana from 1960–2015. *Scientific African*, 10, e00571. https://doi.org/10.1016/j.sciaf.2020.e00571.

Intergovernmental Panel on Climate Change (IPCC). 2014. Climate Change 2014: Synthesis Report. Contribution of Working Groups I, II and III to the Fifth Assessment Report of the Intergovernmental Panel on Climate Change [Core Writing Team, Pachauri RK and Meyer LA (Eds)]. IPCC, Geneva, Switzerland, pp. 151.

Intergovernmental Panel on Climate Change (IPCC). 2007. Climate change 2007: The physical science basis: contribution of Working Group I to the Fourth Assessment Report of the Intergovernmental Panel on Climate Change. Cambridge; New York: Cambridge University Press.

Jamshadali, V. H., Reji, M. J. K., Varikoden, H. and Vishnu, R. 2021. Spatial variability of south Asian summer monsoon extreme rainfall events and their association with global climate indices. *Journal of Atmospheric and Solar-Terrestrial Physics*, 221, 105708.

Jellason, N. P., Conway, J. S., Baines, R. N. and Ogbaga, C. C. 2020. A review of farming challenges and resilience management in the Sudano-Sahelian drylands of Nigeria in an era of climate change. *Journal of Arid Environments*, 186, 104–398. https://doi.org/10.1016/j.jaridenv.2020.104398.

Kumar, A., Giri, R. K., Taloor, A. K. and Singh, A. K. 2021. Rainfall trend, variability and changes over the state of Punjab, India 1981–2020: a geospatial approach. *Remote Sensing Applications: Society and Environment*, 23, 100595. https://doi.org/10.1016/j.rsase.2021.100595.

Mishra, A. P., Khali, H., Singh, S., Pande, C. B., Singh, R. and Chaurasia, S. K. 2021. An assessment of in-situ water quality parameters and its variation with landsat 8 level 1 surface reflectance datasets. *International Journal of Environmental Analytical Chemistry*, 103, 18. https://doi.org/10.1080/03067319.2021.1954175.

Nicholson, S. E. and Grist, J. P. 2001. A conceptual model for understanding rainfall variability in the West African Sahel on interannual and interdecadal timescales. *The International Journal of Climatology*, 21(14), 1733–1757.

Oguntunde P. G., Abiodun, B. J. and Lischeid, G. 2011 Rainfall trends in Nigeria, 1901–2000. *Journal of Hydrology*, 411(3–4), 207.

Ogunrinde, A. T., Oguntunde, P. G., Olasehinde, D. A., Fasinmirin, J. T. and Akinwumiju, A. S. 2020. Drought spatiotemporal characterization using self-calibrating palmer drought severity index in the northern region of Nigeria. *Results in Engineering*, 5, 100088. https://doi.org/10.1016/j.rineng.2019.100088.

Okonkwo, C. Demoz, B. and Tesfai, S. 2014. Characterization of West African jet streams and their association to ENSO events and rainfall in ERA-interim 1979–2011. *Advances in Meteorology*, 2014, 405617.

Okpara, J. N., Tarbule, A. and Perumal, M. 2013. Study of climate change in Niger River Basin. West Africa reality not a myth. In: Singh, B. R. (Ed.). *Climate Change Realities Over Ice Caps, Sea Level and Risk in Tech Rijeka*. London: IntechOpen. http://doi.org/10.5772/55186.

Olaniran, O. J. 2002. Rainfall Anomalies in Nigeria: The Contemporary Understanding. The 55th Inaugural Lecture. University of Ilorin, Nigeria, pp. 55.

Oloruntade, A. J., Mohammad, T. A. and Wayayo, A. 2017. Rainfall trends in the Niger South Basin, Nigeria, 1948–2008. *Pertanika Journal of Science and Technology*, 25(2), 476–496.

Raji, S. A., Odunuga, S. and Fasona, M. 2020. Simulating future ecosystem services of the Sokoto–Rima Basin as influenced by geo-environmental factors. *Turkish Journal of Remote Sensing*, 1(2), 106–124.

Raji, S. A., Odunuga, S. and Fasona, M. 2021. Quantifying ecosystem service interactions to support environmental restoration in a tropical semi-arid basin. *ActaGeophysica*, 69, 1813–1841. https://doi.org/10.1007/s11600-021-00644-z.

Sahoo, M. and Yadav, R. K. 2021. Teleconnection of Atlantic Nino with summer monsoon rainfall over northeast India. *Global and Planetary Change*, 203, 103550. https://doi.org/10.1016/j.gloplacha.2021.103550.

Shiru, M. S., Shahid, S., Chung, E. and Alias, N. 2019. Changing characteristics of meteorological droughts in Nigeria during 1901–2010. *Atmospheric Research*, 223, 60–73. https://doi.org/10.1016/j.atmosres.2019.03.010.

Usman, M. U., Nichol, J. E., Ibrahim, A. T. and Buba, L. F. 2018. A spatio-temporal analysis of trends in rainfall from long term satellite rainfall products in the SudanoSahelian zone of Nigeria. *Agricultural and Forest Meteorology*, 260–261, 273–286. https://doi.org/10.1016/j.agrformet.2018.06.016.

Vizy, E. K. and Cook, K. H. 2018. Mesoscale convective systems and nocturnal rainfall over the West African Sahel: role of the inter-tropical front. *Climate Dynamics*, 50, 587–614.

World Meteorological Organisation (WMO). 2018. *Guide to Climatological Practices*. Geneva: WMO, WMO-No. pp. 100.

9 Climate Change Awareness, Perception, and Adaptation Strategies for Small and Marginal Farmers in Yobe State, Nigeria

Mohammed Bashir Umar

9.1 INTRODUCTION

Climate change is primarily characterized by the occurrence of severe weather, temperature extremes, and altering rainfall patterns (Voccia, 2012). As a result, attempts to address present climate change consequences will necessitate adaptation and mitigation responses from smallholder farmers (IPCC, 2014; Spires et al., 2014). Risk perception and awareness are critical components in adaptive behavior. The former refers to how well potential hazards connected with climate change are understood, whereas the latter is a subjective assessment of those risks (Lechowska, 2018). A number of studies have identified them as important motivators for management techniques and disaster preparedness (Bamberg et al., 2017; Buchecker et al., 2016; Rufat et al., 2020; van Valkengoed and Steg, 2019). However, a number of empirical researches, most of which focused on flood risk, have delved into the origins and uniqueness of risk awareness and perception (Raška, 2015). But, because of the strong influence of context-specific variables, the relevant results are distinct or even contradictory (Attems et al., 2020). Consequently, risk perception is critical in decision-making since all possible outcomes must be anticipated in order to be effectively addressed (Hardaker et al., 2015). Many socio-economic, technical, and institutional factors have been shown to influence adaptation decisions, for instance, asset and land ownership (Hisali et al. 2011), access to credit and extension information (Fosu-Mensah et al. 2012), investments in rain-fed agriculture, and efficient rainfall utilization (Wani et al. 2009). Because climate change hazards are critical, certain adaptation strategies require the acquisition or possession of real assets. However, before adaptation can begin, climate change must be acknowledged; therefore, the importance of perception as a precursor to adaptation cannot be overstated (Tambo and Abdoulaye 2013).

According to Haider (2019), popular climate change adaption techniques in Nigeria are clarified by numerous features aimed at reducing negative effects of climate change. Farmers were given assistance in modifying their habits. However, smallholders are anticipated to be open to effects they have never seen before, according to the Intergovernmental Panel on Climate Change (IPCC) (2007).

As a result, the purpose of this research is to investigate farmers' climate change awareness, perspectives, and adaption techniques in Yobe State, Nigeria:

i. Identify the socioeconomic and institutional features of farmers.
ii. Investigate farmers' views on climate change.

DOI: 10.1201/9781003303237-9

iii. Determine the factors that influence people's views on climate change.
iv. Explain the adaption strategies used by farmers.
v. Identify the challenges that farmers encounter.

9.2 MATERIALS AND METHODS

The methodologies of investigation are described in the following sections.

9.3 STUDY AREA

The study was conducted in the Yobe State. It has a population of two million, three hundred and twenty-one thousand, five hundred and ninety-one (2,321,591) people and a land area of roughly 45,502 square kilometers (km^2). It lies between the latitudes of 12.00°N and 11.30°E (NPC, 2006). The Agricultural Development Project (ADP) divides the state into Zones I and II. The nine (9) villages of Potiskum in Zone II are Bareri/Bauya Lailai, Bolewa B, Danchuwa, Dogo Nini, Dogo Tebo, Hausawa Asibiti, Mamudo, Ngogi/Alaraba, and Yarimaram.

9.4 DATA COMPILATION TECHNIQUES

For data collection, a structured questionnaire and focus groups were used. The primary source consisted of a cross-sectional survey of respondents. Farmers' socioeconomic and institutional characteristics, awareness, perception, and adaptation strategies as well as the constraints faced by farmers in the study area as a result of climate change were all obtained. Secondary data such as reports and published documents were also used to cross-check information provided by farmers.

9.5 SAMPLING PROCEDURES

This study employed a multi-stage random sampling approach. ADP zone II was carefully picked in the early stages of the investigation. Potiskum was chosen as a Local Government Area in the second round (LGA). Nine (9) communities were picked in the third stage. The key component of the fieldwork in 2021 was the fourth stage, which included in-depth data collection. Six hundred and six (606) respondents from ADP Zone II were chosen as the study's population in the fourth stage of the research. A twenty percent (20%) proportionate random sampling was used in the fifth stage of a multi-stage random sampling strategy that required larger samples, generating a sample size of one hundred and twenty (120) respondents (Table 9.1).

TABLE 9.1
Population and Sample Size of Respondents ($n = 120$)

Villages	Sampling Frame	Sample Size (20%)
Bare Bari bauyalailai	80	16
Bolewa B	63	12
Danchuwa	61	12
Dogonini	71	14
Dogotaibo	67	13
Hausawa asibiti	58	12
Mamudo	60	12
Ngogi/Alaraba	81	16
Yarimaram	65	13
Total	**606**	**120**

9.6 ANALYSIS OF DATA

The tools of evaluation were STATA and the Statistical Programs for Social Sciences (SPSS). In order to achieve the objectives, descriptive statistics and multiple regressions were used to investigate awareness, perception, farmers' adaptation strategies, and constraints faced as a result of climate change in the study area.

9.7 RESULTS AND DISCUSSION

9.7.1 The Socio-economic and Institutional Characteristics of Respondents

The initial purpose of the study was to describe the socio-economic and institutional features of the respondents. Awareness, perception, age, gender, marital status, level of education, farm size, farming experience, household size, extension visits, membership of association, and farm income level were all taken into account. The findings are summarized in Table 9.2. According to the table, farmers had an average of 1 year and 5 years of climate change awareness and perceptions, respectively. According to the findings, the average age was 35 years with a mean degree of education of 15 years. The average household had four (4) members, a mean farm size of six (6) acres, and the average farmer had eleven (11) years of farming experience. The mean number of extension visits was one (1) in the previous season (2020–2021), and the average number of farmers' organization membership was one (1). The mean number of years of climate change experience was 5, and the average income level was N 181, 8,917.00 (Table 9.2).

9.7.2 Farmers' Awareness of Changes in Climate Parameters

Respondents were asked to identify their awareness and perception of changes in five environmental characteristics (temperature, rainfall, wind, relative humidity, and cloud cover), which served as the foundation for explaining their impression of climate change. Seventy-four percent (74%) of the respondents ranked change in temperature as the first parameter, 72% ranked change in rainfall, wind (45%), relative humidity (44%), and cloud cover (38%), respectively. However, Smit et al. (1996) observed that other drivers of climate change (economic, cultural, government, technological, and

TABLE 9.2
Distribution of Respondents According to Socio-Economic and Institutional Characteristics ($n = 120$)

Variables	*N*	Mean	*SD*	Min	Max
Perception of CC	120	5.38333	5.06957	0	28
CC awareness		0.86667	0.34136	0	1
Age		34.8333	10.7714	20	75
Gender		1.125	0.33211	1	2
Marital status		0.66667	0.62622	0	5
Level of education		14.8333	3.28053	6	20
Farm size		3.26083	2.14147	2	15
Farming experience		10.5667	8.07604	1	42
Household size		4.025	4.70734	0	32
Membership. of association		0.13333	0.34136	0	1
Extension visit		0.64167	1.22163	0	1
Farm income		181.892	180.764	20	800

Source: Field Survey (2021).

Table 9.3
Distribution of Farmers According to Changes in Climate Parameters ($n = 120$)

Parameters	Frequency	Ranking
Temperature	74	1st
Rainfall	72	2nd
Wind	45	3rd
Relative humidity	44	4th
Cloud cover	38	5th

Source: Field Work (2021).

TABLE 9.4
Distribution of Selected Climate Change Perception Variables According to the Literature

Perception Variable	Reference
Poor fertility of most soils	Swe et al. (2015)
Decrease in arable yield	Ndamani and Watanabe (2015)
Increased drought	Smit et al. (1996); Okonya et al. (2013)
Change in rainfall pattern	Smit and Skinner (2002); Okonya et al. (2013)
Poor humidity/dryness	Okonya et al. (2013)
Increase in temperature/hot	Smit and Skinner (2002); Okonya et al. (2013)
High sunshine intensity	Mehar et al. (2016)
Increased rate of erosion	Nearing et al. (2004)
Flooding	Okonya et al. (2013)
Pests and diseases	Bryant et al. (2000); Brklacich et al. (2000); (Swe et al. 2015)
Loss of biodiversity	Elisha et al. (2017)
Loss of fish resources	Elisha et al. (2017)
Crop-specific information	Abraham and Fonta (2018)
Climate finance	Agomuo et al. (2015); Chukwuone (2015)
Loss of farm income	Amusa et al. (2015); Assan et al. (2018)
Altered composition of herds	Amusa et al. (2015); Assan et al. (2018)
Reduced livelihoods	Amusa et al. (2015); Assan et al. (2018)

environmental influences) may increase, negate, or otherwise modify the impacts of climate on agricultural systems (Tables 9.3 and 9.4).

9.7.3 Climate Change Information for Farmers

According to the poll, the most important source of climate information in the research area was friends and relatives. Friends and relatives are where 22% of respondents acquire their climate change information. Radio/television (16%), Yobe State Agricultural Development Project (YOSADP) (14%), non-governmental organizations (14%), other farmers (13%), personal experience (9%), fellow extension staff (6%), farmer association (4%), meetings/workshops (2%), and books/journals (1%) were all mentioned as sources of climate information. This agrees with Smith and Leiserowitz (2012), who claim that efforts to illustrate climate change hazards may be futile since climate change risk perceptions are mostly driven by worldview rather than scientific thinking.

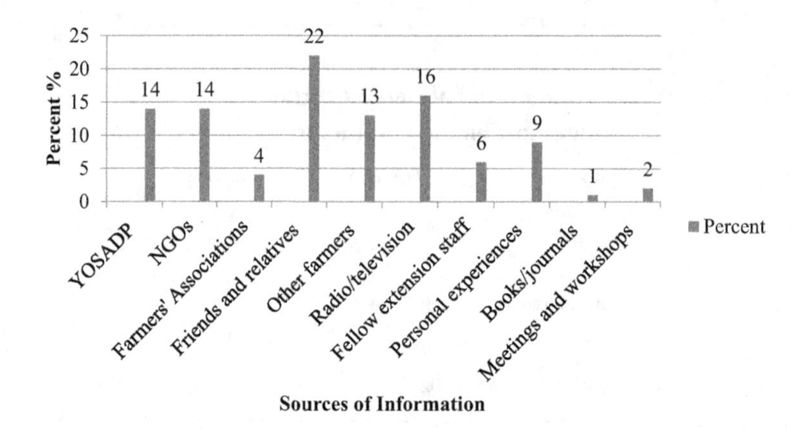

FIGURE 9.1 Utilized sources of climate change information in Potiskum ($n = 120$). YOSADP, Yobe State Agricultural Development Project; NGOs, non-governmental organization.

As a result, if climate change information is spread through these channels, there is a good chance that it will be believed. According to the findings, traditional methods of disseminating information could be employed to enhance climate change awareness. However, in order to lessen the influence of worldviews on how people perceive the hazards associated with climate change in the research region, it is necessary to push the agenda in order to promote climate science in the media (Figure 9.1).

9.8 CLIMATE CHANGE AND FARMERS' PERCEPTIONS

The results showed that 30% of respondents consider that poor soil fertility emanates from climate change in the area. This confirm the findings of Horel et al. (2015) and Pande et al. (2022) that soil adsorption, water, heat, and mass mobility, nutrient delivery, biological activity, and other chemical and biological processes have a significant effect on soil fertility. As a result, this finding supports the theory that soil physical characteristics are significantly linked to their climate change sensitivity. Climate variables such as seasonal temperatures and precipitation intensities have a big impact on soil hydro-physical characteristics. Horel et al. (2014) go on to say that a variety of soil physical processes, such as rising temperature and carbon dioxide (CO_2) concentrations, changes in rainfall patterns, and their interactions, are likely to make soils vulnerable to salinization, decreased water and nutrient availability, altered C and N dynamics, and a loss of soil biodiversity. As a result, climate change is predicted to have an impact on these, as Benbi and Kaur have observed (2009). Mills et al. (2014) also discovered that soil moisture stress affects plant development by lowering soil performance. According to thirteen percent (13%) of respondents, climate change is the cause of pests and diseases in the area. This is in line with the findings of Stöber et al. (2017) and Orimoloye et al. (2022), who found that drought and excessive heat promote wilting, attract pests, and diminish yields. This could be due to excessive moisture levels induced by frequent and heavy rainstorm events, which provide ideal conditions for diseases to thrive, damaging crops and causing nutrient leaching (Stöber et al., 2017). Also, according to Datta (2013), more harsh weather conditions may make certain illnesses (such as rust and powdery mildews) more sporadic and stimulate those that proliferate quickly in warm settings.

Farmers in the study area (8%), for example, have noticed that temperatures have gone up. This finding is consistent with a study that suggests farmers have observed a temperature increase (Fadina and Barjolle, 2018). Similarly, since the 1980s, Nigerian temperatures have been well above average, with record highs in 1973, 1987, and 1998 (Federal Ministry of Environment, 2014). In the country's numerous ecological zones, temperature increases of 0.2°C–0.3°C have been reported every decade (Enete, 2014). Further, the minimum temperature in the country has increased slightly quicker than the national average (Federal Ministry of Environment, 2014).

As a result, 8% of farmers examined in the area reported seeing a change in rainfall patterns. Many investigations, including those by Dhanya and Ramachandran (2016), support similar findings. According to these authors, farmers have faced less rain, shorter rainy seasons, and unexpected and changeable rainfall patterns. Several studies (Ochenje et al., 2016) have found comparable results, corroborating farmers' perceptions of rain being delayed, reduced, and ending early. According to the survey, 8% of respondents have noticed an increase in dry spells. This finding is consistent with farmer sentiments in Zambia and Benin, according to Fadina and Barjolle (2018), who found that farmers were experiencing lengthy and frequent dry times. Furthermore, 6% of farmers in the study area attribute the drought to climate change. This research backs up farmers' long-held concerns about climate change in the Sahel and Benin. Similarly, Sanogo et al. (2016) and Mkonda and He (2017) found that extreme droughts have been more often in recent years. Crop yields had decreased in roughly 5% of the cases. According to Thornton et al. (2013), this can be accomplished by changing agricultural inputs and how farmers use them. Adaptation tactics may be able to compensate for productivity decreases that are expected. As a result, reduced yields are caused by high temperatures and low soil moisture, which disrupt several physiological and biochemical processes.

Flooding has been witnessed in the study region by 5% of the respondents. This is in line with previous studies that farmers in nations like Ethiopia, Zambia, Senegal, and Nigeria have also reported hail stones, flooding, and regular droughts (Mengistu, 2011; Nyanga et al., 2011). Farmers are experiencing more floods, excessive rainfall, strong showers, and hailstorms, according to authors. According to four percent (4%) of the respondents, soil erosion has increased in the area. This supports the findings of Angima et al. (2003), who found that soil erosion accounts for more than eighty (80%) percent of current agricultural land degradation. Consequently, rapid erosion increases the risk of soil depletion, endangering natural resources and the ecosystem (Rahman et al., 2009). In the study area, three percent (3%) had seen a decline of biodiversity. This supports the findings that species exploitation, land degradation, nitrogen deposition, pollution, introduction of invasive or alien species, water diversion, landscape fragmentation, urbanization, and industrialization are all examples of man's impact on biodiversity (Penuelas et al., 2020). Also, anthropogenic global shifts in biospheric N and P concentrations and ratios and their impacts on biodiversity, ecosystem productivity, food security, and human health (Penuelas et al., 2020; Pilling et al., 2020; Webb et al., 2018). However, the interplay of CC with pre-existing dangers to the biota has recently emerged as the most serious and pressing issue. Similarly, two percent (2%) of the respondents are aware of a lot of sunshine (Figure 9.2).

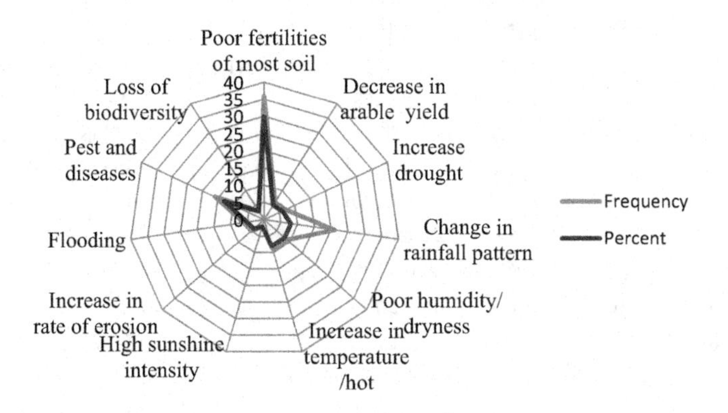

FIGURE 9.2 Web diagram of farmers' perception.

9.9 FACTORS THAT INFLUENCE FARMERS' PERCEPTIONS OF CLIMATE CHANGE

Using a theory-based approach, multiple regression analysis was used to see how socio-economic and institutional factors can explain the perception of climate change. According to regression analysis findings, farmers' perceptions of climate change and variability are strongly influenced by four characteristics (awareness 4.75428, level of education –0.2136, farming experience 0.18332, and household size 0.53439) (Table 9.5). Farmers' awareness is expected to rise as a result of a positive link between climate change and perceptions of variability. The data demonstrate a positive and strong correlation between climate change awareness and perception in the research area (Coef. 4.75428). This is consistent with the findings of a study by Lechowska (2018) that looked at the factors that influence risk awareness and perception (with a focus on flood hazards). The relevant results are varied or even contradictory due to the significant influence of context-specific factors (Attems et al., 2020; van Valkengoed and Steg, 2019). It is assumed that perceptions of climate change and variability will improve if there is a positive relationship between farmers' level of education and perceptions of climate change and variability. In the research area, the findings show a negative and significant relationship between educational achievement and climate change perception (Coef. –0.2136, Sig. 0.041**). But, in some studies (Jamshidi et al., 2018; Liu et al., 2018; Qasim et al., 2018), higher risk perception is linked to higher levels of education, while in others, it is linked to lower levels of education (Jamshidi et al., 2018; Liu et al., 2018; Barret and Bosak, 2018;

TABLE 9.5
Multiple Regression Analysis of Relationship between Socio-Economic and Institutional Characteristics of Farmers with Climate Change Perception (n = 120)

						Lower	Upper
Dependent Variable	***N***	**Coef.**	**Std. Err.**	***T*-Value**	***P*>|*t*|**	**Limit**	**limit**
Perception of CC	120	3.13569	2.29482	1.37	0.175	–1.414	7.68539
			Independent variables				
CC awareness		4.75428	0.99623	4.77	0***	2.77917	6.7294
Age		–0.0259	0.06084	–0.43	0.672ns	–0.1465	0.09476
Gender		–0.5174	1.03175	–0.5	0.617ns	–2.563	1.52814
Marital status		–0.1737	0.61888	–0.28	0.78 ns	–1.4007	1.05329
Level of education		–0.2136	0.10304	–2.07	0.041**	–0.4179	–0.0093
Farm size		–0.159	0.15866	–1	0.319 ns	–0.4735	0.15561
Farming experience		0.18332	0.06954	2.64	0.01**	0.04544	0.3212
Household size		0.53439	0.10786	4.95	0***	0.32055	0.74822
Membership. of association		–0.3269	0.97711	–0.33	0.739 ns	–2.2642	1.6103
Extension visit		–0.0747	0.402	–0.19	0.853 ns	–0.8717	0.7223
Farm income		–0.0006	0.00201	–0.29	0.772 ns	–0.0046	0.00339

Source	SS	df	MS		
Model	1754.5	13	134.962	Number of obs	120
Residual	1303.86	106	12.3006	F(13, 106)	10.97
Total	3058.37	119	25.7006	Prob > F	0
				R-squared	0.5737
				Adj R-squared	0.5214
				Root MSE	3.5072

*$p<0.05$, **$p<0.01$, ***$p<0.001$ (*ns* = Not Significant).

Roder et al., 2016). Also, a positive association between farming experience and perceptions of climate change and variability is likely to improve perceptions of climate change and variability. The analysis shows a positive and robust relationship between farming experience and climate change perception (Coef. 0.18332, Sig.0.01**) in the study area. A prior study revealed that elderly farmers are uninterested in and unwilling to adapt to climate change in general, lending credence to this claim (Uddin et al., 2014). Direct personal experience of harm caused by a natural-hazard event, on the other hand, raises risk perception for rapid-onset disasters (Sherpa et al., 2019).

A positive relationship between farmers' perceptions of climate change unpredictability and family size is likely to improve farmers' awareness of climate change and its unpredictability. Household size and climate change attitudes have a positive and substantial relationship in the studied region (Coef. 0.53439, Sig. 0***). This is in line with earlier research, which implies that larger households' apparent proclivity to adapt to climate change is due to their greater labor endowment (Oyekale and Oladele, 2012).

9.10 FARMERS' CLIMATE CHANGE ADAPTATION

Despite the lack of a defined definition, some authors define climate change adaptation as changes to industrial processes or practices in order to reduce, protect, or resist dangers (Carman and Zint, 2020; Shahid et al. 2021). As a result, the literature offers a variety of adaptation tactics and practices, such as shifting sowing and harvesting dates, switching crops, introducing effective irrigation equipment, applying new temperature control technologies, and migrating businesses, to name a few (FAO, 2017). Adaptation techniques have also been divided into numerous groups. To separate reactive (response after the shock) from proactive (anticipatory) adaptation, Robert, Thomas, and Bergez (2016) use time, temporal, and geographical scopes. Crop diversification, the use of drought-tolerant and early maturing crop varieties, and crop cover are only a few of Nigeria's agricultural activities for various ecological zones (Amadi and Udo, 2015). According to studies, Nigerian farmers are already adapting to climate change on their own and with the help of government and other intervention organizations by using these and other strategies (Ifeanyi-obi and Nnadi, 2014). On the other side, 28 percent of respondents used good practices such as using improved crop varieties. This was consistent with findings demonstrating that smallholder farmers in Nigeria, Senegal, Burkina Faso, and Ghana had looked into drought-resistant crop varieties as climate change adaptation strategies (Ngigi, 2009).

Climate change awareness campaign/workshops were used to increase adoption of existing and new technology for adapting to climate change and variability (18%). Furthermore, agricultural extension services are critical for increasing agricultural output by supplying farmers with relevant farming and weather-related information as well as skill training (Federal Ministry of Environment, 2014). In Nigeria, the existing irregularity of extension services is a barrier to agricultural adaptation (Federal Ministry of Environment, 2014). Numerous studies, including those conducted in India, Kenya, and Burkina Faso, have found substantial evidence linking extension services with agricultural practice awareness and knowledge, according to Evenson (1997). Another approach for adaptation is to reduce deforestation (11%). This is in accordance with the fact that the country's forest reserves are not immune to the threat posed by climate change. It is worth noting that 11 of the 36 states in the country, referred to as frontline states, are being gradually engulfed by desertification. Deforestation is one of the world's most significant issues today, affecting human survival, welfare, and development (Abdullahi et al., 2017). Deforestation is one of the world's most pressing development issues. It is, in particular, the world's most critical long-term environmental concern, and Nigeria is no exception (Ogunwale, 2015). Nigeria has the world's greatest yearly primary forest destruction rate of 55.7%. Nigeria has one of the biggest annual natural forest losses in Africa. With an annual rate of 11.1%, the country is on track to lose practically all of its main forest in the next years. It is a significant issue that affects many sections of the country, with the north bearing the brunt of it (Abdullahi et al., 2017).

Farmers used an integrated farming strategy and promoted biodiversity (8% each). This supports the findings that an integrated farming system (IFS) blends animal and crop agricultural systems in which animals eat agrarian by-products and their bodies are used to cultivate land and provide manure for fertilizer and fuel (Jayanthi et al., 2000; Pande et al. 2021). According to Radhamani et al. (2003), IFS strives to increase production and profits while lowering risks by utilizing organic waste and crop leftovers properly. Traditional agricultural techniques, according to Edwards (2003), have resulted in economic difficulties such as crop exploitation, increased energy-related input prices, and reduced farm profitability. It also resulted in ecological issues such as a lack of diversity, contamination of soil and water, and soil erosion. As a result, implementing integrated agricultural production systems that utilize fewer inputs like fertilizers, chemicals (pesticides), and cultivations can help solve these economic and environmental concerns. The IFS, according to Agbonlabor et al. (2003), is a mixed farming system in which agricultural and animal components are integrated in a complimentary and supplementary way.

Biological diversity on the other hand is essential for the survival of ecosystem processes that provide a variety of services that keep life on Earth afloat (Manley, 2008). Biological diversity and ecosystems are both impacted by climate change. Climate change is already wreaking havoc on the world's biodiversity, forcing managers and planners to reconsider how they plan for species and ecosystem protection (Rose and Burton, 2009). Because a natural ecosystem is usually specialized to certain climatic conditions in a specific place, changes in the natural ecosystem in a changing environment are immediately noticeable. Ecosystems are key parts of the climate system because they help regulate the temperature by sequestering greenhouse gases (mitigation through emission reductions) and regulating water flow which can help with flood and drought adaptation. Therefore, ecosystems should be a fundamental component of climate change initiatives (Doswald and Osti, 2011). According to studies (UNEP/CMS, 2006), variations in local or regional climate, weather patterns, and subsequent changes in vegetation and habitat quality all affect changes in the distribution and behavior of a large number of species.

As a result of the expected climate changes, the extinction of rare and endangered plant and animal species would increase. Forest crops, economic trees, and animal habitat could all be destroyed as a result of adverse environmental change. This could have an impact on people's livelihoods, as many villages in the region rely on forest resources for survival (Dadiowei, 2009). Climate change is already hurting species and ecosystems, according to global research, and it will continue to do so in the future (Yates et al., 2009). Eight percent (8%) of farmers utilized conservation agriculture as a climate change adaptation strategy in the study area. Previous findings of Lema and Majule (2009) confirm this result that farmers time various farming activities properly, such as incorporate crop residues to augment soil fertility, burning crop residues to improve quick release of nutrients, and let livestock to graze on farmlands to improve the soil organic matter after harvests.

In Senegal and Burkina Faso, farmers increased the tree density by using traditional trimming and fertilization procedures, enhancing their adaptability. This contributes to soil stabilization and desertification reversal. According to Nyong et al. (2007), local farmers in the Sahel conserve carbon in soils by utilizing zero-tilling methods in cropping, mulching, and other soil management strategies. Natural mulches help to regulate soil temperatures and extremes, manage disease and pests, and save moisture. Prior to the advent of commercial fertilizers, local farmers relied mainly on organic farming, which helps to reduce GHG emissions. Furthermore, 4percent of the farmers in the study area practiced timely weeding. Crop yields are harmed by climate change because long-term variations in rainfall patterns influence cropping patterns and operation schedules. Some were sowed a few days after the first rains, while others were planted right after the first rains. Tilling of the land begins in fields that were planted prior to cultivation in the third week after the onset of rain, allowing early geminating weeds to be eradicated and weeding to be reduced. These were done on purpose to disperse risk by ensuring that any rain that fell on the dry area was used to the maximum extent feasible by the crop planted there (Liwenga, 2003).

Farmers, like everyone else, relocate if they do not have a choice (3%). According to some studies, migration is a prominent mode of labor (seasonal migration) and a major source of money. The role of remittances created by migration provides a significant coping mechanism in both drought and non-drought years, but it is one that can be significantly impacted by periods of climatic shock when fundamental items like food prices are altered by food aid and other acts (Devereux and Maxwell, 2001). Climate migration is (and has always been) an important tool for combating climate change's effects. As a standard response to drought, pastoral civilizations have relocated their animals from water sources to grazing fields. However, it is becoming clear that migration is not limited to nomadic societies as a response to environmental change. According to studies, an older male family member traveling to Khartoum in search of paid labor to tide the family over until the drought passes is one adaptive reaction to drought in western Sudan (McLeman and Smit, 2004). Many places have previously seen temporary migration as an adaptive response to climatic stress. The truth is a little more complicated; one's ability to relocate is determined by mobility and wealth (both financial and social). To put it another way, the people who will be most affected by climate change are not always the ones who will migrate. A recent study in West Africa's Sahel region looked at seasonal migration as a climate change adaptation mechanism. Drought has been a problem in the region for much of the previous three decades, and one way families have dealt with it is by sending their young men and women away after each crop in search of wage labor. However, the extent to which they relocate is determined by the harvest's success (Oli, 2008).

Farmers in the research area occasionally adopt the effective use of genetic resources (3%) as a climate change adaptation technique. A range of farm-level reactions are among the most important and direct current adaptations to climate variability, according to the literature. Crop and livestock variety diversification, including the replacement of plant types, cultivars, hybrids, and animal breeds with new varieties with higher drought or heat tolerance, has been advocated as a way to boost productivity in the face of improved soil management practices and temperature and moisture stresses (Baker and Viglizzo 1998). Seed genetic diversity has long been recognized as an excellent defense against a variety of threats, including disease and insect outbreaks, as well as, most crucially, climate change. Mortimore et al. (2000) discovered that farmers in Nigeria utilized 3–12 types of pearl millet, 6–22 variations of sorghum, and 14–42 varieties of other cultivars in a study of adaptations.

Seed inventories came from a variety of places, including inheritances, personal selections from planted material, and imported varieties with known climate benefits over local types. Direct seed transfers from extension agents were unusual, according to the scientists; however, some were traceable to those developed in agricultural stations in Nigeria or adjacent Niger. Out-crossing appears to have been the predominant means of dispersal, with farmers selecting from kinds produced on neighboring fields or even in the wild and storing the seeds for planting in subsequent years. Farmers manage their own genetic pool by picking and saving the best seeds from each year's crop, according to the evidence from millet seed selection.

Farmers employ a strategy of planting drought-resistant crop varieties (3%). This was in line with results indicating smallholder farmers in Nigeria, Senegal, Burkina Faso, and Ghana had explored drought-resistant crop types as climate change adaptation approaches (Ngigi, 2009). Nomadic pastoralists living on Kenya's desert fringes have also embraced drought-prevention methods (Langill and Ndathi, 1998). Land usage is influenced in Southern Africa, for example, leading to land conversion, such as the move from cattle to game farming (Ziervogel et al., 2008). Food crops have replaced cash crops in Western Sudan's Kordofan and Drafur states, and hardier crop varieties have been introduced (DFID, 2004). Farmers in Tanzania diversify their crop kinds to spread risk on the land (Adger et al., 2003). Crop diversity helps protect farmers from unpredictable rainfall. Concerning adopting agroforestry practices, only 2% of the farmers employ it as an adaptation strategy. As a practical land-use technique for establishing a healthy balance of food production and forest preservation, Adesina et al. (1999) pointed that most African rural farmers are planting trees to mitigate climate change. Also, according to Adesina (1988), shade-tolerant crops, such as

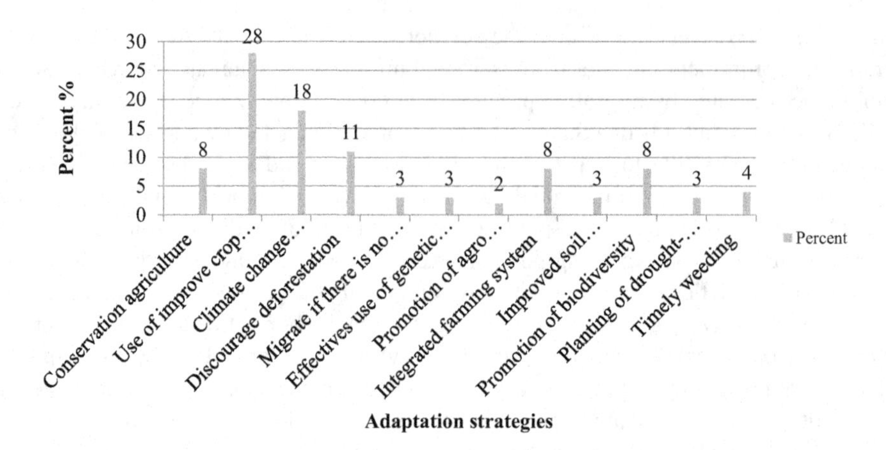

FIGURE 9.3 Farm practices of farmers.

Dioscorea spp. and cocoyam, are produced in an almost permanent forest setting of Nigeria's south western area. Therefore, this technique can be developed to cope with the new conditions of a drier climate, increased human density to help increase the amount of organic matter in the soil, thereby increasing agricultural production and reducing forest strain (Figure 9.3).

9.11 CONSTRAINTS FACING FARMERS AS A RESULT OF CLIMATE CHANGE

As an evolutionary process, adaptation refers to the perception of coping with the repercussions of climate change in order to become more suitable to habitats. The issues of agricultural climate adaptation in Nigeria are discussed in this section. Adaptation tactics used by farmers, on the other hand, are limited by country-specific constraints. According to Satishkumar et al. (2013), farmers encounter individual, institutional, and technical issues. Property ownership fragmentation, low literacy, and a lack of awareness on how to manage or create resilience and traditional attitudes are individuals' hurdles. Lack of access to extension services, information sources, and credit, on the other hand, are institutional impediments. Lack of drought-tolerant varieties, lack of access to weather forecasting data, and dependency on monsoon rains are all contributing factors.

Farmers' actions to mitigate the effects of climate change in Northern Ghana were ranked by Fagariba et al. (2018). According to the paper, the most major hindrance to climate change adaptation strategies is unpredictable weather. Other obstacles included a lack of government support, a lack of access to weather data, property ownership issues, high input costs, insufficient extension personnel, a lack of formal education, and poor soil fertility. Otitoju and Enete (2016) divided the hurdles to food crop farmers employing adaptation strategies in South-western Nigeria into four categories. Land, neighborhood customs, and religious beliefs restrictions; high input costs, technology, and lack of information on early warning systems constraints; and farm distance, insufficient access to climate change information.

Changes in rainfall (29%), poor seed germination (13%), poor land (8%), pests and diseases (8%), deforestation (8%), temperature (8%), labor (6%), erosion (5%), lack of fertilizer (4%), lack of improved seeds (4%), and wind (3%), flooding (3%) were among the top farming problems listed by farmers in the area (3%). This was also reported by Ekwe et al. (2011) and Ozor and Nnaji (2011), who found that between 2006 and 2009, Liberia had more farming problems (including disease incidence, weed infestation, soil infertility, overflowing/drying up of streams/rivers, low farm yields, and landslides) than Nigeria and Sierra Leone. They also noted that insufficient access to information essential to adaptation, a lack of financial resources, poor/low extension services, and a lack of access to weather forecasts were all obstacles in Nigeria, Sierra Leone, and Liberia, respectively (Table 9.6).

TABLE 9.6
Distribution of Farmers Based on Major Constraints Faced Due to Climate Change (n=120)

Type of Problem	Frequency	Percent
Poor germination of seed	15	13
Lack of fertilizer	5	4
Poor land	9	8
Erosion	6	5
Pests and diseases	10	8
Deforestation	10	8
Temperature	10	8
Change in rainfall	35	29
Labor	7	6
Wind	4	3
Flooding	4	3
Lack of improved seeds	5	4
Total	**120**	**100**

Source: Field Work (2021).

9.12 CONCLUSION AND RECOMMENDATION

While socioeconomic and institutional factors are essential determinants of climate change adaptation, this paper indicated that there are other key determinants, such as climate change awareness (4.75428), level of education (–0.2136), farming experience (0.18332), and household size (0.53439) that can compensate for the limitations of other socio-economic and institutional factors. Farmers' answers and field observations show that climatic parameters have been varying, implying that the climate is changing. However, an assessment of present farming household activities revealed some continuous good adaptation behaviors, despite the fact that several critical adaptation activities were underutilized. Using descriptive and multiple regression analysis, the study looked into characteristics related to climate change awareness, perception, and adaptability in the research area. More research is needed to corroborate this, though. Multiple regression models were used to explain the variation (0.5214). The paper's finding has policy and practice consequences. The findings will aid policymakers in designing programs and extension initiatives to encourage people to adapt to climate change, hence reducing climate change's impact on agricultural household livelihoods. Farmers should be made more aware of the importance of employing long-term adaptation strategies that are appropriate for managing conditions in the research area for practitioners.

ACKNOWLEDGMENT

Appreciation to all authors consulted and cited.

REFERENCES

Abdullahi, A., Girei, A.A., Usman, I.S., and Abubakar, M.G. 2017. Assessment of adaptation strategies for deforestation and climate change: implication for agricultural extension system in Nigeria. *International Journal of Innovative Agriculture and Biology Research*, 5, 2, 11–17.

Adesina, F.A. 1988. Developing stable agroforestry systems in the tropics: an example oflocal agroforestry techniques from south western Nigeria. *Discussion Papers in Geography*, 37, 27.

Adesina, F.O., Siyambola, W.O., Oketola, F.O., Pelemo, D.A., Ojo, L.O., and Adegbugbe, A.O. 1999. Potentials of agroforestry for climate change mitigation in Nigeria: somepreliminary estimates. *Global Ecology and Biogeography*, 8, 163–173.

Adger, W.N., Huq S., Brown, K.B., Conway, D., and Hulme, M. 2003. Adaptation to climate change in the developing world. *Programme and Development Studies*, 33, 179–195.

Agbonlabor, M.U., Aromolaran, A.B., and Aiboni, V.I. 2003. Sustainable soil management practices in small farm of Southern Nigeria: a poultry-food crop integrated farming approach. *Journal of Sustainable Agriculture*, 22, 51–62.

Amadi, S.O., and Udo, S.O. 2015. Climate change in contemporary Nigeria: An empirical analysis of trends, impacts, challenges and coping strategies. *IOSR Journal of Applied Physics*, 7, 2, 1–9.

Angima, S.D., Stott, D.E., O'Neill, M.K., Ong, C.K., and Weesies, G.A. 2003. Soil erosion prediction using RUSLE for central Kenyan highland conditions agriculture. *Ecosystems and Environment*, 97(1–3), 295–308.

Attems, M.S., Thaler, T., Genovese, E., and Fuchs, S., 2020. Implementation of property-level flood risk adaptation (PLFRA) measures: choices and decisions. *Wiley Interdisciplinary Reviews: Water*, 7, 1, e1404.

Baker, B., and Viglizzo, E.F. 1998. Rangeland and Livestock In: Feenstra, J.F., Burton, I., Smith, J.B., and Tol, R.S.J. (eds.) *Handbook on Methods for Climate Change impact Assessment and Adaptation Strategies*. Amsterdam: UNEP.

Bamberg, S., Masson, T., Brewitt, K., and Nemetschek, N. 2017. Threat, coping and flood prevention – a meta-analysis. *Journal of Environmental Psychology*, 54, 116–126.

Barrett, K., and Bosak, K. 2018. The role of place in adapting to climate change: a case study from Ladakh, Western Himalayas. *Sustainability*, 1, 898.

Benbi, D.K., and Kaur, R. 2009. Modeling soil processes in relation to climate change. *Journal of Indonesian Society of Soil Science*, 57, 433–444.

Buchecker, M., Ogasa, D. M., and Maidl, E. 2016. How well do the wider public accept integrated flood risk-management? An empirical study in two Swiss Alpine valleys. *Environmental Science and Policy*, 55, 309–317.

Carman, J., and Zint, M.T. 2020. Defining and classifying personal and household climate change adaptation behaviors. *Global Environmental Change*, 61, 102–162.

Dadiowei, T. 2009. Environmental impact assessment and sustainable development in the Niger Delta: the Gbarain oil field experience. *Niger Delta Economies of Violence Working Paper*, 24, 1–50.

Datta, S. 2013. Impact of climate change in Indian horticulture a review. *International Journal of Science, Environment and Technology*, 2, 4, 661–667.

Department for Food and International Development (DFID). 2004. Adaptation toclimate change: the right information can help the poor to cope. Global and localenvironment team, policy division.

Devereux, S., and Maxwell, S. 2001. *Food Security in Sub-Saharan Africa*. ITDG Publishing, London.

Dhanya, P., and Ramachandran, A. 2016. Farmers' perceptions of climate change and the proposed agriculture adaptation strategies in a semi-arid region of South India. *Journal of Integrative Environmental Sciences*, 13(1), 1–18.

Doswald, N., and Osti, M. 2011. Ecosystem-based approaches to adaptation and mitigation – goodpractice examples and lessons learned in Europe. Federal Ministry of Environment, Nature Conservation and Nuclear Safety, Bonn, Germany, pp. 3–4.

Edwards, K. 2003. The importance of integration in sustainability agricultural systems. *Agriculture, Ecosystem and Environment*, 27, 25–35.

Ekwe, A.A., Amadu, F.O., Morlai, T.A., Wollor, E.T., and Cegbe, L.W. 2011. Agricultural innovations for climate change adaptation and food security in West Africa: the case of Nigeria, Sierra Leone and Liberia. African Technology Policy Studies Network Working Paper Series, pp. 61.

Enete, I.C. 2014. Impacts of climate change on agricultural production in Enugu State, Nigeria.

Evenson, J.P., Apurva, S., Kavi, K., James, M., and Stephen, L. 1997. Measuring the Impact of Climate Change on Indian Agriculture. World Bank Technical Paper, 402, Washington, DC.

Fadina, A.M.R. and Barjolle, D. 2018. Farmers' adaptation strategies to climate change and their implications in the Zou department of South Benin. *Environments*, 5, 1.

Fagariba, C.J., Song, S., and Baoro, S.K.G.S. 2018. Climate change adaptation strategies and constraints in northern Ghana: evidence of farmers in Sissala West District. *Journal of Sustainability*, 10(5), 1–18.

Federal Ministry of Environment. 2014. United Nations Climate Change Nigeria. National Communication (NC). NC 2.

Food and Agriculture Organisation (FAO). 2017. Tracking adaptation in agricultural sectors: Climate change adaptation indicators. Available online: http://www.fao.org/3/i8145e/i8145e.pdf (Accessed 2 February, 2022).

Fosu-Mensah, B.Y., Vlek, P.L.G., and MacCarthy, D.S. 2012. Farmers' perceptions and adaptation to climate change: a case study of Sekyeredumase district in Ghana. *Environment, Development and Sustainability*, 14, 495–505.

Haider, H. 2019. Climate change in Nigeria: impacts and responses. K4D Helpdesk Report 675. Brighton, UK: Institute of Development Studies.

Hardaker, J.B., Gudbrand, L., Jock R.A., and Rudd, B.M.H. 2015. *Coping with Risk in Agriculture: Applied Decision Analysis*, 3rd ed. Wikiwand: CAB International.

Hisali, E., Patrick, B., and Faisal, B. 2011. Adaptation to climate change in Uganda: evidence frommicro level data. *Global Environmental Change*, 21, 1245–1261.

Horel, A., Lichner, L., Alaoui, A., Czachor, H., Nagy, V., and Tóth, E. 2014. Transport of iodide in structured clay–loam soil under maize during irrigation experiments analysed using HYDRUS model. *Biologia*, 69, 1531–1538.

Horel, Á., Tóth, E., Gelybó, G.Y., Kása, I., Bakacsi, Z.S., and Farkas, C.S. 2015. Effect of land use and management on soil hydraulic properties. *Open Geoscience*, 1, 42–754.

Ifeanyi-obi, C.C., and Nnadi, F.N. 2014. Climate change adaptation measures used by farmers in Southsouth Nigeria. *Journal of Environmental Science, Toxicology and Food Technology*, 8, 4.

Intergovernmental Panel on Climate Change (IPCC). 2014. Climate change impacts, adaptation, and vulnerability. part b: regional aspects. Contribution of Working Group II to the Fifth Assessment Report of the Intergovernmental Panelon Climate Change, Cambridge University Press, Cambridge, New York, NY, pp. 688.

Intergovernmental Panel on Climate Change (IPCC). 2007. Climate change 2007: Impacts, adaptation and vulnerability. Intergovernmental Panel on Climate Change Fourth Assessment Report. IPCC: Geneva, Switzerland.

Jamshidi, O., Asadi, A., and Kalantari, K. 2018. Perception, knowledge and behavior towards climate change: a survey among agricultural professionals in Hamadan Province, Iran. *Journal of Agricultural Science and Technology*, 20(7), 1369–1382.

Jayanthi, C., Rangasamy, A., and Chinnusamy, C. 2000. Water budgeting for components in lowland integrated farming system. *Agricultural Journal*, 87, 411–414.

Langill, S., and Ndathi, A.J.N. 1998. Indigenous knowledge of desertification: A progress report from the Desert Margins Program in Kenya. People, Land and Water Series Report, 2. Ottawa: International Development Research Centre.

Lechowska, E. 2018. What determines flood risk perception? A review of factors of flood risk perception and relations between its basic elements. *Natural Hazards*, 94(3), 1341–1366.

Lema, M.A., and Majule, A.E. 2009. Impacts of climate change, variability and adaptation strategies on agriculture in semi-arid areas of Tanzania: the case of Manyoni District in Singida Region, Tanzania. *African Journal of Environmental Science and Technology*, 3(8), 206–218.

Liu, D., Li, Y., Shen, X., Xie, Y., and Zhang, Y. 2018. Flood risk perception of rural households in western mountainous regions of Henan Province, China. *International Journal of Disaster Risk Reduction*, 27, 155–160.

Liwenga, E.T. 2003. Food insecurity and coping strategies in semi-arid areas: the Case of Mvumi in Central Tanzania. Ph.D Dissertation No. 11. Stockholm Studies in Human Geography, Stockholm University Stockholm, Sweden.

Manley, P. 2008. Biodiversity and climate change (May 20, 2008). U.S. Department of Agriculture, Forest Service, Climate Change Resource Center. http://www.fs.fed.us/ccrc/topics/biodiversity.shtml. (Accessed 24 February 2022).

McLeman, R., and Smit, B. 2004. Climate change, migration and security. Canadian security intelligence service. *Commentary*, 86, 8.

Mengistu, D.K. 2011. Farmers' perception and knowledge on climate change and their copingstrategies to the related hazards: case study from Adiha, Central Tigray, Ethiopia. *Agricultural Sciences*, 2(2), 138–145.

Mills, R.T.E., Gavazov, K.S., Spiegelberger, T., Johnson, D., and Buttler, A. 2014. Diminished soil functions occur under simulated climate change in a sup-alpine pasture, but heterotrophic temperaturesensitivity indicates microbial resilience. *Science of the Total Environment*, 473, 465–472.

Mkonda, M., and He, X. 2016. Are rainfall and temperature really changing? Farmer's perceptions, meteorological data, and policy implications in the Tanzanian semi-arid zone. *Sustainability*, 9(8), 128–136.

Mortimore, M.J., and Adams, W.M. 2000. Farmer adaptation, change and crisis in the sahel. *Global Environmental Change*, 11, 49–57.

Ngigi, S.N. 2009. Climate change adaptation strategies: Water resources management options for smallholder farming systems in Sub-Saharan Africa. The MDG Centre for East and Southern Africa, the Earth Institute at Columbia University, New York, pp. 189.

Nyanga, P., Johnsen, F., Aune, J., and Kahinda, T. 2011. Smallholder farmers' perceptions of climate change and conservation agriculture: evidence from Zambia. *Journal of Sustainable Development*, 4(4), 73–85.

Nyong, A., Adesina, F., and Osman, E.B. 2007. The value of indigenous knowledge in climate change mitigation and adaptation strategies in the African Sahel. *Mitigation and Adaptation Strategies for Global Change*, 12, 787–797.

Ochenje, I.M., Ritho, C.N., Guthiga, P.M., and Mbatia, O.L.E. 2016. Assessment of farmers' perception to the effects of climate change on water resources at farm level: the case of Kakamega county, Kenya, Invited poster presented at the 5th International Conference of the African Association of Agricultural Economists, Addis Ababa.

Ogunwale, A.O. 2015. Deforestation and greening the Nigerian environment. International Conference on African Development Issues (CUICADI) 2015: Renewable Energy Track, pp. 212–219.

Orimoloye, I.R., Olusola, A.O., Belle, J.A. et al. 2022. Drought disaster monitoring and land use dynamics: identification of drought drivers using regression-based algorithms. *Nat Hazards*, 112, 1085–1106. https://doi.org/10.1007/s11069-022-05219-9.

Oli, B. 2008. Migration and climate change. IOM Migration Research Series, 31: International Organization for Migration, Geneva, pp. 1–68.

Otitoju, M.A., and Enete, A.A. 2016. Climate change adaptation: uncovering constraints to the use of adaptation strategies among food crop farmers in south-west, Nigeria using principal component analysis (PCA). *Cogent Food and Agriculture*, 2, 1.

Oyekale, A.S., and Oladele, O.I. 2012. Determinants of climate change adaptation among cocoa farmers in southwest Nigeria. *ARPN Journal of Science and Technology*, 2, 154–168.

Ozor, N., and Nnaji, C.E. 2011. The role of extension in agricultural adaptation to climate change in Enugu State, Nigeria. *Journal of Agricultural Extension and Rural Development*, 3(3), 42–50.

Pande Chaitanya, B., Kanak, N.M., Sudhir, K.S., Abhay, M.V., Ahmed Elbeltagie, S.F.R., and Khadri, P.C. 2021. Estimation of crop and forest biomass resources in a semi-arid region using satellite data and GIS. *Journal of the Saudi Society of Agricultural Sciences*, 20(5), 302–311.

Pande Chaitanya B., Sunil, A.K., Rajesh, J., Sunil, G., and Mukund, S. 2022. Prediction of soil chemical properties using multispectral satellite images and wavelet transforms methods. *Journal of the Saudi Society of Agricultural Sciences*, 21(1), 21–28.

Penuelas, J., Janssens, I.A., Ciais, P., Obersteiner, M., and Sardans, J. 2020. Anthropogenic global shifts in biospheric N and P concentrations and ratios and their impacts on biodiversity, ecosystem productivity, food security, and human health. *Global Change and Biology*, 26:1962–85.

Pilling, D., Belanger, J., and Hoffmann, I. 2020. Declining biodiversity for food and agriculture needs urgent global action. *Natural Food*, 1, 44–47.

Qasim, S., Qasim, M., Shrestha, R.P., and Nawaz Khan, A. 2018. Socio-economic determinants of landslide risk perception in Murree hills of Pakistan. *AIMS Environmental Science*, 5, 305–314.

Radhamani, S., Balasubramanian, A., Ramamoorthy, K., and Geethalakshmi, V. 2003. Sustainable integrated farming system for dry lands: a Review. *Agricultural Reviews*, 24, 204–210.

Rahman, E.L., Abd, M.A., Ali, R.R, Hussain, M.A., and El-Semey, M. A. 2009. Remote sensing and GIS-based physiography and soils mapping of the Idku-Brullus Area, North Delta, Egypt. *Egyptian Journal of Soil Science*, 49(3), 209–432.

Raška, P. 2015. Flood risk perception in Central-Eastern European members states of the EU: a review. *Natural Hazards*, 79(3), 2163–2179.

Robert, M., Thomas, A., and Bergez, J.E. 2016. Processes of adaptation in farm decision-making models: a review. *Agronomy and Sustainable Development*, 36, 64.

Roder, G., Ruljigaljig, T., Lin, C.-W., and Tarolli, P. 2016. Natural hazards knowledge and risk perception of Wujie indigenous community in Taiwan. *Natural Hazards*, 81, 641–662.

Rose, N.A., and Burton, P.J. 2009. Using bioclimatic envelopes to identify temporal corridors in support of conservation planning in a changing climate. *Forest Ecology and Management*, 258, 64–74.

Rufat, S., Fekete, A., Armaş, I., Hartmann, T., Kuhlicke, C., Prior, T., Thaler, T., and Wisner, B. 2020. Swimming alone? Why linking flood risk perception and behavior requires more. *WIREs Water*, 7(5), e1462.

Sanogo, K., Sanogo, S., and Ba, A. 2016. Farmers' perception and adaption to land use change and climate variability in fina reserve, Mali. *Turkish Journal of Agriculture–Food Science and Technology*, 4, 4.

Satishkumar, N., Tevari, P., and Singh, A. 2013. A study on constraints faced by farmers in adapting to climate change in rain-fed agriculture. *Journal of Human Ecology*, 44(1), 23–28.

Sherpa, S.F., Shrestha, M., Eakin, H., and Boone, C.G. 2019. Cryospheric hazards and risk perceptions in the Sagarmatha (Mt. Everest) National Park and Buffer Zone, Nepal. *Natural Hazards*, 96, 607–626.

Smit, B., McNabb, D., and Smithers, J. 1996. Agricultural adaptation to climatic variation. *Climate Change*, 33, 7–29.

Smith, N., and Leiserowitz, A. 2012. The rise of global warming skepticism: exploring affective image associations in the United States over time. *Risk Analysis*, 32(6), 1021–1032.

Spires, M., Shackleton, S., and Cundill, G. 2014. Barriers to implementing planned community-based adaptation in developing countries: a systematic literature review. *Climate Development*, 6, 277–287.

Stöber, S., Chepkoech, W., Neubert, S., Kurgat, B., Bett, H., and Lotze Campen, H. 2017. Adaptation pathways for African indigenous vegetables' value chains In: Leal Filho, W., Belay, S., Kalangu, J., Menas, W., Munishi, P., and Musiyiwa, K. (Eds), Climate Change Adaptation in Africa, Climate Change Management, pp. 413–433.

Shahid, M., Rahman, K.U., Haider, S. et al. 2021. Quantitative assessment of regional land use and climate change impact on runoff across Gilgit watershed. *Environmental Earth Sciences*, 80**,** 743. https://doi.org/10.1007/s12665-021-10032-x.

Tambo, J.A., and Abdoulaye, T. 2013. Smallholder farmers' perceptions of and adaptations to climate change in the Nigerian savanna. *Regional Environmental Change*, 13, 375–88.

Thornton, P.K., Lipper, L., Baas, S., Cattaneo, A., Chesterman, S., Cochrane, K., de Young, C., Ericksen, P., van Etten, J., de Clerck, F., Douthwaite, B., DuVal, A., Fadda, C., Garnett, T., Gerber, P.J., Howden, M., Mann, W., McCarthy, N., Sessa, R., Vermeulen, S., and Vervoort, J. 2013. How does climate change alter agricultural strategies to support food security? Background paper for the conference 'Food Security Futures:' Research Priorities for the 21st Century, Dublin.

Uddin, M.N., Bokelmann, W., and Entsminger, J.S. 2014. Factors affecting farmers' adaptation strategies to environmental degradation and climate change effects: a farm level study in Bangladesh. *Climate*, 2, 223–241.

United Nations Environmental Program and Convention on the Conservation of Migratory Species (UNEP/CMS). (2006). Migratory species and climate change: impacts of changing environment on wild animals, pp. 8–60.

van Valkengoed, A.M., and Steg, L. 2019. Meta-analyses of factors motivating climate change adaptation behaviour. *Nature Climate Change*, 9, 158–163.

Voccia, A. 2012. Climate change: What future for small, vulnerable states? *International Journal of Sustainable Development and World Ecology*, 19, 101–115.

Wani, S.S., Pralhad, T.K., Rockstrom, J., and Ramakrishna, Y.S. 2009. Rainfed agriculture-past trends and future prospects In: Wani, S.S., Pralhad, T.K., Rockstrom, J., and Oweis, T. (Eds.) Rainfed Agriculture: Unlocking the Potential. Wikiwand: CAB International, pp. 1–35.

Webb, N.P., Marshall, N.A., Stringer, L.C., Reed, M.S., and Chappell, A.H.J. 2018. Land degradation and climate change: building climate resilience in agriculture. *Front Ecology and Environment*, 15, 450–459.

Yates, C.J., Elith, J., Latimer, A.N., Le Maiter, D., Midgley, G.F., Schurr, F.M., and West, A.G. 2009. Projecting climate change impacts on species distributions in mega-diverse South African regions: opportunities and challenges. *Ecological Society of Australia*, 35, 374–391.

Ziervogel, G., Cartwright, A., Tas, A., Adejuwon, J., Zermoglio, F., Shale, M., and Smith, B. 2008. Climate change and adaptation in African agriculture. Stockholm Environment Institute, March 2008, pp. 17–19.

10 Delineation of Groundwater Prospect Zones based on Earth Observation Data and AHP Modeling – A Study from Basaltic Rock Formation

Kanak N. Moharir, Chaitanya B. Pande, Sudhir Kumar Singh, Pandurang Choudhari, and A. M. Varade

10.1 INTRODUCTION

Groundwater resource plays a significant part in the overall development of many areas, *viz.*, agriculture, manufacturing, and urbanization in many countries. Nowadays, India is suffering problems related to sustainable water resources, and this problem has directly impacted human health and crop yield production. However, groundwater quality is continuously changing due to intensive cultivation, climate change, population growth, and other natural processes (Shinde et al., 2020). India is a second major populous country. It contributes 18% of the world's population, but only 4% of the world's renewable and surface water resources are available (National Water Policy of India, 2012). Since the previous three decades, many parts of the world's water level in small dams and aquifer have been dropped significantly in several regions of the country (Biswas et al., 2013). Traditionally, groundwater explorations by the geological, hydrological, drilling, and geophysical strategies are exorbitant and tedious (Razandi et al., 2015). Since the last few decades, remote sensing, GIS, geospatial modeling, and ANN and machine learning models have been effectively used for the development and planning of water resources in the semi-arid region (Biswas et al., 2013; Shekhar and Pandey, 2015; Jasrotia et al., 2016; Choudhari et al., 2018). Nowadays, geospatial technologies are applied in earth science, engineering, and environmental engineering, and these technologies have a significant role in natural resources planning and management (Watkins et al., 1997; Stafford et al., 2008; Goodchild, 1993), and sizeable hydrogeological data simulation and modeling (Loague and Corwin, 1998; Gogu et al., 2001; Gossel et al., 2004; Essay, 2007). Many researchers have identified appropriate groundwater sites, watershed management, natural resource evaluation, and ecosystem management (Kumar et al., 2016; Mogaji, 2017).

Currently, air-borne and ground-based platform data are simple, quickly available, and these data are helpful in the upgradation of baseline map, sub-basin management, groundwater suitable zones, and mapping in the aquifer zones. The real-time satellite images support the generation of numerous maps, namely, drainage network, river, water body, geology, lineaments, land use/land cover, slope, and other layers. These layers regulate and are used to monitor the hydrological processes such as infiltrations, soil erosion, runoff, evaporation, and percolation (Bobba et al., 1992; Meijerink et al., 2000; Patode et al., 2017; Choudhari and Nayak, 2012). However, researchers rarely studied full regulatory parameters (Guru et al., 2017). Many groundwater professionals have applied various methods, namely, frequency ratio (Ozdemir et al., 2011; Al-Abadi et al., 2016; Balamurugan et al., 2017; Das and Pardeshi,

DOI: 10.1201/9781003303237-10

2018), weighted overlay (WOA), analytical hierarchal process (AHP), artificial neural network (ANN), convolution neural network (CNN), multi-influencing factor (MIF), and machine learning algorithm (Nampak et al., 2014; Lee et al., 2015; Saaty, 2008; Singh et al., 2018; Pourtaghi et al., 2014; Naghibi et al., 2015; Golkarian et al., 2018) to delineate groundwater prospect zones in the arid and semi-arid regions. Multi-criteria decision methods (MCDM) play a vital role in the sustainable development and management of soil and water resources and support decision-makers in better decision-making (Jha et al., 2010). Saaty's (1990) method is an MCDM method, which is based on a pairwise comparison matrix. Basalt rock is a type of igneous rock. The study region comprises basalt rock, and the runoff is high and infiltration is less in this region. Therefore, finding the potential groundwater zones is challenging in the study region. The help of various layers and their integration in the GIS environment will facilitate locating potential groundwater zones. However, very few studies have been carried out for groundwater potential zone mapping in the basalt region (Mundalik et al., 2018; Doke et al., 2021). The research aims to delineate the groundwater prospect zones from the basaltic formation using the AHP and weighted overlay method, earth observation datasets, and geoinformatics technique.

10.2 STUDY AREA DESCRIPTION

The area is located between 20°54′59″ N latitude and 76°41′23″ E longitude in the Akola and Buldhana districts of Maharashtra, India (Figure 10.1). The Man River Basin was selected as a study area, and it is a tributary of the Purna River. The elevation ranges between 300 and 528 meter above mean sea level (Pande et al., 2018). The total study area is 449.52 km^2, and the average annual rainfall ranges from 350 to 800 mm. The Maharashtra state is located within the peninsular shield area of the country, with about 94% of its total geographical area underlain by complex rock formations and the remaining 6% with localized occurrences of sedimentary and alluvial deposits. Basalt is a dark-colored, fine-grained, igneous rock composed mainly of plagioclase and pyroxene minerals (Sheth, 2005). It most commonly forms as an extrusive rock, such as a lava flow, but also develops in small intrusive bodies, such as an igneous dike or a thin sill. It has a composition similar to gabbro. About 80% area of the state is covered by basaltic lava flows with overlying alluvium confined to the places in the vicinity of significant rivers and streams (Sethna and Battiwala, 1977). The alluvial deposits of shallow thickness occurred extensively along the stream courses consisting of gravel, sand, and clay mixtures (Pande et al., 2021b).

10.3 MATERIALS AND METHODS

10.3.1 Used Datasets

Satellite imagery, geomorphology, slope, geology, drainage density, lineament density, soil, groundwater fluctuation data, rainfall, and LULC delineate groundwater prospect zones. Cartosat digital elevation model (DEM) of 30 m spatial resolution and LISS-III image having spatial resolution of 23.5 m, which are freely available, was downloaded (acquired on 15-02-2017) from the website (http://bhuvan.nrsc.gov.in) of the National Remote Sensing Centre (NRSC), Hyderabad, India. The Cartosat Dem was used for the preparation of slope and drainage density layers. The LISS-III data were used for the preparation of the LULC map using supervised classification method. The water-level data for pre- and post-monsoon were obtained from Central Groundwater Board (CGWB), India.

10.3.2 Methodology

All the satellite data were pre-processed, and base layers were created. The drainage network map was prepared using DEM data. The groundwater fluctuation map was prepared using the inverse distance weighted (IDW) interpolation method. All layers were prepared in a standard reference

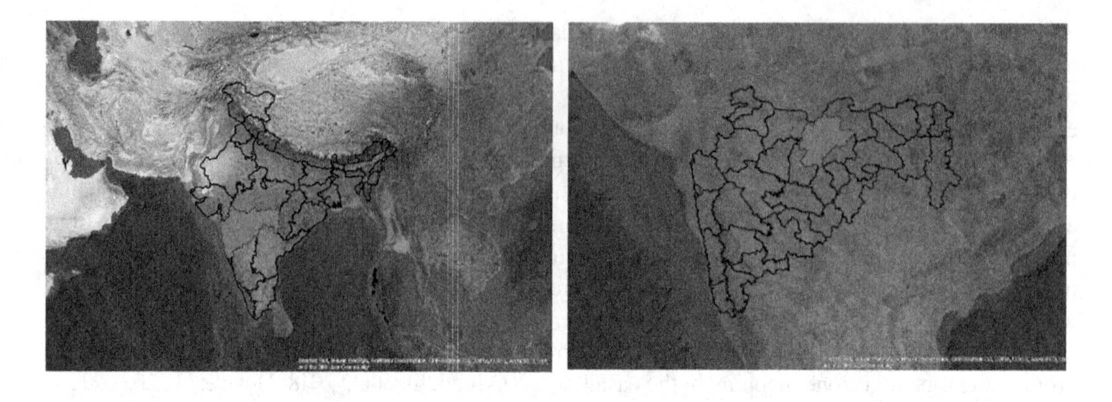

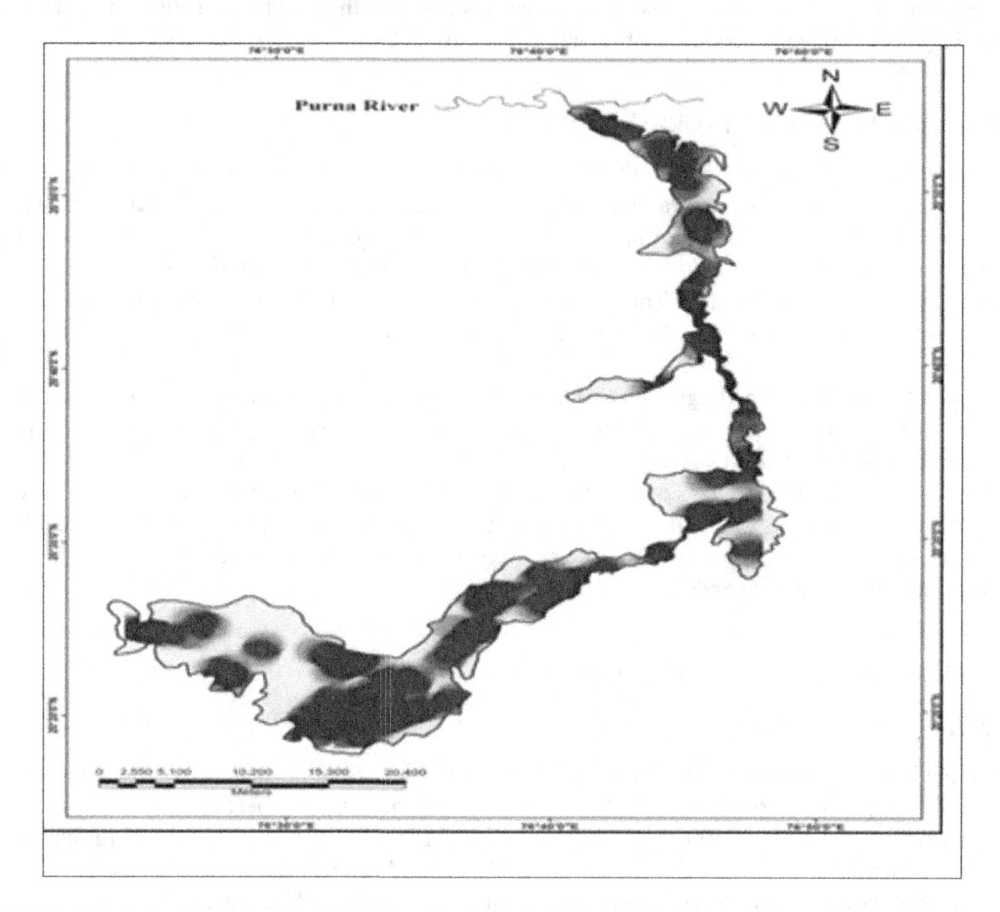

FIGURE 10.1 Location map of the Man River Basin, Maharashtra, India.

system (Datum WGS, 1984, UTM zone 43). The vector layers were converted into raster format in the first step; computed weights were assigned to each thematic layer based on the AHP method. The accuracy of the weights was assessed based on the consistency index (CI) and consistency ratio (CR). The highest importance was given to the most crucial layer and the lowest to the least essential layer. The ranks were assigned to each attributes of thematic layer based on their importance.

All layers were integrated using the weighted overlay analysis method in the GIS environment, and the groundwater potential zones were created. The applied procedure of methodology is illustrated in Figure 10.2.

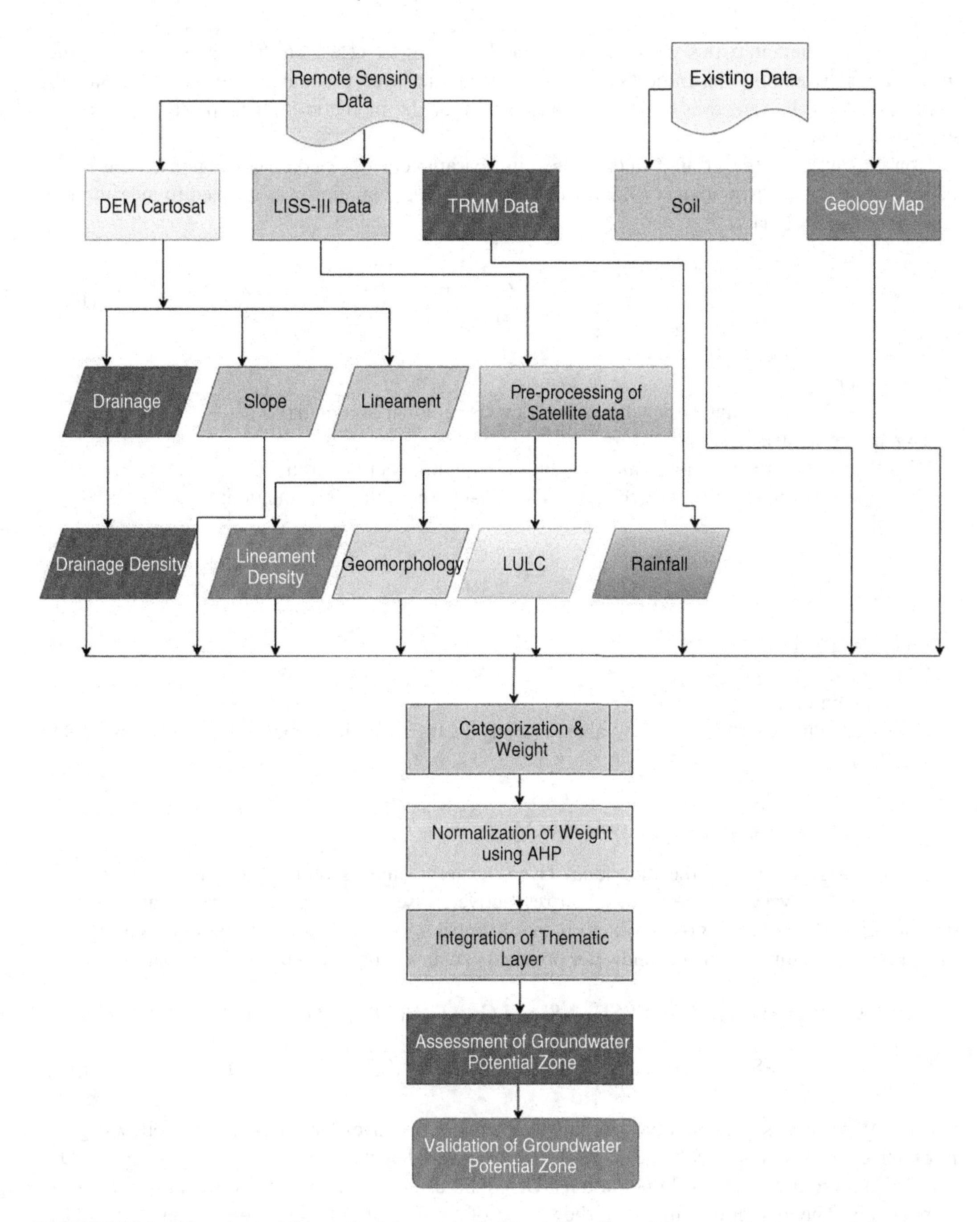

FIGURE 10.2 Flowchart of the methodology adopted for delineation of the groundwater potential zones.

10.3.3 AHP Method and Weight Normalization

AHP is the most commonly used method with GIS for delineating groundwater prospect zones (Agarwal et al., 2013; Razandi et al., 2015; Murmu et al., 2019). There are nine thematic layers, and these layers control the storage and flow of water in the study area. This method was used to solve the problem in a hierarchical structure. A character with a high weight demonstrates a significant influence, and characteristics with a low weight show a minor effect on the suitable zones of the groundwater potential. The importance of each thematic layer was calculated based on Saaty's scale (1–9), and the rank was given to each attribute of the thematic layer as per their vital and water-holding volume.

Consequently, all the groundwater-influencing parameters were compared against each other in a pairwise comparison matrix (Table 10.1). The following equation was utilized to measure the consistency index (CI).

$$CI = \frac{(\lambda_{max} - n)}{(n-1)} \tag{10.1}$$

CI = consistency index,

λ_{max} = it is the significant eigenvalue of the pairwise comparison matrix,

n = number of variables,

RI is the random index whose value depends on the influencing parameter.

CR is the consistency ratio, which is calculated using the following equation:

$$CR = \frac{CI}{RI} \tag{10.2}$$

CR = consistency ratio,

CI = consistency index,

RI = random index.

If the CR value is ≤0.10, it is acceptable for analysis. If CR = 0, it is a perfect level of consistency (Table 10.1).

10.3.4 Weighted Overlay Analysis (WOA)

Initially, reclassification of the individual layers (criteria) and assignment of ranks was carried out accordingly. Every parameter and their respective classes are allocated the ranks and weights depending on their relative groundwater potential contribution (Table 10.2). Equation (10.3) is used in a raster calculator, and the groundwater prospects zone was prepared using this equation:

$$\begin{aligned} GWPZ = (&GM_w * GM_{wi} + SL_w * SL_{wi} + G_w * G_{wi} + DD_w * DD_{wi} + LD_w * LD_{wi} \\ &+ S_w * S_{wi} + GL_w * GL_{wi} + LULC_w * LULC_{wi} + RF_w * RF_{wi}) \end{aligned} \tag{10.3}$$

where GWPZ shows the potential groundwater zones; the layers, namely, are as follows: geomorphology (GM), slope (SL), geology (GG), drainage density (DD), lineament density (LD), soil (S), and groundwater-level fluctuations (GL), land use and land cover (LULC), rainfall (RF), respectively. The weight of a map and the weight of a particular class of map are represented by w and wi.

10.3.5 Groundwater Fluctuation

Groundwater well location was marked with the help of Garmin GPS. The IDW is an important and powerful tool for point data interpolation in the GIS environment. The groundwater fluctuation data were used for describing and determining the suitable zone and helping in water resources management. Deep acquifer depicts immense groundwater potential. In addition, shallow groundwater acquifers display large fluctuations, hence low groundwater prospects.

TABLE 10.1
Pairwise Comparison Matrix of Thematic Layers

Themes	Geomorphology	Geology	Drainage density	Lineament density	Soil	Groundwater-level fluctuation	LULC	Rainfall	Slope	Geometric Mean	Weight
Geomorphology	1	0.5	0.5	0.33	0.33	0.33	0.25	0.16	0.25	0.42	0.03
Geology	2	1	0.25	0.25	0.2	0.5	0.5	0.16	0.16	0.47	0.04
Drainage density	2	4	1	0.5	0.33	0.33	0.33	0.33	0.16	0.72	0.05
Lineament density	3	4	2	1	0.5	0.5	0.33	0.33	0.33	0.96	0.07
Soil	3	5	3	2	1	0.5	0.33	0.33	0.2	1.20	0.09
Groundwater-level fluctuation	3	2	3	2	2	1	0.33	0.5	0.2	1.32	0.10
LULC	4	3	3	3	3	3	1	0.5	0.16	1.99	0.15
Rainfall	6	6	3	3	3	2	2	1	0.5	2.50	0.19
Slope	4	6	6	3	5	5	6	2	1	3.69	0.28
Total	28	31.5	21.75	15.08	15.36	13.16	11.08	5.33	2.98	13.25	1.00

TABLE 10.2
Rank and Weight for Different Categories Derived Using AHP

Sr. No	Thematic Layers	Weightage	Classes	Rank
1	Geomorphology	0.03	Alluvial plain	9
			Denudation Hill	2
			Floodplain	1
			Plateau	2
2	Slope	0.28	1%–3%	5
			3%–5%	4
			5%–10%	3
			10%–15%	2
			15%–35%	1
3	Geology	0.47	Alluvial plain	7
			Plateau	3
4	Drainage density	0.05	0.00–0.11	3
			0.11–0.33	5
			0.33–0.97	7
5	Lineament density	0.07	0.00–0.10	2
			0.10–0.44	3
			0.40–0.90	4
6	Soil	0.09	Clay loam	1
			Clayey	2
			Gravel clay	3
			Gravel clay loam	4
			Gravel sandy clay loam	5
			Gravel sandy loam	6
7	Groundwater level	0.10	0.2–13	7
8.	Rainfall	0.19	350–500	1
			500–650	3
			650–800	4
9.	LULC	0.15	Agriculture land	5
			Forest land	3
			Water bodies land	2
			Wasteland land	5
			Built-up land	5

10.3.6 Accuracy Assessment

The groundwater prospects zone map was delineated in the present study and verified using the available well groundwater fluctuation data of 45 observation wells. The well's data were collected during the pre-post monsoon of 2014. The well's fluctuation points were overlaid on the final groundwater prospect map to check the accuracy of the present work in the various groundwater prospective zones. We also compared the resultant groundwater prospect map with groundwater water fluctuation data of the published report by Central Ground Water Development (CGWB), Govt. of India.

10.4 RESULTS AND DISCUSSION

The movement and storage of groundwater based on different parameters such as lineaments, lineament density, drainage map, groundwater level changes, geology, slope, soil, rainfall, and hydrological situation of a particular region and the associates among all parameter's factors (Singh et al., 2011; Pande et al., 2021a; Rajesh et al., 2021).

10.4.1 LULC

LULC analysis was carried out based on satellite images with the help of image interpretation methods and field checks to verify the classes. Based on supervised classification method five land use land cover classes were classified, namely, agriculture, built-up, forest, waste land, water body, and their occupied area are 246.02 km^2, 4.27 km^2, 65.83 km^2, 113.95 km^2, and 19.45 km^2, respectively. The LULC accuracy was 86.23% (Figure 10.3).

10.4.2 Landforms and Geomorphological Features

Landforms and geomorphological features are essential for groundwater prospects. The denudational origin with pediment-peneplain complex comprises gently undulating plains covered with weathered

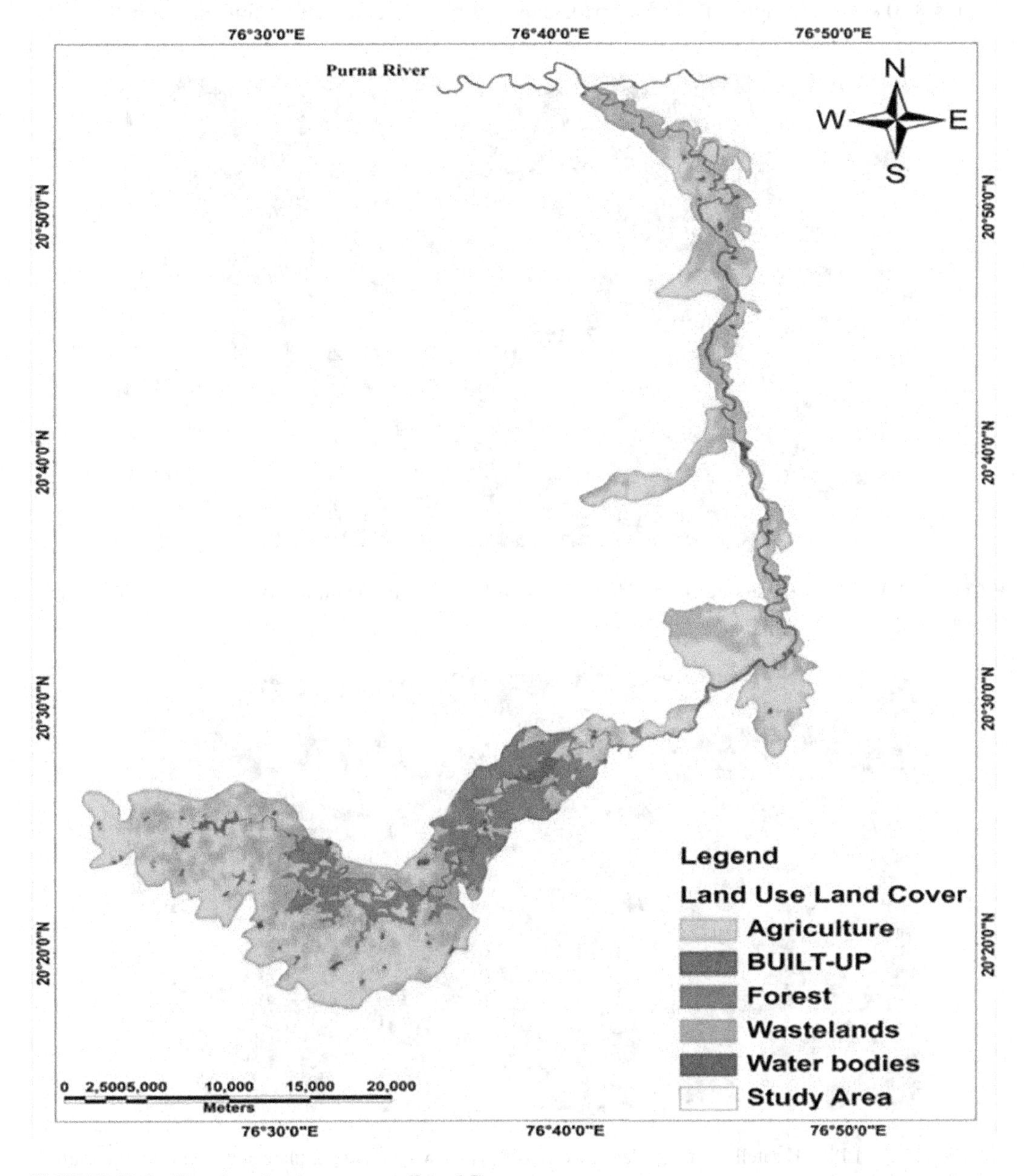

FIGURE 10.3 Land use land cover map (Level-I).

material favorable for groundwater recharge. The investigation of geomorphology features is significant and plays a vital role in identifying and conserving aquifers and groundwater planning in a drought-affected area. The main geomorphological features are alluvial plain, denudational hill, flood plain, and plateau, as illustrated in Figures 10.3–10.5. These features can be improved using the classification approach, but we have extracted them through visual image interpretation.

Alluvial plain – 87.05, denudation hill – 0.062, flood plain – 1.109, habitation mask – 4.313, plateau – 337.54, water body – 19.454 (Figure 10.6).

10.4.3 Lineament Density

Lineament characteristics, *viz.*, joints, fractures, and faults, have essential roles in the hydrology and groundwater development. The maximum lineament density was 0.90 km/km^2, and the minimum is 0.10 km/km^2 (Figure 10.7). Its performance as a channel to aquifer flow movement results

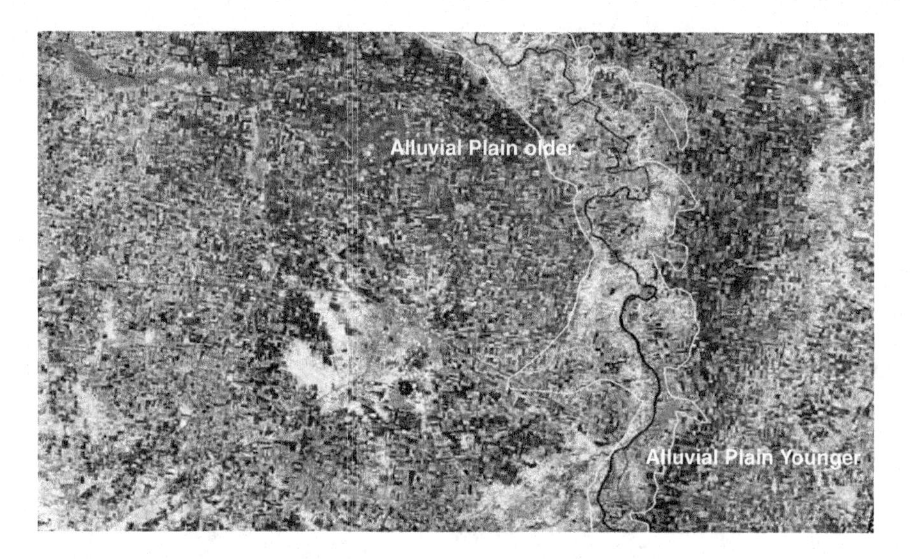

FIGURE 10.4 LISS-III satellite image (date:15-02-2017) is showing plain alluvial landforms of Basaltic rock area (band combinations 2, 3, and 4).

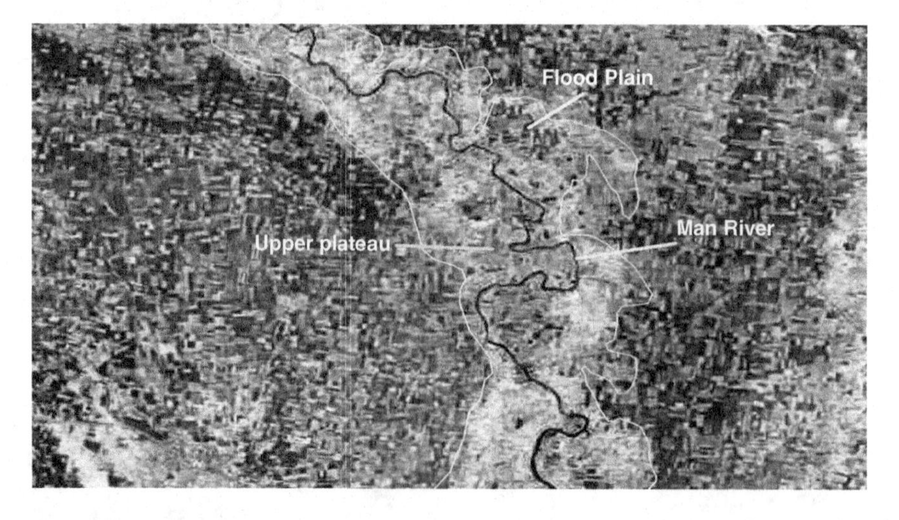

FIGURE 10.5 LISS-III satellite image (date:15-02-2017) is showing an upper plateau which is another landform exposed at Man River Basin (band combinations 2, 3, and 4).

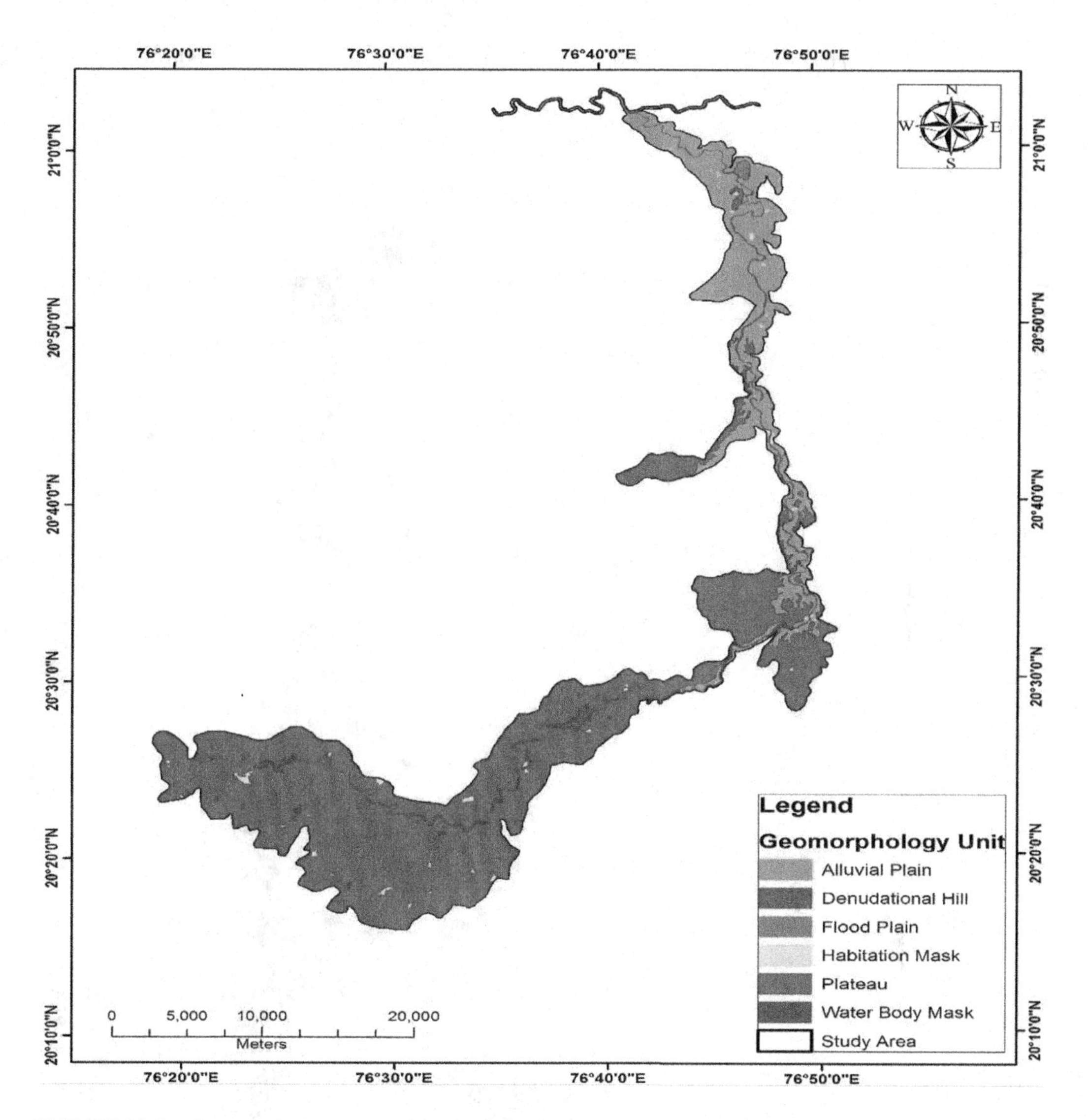

FIGURE 10.6 Geomorphology map of the study area.

in a better rising porosity, enhancing groundwater level and therefore helps as a groundwater prospects zone (Obi Reddy et al., 2000). The intersections of lineament in surrounding zones are suitable for groundwater accumulation. Lineament has been found in the northern tip of the study area. Rock faulting was identified in some segments of the contour. The analysis of the feature was restricted from the satellite data with the geological map. Lineaments have been structurally measured in two types: linear classes, which have been obtained from the satellite images by them comparatively with linear alignments. These lineament structures defined the land slope of the fundamental physical elements. The lineament density (Ld) study is a quantifiable quantity resulting from using the structure map. In satellite image data, lineaments line is interpreted using visual interpretation techniques and ground verification data key which was found mainly depending on linear nature, land-water condition, the configuration of plants, direct drainage parts, and numerous soil character. It is shown that rock fracture and poor aquifer zones ensuing in enlarged secondary porosity and permeability. All these parameters are hydrologically significant to given pathways for water drive. Lineament density can circuitously disclose the suitable groundwater area (Haridas et al., 1998).

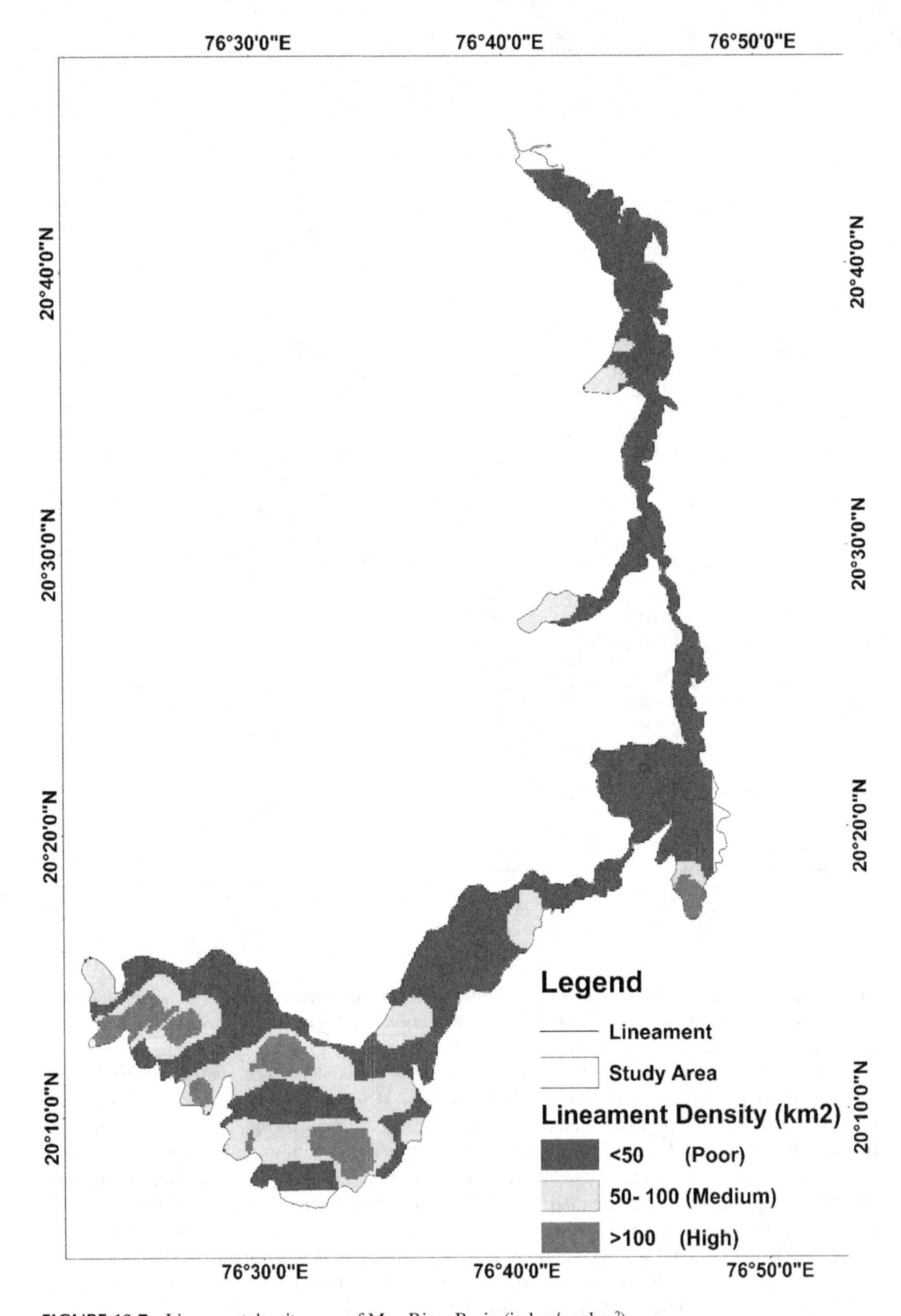

FIGURE 10.7 Lineament density map of Man River Basin (in km/per km^2).

10.4.4 Drainage Map

The drainage density in the study area ranges from 0.5 to 1.5 km/km^2. Three zones were created, namely, high, moderate, and low for drainage density. The river during the monsoon becomes flooded and dry in summer. Cartosat DEM data are helpful for the marked drainage system and give significant hydrology-related information (Saraf et al., 2007). The morphometric characteristics are measured with the reference of the Strahler scheme (Strahler, 1954) (Figure 10.8).

10.4.5 Groundwater Fluctuation

Groundwater fluctuations map was prepared based on pre- and post-monsoon data for the year 2013-13. The groundwater fluctuation map was categorized into 10 zones (Figure 10.9).

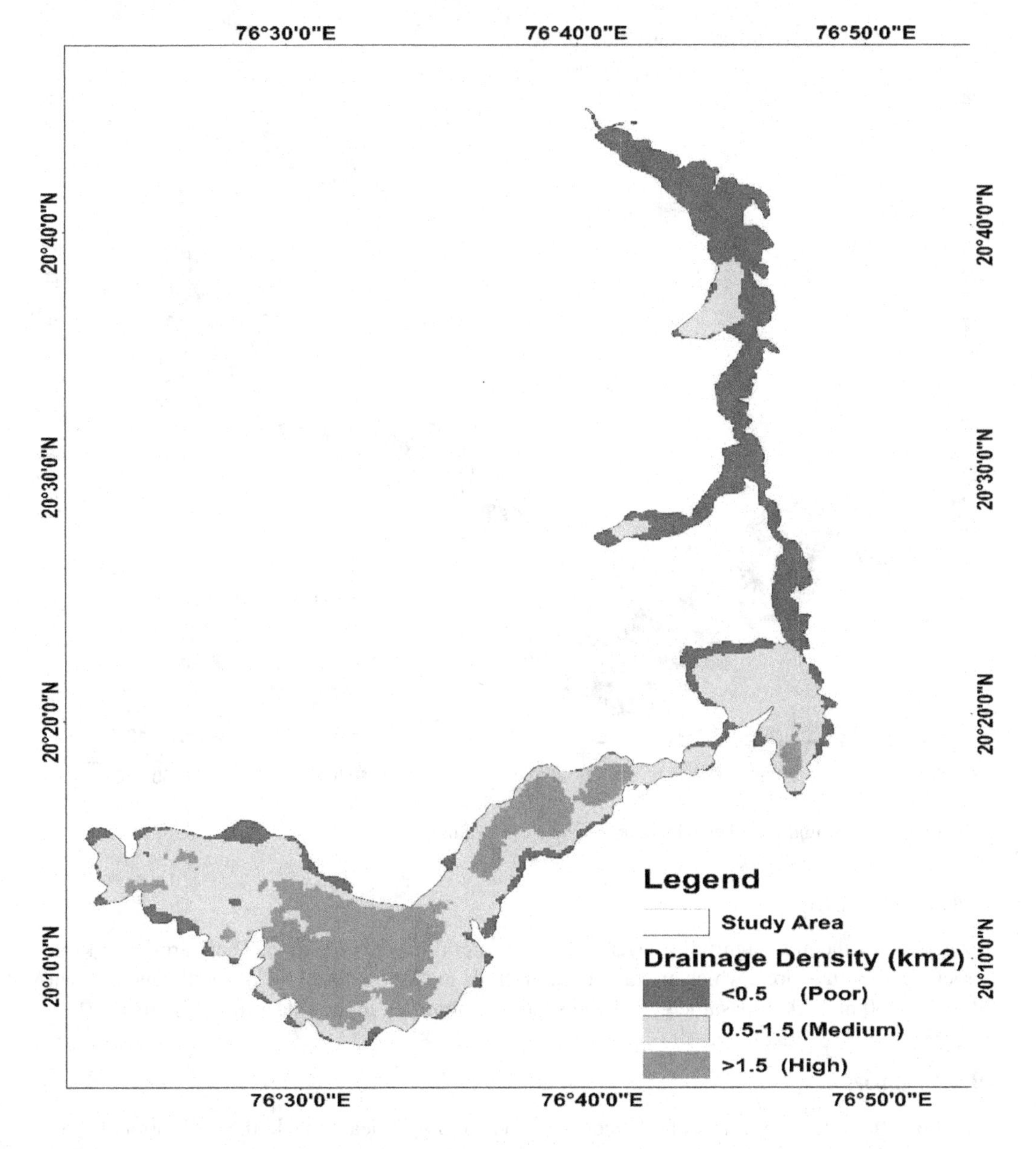

FIGURE 10.8 Drainage density map of the study area.

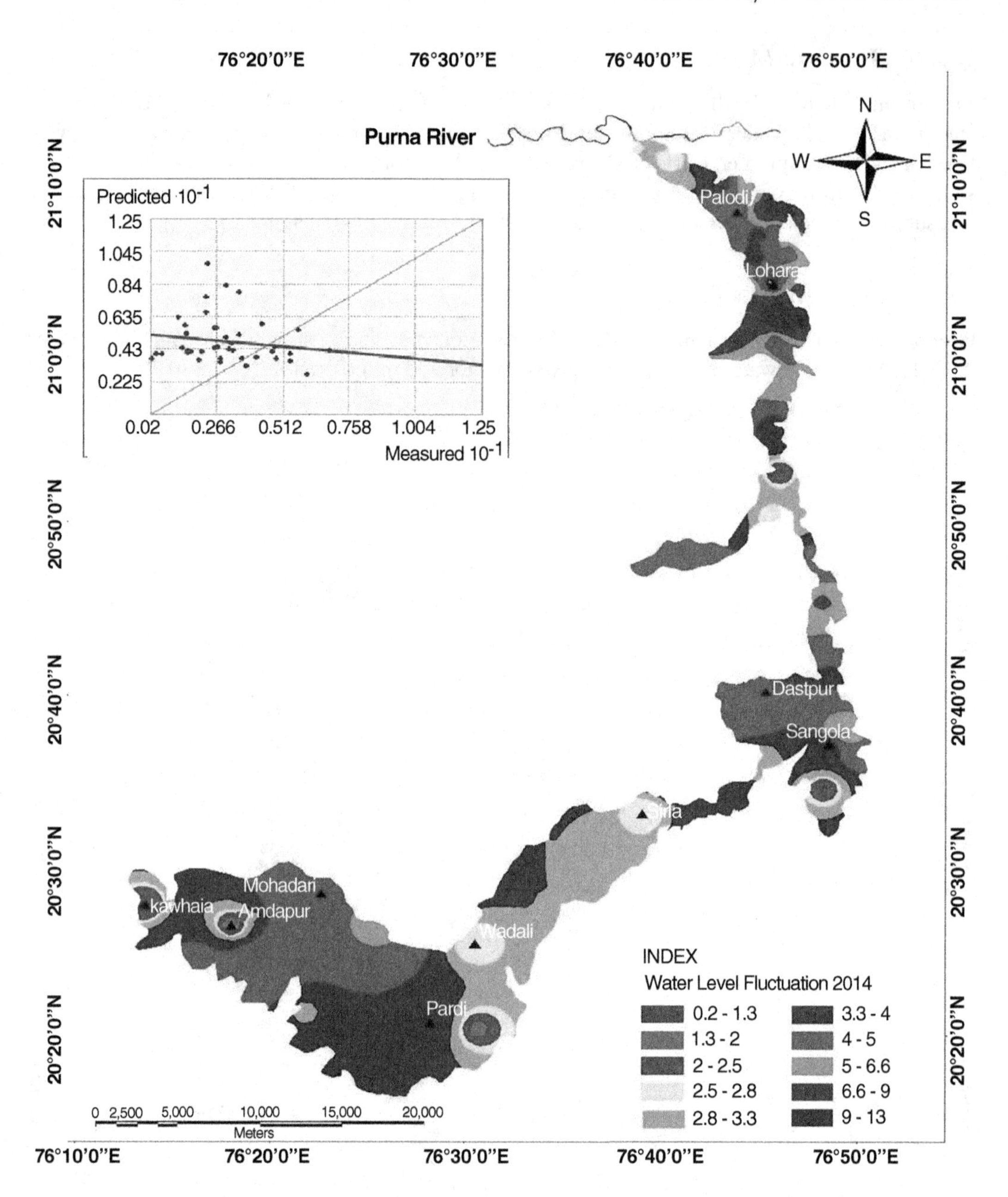

FIGURE 10.9 Groundwater level fluctuation map (pre-post monsoon).

10.4.6 Geology

The geology is the most significant layer in groundwater prospects zonation. There are many kinds of rocks in the study area, which are low- to high-grade metamorphosed rocks. On the more downside, 200–250 m thick of basaltic gravel and sand was observed in the study area (Figure 10.10).

10.4.7 Slope

The slope affects the movement of surface water and has a significant role in the infiltration of rainwater. The slope map was prepared using Cartosat data (32 m) resolution, and it observed 1%–35%.

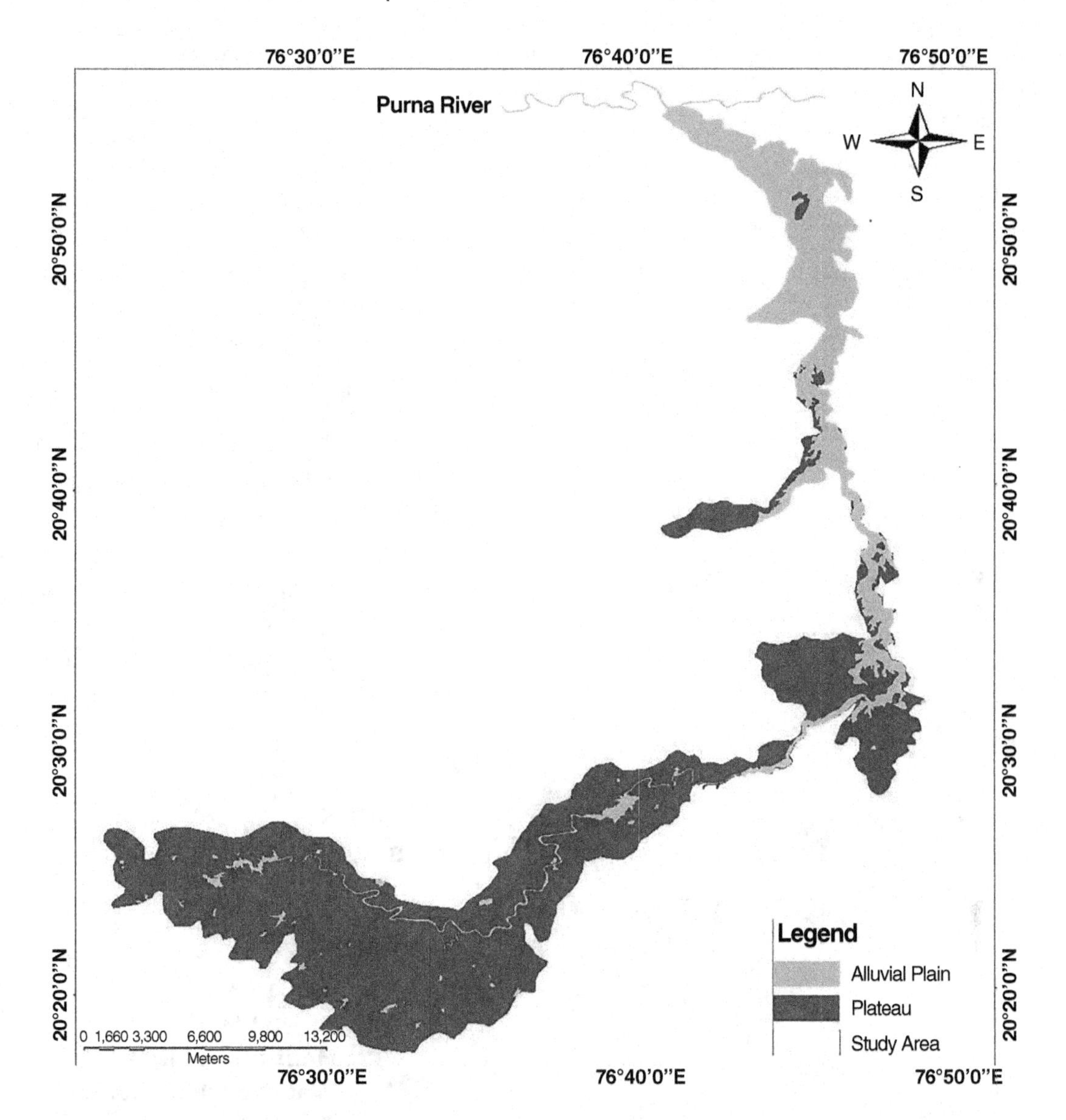

FIGURE 10.10 Geological map of the study area.

The run-off rate and soil erosion are very high on a steep slope. Few study areas have a moderate incline. Hence the runoff is slow, so more time is required for precipitation to infiltrate and water conservation. However, the vice versa at high slope areas where high runoff allows less period to rainfall and relatively small infiltration (Figure 10.11).

10.4.8 Soil

Surface soils are sandy loam in texture, and about 65% of fine sand and soil texture become gradually more refined with depth. The soil layers below 1 m and up to a depth of 5.2 m are loamy in texture, and below 5.2 m depth, the soil is silty clay. Broadly, the area's soil is divided into three classes, viz., coarse shallow, found in the south at a higher level. The Purna River area has maximum land under the moderate black soil, and the deep black soil has been observed in the river part.

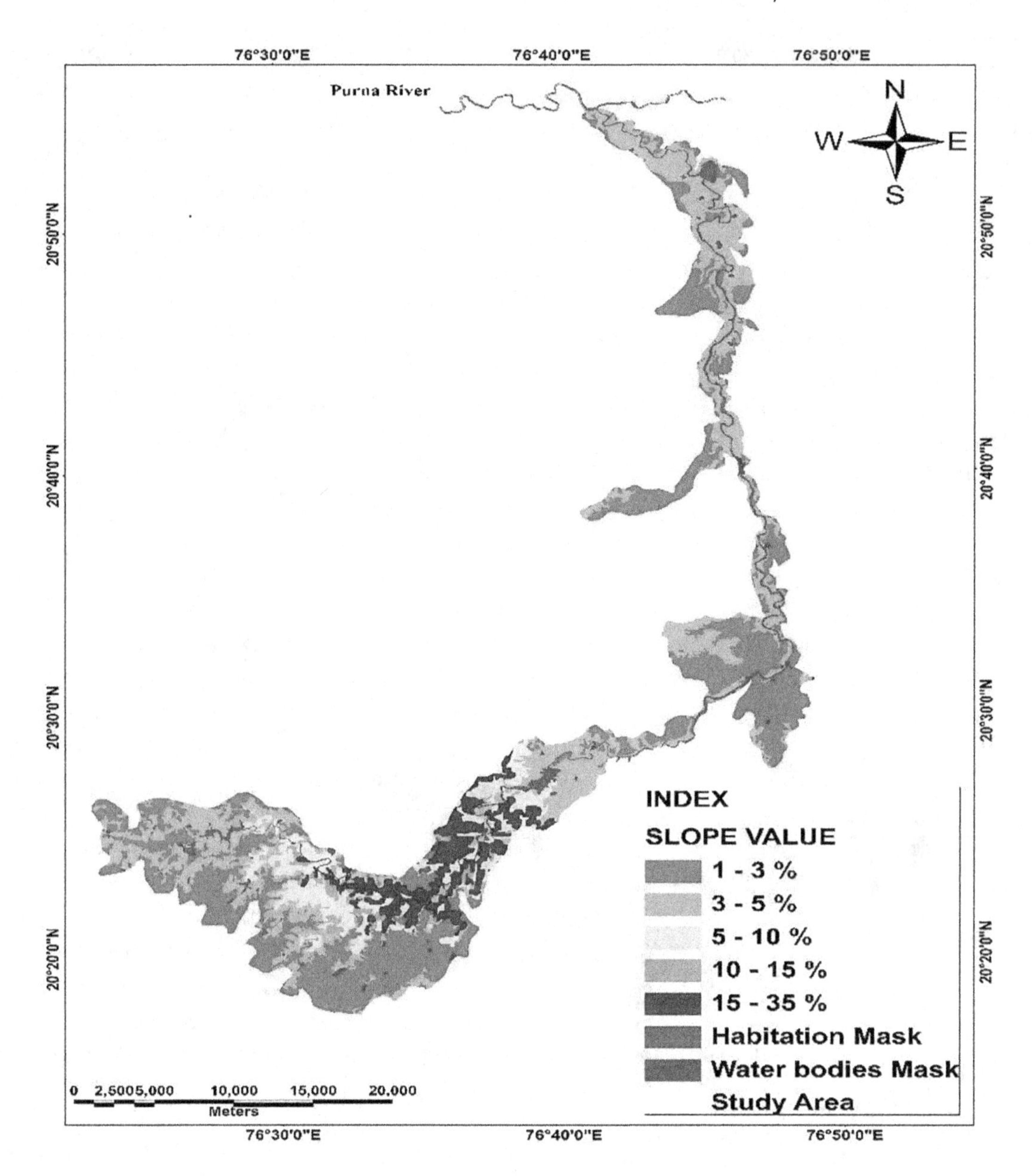

FIGURE 10.11 Slope map of the study area.

The soil depth changes from the north and central part of the basin; it is minimum in the basin's southern part. Again, it shows the rise in thickness of soil, and the soil in the area shows good layering characteristic in the northern region compared to the southern part of the basin. The soil depth map of the room was prepared, and different depth zones are shown; the deep soil zone is found in the southern and an eastern region of the basin; the intense site lies in most part of the basin, while the moderately deep zone is occurred in the central-western part as well as at some places in the southern region. The shallow soil zone appears in the basin's central and marginal or peripheral parts (Figure 10.12).

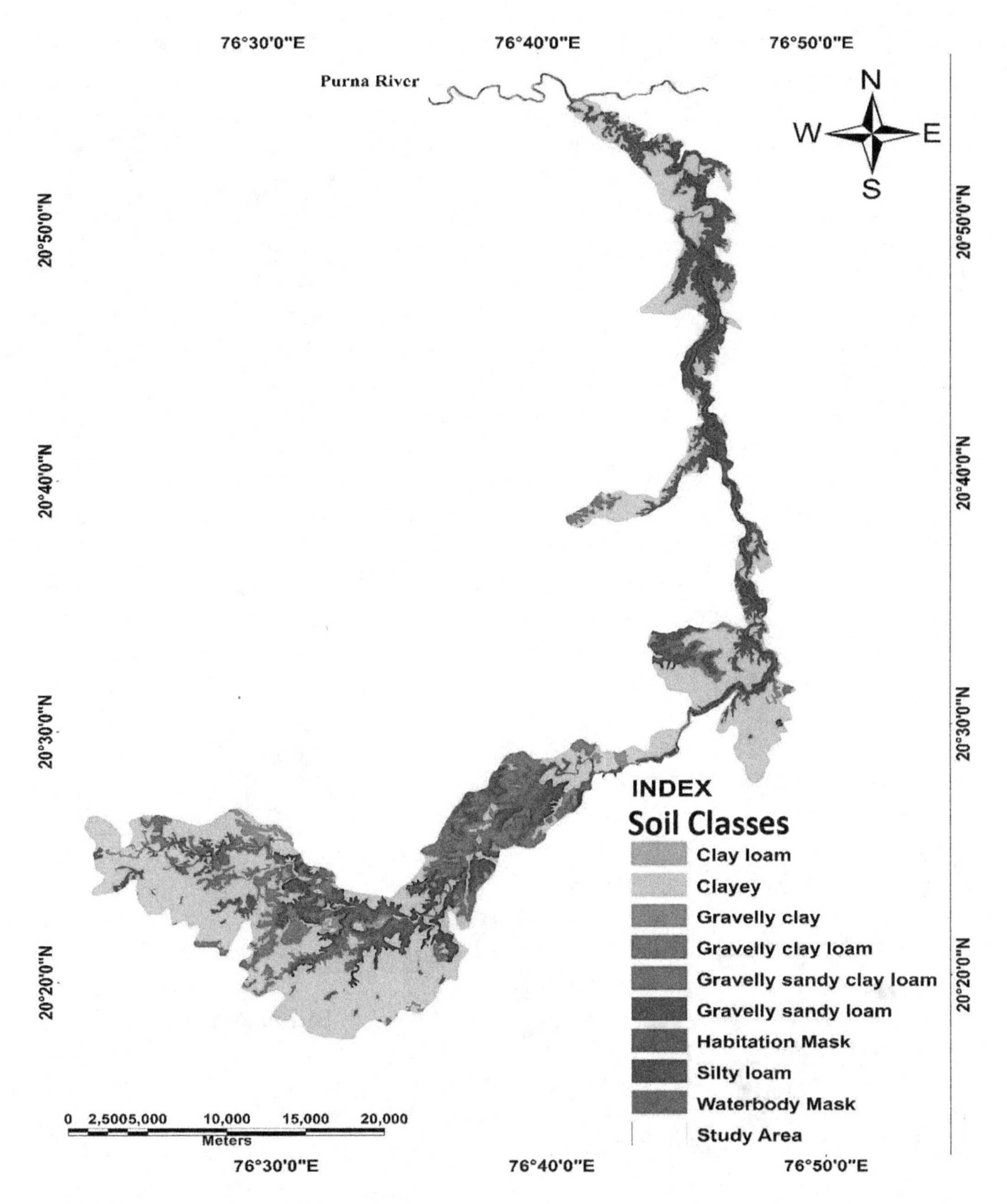

FIGURE 10.12 Soil map of the study area.

10.4.9 Rainfall

The annual rainfall is 350–800 mm in the river valley (*Source*: https://power.larc.nasa.gov) with a spatial resolution of 0.5° latitude by 0.5° longitude. The possibility of the rate of severe drought was the maximum at Akola (4%). Thus, the study reveals that extreme northern and north-eastern parts of the district, which experience moderate and severe drought conditions near about 20% of the years, can be separated as drought areas (Figure 10.13).

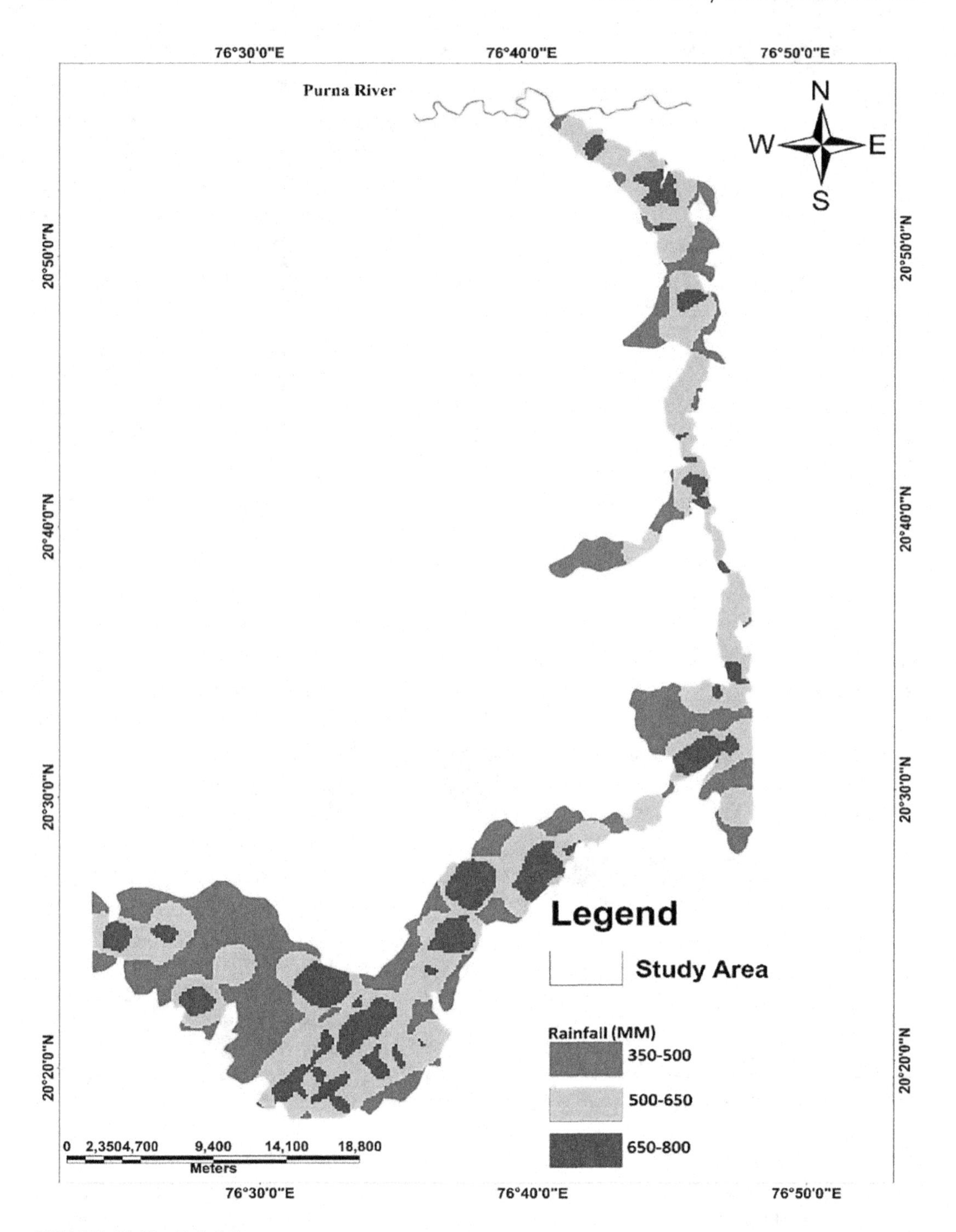

FIGURE 10.13 Rainfall pattern map.

10.5 GROUNDWATER PROSPECTIVE ZONE MAPPING

It is easy to recognize with the assistance of many factors such as geological structures, geomorphology, slope, and drainage (Sisay, 2007; Dev, 2015; Rose and Krishnan, 2009). Groundwater rises within different hydrogeological environments and investigation settings that control the groundwater-level distribution and development for other purposes (Tesfaye, 2010; Ayele et al., 2014). For this research, the prospective groundwater map of the area was completed by integrating thematic layers by using a weighted overlay method (Pande et al., 2019). The study area has excellent areas (97.73 km^2), very good (4.31 km^2), moderate (316.66 km^2), and poor (11.37 km^2) (Figure 10.14). The area details of groundwater prospect zones have been given in Table 10.3.

The excellent groundwater potential regions are present, where the hydrogeological setting has a flat topography, lower slope, high porosity, and high infiltration capacity. The site, which is under the agricultural influence, shows a substantial impact on groundwater holding capacity, and there are excellent potential regions. Inadequate and inferior groundwater water potential regions are located in mountainous areas, ridges, and denudation hills with steeper slopes, high run-off potential resulting in less water infiltration and storage in areas. The low groundwater prospects areas also occur at the lithological formations where lower porosity/permeability and present at high vertical slopes and high drainage density areas (Pande et al., 2017).

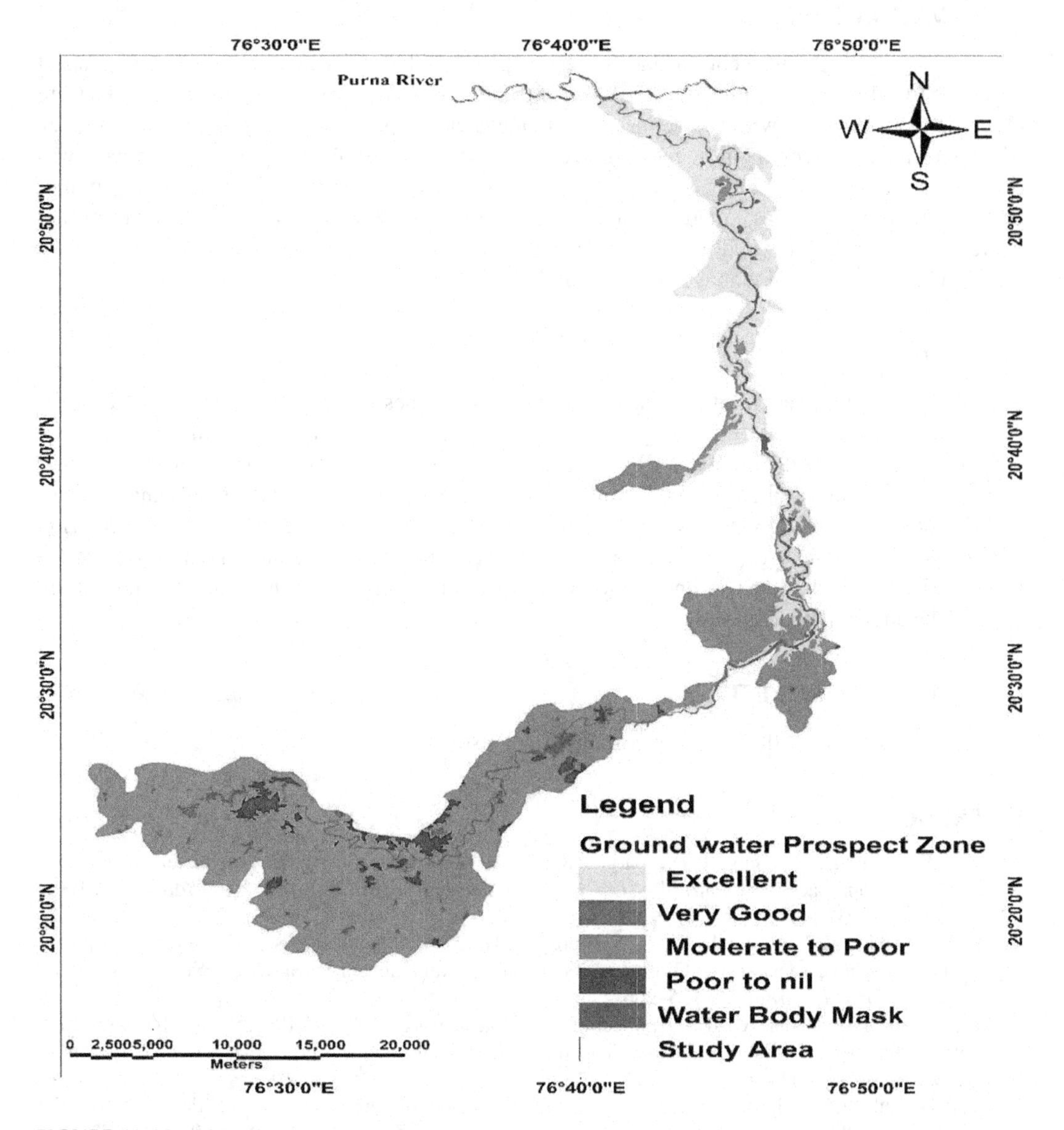

FIGURE 10.14 Map of groundwater prospect zones.

TABLE 10.3
Area Statistics of Different Groundwater Prospects Zones

Sr. No.	Groundwater Prospects category	Total Area (km²)
1	Excellent	97.73
2	Very good	4.31
3	Moderate to poor	316.66
4	Poor to nil	11.37
5	Water body	19.45
	Total	**449.52**

- *Including water body area*

10.6 VALIDATION

The existing 20 wells' water level data were compared with a delineated groundwater prospect zone map to validate the result (Figure 10.14). Of the 20 locations, nearly 17 observation wells were observed to cross-check with the delineated groundwater prospect zones map. The total accuracy of groundwater prospect mapping by weighted overlay method and AHP techniques approach was observed to be 85% correctness, signifying the consistency of the techniques. The investigation of groundwater prospect areas can give basic strategies to explore new wells, boreholes, and rainwater harvesting structures sites to demarcate vital sustainable groundwater and surface resources in the area and development and management purposes.

10.7 CONCLUSION

This study tries to demarcate the groundwater prospects zones by using the AHP and GIS techniques. Nine thematic maps, namely, geology, geomorphology, drainage density, lineament density, soil, groundwater level fluctuation, LULC, and rainfall, the slope was used to delineate groundwater prospects zones and assigned a suitable rank for every thematic map. The study area's final groundwater's potential map is divided into five categories: excellent, good, moderate, poor, and inferior. Using this groundwater prospect zone map, appropriate preparation and management of groundwater for agriculture and urban use, and improving the irrigation facility and enhancing the agriculture yield of the area will be achieved under climate change.

CONFLICT OF INTEREST

The authors declare that they have no conflict of interest.

REFERENCES

Agarwal, E., Agarwal, R., Garg, R.D., Garg, P.K., 2013. Delineation of groundwater potential zone: an AHP/ANP approach. *The Journal of Earth System Science*, 122(3), 887–898. https://doi.org/10.1007/s12040-013-0309-8.

Ayele. A.F., Addis. K., Tesfamichael. G., Gebrerufael, H. 2014. Spatial analysis of groundwater potential using remote sensing and GIS-based multi-criteria evaluation in Raya Valley, Northern Ethiopia. *Hydrogeological Journal*, 23, 195–206.

Biswas, A., Jana, A., Mandal, A. 2013. Application of remote sensing, GIS and MIF technique for elucidation of groundwater potential zones from a part of Orissa coastal tract, Eastern India. *Research Journal of Recent Sciences*, 2(11), 42–49.

Bobba, A.G., Bukata, R.P., Jerome, J.H. 1992. I digitally processed satellite data as a tool in detecting potential groundwater flow systems. *Journal of Hydrology*, 131(1–4), 25–62. https://doi.org/10.1016/0022-1694(92)90212-E

Choudhari, P.P., Nayak, L. 2012. Application of Remote Sensing GIS and GPS in Roof-Top rainwater harvesting of Chowgule College Campus, Margao: A Micro-Level Analysis. *Climate change*, Mumbai, 9/2012.

Choudhari, P.P., Nigam, G.K., Singh, S.K., Thakur, S. 2018. Morphometric based prioritization of watershed for groundwater potential of Mula river basin, Maharashtra, India. *Geology, Ecology, and Landscapes*, 2(4), 256–267. https://doi.org/10.1080/24749508.2018.1452482

Dev, S.G. 2015. Groundwater potential mapping of Sihu river watershed area of Mahoba District, UP using Remote Sensing and GIS. *IJAR*, 1(10), 241–248.

Doke, A.B., Zolekar, R.B., Patel, H., Das, S. 2021. Geospatial mapping of groundwater potential zones using multi-criteria decision-making AHP approach in a hardrock basaltic terrain in India. *Ecological Indicators*, *127*, 107685. https://doi.org/10.1016/j.ecolind.2021.107685

Essay, L. 2007. Application of remote sensing and GIS for groundwater potential zone mapping in Northern Ada'a plain (Modjo catchment) University/Publisher Addis Ababa University. http://etd.aau.edu.et/dspace/handle/123456789/386

Gogu, R.C., Carabin, G., Hallet, V., Peters, V., Dassargues, A. 2001. GIS-based hydrogeological databases and groundwater modelling. *Hydrogeological Journal*, 9, 555–569. https://doi.org/10.1007/s10040-001-0167-3

Goodchild, M.F. 1993. The state of GIS for environmental problem-solving. In: Goodchild, M.F., Parks, B.O., Steyaert, L.T. (eds.) *Environmental Modelling with GIS*. Oxford University Press, New York, pp. 8–15

Gossel, W., Ebraheem, A.M., Wycisk, P. 2004. A vast scale GIS-based groundwater flow model for the Nubian sandstone aquifer in Eastern Sahara (Egypt, northern Sudan and eastern Libya). *Hydrogeological Journal*, 12(6), 698–713.

Haridas, V.R., Aravindan, S., Girish, G. 1998. Remote sensing and its applications for groundwater favourable area identification. *Quarterly Journal of GARC*, 6, 18e22.

Loague, K., Corwin, D.L. 1998. Regional-scale assessment of non-point source groundwater contamination. *Hydrological Processes*, 12(6), 957–966. https://doi.org/10.1002/(SICI)1099-1085(199805)12:6<957::AID-HYP665>3.0. C.O.;2-J

Meijerink, A.M.J., Schultz, G.A., Engman, E.T. 2000. Remote sensing in hydrology and water management. Springer, Berlin, pp. 305–325. https://doi.org/10.1007/s11269-006-9024-4

Mundalik, V., Fernandes, C., Kadam, A.K., Umrikar, B.N. 2018. Integrated geomorphological, geospatial and AHP technique for groundwater prospects mapping in basaltic terrain. Hydrospatial Analysis, 2(1), 16–27.

Murmu, P., Kumar, M., Lal, D., Sonker, I., Singh, S.K. 2019. Delineation of groundwater potential zones using geospatial techniques and analytical hierarchy Process in Dumka District, Jharkhand, India. *Groundwater for Sustainable Development*, 9, 100239. https://doi.org/10.1016/j.gsd.2019.100239

Pande, C.B., Khadri, S.F.R., Moharir, N.K., Patode, R.S. 2017. Assess potential groundwater zonation of Mahesh River basin Akola and Buldhana districts, Maharashtra, India, using remote sensing and GIS techniques. *Sustainable Water Resources Management*, 4, 965–979. https://doi. org/10.1007/s40899–017–0193–5

Pande, C.B., Moharir, K.N., Pande, R. 2018. Assessment of morphometric and hypsometric study for watershed development using the spatial technology-A case study of Wardha river basin in Maharashtra, India. *International Journal of River Basin Management*, 4(4), 1–36. https://doi.org/10.1080/15715124.2018.1505737

Pande, C.B., Moharir, K.N., Singh, S.K., Varade, A.M. 2019. An integrated approach to delineate the potential groundwater zones in Devdari watershed area of Akola district, Maharashtra, Central India. *Environment, Development and Sustainability*, 22, 4867–4887. https://doi.org/10.1007/s1066 8–019–00409-1

Pande, C.B., Moharir, K.N., Panneerselvam, B. et al. 2021a. Delineation of groundwater potential zones for sustainable development and planning using analytical hierarchy process (AHP), and MIF techniques. *Applied Water Science*, 11, 186. https://doi.org/10.1007/s13201-021-01522-1

Pande, C.B., Moharir, K.N. 2021b. *Groundwater Resources Development and Planning in the Semi-Arid Region*. Springer, Cham, 1, pp. XIV, 571. https://doi.org/10.1007/978-3-030-68124-1

Patode, R.S., Pande, C.B., Nagdeve, B.M., Moharir, K.N., Wankhade, R.M. 2017. Planning of conservation measures for watershed management and development by using geospatial technology- a case study of Patur watershed in Akola district of Maharashtra. *Current World Environment*, 12(3), 708–716.

Rajesh, J., Pande, C.B., Kadam, S.A. et al. 2021. Exploration of groundwater potential zones using analytical hierarchical process (AHP) approach in the Godavari river basin of Maharashtra in India. *Applied Water Science*, 11, 182. https://doi.org/10.1007/s13201-021-01518-x

Razandi, Y., Pourghasemi, H.R., Neisani, N.S., Rahmati, O., 2015. Application of analytical hierarchy process, frequency ratio, and certainty factor models for potential groundwater mapping using GIS. *Earth Science Informatics*, 8(4), 867–883. https://doi.org/10.1007/s12145-015-0220-8

Rose, R.S., Krishnan, N. 2009. Spatial analysis of groundwater potential using remote sensing and GIS in the Kanyakumari and Nambiyar basins. India. *The Journal of the Indian Society of Remote Sensing*, 37(4), 681–692. https://doi.org/10.1007/s12524-009-0058-y

Saaty, T.L. 2008. Decision making with the analytic hierarchy process. *International Journal of Services Sciences*, 1, 1–83.

Saaty, T.L. 1990. *Decision Making for Leaders: The Analytic Hierarchy Process for Decisions in a Complex World*. RWS publications, Pittsburgh, PA.

Saraf, A.K., Jasrotia, A.S., Kumar, R. 2007. Delineation of groundwater recharge sites using integrated remote sensing and GIS in Jammu district, India. *International Journal of Remote Sensing*, 28(22), 5019–5036. https://doi.org/10.1080/01431160701264276

Sethna, S.F., Battiwala, H.K. 1977. Chemical classiµcation of the intermediate and acid rocks (Deccan Trap) of Salsette Island, Bombay. *Journal of the Geological Society of India*, 18, 323–330.

Shekhar, S., Pandey, A.C. 2015. Delineation of groundwater potential zone in hard rock the terrain of India using remote sensing, geographical information system (GIS) and Analytic hierarchy process (AHP) techniques. *Geocarto International*, 30(4), 402–421. https://doi.org/10.1080/10106049.2014.894584

Sheth, H.C. 2005. Were the Deccan flood basalts derived in part from ancient oceanic crust within the Indian continental lithosphere?. *Gondwana Research*, 8(2), 109–127. https://doi.org/10.1016/S1342-937X(05)71112-6

Shinde, S., Choudhari, P.P., Popatkar, B. 2020. Assessment of groundwater quality using GIS in Thane Municipal Corporation, Maharashtra, India. *Modeling Earth Systems and Environment*, 7, 1739–1751. https://doi.org/10.1007/s40808-020-00906-7

Singh, C.K., Shashtri, S., Singh, A., Mukherjee, S. 2011. Quantitative modelling of groundwater in Satluj River basin of Rupnagar district of Punjab using remote sensing and geographic information system. *Environmental Earth Sciences*, 62(4), 871–881. https://doi.org/10.1007/s12665-010-0574-7

Stafford, K.W., Rosales-Lagarde, L., Boston, P.J. 2008. Castile evaporates karst potential map of the Gypsum Plain, Eddy County, New Mexico and Culberson County, Texas: a GIS methodological comparison. *The Journal of Cave and Karst Studies,* 70(1), 35–46.

Strahler, A.N. 1954. Quantitative geomorphology of erosional landscapes. *C-R 19th International Geological Congress Algiers* 1952 sec.13 (pt. 3), pp. 341–354.

Watkins, B., Lu, Y.C., Hart, G., Daughtry, C. 1997. The current state of precision farming. *Food Research International*, 13(2), 141–162. https://doi.org/10.1080/87559129709541104

11 Hydro-geochemical Evaluation of Phreatic Groundwater for Assessing Drinking and Irrigation Appropriateness – A Case Study of Chandrapur Watershed

Murkute YA

11.1 INTRODUCTION

Water is the fundamental need of life, and hence, its quality has turned out to be the vital issue worldwide. In situ geogenic sources, which are primarily responsible for higher concentration and intense mineralization, increase the contamination of groundwater. To come to a decision on the quality and suitability of groundwater for distinct purposes, lot of work has been carried out on the type of mineralization and physicochemical characterization of groundwater. Such studies have been found to be necessary to ascertain the processes responsible for physical characters of mineral species and groundwater chemistry (Eaton, 1950; Gupta, 1983). The poor quality of groundwater has direct adverse effects on the human health as well as on the growth of plants (US Salinity Laboratory Staff, 1954; Davis and DeWeist, 1966; Wilcox, 1955; Teotia and Teitia, 1988; Hem, 1991; Loizidou and Kapetanious, 1993; Deshmukh et al., 1995; Deshmukh and Vali, 1995; Todd, 1995; Canter, 1997; Apambire et al., 1997; Jha and Verma, 2000; Ekambaram, 2001; Sanhez-Perez and Tremolieres, 2003; Sunitha et al., 2005; Subba Rao, 2011; Nazzal et al., 2014; Murkute, 2014, 2022; Voutsis et al., 2015; Kumar et al., 2018; Laxmankumar et al., 2019; Panneerselvam et al., 2022).

Fluorite is a halide mineral of uncommon incidence in the soils; nevertheless, its mineralization in aquifer-rock poses threat to the physicochemical environment of the groundwater regime. The soluble fluoride-bearing minerals having extended residence time are the sole sources of fluoride (F^-) in natural water. The present investigation is an attempt to unravel the impact of fluorite mineralization on the physicochemical environment of the groundwater from fluoride deposit area, WRD watershed (Lat. 20°15′15″–20°21′05″: N and Long. 79°53′55″–79°03′35″: E), Chandrapur District, Maharashtra, Central India (Figure 11.1).

11.2 FLUORITE (CAF_2) MINERALIZATION IN GEOLOGICAL ENVIRONMENT

11.2.1 PHYSICAL CHARACTERISTICS OF FLUORITE (CAF_2)

Fluorite is a halide mineral of uncommon occurrence in soils having hardness of 4 in the Mohs Hardness Scale with 3.2 specific gravity. It belongs to the cubic crystal system with a body-centered structure, often with penetration twins, and has four directions of perfect cleavage, habitually

DOI: 10.1201/9781003303237-11

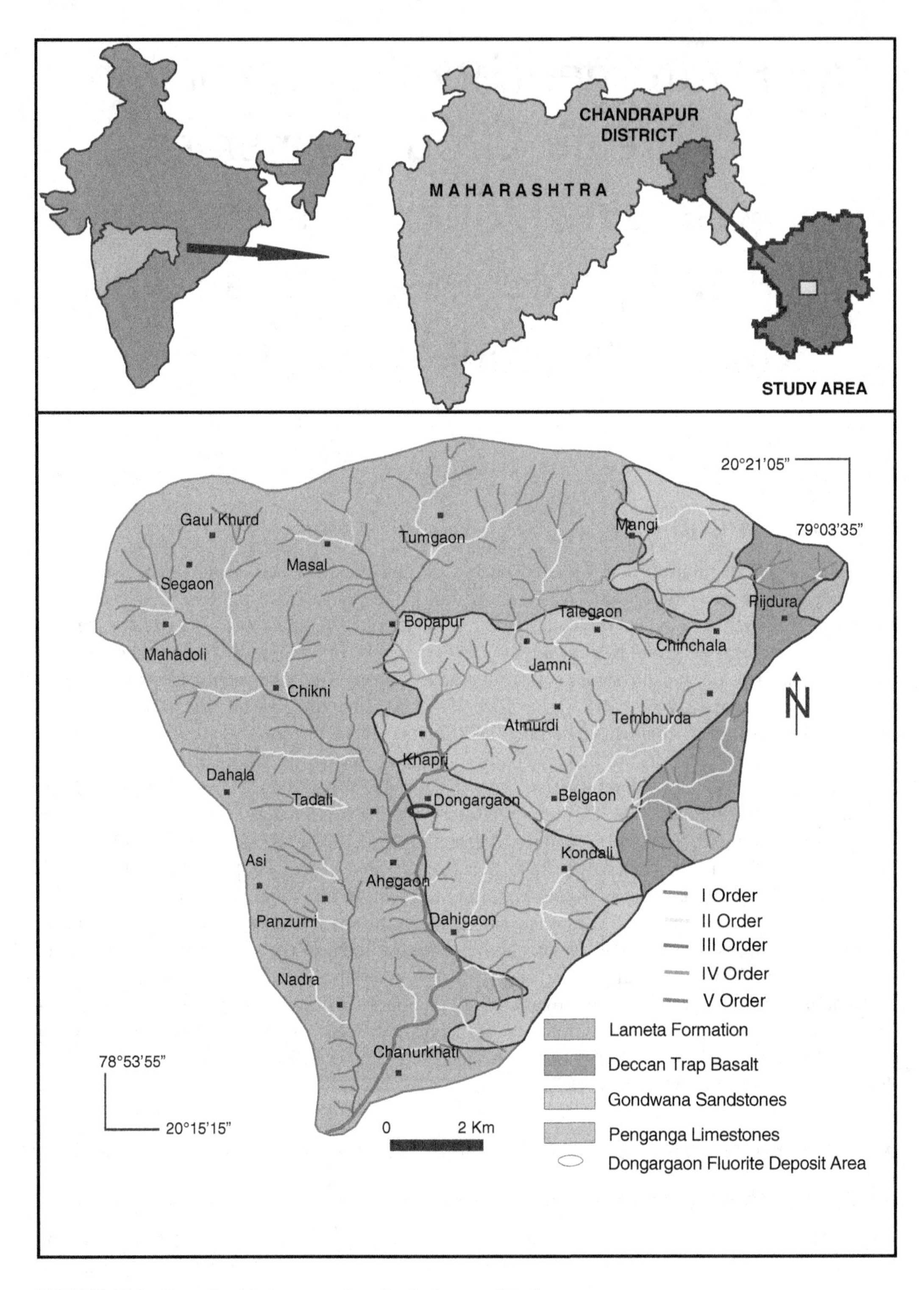

FIGURE 11.1 Location, drainage, and geological map of study area.

breaking into pieces with the shape of an octahedron. It also occurs in both massive and earthy forms and as crusts or globular aggregates with radial fibrous texture. Pure fluorite is transparent, both in visible and ultraviolet light, but impurities usually make it a colorful mineral and the stone has ornamental and lapidary uses. Industrially, fluorite is used as a flux for smelting, and in the production of certain glasses and enamels.

11.2.2 High Concentration of F^- in Groundwater

The sources of fluoride (F^-) in natural fresh water are the soluble fluoride-bearing minerals having longer residence time. Many times percolating rain water carries soluble fluoride compounds from soil to the deeper depths. Fluoride has both positive as well as negative impacts on human body (Richards, 1954; Wood, 1974; Apambire et al., 1997; Datta et al., 2000; BIS, 2003). Similarly, fluoride absorption in different food-bearing plants is different and the matter of the fact is that the fluoride percentage in such plants is always higher in endemic areas. Fluoride ingestion is essential for human health because lack of it causes the dental carries, and on the other hand, excess ingestion leads to dental and skeletal fluorosis (WHO, 1984; Dev Burman et al., 1995; Subba Rao, 2003). Fluorosis is manifested in three main types (Kharb and Susheela, 1994; Susheela, 2001). The dental fluorosis causes pitting of teeth, opaque patches, chalkiness, staining, chipping of enamel while the skeletal fluorosis causes pain in the neck, joints, back, etc. The skeletal fluorosis is also associated with rigidity and restricts movement of cervical and lumbar spine, shoulder joints as well as knee and pelvic joints (WHO, 1994; Teotia and Teotia, 1988; Susheela, 2001; Panneerselvam et al., 2021). The non-skeletal fluorosis is apparent in the form of neurological, muscular, allergic, gastrointestinal ailments as well as headache and urinary tract infections (Ozha and Mathur, 2001).

11.2.3 Genesis of Fluorite (CaF_2) Mineralization in Study Area

Fluorite (CaF_2) mineralization has been hosted in brecciated and silicified Penganga calcareous sediments in the southwest of Dongargaon village of Warora Tehsil, Chandrapur district. The mineralization has been attained in three distinct tectogenetic episodes, wherein the Limestones of Penganga were enriched in fluoride and barite representing a syngenetic sedimentary mineralization at the first phase. Subsequently, calcareous sediments were pervasively silicified, and in the last phase, the rocks were intensely sheared, brecciated, and mineralized with fracture filling-type fluorite (Deshmukh et al., 1995).

11.3 STUDY AREA DETAILS

The study area representing WRD watershed of 282.82 km^2 covers 25 villages (Lat. 20°15′15″–20°21′05″: N and Long. 79°53′55″–79°03′35″: E) from Warora tehsil, Chandrapur district, Maharashtra state, Central India. The area falls under tropical dry sub-humid climatic zone, where in summer, months are very hot (max. temp. ~46°C), while winter is cool (min. temp. ~12°C). The average annual precipitation falls in the study area is 1,089 mm by the southwest monsoon (June to September) and has a relative humidity of 71% during the monsoon season. The WRD watershed area represents moderately dissected plateau topography with a few isolated hills. The study area has a general slope toward south direction. The Dahiwal River and its tributaries drain the entire watershed area in the south.

11.4 MATERIALS AND METHODS

In the present work, the standard methods prescribed by American Public Health Association (APHA, 1995) have been followed for the field as well as laboratory analysis of groundwater samples. In total, 25 groundwater samples were collected from 25 villages spread over the watershed area. In additional, usual procedures prescribed by the Hem (1991), Handa (1974, 1975), Trivedy

TABLE 11.1
Analytical Units and Formulae Followed

Sr. No	Parameters	Characteristics	Unit
1	Physical	Temperature	°C
2		pH	–
3		Electrical conductivity (EC)	µS/cm
4		Total dissolved solids (TDS)	mg/L
5		Total hardness (as $CaCO_3$)	mg/L
6	Major cations	Calcium (Ca^{++})	mg/L
7		Magnesium (Mg^{++})	mg/L
8		Sodium (Na^+)	mg/L
9		Potassium (K^+)	mg/L
10	Major anions	Carbonate (CO_3^{--})	mg/L
11		Bicarbonate (HCO_3^-)	mg/L
12		Nitrate (NO_3^-)	mg/L
13		Sulphate (SO_4^{--})	mg/L
14		Chloride (Cl^-)	mg/L
15		Fluoride (F^-)	mg/L
16	Irrigation water use (by calculations)	Residual sodium carbonate (RSC) $RSC=(HCO_3^-+CO_3^-)-(Ca^{++}+Mg^{++})$	meq/L
17		Residual sodium bicarbonate (RSBC) $RSBC=HCO_3^- - Ca^{++}$	meq/L
18		Soluble sodium percentage (SSP) $SSP=[(Na^++K^+)/(Ca^{++}+Mg^{++}+Na^++K^+)]\times 100$	meq/L
19		Mg ratio (MR) $MR=(Mg^{++}\times 100)/(Ca^{++}+Mg^{++})$	meq/L
20		Kelley's ratio (KR) $KR=Na^+/(Ca^{++}+Mg^{++})$	meq/L

and Goyal (1986), and Singh et al. (2010) were also considered for the analysis. The guidelines given by WHO (1997) and BIS (1991) were employed for the standards correlation of the chemical data. The overall relative standard deviation (RSD) was less than 10%, and the results in respective units as well as formulae were noted as per Table 11.1. The mechanism of rock–water interaction was established by following Gibbs variation diagrams (1970).

Various modeling methods/software were used to understand and evaluate the proper understanding of hydro-geochemical interrelationship and rock-water interface. The statistical analyses of hydro-geochemical data of all the groundwater samples were performed by using the Microsoft Office Excel 2007. This software was also used to construct bivariate inter-relationship diagrams of individual cation versus anion or even in combinations/group because such diagrams are extremely helpful to distinguish hydro-geochemical reactions between groundwater and rock-minerals interface. The Piper trilinear diagram as well as the US Salinity diagram were generated through AquaChem 4.0 software just to understand the various combinations of water types and to recognize the control over the groundwater pollution.

11.5 PHYSICOCHEMICAL CHARACTERISTICS OF GROUNDWATER

11.5.1 Physical Parameters

The temperature of the groundwater samples has been noted to vary from 26.5°C to 29.2°C and pH values grade between 7.1 and 8.5, suggesting dominantly alkaline nature (Table 11.2). The high soil-CO_2 ultimately accelerates the dissolution of silicate minerals, particularly feldspars and the

TABLE 11.2
Analytical Data of 25 Groundwater Samples Collected from Study Area

	pH	EC	TDS	TH	Ca^{++}	Mg^{++}	Na^{+}	K^{+}	NO_3^-	CO_3^{--}	HCO_3^-	SO_4^{--}	Cl^-	F^-
Min	7.1	809.0	517.8	305.2	47.8	26.4	44.0	2.9	16.9	16.7	161.5	65.4	56.5	0.3
Max	8.5	1,715.0	1,097.6	1,718.7	258.3	261.7	474.2	145.8	188.4	157.9	580.1	528.4	851.4	4.4
Avg	8.0	1,142.0	730.9	701.4	115.4	100.7	232.9	29.4	71.4	40.1	337.1	231.0	367.2	2.0
Sd	0.5	321.2	205.6	398.6	59.4	69.8	130.8	39.6	47.4	35.2	106.3	149.9	238.1	1.6
CoV	6.0	28.1	28.1	56.8	51.5	69.4	56.2	134.7	66.4	87.8	31.5	64.9	64.8	78.8

Min, minimum; Max, maximum; Avg, average; Sd, standard deviation; CoV, coefficient of variance.
Except EC (μS/cm), other parameters are presented in mg/L.

carbonate minerals, leading the high alkalinity in groundwater (Roy et al., 2018). The values of electrical conductivity (EC) were found to range from 809 to 1,715 μS/cm. All the groundwater samples have TDS values, ranging from 517.8 to 1,097.6 mg/L, exceeding the desirable limit of 500 mg/L recommended by WHO (1997). The spatial variation in TDS values may be attributed to variations in lithology and hydrological processes (Singh et al., 2010; Pande et al., 2020). The 80% groundwater samples have TH values more than the desirable limit of 500 mg/L (305.2–1718.7 mg/L) as suggested by WHO (1997).

11.5.2 Cation-Anion Chemistry

Among the cations, Ca^{++} is the major constituent, which ranges from 47.8 to 258.3 mg/L and the concentration of Mg^{++} ranges from 26.4 to 261.7 mg/L. The sources of Mg^{++} in the natural waters are the magnesium-bearing minerals present in the rocks. The domestic as well as industrial wastes also contribute to increase Mg^{++} concentration. The Na^+ and K^+ contents in shallow aquifer range between 44–474.2 mg/L and 2.9–145.8 mg/L, respectively. The positive interrelationship between Na^+ and K^+ corresponds to minerals like sodium plagioclase, potash plagioclase along with anthropogenic sources like domestic and animal waste. The inverse inter-relationship of increased Na^+ content with decreased Ca^{++} concentration in the present investigation is usual phenomenon in alkaline groundwater conditions (Handa, 1975; Jacks et al., 2005; Murkute, 2014).

The CO_3^{--} contents from the study area grade from 16.7 to 157.9 mg/L, and the HCO_3^- contents vary from 161.5 to 580.1 mg/L. CO_3^{--} and HCO_3^- contents in the groundwater are generally derived from rain water containing dissolved CO_2. When rain water infiltrates the soil, it dissolves more CO_2 from decaying organic matter and the carbonate minerals like calcite and dolomite present in the soil and finally forms HCO_3^- (Karanth, 1987). In the present study, 62% of groundwater samples have HCO_3^- concentration above 300 mg/L as prescribed by WHO (1997) and BIS (1991). NO_3^- content of groundwater samples grades from 16.9 to 188.4. NO_3^- is the most widespread contamination of groundwater (Hallberg and Keeney, 1993), and it is a non-lithological source of contamination (Ritzi et al., 1993). Decomposition of organic matter in soils, leaching of soluble chemical fertilizers, human and animal excreta, and untreated effluents of nitrogenous industries and sewage disposal are potential sources of nitrate contamination in groundwater. The SO_4^{--} concentration of groundwater from the study area is within the potable limit (BIS, 1991) (65.4–528.4 mg/L). The concentration of Cl^- in the groundwater samples from the study area ranges from 56.5 to 851.4 mg/L. The high concentration of chloride (Cl^-) in groundwater is derived from weathering of minerals like halite. Excess of Cl^- in the groundwater strongly points out toward the pollution and is considered as tracer for groundwater contamination (Loizidou and Kapetanions, 1993).

11.5.3 Rock–Water Interaction

The hydrogeochemistry of groundwater to the large extent depends on the aquifer mineralogy, climatic setup of the area, general slope characters, drainage characteristics, longer time of residence in groundwater, and anthropogenic activities influencing that area. Gibbs (1970) has propounded the mechanism of controlling the chemical composition of water as the proximity of chemical composition of water with aquifer lithology. In the present study, the Gibbs plots are constructed by plotting ratio against the TDS by dominant anions [$(Na^{+}+K^{+})/(Na^{+}+K^{+}+Ca^{++})$] (Figure 11.2a) and dominant cations [$(Cl^{-}/Cl^{-}+HCO_3^{-})$] (Figure 11.2b). The sample points in Gibbs diagram reveal that chemical weathering of rock forming minerals is the main causative factor in the evolution of chemical composition of groundwater occurring in all the lithological domains of study area, which is afterward influenced by anthropogenic activities (Gibbs, 1970; Ravikumar et al., 2010).

The interrelationship of $Ca^{++}+Mg^{++}$ *vs* $SO_4^{-}+HCO_3^{-}$ has been used to interpret the rock–water interaction which points out toward the silicate weathering as well as sources of calcium and bicarbonate (Lakshmanan et al., 2003), where in the present investigation, it reveals that the majority of samples fall below the equiline suggesting excess of $SO_4^{-}+HCO_3^{-}$ values. The dominance of $SO_4^{-}+HCO_3^{-}$ reveals silicate weathering representing dissolution of silicate minerals, thereby, decreasing calcium content in groundwater (Datta and Tyagy, 1996). The points which lie above the equiline indicate excess of $Ca^{++}+Mg^{++}$. Such excess of $Ca^{++}+Mg^{++}$ in the groundwater regime of the study area is attributed to carbonate weathering (Lakshmanan et al., 2003). The molar ratios of Na^{+}/Cl^{-} in the maximum number of the groundwater samples are found to be more than 1, which points out the sources for Na^{+} in the study area as a result of silicate weathering reactions (Meyback, 1987; Murkute, 2014). The Na^{+} *vs* HCO_3^{-} is also a measure of silicate weathering. The scatter plot of such interrelationship shows that HCO_3^{-} is comparatively more in concentration than Na^{+} content. The increased concentration of HCO_3^{-} than Na^{+} indicates the prevalence of silicate weathering, which is due to the reduction of the sodium concentration because of ion exchange process.

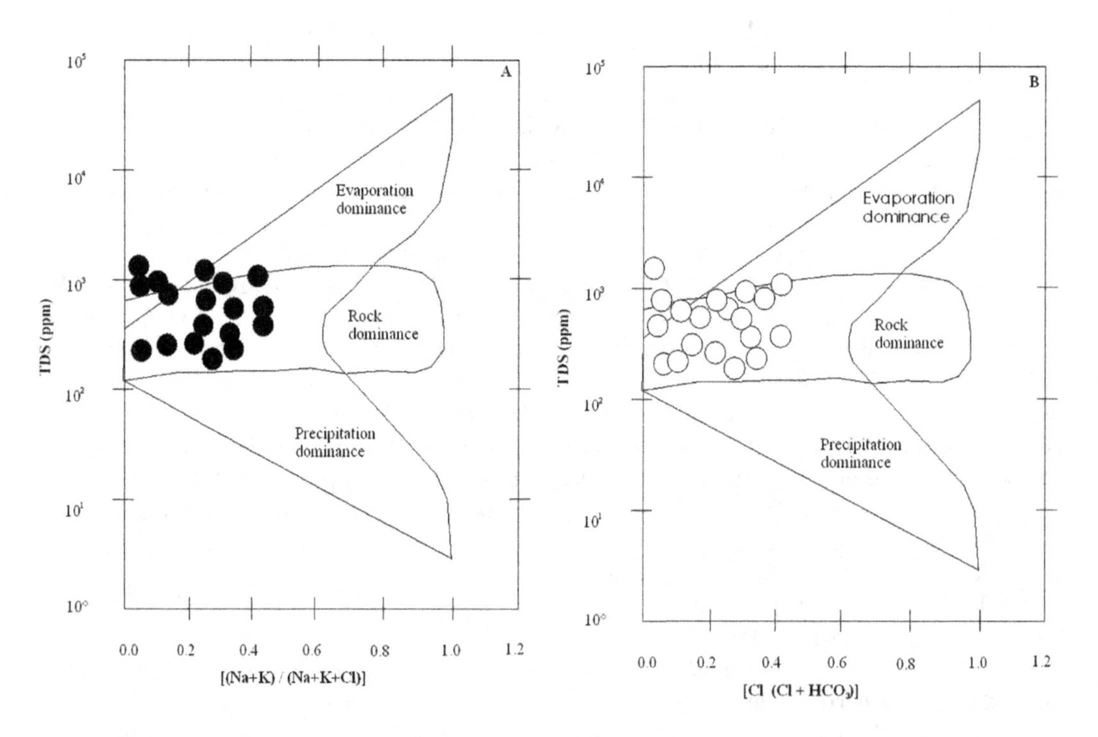

FIGURE 11.2 Gibbs diagrams for study area. (a) cation category and (b) anion category.

The fertilizers and soil amendments for agricultural purpose, animal wastes, industrial as well as municipal sewage are the sources of Cl^-, SO_4^{--}, NO_3^-, and Na^+ ions (Jalali, 2009). The higher concentration of Cl^-, SO_4^{--}, NO_3^-, and Na^+ contaminates the groundwater sources, which lie nearby. The correlation of these ions with TDS clearly indicates the influence of human activity on the water chemistry (Han and Lui, 2004; Jalali, 2009). A positive correlation between Cl^- and NO_3^- in the present study (Table 11.3) also points out anthropogenic input (Back and Hanshaw, 1966; Piskin, 1973; Ritter and Chrinside, 1984).

The major cations–anions have been plotted in the Piper's trilinear diagram. The data shows that Ca^{2+} is the dominant cation followed by Na^+ and Mg^+. HCO_3^- is the most dominant anion followed by Cl^- and SO_4^{2-}. The diagram also shows that groundwater from the study area belongs to six hydro-chemical types. In the study area, in general, alkaline earths ($Ca^{++}+Mg^{++}$) exceed the alkalis (Na^++K^+); however, in 24% samples, alkalis (Na^++K^+) exceed alkaline earths ($Ca^{++}+Mg^{++}$). 80% samples from study area have strong acids ($SO_4^{--}+Cl^-+F^-$) exceed weak acid ($CO_3^{--}+HCO_3^-$) while remaining 20% samples show weak acids ($CO_3^{--}+HCO_3^-$) exceed strong acids ($SO_4^{--}+Cl^-+F^-$).

11.6 DISCUSSION ON GROUNDWATER APPROPRIATENESS

The hydro-geochemical data have been synthesized to deduce appropriateness of groundwater from the study area for drinking and general domestic use as well as irrigation purposes.

11.6.1 Drinking and General Domestic Use

The suitability of groundwater for drinking and public health purposes was assessed using prescribed limits of WHO (1997) and Indian drinking water standards (BIS, 1991). The groundwater use for drinking purpose on the basis of TDS range (Davis and De Wiest, 1966) has following class limits, namely, up to 500 mg/L (desirable for drinking), 500–1,000 mg/L (permissible for drinking), and up to 3,000 mg/L (useful to agriculture). According to this classification, 83.66% of groundwater samples from the entire study area are not potable. The 58.33% groundwater samples have TH value above the desirable limit suggested by WHO (1997) and hence not suitable for drinking purpose and even for domestic use. As per this classification given by Dufor and Becker (1964), the groundwater of the study area is hard to very hard, suggesting non-suitability for domestic purpose. The 23% of groundwater samples have SO_4^{--} concentration exceeding the value of permissible

TABLE 11.3
Correlation Matrix

	pH	EC	TDS	TH	Ca^{2+}	Mg^{2+}	Na^+	K^+	NO_3^-	HCO_3^-	SO_4^{2-}	Cl^-
pH	1.00											
EC	−0.29	1.00										
TDS	−0.31	1.00	1.00									
TH	−0.38	0.66	0.55	1.00								
Ca^{2+}	−0.39	0.76	0.83	0.95	1.00							
Mg^{2+}	−0.28	0.32	0.42	0.83	0.64	1.00						
Na^+	−0.05	0.66	0.74	−0.25	0.37	0.19	1.00					
K^+	−0.27	0.58	0.68	−0.29	0.24	0.22	0.52	1.00				
NO_3^-	0.36	0.64	0.74	−0.36	0.65	0.58	0.47	0.24	1.00			
HCO_3^-	0.61	0.41	0.39	−0.31	0.17	−0.17	0.41	0.42	−0.19	1.00		
SO_4^{2-}	−0.24	0.75	0.83	0.78	0.82	0.52	0.72	0.41	0.18	0.64	1.00	
Cl^-	−0.41	0.74	0.85	0.22	0.68	0.28	0.85	0.52	0.61	0.19	0.37	1.00

limit of 400 mg/L (BIS, 1991). The 30% of the samples have SO_4^{--} content within the limit of 200–400 mg/L, which has a bitter taste. According to Singh et al. (2010), such water may also cause corrosion of metals, if used for longer time, particularly in water with low alkalinity.

11.6.2 Irrigation Use

Various irrigation quality indicators have been applied to evaluate the appropriateness of groundwater samples from the study area (Table 11.4). Sodium adsorption ratio (SAR): It is the measure of soil permeability, where in Na^+, Ca^{++}, and Mg^{++} cations are involved. The calculated SAR values of the groundwater samples from the study area range from 0.51 to 3.63 meq/L and infer excellent quality of groundwater for irrigation purpose. The SAR values in terms of sodium hazard are compared with the salinity hazard in the US Salinity Laboratory's diagram (US Salinity Laboratory Staff, 1954). The main cluster of plots of the groundwater samples spreads in C_3–S_1 type specifying the water of medium-to-high salinity – medium sodium type. The groundwater

TABLE 11.4
Irrigation Suitability Indices for Groundwater of Study Area

Sr. No	Irrigation parameter	Min	Max	Classification	Water quality	No. of Samples	% of Samples
1	Sodium adsorption ratio (SAR)	0.51	3.63	< 10	Excellent	40	100
				11–18	Good	–	–
				19–26	Doubtful	–	–
				> 27	Unsuitable	–	–
2	% Sodium	7.73	122.0	< 20	Excellent	28	65
				21–40	Good	10	25
				41–60	Permissible	1	5
				61–80	Doubtful	–	–
				> 80	Unsuitable	1	5
3	Residual sodium carbonate (RSC)	0.03	2.67	< 1.25	Good	17	42.5
				1.25–2.5	Doubtful	19	47.5
				> 2.5	Unsuitable	4	10
4	Residual sodium bicarbonate (RSBC)	–7.4	23.8	< 5	Satisfactory	20	52
				5–10	Marginal	14	35
				> 10	Unsatisfactory	6	13
5	Soluble sodium percentage (SSP)	11.9	54.2	≤ 50	Good	36	90
				> 50	Unsuitable	4	10
6	Mg ratio (MR)	29.7	84.8	< 50	Suitable	21	52.5
				> 50	Harmful	19	47.5
7	Corrosivity ratio (CR)	0.03	0.31	≤ 1	Good	40	100
				> 1	Corrosive	–	–
8	Kelley's ratio (KR)	0.1	1.2	< 1	Good	37	92.5
				> 1	Unsuitable	3	7.5
9	Potential salinity index (PS)	1.4	7.8	< 3	Excellent to Good	26	65
				3–5	Good to Injurious	9	22.5
				> 5	Injurious to Unsatisfactory	5	12.5
10	Synthetic harmful coefficient (*K*)	1.3	11.6	< 25	Excellent	40	100
				26–36	Good	–	–
				37–44	Injurious	–	–
				> 44	Unsuitable	-	-

points corresponding to the C_3–S_2 type (20%) represent the medium salinity – medium sodium type. Such groundwater for the above two categories may be utilized for irrigation with a slight threat of exchangeable sodium. The other very minor clusters of points are in C_2–S_1 (10%) and C_4–S_1 (5%). Such groundwater for the above two categories may be utilized for salt-tolerant or semi-tolerant crops under the favorable drainage conditions. The surplus salinity decreases the osmotic activity of plants, moreover impedes the absorption of water and nutrients from the soil (Saleh et al., 1999) as well as affects the chemical as well as the physical setup of agricultural soil and consequently directly or indirectly affects the soil structure, permeability, and aeration (Subba Rao, 2003, 2006; Kouadri et al., 2022; Elbeltagi et al., 2022).

Percent Sodium (%Na): The higher % Na^+ in irrigation water commences the exchange of sodium in water, and exchange calcium as well as magnesium concentrations in the soil, having deprived inner drainage (Simsek and Gunduz, 2007; Murkute, 2014; Chacha et al., 2018). The groundwater having higher EC as well as high sodium percentage (% Na), due to altered soil structures, eventually kills the plant (Islam et al., 2017). Only one sample is found unsuitable while the others are either of excellent (65%) to good (25%) category of the %Na in the study area.

Residual Sodium Carbonate (RSC): The RSC involves the combination of HCO_3-, Ca^{2+}, and Mg^{2+}. The RSC value >2.5 meq/L is considered harmful to the growth of plants, and thus, the groundwater is unsuitable for irrigation. In the present investigation, the RSC values grade from 0.03 to 2.67. All the positive RSV values, as noted in the present investigation, result in alteration of the soil structure (Zhang et al., 2019) and obstruct the plant growth due to sodium-induced calcium deficiency and sodium toxicity (Adimalla and Quian, 2019).

Residual Sodium Bicarbonate (RSBC): Though the RSBC value represents the simple mathematical difference between the HCO_3^- anion and Ca^{2+} cation, the concentration of these ions in groundwater is largely controlled by rock-water interaction. It imparts geogenic alkalinity to the groundwater and hence does not support the alkaline-intolerant type of crops. It has a range from <5 meq/L (satisfactory: 52%) to >10 meq/L (unsatisfactory: 13%) with marginal utilization of groundwater (35%) for irrigation.

Soluble Sodium Percentage (SSP): The higher soluble sodium percentage has undeserved consequences in irrigation water due to the process of base exchange, which replaces the calcium cation primarily by the sodium cation in the soil, which in turns lowers the soil permeability. The groundwater having SSP values ≤50 represents good quality, whereas values >50 are inapt for irrigation utility. 90% of the groundwater samples of the study area have SSP ≤ 50, and hence found to be suitable for irrigation purpose.

Mg Ratio (MR): The alkaline earth metals (Ca^{2+} and Mg^{2+}) are naturally the sole contributors to chemistry of groundwater at rock–water interface. These earth metals have naturally maintained equilibrium which when deviates causes the reduction in the infiltration capacity of soil, and consequently, the crop yield is reduced. The MR values < 50 are suitable, and the MR values > 50 are unsuitable for the irrigation purpose. In totality, 21 groundwater samples from the study area have suitability, while 19 groundwater samples are harmful for irrigation purpose.

Corrosivity Ratio (CR): The CR is crucial parameter because it involves probably the anthropogenically generated anions (Cl^- and SO_4^{--}) in equilibrium with geogenically generated ion HCO_3^- (and/or with CO_3^{--}, if present at aquifer–water interface). The groundwater is good for irrigation purpose if the CR value is ≤ 1; if the CR value is > 1, then such groundwater is considered as corrosive in nature, hence such water is not suitable for irrigation purpose. All the groundwater samples from the study area are non-corrosive, and hence, water can be transported to longer distances for irrigation purpose.

Kelley's Ratio (KR): The KR involves the mathematical relationship of divalent cations (Ca^{++} and Mg^{++}) and univalent cation (Na^+). The equilibrium between the divalent cations and the univalent cation gets disturbed with the enrichment of univalent cation. The maintained equilibrium in the groundwater hydrochemistry is represented by KR < 1, while disturbed equilibrium in the groundwater hydrochemistry is represented by KR > 1. The Kelley's ratio (KR) for the groundwater of the study area ranges from 0.4 to 1.4 meq/L. The KR value for the 92.5% of the groundwater samples from the study area is less than 1, and hence, such water is suitable for irrigation purpose.

Potential Salinity index (PS): Mathematically, the total sum of Cl^- with half concentration of SO_4^{--} represents the potential salinity index. Cl^- presence in the groundwater system is non-geogenic and clearly corresponds with anthropogenic activities in that region. The SO_4^{--} content in groundwater primarily connotes geogenic gypsum dissolution or even through irrigation runoff (Rao et al., 2017). The 65% of the groundwater samples in the present investigation have PS values representing the excellent to good water type (PS < 3); 22.5% good to injurious water type (PS: 3–5); and 12.5% injurious to unsatisfactory (PS > 5).

Synthetic Harmful Coefficient (*K*): A very distinct combination of TDS and SAR values represents the synthetic harmful coefficient. Xu et al. (2018) and Zhou et al. (2020) have used the *K* value to evaluate high salt presence and alkali hazards in the groundwater to infer its suitability for irrigation purpose. Since the *K* value of all the groundwater samples from the study area is less than 25 (*K* range: 1.19–12.1 meq/L), it is suitable for irrigation purpose.

11.7 F^- CONTENTS FROM STUDY AREA

The F^- content in groundwater from the study area grades from 0.3 to 4.4 ppm, where in more than 58 % of samples exceeds the permissible limit of WHO (1997). The F^- concentration is higher in limestone as well as sandstones than the basalt flows in the study area. The inverse relationship has been noted between F^- with Ca^{++} and Mg^{++} for the groundwater samples from the study area. The higher concentration of F^- is ascribed to the presence of fluoride bearing minerals like fluorite and apatite in limestones and the clays, which absorb fluorine by F^- to OH^- replacement (Teotia and Teotia, 1992; Murkute and Badhan, 2011). The fluoride concentration is the result of groundwater which flows through rocks and also increases along with the depth (Dev Burman et al., 1995). In general, higher alkalinity of the water promotes the leaching of F^- and thus increases the F^- concentration (Saxsena and Ahmed, 2001). The inverse correlation of F^- and HCO_3^- from the study area has been attributed to alkaline nature of water, where carbonates and hydroxyl ions are dominant as compared to HCO_3^- (Hem, 1991; Karanth, 1987; Apambire et al., 1997; Shivanna and Mahokar, 2003). In the study area, it has been observed that due to higher concentration of fluoride content in groundwater, the safe limit for fluoride ingestion has already been exceeded (Figure 11.3) and such water is most unsuitable for drinking purpose.

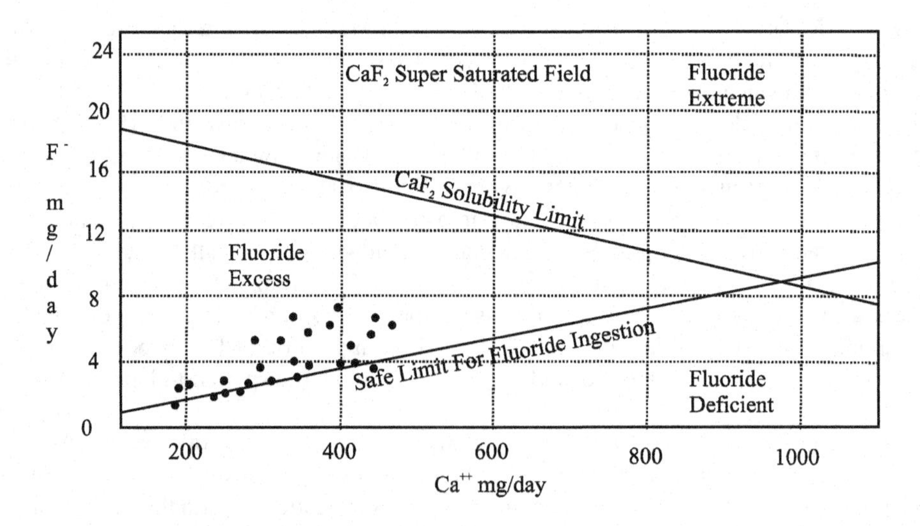

FIGURE 11.3 Safe limit fluoride ingestion diagram for study area.

11.8 CONCLUSIONS

1. The ion exchange and solubilization processes existing within the aquifer are accountable for the higher EC in groundwater of study area. The outsized variation in EC may also be ascribed to both anthropogenic activity as well as the geochemical processes through the rock–water interaction. The spatial variation in TDS values may be attributed to variations in lithology and hydrological processes.
2. The strong positive association between Na^+ and K^+ unravels the weathered rock forming minerals like sodium plagioclase, potash plagioclase along with anthropogenic sources like domestic and animal waste, while the inverse interrelationship of increased Na^+ content with decreased Ca^{++} concentration has been found to be the usual phenomenon in alkaline groundwater conditions.
3. The chemical weathering of rock forming minerals is the main causative factor in the evolution of chemical composition of groundwater occurring in all the lithological domains of study area.
4. The groundwater from the study area is not at all suitable for drinking and even for domestic use. The groundwater from the study area is of medium-to-high salinity – medium sodium type and thus be utilized for irrigation with little danger of exchangeable sodium; hence marginally suitable for irrigation, but not for longer time period.
5. The higher concentration of F^- is ascribed to the presence of fluoride bearing minerals like fluorite and apatite in limestones and the clays, which absorb fluorine by F^- to OH^- replacement. The safe limit for fluoride ingestion has already been exceeded in the study area, and hence, such water is unsuitable for drinking purpose.

REFERENCES

Adimalla, N., Qian, H. 2019. Groundwater quality evaluation using water quality index (WQI) for drinking purposes and human health risk (HHR) assessment in an agricultural region of Nanganur, south India. *Ecotoxicology and Environmental Safety*, 176, 153–161.

Apambire, W.B., Boyle, D.R., Michel, F.A. 1997. Geochemistry, genesis and health implications of fluoriferous groundwater in the upper region of Ghana. *Environmental Geology*, 33(1), 13–24.

APHA. 1995. Standard method for examination for the examination of water and waste water, 19th edn. Americal Health Association, Washington DC, USA.

Back, W., Hanshaw, B., 1966. Hydrochemistry of Northern Yucatan Peninsula, Mexico, with a section on Mayan water practices. In Wiedie, A.E. (ed.) *Field Seminar on Water Carbonate Rocks of the Yucatan Peninsula*. Mexico, New Orleans Geological Society, pp. 45–77.

BIS. 1991. Bureau of Indian standard specification for drinking water. IS:10500, Indian Standard Institute, pp. 1–5.

BIS. 2003. Indian standard drinking water specifications. IS: 10500, Edition 2.2 (2003–2009). Bureau of Indian Standards, New Delhi.

Canter, L.W. 1997. *Nitrates in Groundwater*. Lewis Publisher, New York.

Chacha, N., Njau, K.N., Lugomela, G.V., Muzuka, A.N.N., et al. 2018. Hydro-geochemical characteristics and spatial distribution of groundwater quality in Arusha well fields, northern Tanzania. *Applied Water Science*, 8, 1–23.

Datta, D.K., Gupta, L.P., Subramanian, 2000. Dissolved fluoride in the lower Ganges-Brahmaputra-Meghna river system in the Bangal Basin, Bangladesh. *Environmental Geology*, 39(10), 1163–1168.

Davis, S.N., De Wiest, R.J.M. 1966. *Hydrogeology*. John Wiley and Sons, Hoboken, pp. 463.

Deshmukh, A.N., Vali, S. 1995. Biochemical impact of fluorides in drinking water on development of dental and skeletal fluorosis. *The Gondwana Geological* Society, 9, 139–150.

Deshmukh, A.N., Wadaskar, P.M., Malpe, D.B. 1995. Fluorine in environment: a review. *The Gondwana Geological* Society, 9, 1–20.

Dev Burman, G.K., Singh, B., Khatri, P., 1995. Hydro-geochemical study of groundwater having high fluoride content in Chandrapur District of Vidharbha region, Maharashtra. *The Gondwana Geological* Society, 9, 71–80.

Dufor, C.N., Becker, E. 1964. Public water supplies of the 100 largest cities of in the United States. US Geological Survey Water Supply Paper 1812, pp. 364.

Eaton, F.M. 1950. Significance of carbonates in irrigation waters. *Soil Science*, 39, 23–133.

Ekambaram, V. 2001. Calcium preventing locomotor behavioural and dental toxicities of fluoride by decreasing serum fluoride levels in rats. *Environmental Toxicology and Pharmacology*, 9, 141–146.

Elbeltagi, A., Pande, C.B., Kouadri, S., et al. 2022. Applications of various data-driven models for the prediction of groundwater quality index in the Akot basin, Maharashtra, India. *Environmental Science and Pollution Research*, 29, 17591–17605. https://doi.org/10.1007/s11356-021-17064-7

Gibbs, R.J. 1970. Mechanism controlling world water chemistry. *Science*, 17, 1088–1090.

Gupta, I.C. 1983. Concept of residual sodium carbonate in irrigation water in relation to sodic hazard in irrigated soils. *Current Agriculture Science*, 7, 97–113.

Hallberg, G.R., Keeney, D.R. 1993. Nitrate. In: *Regional Groundwater Quality*. Van Nostrand Reinhold, New York, pp. 279–322.

Han, G., Lui, C.Q. 2004. Water geochemistry controlled by carbonate dissolution a study of the river water draining Karst-dominated terrain, Guizhou province. *China Chemical Geology*, 204, 1–21.

Handa, B.K. 1974. Methods of collection and analysis of water samples and interpretation of water analysis data. Govt of India. pp. 365.

Handa, B.K. 1975. Geochemistry and genesis of fluoride containing groundwater in India. *Groundwater*, 13(3), 275–281.

Hem, J.D. 1991. *Study and Interpretation of the Chemical Characteristics of Natural Water*. pp. 2254, 3rd edn. Scientific Publishers, Jodhapur, India.

Islam, A.M., Zahid, A., Rahman, M.M., Rahman, M.S., Islam, M.J., Akter, Y., Shammi, M., Bodrud-Doza, M., Roy, B. 2017. Investigation of groundwater quality and its suitability for drinking and agricultural use in the south central part of the coastal region in Bangladesh. *Exposure and Health*, 9, 27–41.

Jacks, G., Bhattachrya, P., Chaudhary, V., Singh, K.P. 2005. Controls on the genesis of high- fluoride groundwaters in India. *Applied Geochemistry*, 20, 221–228.

Jalali, M. 2009. Geochemistry characterization of groundwater in an agriculture area of Razan, Hamadan, Iran. *Environmental Geology*, 56, 1479–1488.

Jha, A.N. Verma, P.K. 2000. Physico–chemical property of drinking water in town area of Godda district under Santal Pargana (Bihar), India. *Pollution Research*, 19(2), 75–85.

Karanth, K.R. 1987. *Groundwater Assessment Development and Management*. Tata McGraw Hill Publishing Company Ltd. New Delhi.

Kharb, P., Susheela, A.K. 1994. Fluoride investigation in excess and its effect on organic and certain inorganic constituents of soft tissues. *Medical Science Research*, 22, 43–44.

Kouadri, S., Pande, C.B., Panneerselvam, B. et al. 2022. Prediction of irrigation groundwater quality parameters using ANN, LSTM, and MLR models. *Environmental Science* and *Pollution Research*, 29, 21067–21091. https://doi.org/10.1007/s11356-021-17084-3

Kumar, S. Venkatesh, A.S. Singh, R. Udayabhanu, G., Saha, D. 2018. Geochemical signatures and isotopic systematics constraining dynamics of fluoride contamination in groundwater across Jamui district, indo-Gangetic alluvial plains, India. *Chemosphere*, 205, 493–505.

Lakshmanan, E., Kannan, R., Senthilkumar, M. 2003. Major ion chemistry and identification of hydrogeochemical processes of groundwater in a part of Kancheepuram District, Tamil Nadu, India. *Environmental Geosciences*, 10(4), 157–166.

Loizidou, M., Kapetanious, E.G. 1993. Effect of leachate from landfills on underground water quality. *Science Environmental*, 128, 69–81.

Marghade, D., Malpe, D.B., Zade, A.B. 2011. Geochemical characterization of groundwater from northeastern part of Nagpur urban, Central India. *Environmental Earth Sciences*, 62(7), 1419–1430.

Meybeck, 1987. Global chemical weathering of surfacial rocks estimated from river dissolved loads. *American Journal* of *Science*, 287, 401–428.

Murkute, Y.A. 2014. Hydro-geochemical characterization and quality assessment of groundwater around Umrer Coal Mine area, Nagpur District, Maharashtra, India. *Environmental Earth Sciences*, 72, 4059–4073.

Murkute, Y.A. 2022. Major ice chemistry and assessment of groundwater quality around Gangapur village, Nagpur district, Maharashtra, India. *Journal* of *Geosciences Research*, 7(1), 112–120.

Murkute, Y.A., Badhan, P.P. 2011. Fluoride Contamination in Groundwater from Bhadravati Tehsil, Chandrapur District, Maharashtra. *Nature Environment and Pollution Technology*, 10(2), 255–260.

Nazzal, Y., Zaidi, F.K., Ahmed, Z., Ghrefat, H., Naeem, M., Nassir, S.N., Saeed, A.A., Khaled, M.A. 2014. The combination of principal component analysis and geostatistics as a technique in assessment of groundwater hydrochemistry in arid environment. *Current Science*, 108(6), 1138–1145.

Ozha, D.D., Mathur, S.B. 2001. Socio-economic losses to human and livestock health in Rajasthan due to high fluoride bearing groundwater and its mitigation. International workshop on 'Fluoride in drinking water: Strategies, Management and Mitigation', Bhopal, pp. 162–170

Pande, C.B., Moharir, K.N., Singh, S.K. et al. 2020. Groundwater evaluation for drinking purposes using statistical index: study of Akola and Buldhana districts of Maharashtra, India. *Environment, Development and Sustainability*, 22, 7453–7471. https://doi.org/10.1007/s10668-019-00531-0

Panneerselvam, B., Muniraj, K., Pande, C. et al. 2021. Geochemical evaluation and human health risk assessment of nitrate-contaminated groundwater in an industrial area of South India. *Environmental Science* and *Pollution Research*, 29(57), 86202–86219. https://doi.org/10.1007/s11356-021-17281-0

Panneerselvam, B., Muniraj, K., Duraisamy, K. et al. 2022. An integrated approach to explore the suitability of nitrate-contaminated groundwater for drinking purposes in a semiarid region of India. *Environmental Geochemistry and Health*, 45, 647–663. https://doi.org/10.1007/s10653-022-01237-5

Piskin, R. 1973. Evaluation of nitrate content of groundwater in Hall County, Nebraska. *Ground Water*, 11(6), 4–13.

Ravikumar, P., Venkatesharaju, K., Prakash, K.L., Somashekhar, R.K. 2010. Geochemistry of groundwater and groundwater prospects evaluation, Anekal Taluk, Bangalore urban district, Karnataka, India. *Environmental Monitoring and Assessment*, 179, 93–112. https://doi.org/10.1007/s10661-010-1721-z

Richards, L.A. 1954. *Diagnosis and Improvement of Saline and Alkali Soils*. US department of agriculture handbook. Washington DC, USA.

Ritter, W.F., Chrinside, A.E.M. 1984. Impact of land use on groundwater quality in southern Delaware. *Ground Water*, 22(1), 38–47.

Ritzi, R.W., Wright, S.L. Mann, B., Chen, M. 1993. Analysis of temporal variability in hydrogeochemical data used for multivariate analyses. *Ground Water*, 31, 221–229.

Roy, A. Keesari, T., Hemant Mohokar, U.K., Bitra, S. et al. 2018. Assessment of groundwater quality in hard rock aquifer of central Telangana state for drinking and agriculture purposes. *Applied Water Science*, 8, 124. https://doi.org/10.1007/s13201-018-0761-3

Saleh, A., Al-ruwih, F., Shehata, M. 1999. Hydrochemical process operating within the main aquifers of Kuwait. *Journal of Arid Environments*, 42, 195–209.

Sanhez-Perez, J.M., Tremolieres, M. 2003. Change in groundwater chemistry as a consequence of suppression of floods: the case of Rhine floodplains. *Journal of Hydrology*, 270, 89–104.

Saxsena, V.K., Ahmed, S. 2001. Dissolution of fluoride in groundwater: a water rock interaction study. *Enviornmental Geology*, 40(9), 1084–1087.

Shivanna, K.T., Mahokar, H.V. 2003. Isotope hydrochemical approach to study the fluoride contamination in groundwater of Ilkal area, Baglkot District Karnataka. Proceeding of the International Conference on water and environment, Bhopal, India. Allied Pub. Pvt Ltd. pp. 332–346.

Simsek, C., Gunduz, O. 2007. IWQ index: a GIS-integrated technique to assess irrigation water quality. *Environmental Monitoring and Assessment*, 128, 277–300.

Singh, A.K., Mahato, M.K., Neogi, B., Singh, K.K. 2010. Quality assessment of mine water in the Raniganj Coalfield area, India. *Mine Water and Environmental*, 29, 248–262.

Subba Rao, N. 2003. Groundwater quality: focus on fluoride concentration in rural parts of Guntur District, Andhra Pradesh, India. *Hydrological Sciences Journal*, 48(5), 835–847.

Subba Rao, N. 2011. High- fluoride groundwater. *Environmental Monitoring and Assessment*, 176, 637–645.

Sunitha, V., Sudarshan, V., Reddy, R.B. 2005. Hydrochemistry of groundwater, Gooty area, Anantpur district, Andhra Pradesh, India. *Pollution and Research*, 24(1), 217–244.

Susheela, A.K. 2001. Sound planning and implementation of fluoride and fluorosis mitigation programme in an endemic village. International workshop in Fluoride in drinking water: Strategies, Management and Mitigation, Bhopal, pp. 22–24.

Teotia, S.P.S, Teitia, M. 1988. Endemic skeletal fluorosis clinical and radiological variant. *Fluoride*, 21, 39–44.

Teotia, S.P.S., Teitia, M. 1992. *Edemic Fluoride: Bones and Teeth-Uptake; Report on Fluorosis in India*. Institute of Social Science, New Delhi, pp. 50–61.

Todd, D.K. 1995. *Groundwater Hydrology*. John Wiley and Sons, Singapore.

Trivedy, R.K., Goel, P.K. 1986. Chemical and biological methods for waste pollution studies. *Environmental Publication*, 5, 35–96.

US Salinity Laboratory Staff. 1954. *Diagnosis and Improvement of Saline and Alkali Soils*. U.S. Dept Agriculture, Washington, DC.

Voutsis, N., Kelepertzis, E., Tiritis, E., Kelepertsis, A. 2015. Assessing the hydrogeochemistry of ground waters in ophiolite areas of Euboea Island, Greece, using multivariate statistical methods. *Journal of Geochemical Exploration*, 159, 79–92.

WHO. 1984. *The Guideline for Drinking Water Quality Recommendations*. World Health Organization, Geneva.

WHO. 1997. *The Guideline for Drinking Water Quality, Health Criteria and Other Supporting Information*. World Health Organization, Geneva.

Wilcox, L.V. 1955. The quality of water for irrigation use, U.S. Department Agriculture. *Technical Bulletin*, 962, 40

Wood, J.M. 1974. Biological cycles for toxic elements in the environment. *Science*, 183, 1049–1052.

Xu, Y., Dai, S, Meng, K., Wang, Y., Ren, W., Zhao, L., Christie, P. Teng, Y. 2018. Occurrence and risk assessment of potentially toxic elements and typical organic pollutants in contaminated rural soils. *Science of the Total Environment*, 630, 618–629.

Zhang, W., Ma, L., Abuduwaili, J., Ge, Y., Issanova, G., Saparov, G. 2019. Hydro-chemical characteristics and irrigation suitability of surface water in the Syr Darya River, Kazakhstan. *Environmental Monitoring and Assessment*, 191, 572. https://doi.org/10.1007/s10661-019-7713-8

Zhou, Y., Li, P., Xue, L., Dong, Z., Li, D. et al. 2020. Solute geochemistry and groundwater quality for drinking and irrigation purposes: a case study in Xinle City, North China. *Geochemistry*, 80, 125609. https://doi.org/10.1016/j.chemer.2020.125609

12 Variability of Ground Water Quality in Quaternary Aquifers of the Cauvery and Vennar Sub-basins within the Cauvery Delta, Southern India

Aswin Kokkat, N. C. Mondal, P. J. SajilKumar, and E. J. James

12.1 INTRODUCTION

In general, groundwater has been regarded as better than the surface in terms of quality. The shortage of surface water leads to increased dependence on groundwater (Kokkat et al., 2016). Groundwater sources are susceptible to contamination caused by various natural and anthropogenic activities (Mondal and Singh, 2011; Mondal et al., 2016; Rahman et al., 2021). A high level of chemical pollution is a matter of serious concern across the globe. Often, water containing agrochemicals joins the drinking water sources, thereby causing a high threat to human health (Bu et al., 2010). Therefore, groundwater quality assessment is essential for deciding on its utility for human needs (Sajil Kumar, 2016). In recent years, several agencies worldwide have been involved in monitoring the pollutants both at the local and regional scales (Lopez et al., 2015). Quality monitoring at regular intervals is critical for efficient management of groundwater and also to keep track of quality standards for safe use (Masoud et al., 2018). Numerous studies have been conducted across the globe on groundwater quality. Groundwater is extensively used for domestic and agricultural purposes in India (Jacks et al., 2008). Several researchers have studied the quality of groundwater from various locations in India, such as Tuticorin and Kalpakkam, in Tamil Nadu (Mondal et al., 2010a; Mondal et al., 2011; Selvam et al., 2014), Uttar Pradesh (Kumari et al., 2014), Kerala (Satish Kumar et al., 2016), Jammu and Kashmir (Zeeshan and Azeez, 2016), Andhra Pradesh (Saxena et al., 2003, 2004; Mondal et al., 2010b; Nageswara Rao et al., 2017), and Rajasthan (Rahman et al., 2021). The non-availability of surface water for irrigation purposes compels the farmers to resort to groundwater for irrigation (Madramootoo, 2012). This trend calls for the investigation of the suitability of groundwater for irrigation purposes. These investigations make use of indices like sodium percentage (Na%), sodium adsorption ratio (SAR), residual sodium carbonate (RSC), magnesium adsorption ratio (MAR), and Kelly's ratio (KR) (Arumugam and Elangovan, 2009; Sarath Prasanth et al., 2012; Jagadeshan and Elango, 2015; Mondal et al., 2016).

The present study was taken up at the tailend of the Cauvery delta in Southeast India. Agriculture is the mainstay of 70% of the people living in this delta, which is known as the 'rice bowl' of Tamil Nadu. In the delta, rice is the main crop. The Point Calimere wetland, the only Ramsar site in Tamil Nadu, is located within the study area (Kokkat et al., 2016). The evaluation of the suitability of water for drinking and irrigation purposes was previously attempted by some researchers for the entire Nagapattinam and Karaikal districts

DOI: 10.1201/9781003303237-12

(Venkatramanan et al., 2013, 2015). A few researchers studied the groundwater quality in some pockets within Nagapattinam and Karaikal districts; a part of Central Karaikal district (Vetrimurugan et al., 2013; Vetrimurugan and Elango, 2015), the southern part of Vedaraniyam (Krishnakumar et al., 2014), the southern part of Nagapattinam district covering the Vennar sub-basin (Gnanachandrasamy et al., 2015), and the northern part of Nagapattinam district and Karaikal region (Gopinath et al., 2015). The earlier studies were either limited to small pockets in the delta or the entire Cauvery and Vennar sub-basins combined as one unit. In the present study, two distinct sub-basins, namely, the Cauvery and Vennar, were considered separate units. The results reflect the particular characteristics of these two separate sub-basins. Moreover, the earlier studies considered the samples for only two seasons, while the present study has made use of the samples for four seasons. A detailed, systematic, and periodic monitoring system is required to estimate the groundwater quality. This helps in understanding the various factors that affect the groundwater chemistry of the region. Hence, the present investigation on groundwater quality was planned taking heterogeneous hydrogeological characteristics into consideration, which were not explored earlier. The present study focused on groundwater quality's spatial and seasonal variations for drinking and irrigation over the four seasons in the Cauvery and Vennar sub-basins within the Cauvery delta in Southeast India. The geographical information system (GIS) was used to prepare spatial maps of various water quality parameters and WQI. The chapter also discusses the anthropogenic and hydrogeological setting of the sub-basins that impact the groundwater quality in the shallow aquifers in both sub-basins in the Cauvery delta, Southern India.

12.2 MATERIALS AND METHODS

12.2.1 Study Area

The present study was carried out in the coastal districts of Nagapattinam and Karaikal, covering an area of 2,757 km^2 in the Cauvery delta, as shown in Figure 12.1. These districts are located between latitudes: 10°15′ and 11°30′ N and longitudes: 79°30′ and 79°55′ E and have a coastline of 190 km. The region experiences a hot, sub-humid to semi-arid climate. The mean annual rainfall based on the data from 1951 to 2019 is around 1,330 mm. The Nagapattinam and Karaikal districts are influenced by the southwest monsoon during the June-September months, contributing to about 69% of the annual rainfall. The northeast monsoon during October-December contributes 18% of the annual rainfall (Dhinagaran, 2008).

12.2.2 Geology and Hydrogeology

Two-tiered aquifer systems exist in the study area. The lower miocene deep aquifer system consists of the Aquitanian and Burdigalian aquifers, and the upper Pliocene-Quaternary shallow aquifer system consists of the Pliocene and Quaternary aquifers. The present study mainly deals with the Quaternary aquifer. The Quaternary deposit, having a fluvial and semi-marine origin, has a thickness ranging from around 12 m on the western side to around 40 m on the eastern side (UNDP, 1973). The thickness of the shallow aquifer is about 110 m in the coastal region. The groundwater occurs under phreatic conditions, is semi-confined, and is confined in formations consisting of sand, silt, and clay (Dhinagaran, 2008). These deposits consist of sand intercalated with brown and black clay, with thickness varying from 1 to 50 m (UNDP, 1973). The thickness of the phreatic aquifer in the Quaternary deposits ranges from 5 to 25 m, and groundwater occurs at a depth of 2–3 m below ground level (bgl). In the areas located south of the Vettar river, the transition from Quaternary to Pliocene is marked by a bed of red-mottled gray clay at a depth range of up to 35 m, bgl. The recharge of the shallow aquifer usually occurs by precipitation and return flow from irrigation (UNDP, 1973). The net groundwater availability for Nagapattinam district in 2011 was

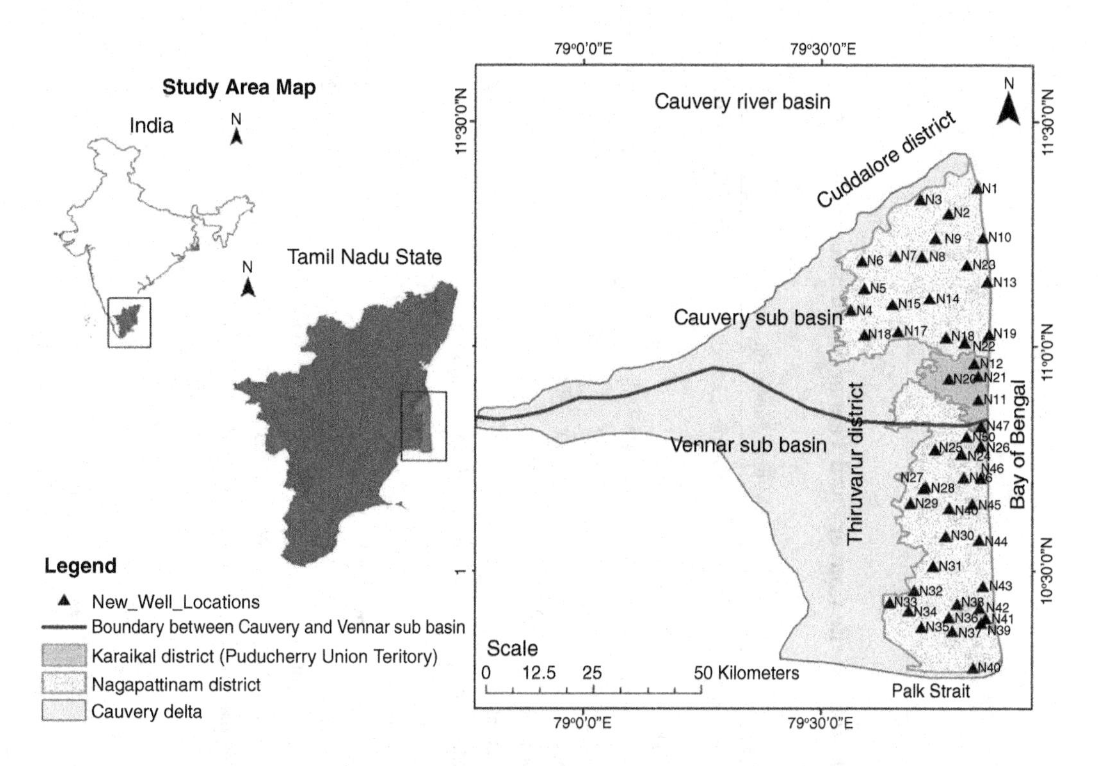

FIGURE 12.1 Map showing the location of the study area along with the sampling points in the Cauvery and Vennar sub-basins, Cauvery delta, Southeast India.

15,916 hectare million (ham), and in 2013, it was 16,544 ham. The gross groundwater draft was 16,288 ham for the year 2011 and 16,724 ham for 2013 (CGWB, 2020).

12.2.3 Groundwater Sampling and Laboratory Analysis

Dug/open wells and handpumps were used for collecting groundwater samples from shallow, unconfined aquifers in the Cauvery and Vennar sub-basins within the Cauvery delta. Of the 50 samples, 23 were from the northern part of the Nagapattinam and Karaikal districts (Cauvery sub-basin), and 27 were from the southern part of the Nagapattinam district (Vennar sub-basin), as shown in Figure 12.1. Groundwater samples were collected from 16 open wells (OW) and 34 suction handpumps (HPs). The suction HPs in the eastern and central parts of the study area have a depth range of 3–4 m, and on the western side, the depth goes up to 6 m. The depth of groundwater levels mainly varied from 1.2 to 3.1 m in the Cauvery sub-basin and 0.9–1.9 m in the Vennar sub-basin (Table 12.1). For homogeneity of groundwater samples, water was pumped for 10–15 mintues before the sampling. The groundwater samples were collected for all four seasons, i.e., the pre-monsoon (PRM), southwest monsoon (SWM), northeast monsoon (NEM), and post-monsoon (POM). The PRM season sampling was carried out during March-May, SWM in June-September, NEM in October-December, and POM in January-February. A handheld GPS (Gramin 76CS×) was used to record the location of each sampling station. The locations of stations were plotted on a map using ArcGIS 10.1. The samples were collected in polyethene bottles with a capacity of 1 L; after the collection of samples, the bottles were sealed and transferred to the laboratory for analysis. A Hach HQ40d instrument was used in the field to measure the pH and EC. The TDS was estimated from the EC value with a factor of 0.64 (Rainwater and Thatcher, 1960). The protocol mentioned by APHA (2012) was employed in the present study. The titration method was used to estimate Ca^{2+}, Mg^{2+}, CO_3^{2-}, Cl^-, and HCO_3^-, flame photometry for Na^+ and K^+, and sulfate by the turbidimetric method.

TABLE 12.1
Details of Well Inventory, Groundwater Level, and Well Depth of Sampling Stations in the Cauvery and Vennar Sub-basins within the Cauvery Delta, South India

In Cauvery Sub-basin							In Vennar Sub-basin						
Well No	Location	Well Type	Latitude	Longitude	Depth of GW level (m, bgl)	Well depth (m, bgl)	Well No	Location	Well Type	Latitude	Longitude	Depth of GW level (m, bgl)	Well depth (m, bgl)
N1	Pudupattinam	HP	11°21'	79°49'	1.4	2.5	N24	Sikkal	HP	10°45'	79°47'	1.4	2.1
N2	Madhanam	HP	11°17'	79°46'	1.3	2.1	N25	Kizhavelur	HP	10°46'	79°44'	1.5	2.1
N3	Kollidam	HP	11°19'	79°42'	2.2	3.1	N26	Nagapattinam	HP	10°46'	79°50'	1.7	2.3
N4	Kuttalam	HP	11°04'	79°33'	1.4	2.5	N27	SathiyaGudi	HP	10°40'	79°43'	1.3	2.0
N5	Pandanellur	HP	11°08'	79°31'	2.8	5.5	N28	Killugudi School	HP	10°41'	79°43'	1.4	2.4
N6	Manalmedu	HP	11°11'	79°35'	3.1	4.1	N29	Valivalam	HP	10°38'	79°41'	1.5	2.7
N7	Thalainayar	HP	11°11'	79°39'	2.9	3.8	N30	Manakudi	HP	10°34'	79°45'	1.4	2.6
N8	VaitheswaranKoil	HP	11°11'	79°42'	2.3	3.7	N31	Upalacheri	HP	10°30'	79°44'	1.6	2.6
N9	Sirkazhi	HP	11°14'	79°44'	2.7	3.8	N32	Manakadu	OW	10°27'	79°41'	1.8	2.4
N10	Thirumulaivasal	HP	11°14'	79°50'	1.9	3.0	N33	Thulasiyapattinam	OW	10°25'	79°38'	1.9	2.7
N11	Periyapalli	OW	10°52'	79°49'	1.4	2.5	N34	Voimedu	HP	10°24'	79°41'	1.7	2.5
N12	Kottucherry	HP	10°57'	79°49'	1.5	2.4	N35	Panchanadikulam	OW	10°22'	79°42'	1.5	2.2
N13	Pompuhar	HP	11°08'	79°51'	1.6	2.5	N36	Ayakarapulam	OW	10°23'	79°46'	1.4	2.6
N14	Sembiarkoil	HP	11°06'	79°43'	1.7	2.7	N37	Siruthalakadu	OW	10°21'	79°46'	1.7	2.3

(*Continued*)

TABLE 12.1 (*Continued*)
Details of Well Inventory, Groundwater Level, and Well Depth of Sampling Stations in the Cauvery and Vennar Sub-basins within the Cauvery Delta, South India

In Cauvery Sub-basin							In Vennar Sub-basin						
Well No	Location	Well Type	Latitude	Longitude	Depth of GW level (m, bgl)	Well depth (m, bgl)	Well No	Location	Well Type	Latitude	Longitude	Depth of GW level (m, bgl)	Well depth (m, bgl)
N15	Mayiladuthurai	HP	11°05’	79°39’	1.7	3.0	N38	Kuravapalam	OW	10°25’	79°47’	1.2	2.0
N16	Komal	HP	11°01’	79°35’	1.3	2.3	N39	Vedaranyam 1	OW	10°22’	79°50’	1.4	2.6
N17	Manganellur	HP	11°01’	79°39’	1.8	2.5	N40	Point Calimere	OW	10°16’	79°49’	1.2	2.4
N18	SankarnPandal	HP	11°01’	79°45’	1.2	2.4	N41	Vedaranyam 2	OW	10°23’	79°50’	1.3	2.8
N19	Tharangambadi	OW	11°01’	79°51’	1.6	2.9	N42	Thettakudi	OW	10°24’	79°50’	1.3	2.7
N20	Thirunallar	HP	10°55’	79°46’	1.3	2.6	N43	Pushpavanam	OW	10°27’	79°50’	1.5	3.1
N21	Karaikal	OW	10°55’	79°49’	1.5	2.5	N44	Vettaikaraniruppu	HP	10°34’	79°50’	1.2	2.8
N22	Thirukalancherry	HP	11°00’	79°48’	2.0	3.4	N45	Thirupondi	HP	10°38’	79°49’	1.2	2.8
N23	Tiruvenkandu	OW	11°10’	79°48’	1.8	3.0	N46	Paravi	HP	10°42’	79°50’	1.1	2.9
							N47	Nagore	OW	10°49’	79°50’	1.3	2.5
							N48	Menamanallur	HP	10°38’	79°46’	0.9	2.8
							N49	Nirtharamangalam	HP	10°42’	79°48’	1.1	2.9
							N50	Perumkadambanoor	HP	10°47’	79°48’	1.1	2.8

HP, Hand Pump (suction hand pump); OW, Open well (dug well); bgl, below ground level.

12.2.4 Drinking Water Quality

The measured concentrations of ions were compared with the drinking water quality standards recommended by the WHO (2011) and the BIS (2012). The spatial interpolation maps of various physicochemical parameters were also prepared using ArcGIS 10.1 software based on WHO and BIS standards.

12.2.4.1 Spatial Mapping Using Inverse Distance Weighting

Spatial variation maps are used to visualize and interpret the distribution of ions in different locations. Several interpolation techniques are available in the spatial analyst tool in ArcGIS. Inverse distance weighting (IDW) is considered to be an effective method, which was used in this study. The IDW interpolation technique is based on the assumption that when several known points are distributed in space, they are closer to the points that are most similar, when compared to the points that are farther away. The main advantage is that the method is logical and efficient. Since groundwater sampling locations were evenly distributed, the IDW method was adopted. The weights for unknown points were determined using the equation (Burrough and McDonnell, 1998):

$$\lambda_i = \frac{D_i^{-\alpha}}{\sum_{i=1}^{n} D_i^{-\alpha}} \tag{12.1}$$

12.2.4.2 Water Quality Index (WQI)

It is a dimensionless number obtained by combining multiple water quality factors into a single number. It serves as a significant parameter for the evaluation and assessment of groundwater. The different samples collected for water quality determination can be compared to the index value of each sample (Abbasi and Abbasi, 2012). The WQI calculation reduces a large number of parameters to a single value, as provided by

$$\text{WQI} = \sum_{i=l}^{n} Q_i \times W_i \tag{12.2}$$

$$Q_i = \left(\frac{C_i}{S_i}\right) \times 100 \tag{12.3}$$

Assigned weights and relative weights are presented in Table 12.2.

12.2.4.3 Irrigation Water Quality

In the Cauvery delta, groundwater is widely used for agriculture, and its suitability depends upon the mineral constituents of the groundwater, as these have an impact on both the plant and the soil. The indices SAR, Na (%), MAR, RSC, and KR were utilized to understand the suitability of groundwater for irrigation.

The SAR is an indicator of the degree to which cation exchange reactions in the soil tend to enter irrigation water. The SAR is estimated using the formula given by Richards (1954) and expressed as

$$\text{SAR} = \frac{\text{Na}^+}{\sqrt{\frac{\text{Ca}^{2+} + \text{Mg}^{2+}}{2}}} \tag{12.4}$$

where the concentrations are represented in meq/l. A SAR value of less than 10 is regarded as excellent, between 10 and 18 as good, and between 18 and 26 as fair.

TABLE 12.2
Details of Relative Weights for WQI Calculation

Chemical Parameters	WHO Standards (2011)	Weight (w_i)	Relative Weight $W_i = \frac{W_i}{\sum_{i=1}^{n} W_i}$
pH	6.5–8.5	4	0.1290
Electrical Conductivity (µS/cm)	1,500	5	0.1613
Bicarbonate	500	1	0.0323
Chloride	250	5	0.1613
Sulfate	250	4	0.1290
Calcium	75	3	0.0968
Magnesium	50	3	0.0968
Sodium	200	4	0.1290
Potassium	12	2	0.0645
		$\Sigma W_i = 31$	$\Sigma W_i = 1.0000$

Wilcox (1955) estimated the suitability of water for irrigation by making use of the following formula

$$Na\% = \frac{Na^+ + K^+}{Ca^{2+} + Mg^{2+} + Na^+ + K^+} \times 100 \quad (12.5)$$

The Na (%) was computed for all the samples from 50 wells during the four different seasons. Based on the classification suggested by Wilcox (1955), the percent value of Na (%) of less than 20 was classified as excellent, between 20 and 40 as good, between 40 and 60 as permissible, between 60 and 80 as doubtful, and greater than 80 as unsuitable.

In irrigation waters, a high amount of bicarbonates can make calcium and magnesium precipitate as carbonate, and RSC are indices to measure this process (Kumar et al., 2007). The RSC was calculated using the method suggested by Richards (1954) as:

$$RSC = \left(HCO_3^- + CO_3^-\right) - \left(Ca^{2+} + Mg^{2+}\right) \quad (12.6)$$

The alkali hazard is studied using KR for irrigation. It was calculated with the help of the following equation (Kelly, 1940).

$$KR = \frac{Na^{2+}}{Ca^{2+} + Mg^{2+}} \quad (12.7)$$

The samples with a KR less than one are suitable, and those greater than one are unsuitable for irrigation.

The MAR was determined by using the equation given by Sappa et al. (2014),

$$MAR = \frac{Mg^{2+}}{Ca^{2+} + Mg^{2+}} \times 100 \quad (12.8)$$

12.2.4.4 Hydrogeochemical Processes

The analysis helps in geochemically categorizing groundwater under a particular 'water type', which is useful for understanding the groundwater chemistry mechanisms. The Gibbs plot is widely used by many researchers to understand the factors regulating the hydrogeochemistry of the region.

The Gibbs plot is based on the ratios of $Na^{+}+K^{+}$: $Na^{+}+K^{+}+Ca^{2+}$ *vs* TDS and Cl–: $Cl^{-}+HCO_3^{-}$ *vs* TDS. This plot was used to study the influence of precipitation, evaporation, and rock formations on the groundwater system. The Chadha diagram (Chadha, 1999) was used to interpret the various hydrogeochemical processes taking place in the coastal aquifers of the Cauvery delta. Here, the major ions were expressed in meq/L, and the percentage differences between the alkaline earths ($Ca^{2+}+Mg^{2+}$) and alkaline metals ($Na^{+}+K^{+}$) were plotted on the *X*-axis of the diagram. On the other hand, the change in percentage between weak acids ($CO_3^{2-}+HCO_3^{-}$) and strong acids ($SO_4^{2-}+Cl^{-}$) was plotted on the *Y*-axis. A total of four distinct fields were included in the diagram, and each of them showed different hydrochemical characteristics of water. Field 5 shows Ca-HCO_3 type of water representing recharging waters, field 6 indicates Ca-Mg-Cl water type showing reverse ion exchange in groundwater, field 7 represents Na-Cl water type (seawater), in which strong acid exceeds weak acid, and field 8 indicates Na-HCO_3 type of water, dominated by the base ion exchange.

12.3 RESULTS AND DISCUSSION

A statistical overview of the hydro-physical and chemical parameters of groundwater as well as a comparison with the standards of WHO (2011) and BIS (2012) are presented in Table 12.3a. The basic statistics and chemical variation of individual water quality parameters are represented graphically using a box and whisker plot for all seasons considered (Figure 12.2). The ionic balance error of each sample was estimated and found to be within the standard limits. For the four seasons, the order of dominance of cations was $Mg^{2+}>Ca^{2+}>Na^{+}>K^{+}$, $Mg^{2+}>Ca^{2+}>Na^{+}>K^{+}$, $Ca^{2+}>Na^{+}>Mg^{2+}>K^{-}$, and $Ca^{2+}>Mg^{2+}>Na^{+}>K^{+}$ during the PRM, SWM, NEM, and POM seasons, respectively, whereas the anions were $Cl^{-}>SO_4^{2-}>HCO_3^{-}>CO_3^{2-}$, $Cl^{-}>HCO_3^{-}>CO_3^{2-}>SO_4^{2-}$, $Cl^{-}>HCO_3^{-}>SO_4^{2-}>CO_3^{2-}$, and $Cl^{-}>HCO_3^{-}>SO_4^{2-}>CO_3^{2-}$. This indicates that there were several hydrochemical processes involved in changing the quality of the groundwater.

12.4 DRINKING WATER QUALITY

12.4.1 pH

The permissible pH range is 6.5–8.5 and the maximum allowable limit is 9.2 (WHO, 2011). In the study site, pH during PRM ranged from 6.48 to 8, with an average value of 7.28, pH during the SWM season ranged from 6.53 to 8.12, with an average of 7.40, pH during the SWM season varied between 6.53 and 8.12, with an average of 7.40. During the NEM, pH ranged between 6.60 and 8.10, with an average of 6.90, and in the POM, the values ranged from 6.90 to 8.70, with an average of 7.69. During the seasonal sampling, the observed pH was within the permissible range.

12.4.2 Electrical Conductivity

In general, the acceptance of groundwater for consumption depends upon its taste. The taste of water is a major factor based on which users decide its potability. The permissible limit of electrical conductivity (EC) is 1,500 μS/cm (WHO, 2011). In the Nagapattinam-Karaikalbelt, the EC values ranged from 289 to 10,540 μS/cm in the PRM season, with an average value of 2,307 μS/cm. In the SWM, it ranged from 399 to 12,450 μS/cm, with an average value of 2,469 μS/cm. During the NEM season, the EC values varied from 372 to 10,600 μS/cm, with an average of 2,208 μS/cm, and in the POM, the EC ranged from 299 to 6,312 μS/cm, with an average value of 2,160 μS/cm. In the Nagapattinam-Karaikalbelt, during the PRM, SWM, NEM, and POM seasons, about 54%, 48%, 48%, and 52% of the samples were found to be within the allowed range. During all four seasons, the samples from the Vennar sub-basin showed higher EC concentrations than the samples from the Cauvery sub-basin. A groundwater quality map was generated considering the EC values based on the WHO drinking water quality standard (2011), as shown in Figure 12.3. The Vennar sub-basin

TABLE 12.3A
Statistics of Groundwater Quality for Drinking Purposes Compared with the WHO and BIS Standards

	WHO (2011)	BIS (2012)		Pre-monsoon (PRM)			Southwest monsoon (SWM)			Northeast monsoon (NEM)			Post-monsoon (PRM)		
Parameters (in mg/L)	Permissible Limits	Most Desirable Limits	Maximum Allowable Limits	Min	Max	Mean	Min	Max	Mean	Min	Max	Mean	Min	Max	Mean
pH	6.5–8.5	6.5	8.5	6.5	8.0	7.3	6.5	8.1	7.4	6.6	8.1	7.3	6.9	8.7	7.7
EC (µS/cm)	1,500	–	–	289	10,540	2,307	399	12,450	2,469	372	10,600	2,208	299	6,312	2,160
TDS	500	500	2,000	185	6,745	1,476	255	7,968	1,580	238	6,784	1,413	191	4,039	1,382
Ca^{2+}	75	75	200	48	896	302	24	1,040	255	48	1,096	237	16	800	193
Mg^{2+}	50	30	100	62	1,809	244	19	1,209	217	19	350	96	24	561	114
Na^{+}	200	–	–	18	1,425	267	29	1,110	260	34	1,415	242	10	865	248
K^{+}	12	–	–	6	256	45	5	266	53	12	347	61	2	257	57
Cl^{-}	250	250	1,000	20	3,313	468	60	3,764	524	60	3453	481	50	1,892	445
CO_3^{-}	–	–	–	ND	60	21	40	420	112	ND	160	42	ND	320	111
HCO_3^{-}	500	200	600	30	140	65	120	770	261	220	1,300	604	130	850	383
SO_4^{2-}	250	200	400	16	956	152	14	476	75	15	583	130	26	669	200
TH	–	200	600	480	8,080	1,775	420	6,120	1,546	280	3,640	996	180	3,080	935

All ions, TDS, and TH: in mg/L except pH, ND, Not detectable.

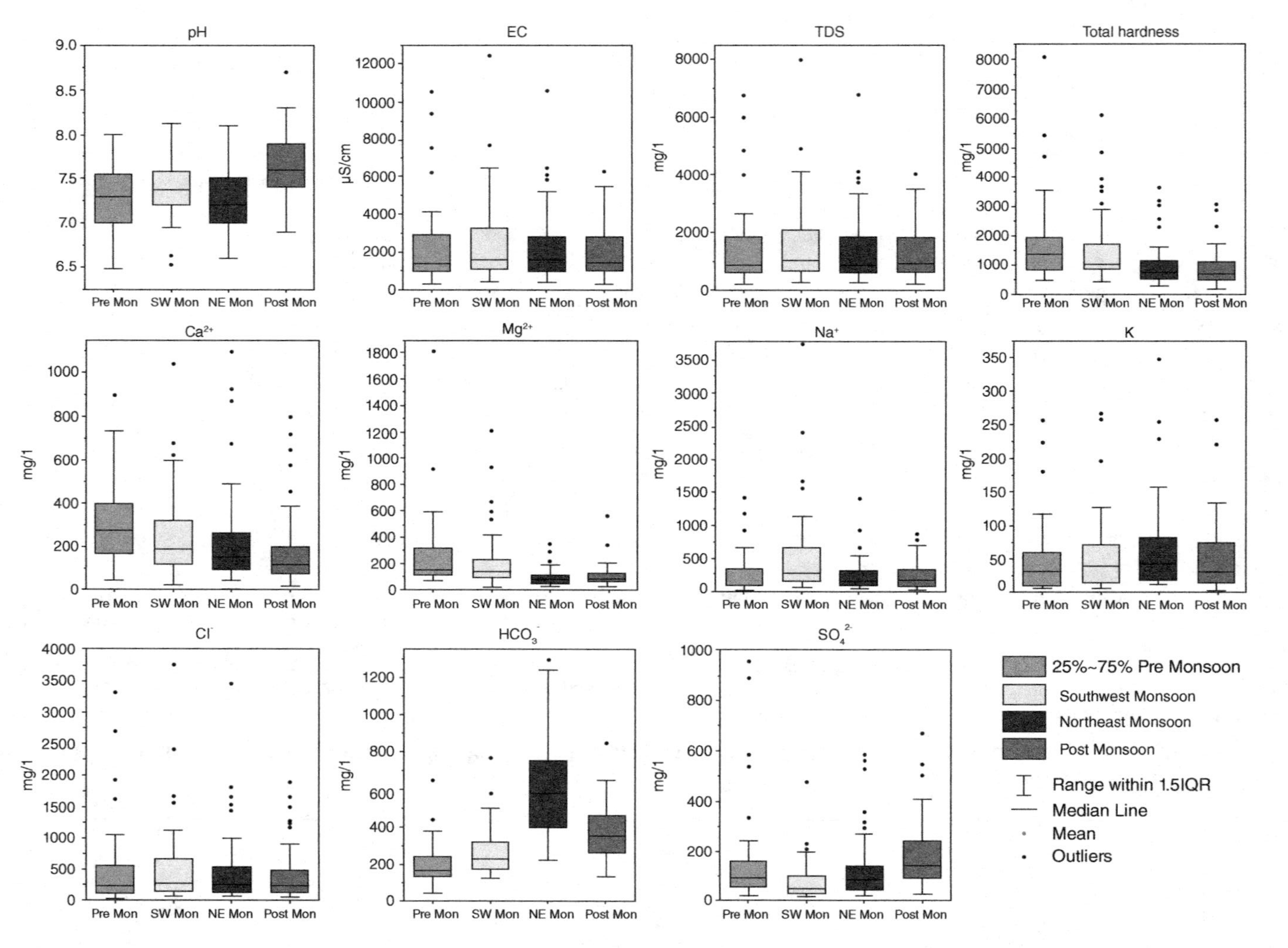

FIGURE 12.2 Statistical and chemical variation of water quality parameters represented using Box and Whiskerplot for all four seasons.

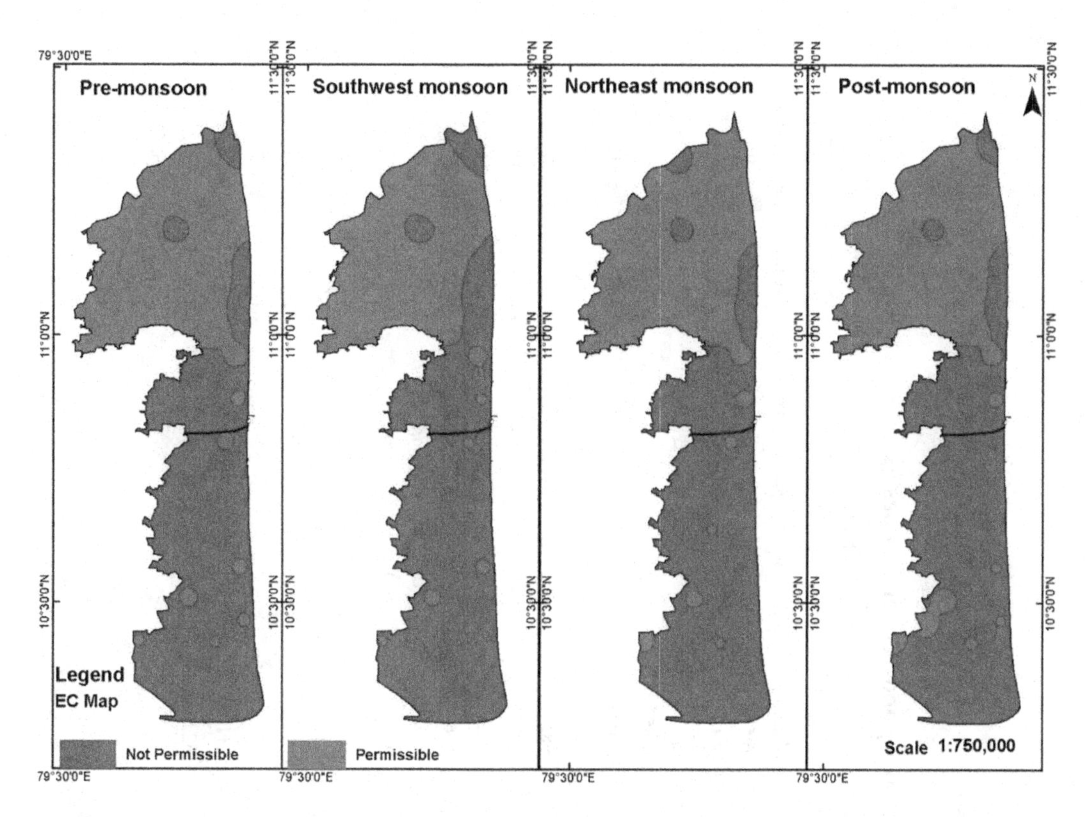

FIGURE 12.3 Spatial interpolation map showing the seasonal variation of EC in the Cauvery and Vennar sub-basins.

showed excessive EC values (Table 12.3b). The high EC values in the eastern part of the study area were due to the influence of seawater intrusion. Higher concentrations of it were reported by Vetrimurugan et al. (2013) in the coastal regions of Karaikal in the Nagapattinam district. During the flood irrigation, paddy was intensively cultivated throughout the year in the study area. In the agriculture fields, about half to two-thirds of the irrigation water applied was lost as evapotranspiration, and the rest infiltrated and joined the groundwater. This process is known as return flow from irrigation. During evapotranspiration, the dissolved salts in irrigation waters get concentrated and eventually leached out from the soil layer to join groundwater. This has a profound effect on the quality of groundwater. This process of groundwater salinization due to the irrigation return flow was widely observed in the semi-arid regions (Vetrimurugan and Elango, 2015; Foster et al., 2018). Return flow from irrigation was also associated with higher EC in the groundwater (Vetrimurugan et al., 2013). A detailed EC-based classification of groundwater as suggested by Wilcox (1955) is presented in Table 12.4.

12.4.3 Total Dissolved Solids

The most desirable limit of total dissolved solids (TDS) is 500 mg/L, and the maximum allowed value is 2,000 mg/L, as per the guidelines of BIS (2012). In the study sites, TDS varies from 185 to 6,745 mg/L in the PRM, with an average of 1,476 mg/L, and in the SWM, the values ranged from 255 to 7,968 mg/L, with an average value of 1,580 mg/L. During the NEM season, values were observed to be in the range of 238–6,784 mg/L, with an average of 1,413 mg/L, and in the POM, the values fall within the range of 191 and 4,039 mg/L, with an average value of 1,382 mg/L. Groundwater

TABLE 12.3B
Comparison of Water Quality Parameters for Samples from the Cauvery and Vennar Sub-basins

Water Quality Parameters (mg/L)		PRM		SWM		NEM		POM	
		Cauvery Sub-basin	Vennar Sub-basin	Cauvery Sub-basin	Vennar Sub-basin	Cauvery Sub-basin	Vennar Sub-basin	Cauvery Sub-basin	Vennar Sub-basin
pH	Min	6.5	6.6	6.6	6.5	6.6	6.6	7.1	6.9
	Max	7.9	8.0	8.1	8.1	8	8.1	8.3	8.7
	Mean	7.2	7.3	7.3	7.4	7.3	7.3	7.7	7.7
EC	Min	290	664	400	1065	372	861	299	823
	Max	3,556	10,540	4,034	12,450	4,600	10,600	4,762	6,312
	Mean	1,256	3,202	1,436	3,349	1,264	3,013	1,284	2,709
TDS	Min	185	425	256	682	238	551	192	527
	Max	2,276	6,746	2,582	7,968	2,944	6,784	3,048	4,040
	Mean	804	2,050	919	2,144	809	1,928	822	1,733
Ca^{2+}	Min	48	144	24	48	48	64	16	40
	Max	512	896	480	1,040	384	1,096	224	800
	Mean	207	383	169	329	129	329	93	256
Mg^{2+}	Min	62	77	19	43	19	38	34	24
	Max	192	1,810	302	1,210	221	350	206	562
	Mean	116	354	123	299	67	123	74	132
Na^{+}	Min	19	71	29	98	34	52	11	60
	Max	656	1,425	760	1,110	669	1,415	698	866
	Mean	157	361	189	322	153	319	162	295

(Continued)

TABLE 12.3B (*Continued*)
Comparison of Water Quality Parameters for Samples from the Cauvery and Vennar Sub-basins

Water Quality Parameters (mg/L)		PRM		SWM		NEM		POM	
		Cauvery Sub-basin	Vennar Sub-basin	Cauvery Sub-basin	Vennar Sub-basin	Cauvery Sub-basin	Vennar Sub-basin	Cauvery Sub-basin	Vennar Sub-basin
K^+	Min	7	7	5	8	12	13	3	9
	Max	118	256	109	267	157	348	135	257
	Mean	31	58	37	67	48	74	45	59
Cl^-	Min	20	100	60	100	60	100	50	100
	Max	601	3,314	911	3,764	1,001	3,454	481	1,892
	Mean	193	703	239	768	211	712	174	612
CO_3^-	Min	0	0	40	40	0	0	20	0
	Max	60	40	280	420	140	160	260	320
	Mean	21	22	109	115	41	44	108	107
HCO_3^-	Min	30	40	120	120	220	300	130	230
	Max	140	130	580	770	1,300	1,240	850	650
	Mean	53	75	257	265	512	682	330	416
SO_4^{2-}	Min	16	31	15	15	16	19	27	76
	Max	199	956	128	477	199	583	393	670
	Mean	75	219	42	104	64	187	114	253
TH	Min	480	740	420	420	280	480	180	360
	Max	1,900	8,080	2,020	6,120	1,300	3,640	1,240	3,080
	Mean	1,001	2,436	934	2,067	599	1,335	541	1,191

classification based on TDS concentration, employing the method suggested by Davis and de Wiest (1966), is given in Table 12.4. According to this classification, only 16% of the water samples were within the desirable limit during the PRM; about 62% of the samples fell under the allowable limit, and 25% of the samples were not suitable. Only 8% of the samples were within the desirable limit during the SWM. The majority of the samples (~66%) fall within the category of allowable limit and about 26% fall under the category of not suitable. In the NEM, only 12% of the samples were in the desirable limit, around 66% were under the allowable limit, and about 22% were in the not suitable category. In the POM, about 14% of samples were in the desirable limit, around 64% were under the category of allowable limit, and about 22% were in the not suitable category.

12.4.4 Total Hardness

According to BIS (2012), the most desirable limit of total hardness (TH) is 200 mg/L, and the maximum allowable limit is 600 mg/L. In the Nagapattinam-Karaikalbelt, TH in the PRM season varied from 480 to 8,080 mg/L, with an average value of 1,775 mg/L; in the SWM season, the values ranged from 420 to 6,120 mg/L, with an average value of 1,546 mg/L. During the NEM, the values were between 280 and 3,640 mg/L, with an average of 996 mg/L. During the POM, the values ranged from 180 to 3,080 mg/L, with an average value of 935 mg/L. Based on BIS standards (2012), about 94%, 88%, 72%, and 62% of samples were unsuitable for drinking purposes during the PRM, SWM, NEM, and POM seasons, respectively. The hardness of water has a certain impact on heart disease (Schroeder and Palmer, 1960). It may also cause kidney failures (Vasanthavigar et al., 2010).

TABLE 12.4
Groundwater Classification Based on EC, TDS, and TH Parameters

		PRM		SWM		NEM		POM	
Indices	**Classification/ Category**	**No. of Samples**	**%**	**No. of Samples**	**%**	**No. of Samples**	**%**	**No. of Samples**	**%**
EC (uS/cm)									
250	Excellent	ND	ND	ND	ND	ND	ND	ND	ND
250–750	Good	8	16	16	32	6	12	7	14
750–2,000	Permissible	23	46	21	42	26	52	26	52
2,000–3,000	Doubtful	7	14	9	18	6	12	10	20
3,000	Unsuitable	12	24	4	8	12	24	7	14
TDS (mg/L)									
500	Desirable for drinking water	8	16	4	8	6	12	7	14
500–1,000	Permissible for drinking water	20	40	20	40	18	36	20	40
1,000–3,000	Useful for irrigation	17	35	22	44	21	42	18	36
3,000	Unfit for drinking and irrigation	5	10	4	8	5	10	5	10
Total hardness (mg/L)									
75	Soft	ND	ND	ND	ND	ND	ND	ND	ND
75–150	Moderately hard	ND	ND	ND	ND	ND	ND	ND	ND
150–300	Hard	ND	ND	ND	ND	2	4	4	8
300	Very hard	100	100	100	100	48	96	46	92

ND, Not detectable.

Groundwater with high TH used for industrial purposes may induce scaling in pipes and boilers. Table 12.4 gives a detailed classification of TH, as suggested by Sawyer and McCarty (1967).

12.4.5 Calcium

The most desirable limit of Ca^{2+} is 75 mg/L as per the standards stipulated by the BIS (2012), and the highest permissible limit is 200 mg/L. In the Nagapattinam-Karaikal belt, Ca^{2+}during the PRM season ranged from 48 to 896 mg/L, with an average value of 302 mg/L. In the SWM, it ranged between 24 and 1,040 mg/L, with an average value of 255 mg/L. During the NEM, the values were between 48 and 1,096 mg/L, with an average of 237 mg/L. During the POM, the values were between 16 and 800 mg/L, with an average value of 193 mg/L. The dissolution of carbonate minerals in the coastal area results in a high concentration of Ca^{2+} (Mondal et al., 2010a). Figure 12.4 shows a spatial interpolation map of Ca^{2+} concentration in the study site. It indicates that the Vennar sub-basin is not at all suitable for drinking purposes in all seasons throughout the year.

12.4.6 Magnesium

The increased application of magnesium fertilizer ($MgSO_4$) used for cultivation may result in return flow into the well (Kelly et al., 1996). A high concentration of magnesium (Mg^{2+}) has a toxic effect on the crop and thereby negatively impacts crop yields (Krishnakumar et al., 2014). The most desirable value for Mg^{2+} is 30 mg/L as per the standards stipulated by BIS (2012), and the maximum allowable is 100 mg/L. In the Nagapattinam-Karaikal belt, Mg^{2+} in the PRM ranged from 62 to 1,809 mg/L, with an average value of 244 mg/L. In the SWM, the values were between 219 and

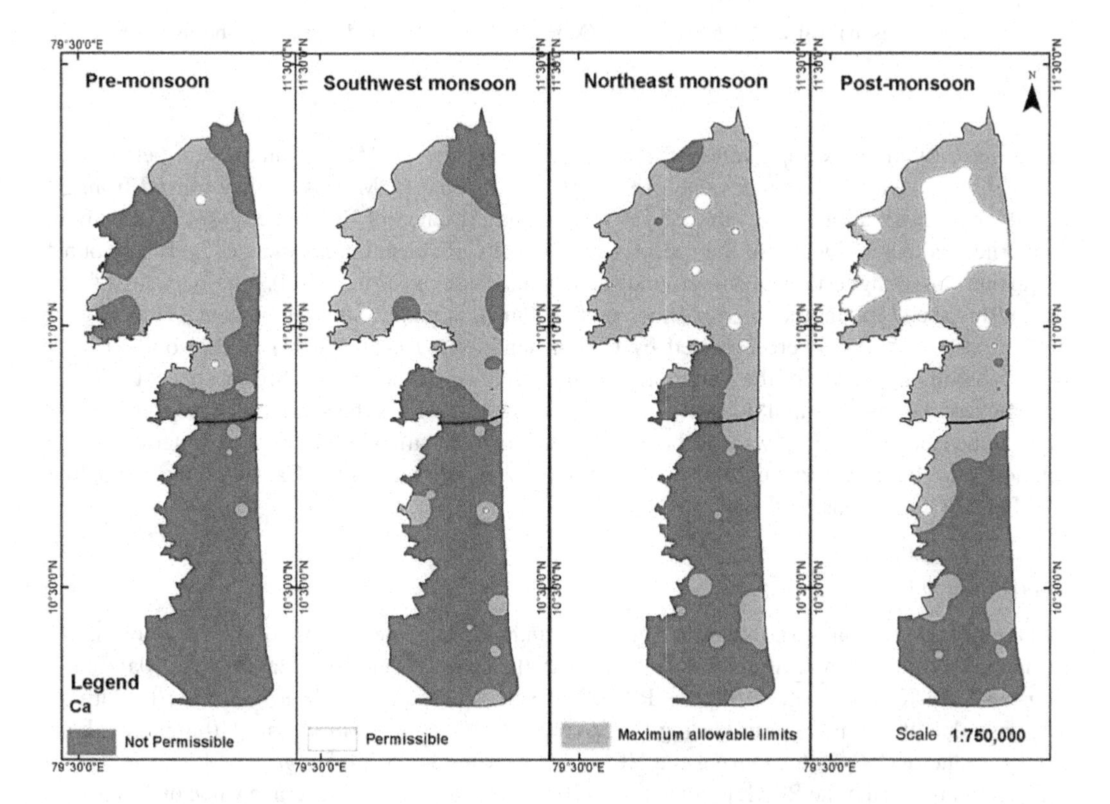

FIGURE 12.4 Spatial interpolation map showing the seasonal variation of calcium (Ca^{2+}) in the Cauvery and Vennar sub-basins.

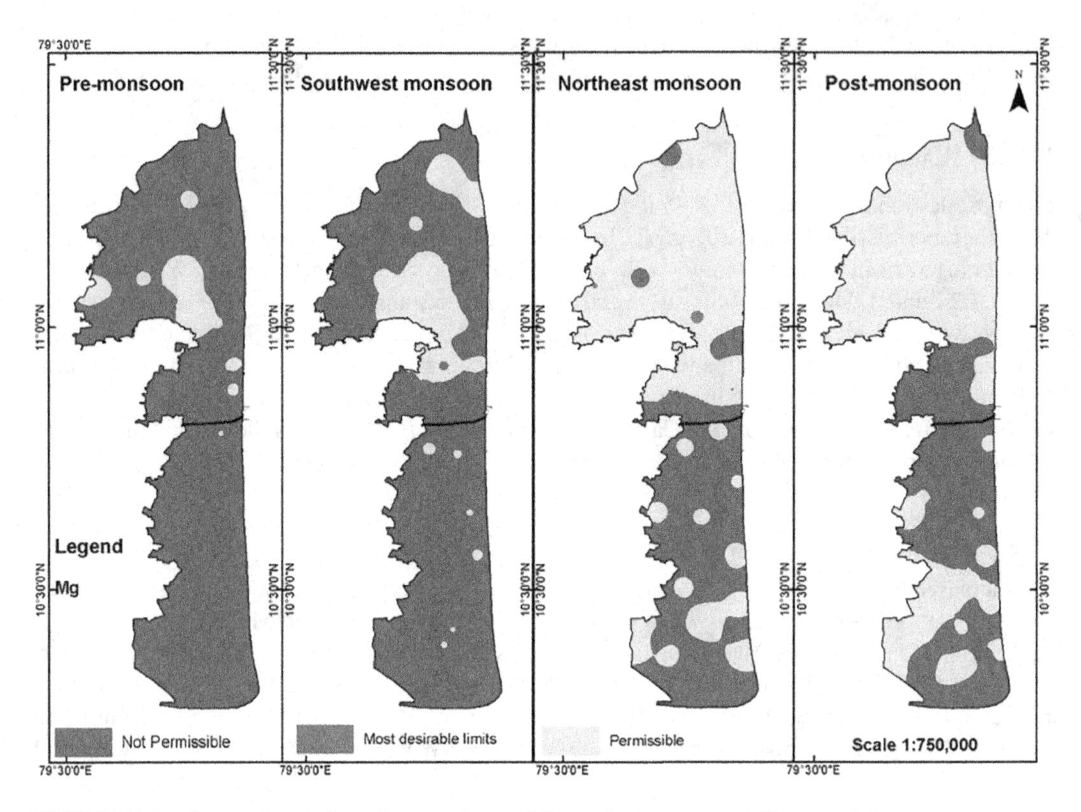

FIGURE 12.5 Seasonal variation of magnesium (Mg^{2+}) in the Cauvery and Vennar sub-basins.

1,209 mg/L, with an average value of 217 mg/L. During the NEM, the values were between 19 and 350 mg/L, with an average value of 96 mg/L, and in the POM season, they ranged from 24 to 561 mg/L, with an average value of 114 mg/L. The presence of limestone, gypsum, dolomite, anhydrite, and clay minerals in the sedimentary deposits of coastal areas induces higher amounts of calcium and magnesium in the groundwater (Chandrasekar et al., 2014). The deposits of fluvio-marine shells are another possible source of calcium and magnesium (Hounslow, 1995). Higher concentrations of Mg^{2+} were reported by Gnanachandrasamy et al. (2015) in the southern part of Nagapattinam district. In the Vedaranayam area on the southern part, a higher concentration of Mg was reported by Krishnakumar et al. (2014). Figure 12.5 shows the spatial interpolation map of Mg^{2+} at the study sites. It shows that the groundwater was not suitable for drinking water use in the southern part, but it was in a useable condition in the northern part of the Cauvery delta during the NEM and POM seasons.

12.4.7 Sodium

The high concentration of sodium (Na^+) in the groundwater may be because of the cation exchange at the interface between seawater and freshwater in the coastal areas (Mondal et al., 2010a). In the Nagapattinam-Karaikalbelt, during the PRM, Na^+ values ranged from 18 to 1,425 mg/L, with an average value of 267 mg/L, and during the SWM, values ranged from 29 to 1,110 mg/L, with an average value of 260 mg/L. During the NEM, it was from 34 to 1,415 mg/L, with an average of 242 mg/L, and during the POM, it ranged from 10 to 865 mg/L, with an average value of 248 mg/L.

FIGURE 12.6 Spatial interpolation map showing the seasonal variation of sodium (Na^+) in the Cauvery and Vennar sub-basins.

The spatial distribution map for Na^+ variation is presented in Figure 12.6. This map reveals high Na^+ values during all the seasons, with lower Na^+ values in the northern part and higher values in the southern, central, and coastal areas of the study area. Multiple factors, like the dissolution of salts from the soil, the impact of marine sources, and anthropogenic influences, contribute to the Na^+ concentration in the groundwater (Nageswara Rao et al., 2017).

12.4.8 Potassium

In natural waters, potassium (K^+) is commonly found as a secondary ion; the allowable limit of potassium in natural waters is 12 mg/L (WHO, 2012). In a study from the coastal Cauvery delta, an abundance of K^+ due to mineralization and biogenic transformation was reported (Asokan and Hameed, 1992). Based on a study in the Karaikal area of the Cauvery delta (Sukhija et al., 1996), it was suggested that high K^+ in groundwater is an indication of recent intrusion of estuarian water due to overexploitation. The study area was intensively cultivated by rice, and the application of potassium fertilizers - potash and nitrogen-phosphorus-potassium (NPK)-mixed fertilizers influenced the groundwater quality in the southern part, especially Vedaranyam in Nagapattinam district (Krishnakumar et al., 2014). The dissolution of pyroxene minerals and the introduction of high K^+ and fertilizers with Mg^{2+} containing nutrients in agricultural lands might be a possible source of Mg^{2+} and K^+ in the region (Härdter et al., 2004). In addition, intrusion of saline water and anthropogenic activities are considered causes of higher K^+ concentrations in groundwater (Kumar et al.,

2009). In the study area, K^+ ranged from 6 to 256 mg/L, with an average value of 45 mg/L during the PRM; during the SWM, it ranged between 5 and 266 mg/L, with an average value of 53 mg/L. During the NEM, it was between 12 and 347 mg/L, with an average value of 61 mg/L. During the POM, ranging from 2 to 257 mg/L, with an average value of 57 mg/L. During all four seasons, Sikkal (N24) showed higher concentrations in the central part of the study area.

12.4.9 Chloride

Leaf burn and leaf necrosis are usually caused when chloride (Cl^-) is absorbed by the plant roots (Hosseinifard and Mirzaei-Aminiyan, 2015). The most desirable limit of Cl^- is 250 mg/L and the maximum allowable limit is 1,000 mg/L (BIS, 2012). In the study area, Cl^- ranged between 20 and 3,313 mg/L in the PRM, with an average value of 468 mg/L. During the SWM season, it varied from 60 to 3,764 mg/L, with an average value of 524 mg/L. During the NEM, it was between 60 and 3,453 mg/L, with an average of 481 mg/L. During the POM, it ranged from 50 to 1,892 mg/L, with an average value of 445 mg/L. A higher concentration of Cl^- during the POM was due to the leaching of the soil derived from the industrial and domestic sources and to the dry climatic conditions (Srinivasamoorthy et al., 2008). Figure 12.7 shows the spatial distribution map of Cl^- in the study area generated based on the BIS (2012) on groundwater quality. The spatial interpolation map indicates that out of the total area of 2,757 km^2, about 7%, 8%, 10%, and 4% of the area falls within the not-permissible limits of Cl^- (> 1,000 mg/L) during the PRM, SWM, NEM, and POM, respectively, especially in the Vennar sub-basin.

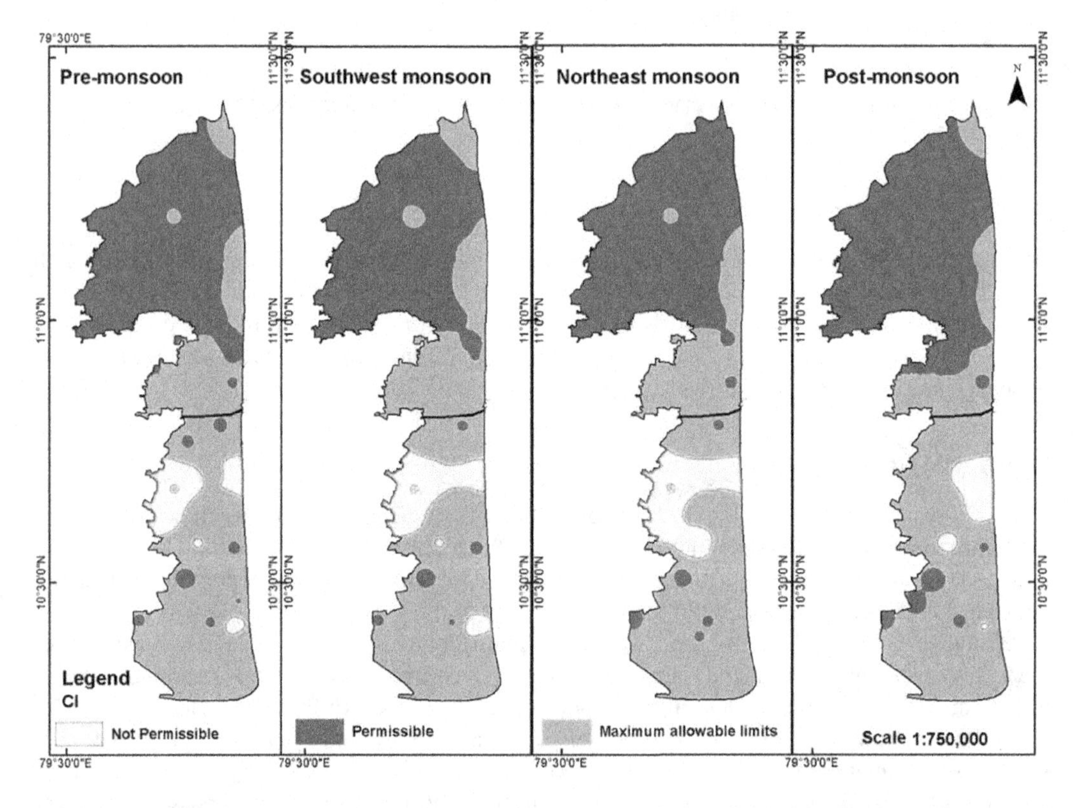

FIGURE 12.7 Spatial interpolation map showing the seasonal variation of chloride (Cl^-) in the Cauvery and Vennar sub-basins.

12.4.10 Carbonate

Carbonate (CO_3^-) during the PRM was found to be a maximum of 60 mg/L, with an average value of 21 mg/L in the study area. In the SWM, it was between 40 and 420 mg/L, with an average of 112 mg/L. In the NEM, the average value of CO_3^- was found to be 42 mg/L, whereas it was 111 mg/L during the POM. It indicates that the carbonate dissolution took place after precipitation.

12.4.11 Bicarbonate

The higher bicarbonate (HCO_3^-) concentrations in natural waters indicate the dominance of mineral dissolution (Stumm and Morgan, 1996). Dissolution of carbonate minerals in coastal areas results in a high concentration of HCO_3^- (Mondal et al., 2010a). In the study area, HCO_3^- concentration during the PRM season varied from 30 to 140 mg/L, with an average of 65 mg/L. In the SWM season, the values varied between 120 and 770 mg/L, with an average of 261 mg/L. During the NEM season, HCO_3^- values ranged between 220 and 1,300 mg/L, with an average value of 604 mg/L. In the POM season, the values were in the range of 130 and 850 mg/L, with an average of 383 mg/L. The dissolution of silicate and calcic minerals in the carbonic acid of organic materials might be the main cause of HCO_3^- (Drever and Stillings, 1997).

12.4.12 Sulfate

The presence of sulfate (SO_4^{2-}) in drinking water with concentrations of more than 250 mg/L may cause a bitter taste for the drinking water and cause corrosion of pipelines used to distribute water (Hosseinifard and MirzaeiAminiyan, 2015). In the study area, SO_4^{2-} in the PRM season varied from 16 to 956 mg/L, with an average of 152 mg/L. In the SWM season, values varied between 14 and 476 mg/L, with an average of 75 mg/L. During the NEM season, SO_4^{2-} values ranged between 15 and 583 mg/L, with an average value of 130 mg/L; in the POM season, it was in the range of 26 and 669 mg/L, with an average of 200 mg/L. In groundwater, higher values of sulfate are caused by reduction, precipitation, and solution action during the interaction between groundwater and formations like gypsum or anhydrite (Hounslow, 1995). High values of sulfate might also be attributed to the leaching and anthropogenic activities in the metamorphic environment caused by the release of sulfur gases from industries and the oxidization of urban utilities and their entrance into the groundwater system (Saxena et al., 2004). Brown and Schoonen (2004) also reported a high concentration of SO_4^{2-} (~600 mg/L) in coastal aquifers of the eastern US caused by saltwater intrusion. It indicates that the area is prone to saltwater intrusion.

12.5 WATER QUALITY INDEX

The WQI was estimated for all the seasons, as shown in Figure 12.8. According to the WHO standards, groundwater can be classified into five types, such as excellent, good, poor, very poor, and unsuitable for drinking use. Most of the study area was in the poor water category with a WQI range of 100–200, followed by good water with a range of 51–100, and very poor water with a range of 200–300. The central part, with a WQI> 300, falls within the unsuitable category for drinking purposes. During the PRM season, out of the total area of 2,757 km^2, about 7, 807, 953, 661, and 329 km^2 areas showed excellent, good, poor, very poor, and unsuitable categories, respectively. In the SWM season, about 4, 667, 1,146, 636, and 304 km^2 areas showed excellent, good, poor, very poor, and unsuitable categories for drinking, respectively. In the NEM season, about 29, 995, 1,002, 434, and 297 km^2 areas showed excellent, good, poor, very poor, and unsuitable categories, respectively.

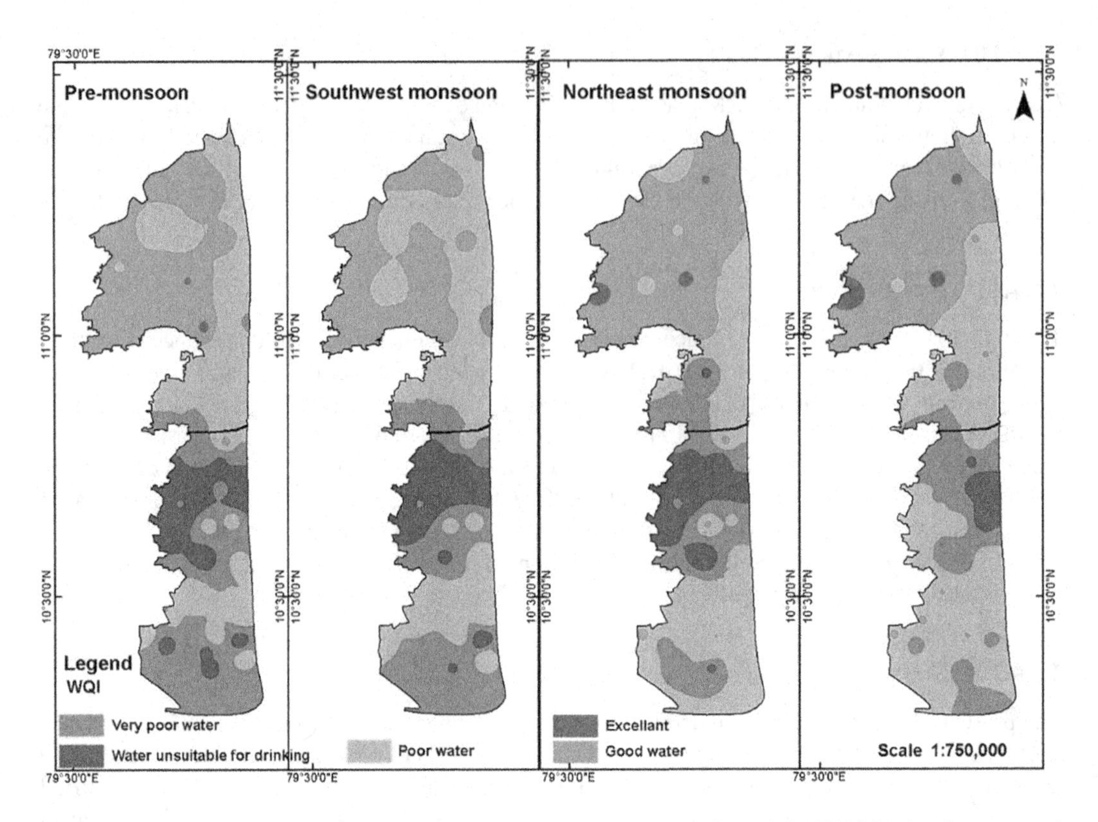

FIGURE 12.8 Spatial interpolation map showing the seasonal variation of the WQI in the Cauvery and Vennar sub-basins.

Whereas about 44, 889, 1,287, 449, and 88 km^2 areas showed excellent, good, poor, very poor, and unsuitable categories, respectively, in the POM season, water quality improved due to the impact of recharge compared to the other seasons. The detailed classification of groundwater based on the WQI and its spatial influence is given in Table 12.5a. The spatial comparison of the WQI between the Cauvery and Vennar sub-basins shows that the groundwater quality in the Vennar sub-basin was inferior to the Cauvery sub-basin (Table 12.5b). From the results of hydrogeochemical analyses and the WQI, it is revealed that the majority of samples from the northern Cauvery sub-basin were found to be within the permissible limits. In contrast, in the southern Vennar sub-basin, most of the samples exceeded the permissible limits as stipulated by WHO (2011) and BIS (2012). The trends were also observed on the spatial maps of the physical parameter EC along with the chemical parameters of Cl^- and Na^+ concentrations of groundwater. Further, this trend was reflected in the WQI map. On studying the hydrogeology of both sub-basins in detail, some factors that cause the variation in the groundwater water quality were identified. They are mainly overexploitation of groundwater and a reversal hydraulic gradient.

A semi-confined/ unconfined layer of sandy silt in the shallow aquifer impacts the recharge in the Cauvery delta (UNDP, 1973; Dhinagaran, 2008). In the northern Cauvery sub-basin, the average thickness of this layer was from 3 to 5 m, and over the southern Vennar sub-basin, it was from 6 to 10 m. As irrigation water is released to the deltaic region from the Mettur dam, the shallow aquifer in this sub-basin generally gets recharged to its full capacity in about 2 months. The fast recharge was due to the water flow in the river channels cutting into 3–5 m of this layer of deposits; this water recharged the shallow aquifer directly by lateral infiltration. The semi-confined, unconfined deposits in the Vennar sub-basin are located much deeper compared to the Cauvery sub-basin. The channels, therefore, are not in direct hydraulic contact with the shallow aquifer;

TABLE 12.5A
Groundwater Classification Based on WQI in the Entire Study Area

		PRM			SWM			NEM			POM		
WQI	Type of water	Samples	%	Km²	Samples	%	Km²	Samples	%	Km²	Samples	%	Km²
<50	Excellent water	2	4	7	2	4	4	4	4	29	3	6	44
51–100	Good water	10	20	807	11	22	667	13	26	995	17	34	889
100–200	Poor water	23	46	953	21	42	1,146	21	42	1,002	17	34	1,287
200–300	Very poor water	6	12	661	6	12	636	3	6	434	8	16	449
>300	Unsuitable for drinking	9	18	329	10	20	304	9	18	297	5	10	88

TABLE 12.5B
Groundwater Classification Based on WQI in the Cauvery and Vennar Sub-basins

	PRM				SWM				NEM				POM			
	Cauvery Sub-basin		Vennar Sub-basin		Cauvery Sub-basin		Vennar Sub-basin		Cauvery Sub-basin		Vennar Sub-basin		Cauvery Sub-basin		Vennar Sub-basin	
Category of Water	km²	%	km²	%	km²	%	km²	%	km²	%	km²	%	km²	%	km²	%
Excellent water	7	0.45	0	0.00	4	0.26	0	0.00	30	1.95	0	0.00	45	2.92	0	0.00
Good water	804	52.17	3	0.25	667	43.28	1	0.08	984	63.81	11	0.90	878	56.98	10	0.82
Poor water	672	43.61	280	23.03	779	50.55	367	30.18	401	26.01	600	49.34	545	35.37	742	61.02
Very poor water	58	3.76	604	49.67	90	5.84	545	44.82	123	7.98	312	25.66	73	4.74	376	30.92
Water unsuitable for drinking	0	0.00	329	27.06	1	0.06	303	24.92	4	0.26	293	24.10	0	0	88	7.24
Total	1,541	100	1,216	100	1,541	100	1,216	100	1,541	100	1,216	100	1,541	100	1,216	100

the lateral infiltration from the channels that carried irrigation water was very negligible. The significant recharge of the shallow aquifer was due to the infiltration from the fields, which was the highest during the peak NEM season (UNDP, 1973). The Quaternary deposits underlie the entire Nagapattinam district and are of fluvial and semi-marine origin. In the Vennar sub-basin, the Quaternary was deposited during the marine transgression and regression (Das, 1991). The Quaternary deposits of marine origin in the Vennar sub-basin may cause groundwater quality deterioration. On the other hand, in the Cauvery sub-basin, Quaternary deposits are mostly fluvial in origin (Kokkat, 2018).

12.6 IRRIGATION SUITABILITY

Since groundwater is widely used for agriculture in the study area, its suitability for irrigation was analyzed. The mineral contents of groundwater determine the appropriateness of water for irrigation to a great extent, as these have an equal impact on both the plant and the soil. Enhancing crop productivity and maintaining healthy soil are critical considerations. The irrigation suitability had been explored and also discussed using the parameters of SAR, Na (%), MAR, RSC, and KR.

As shown in Table 12.6, the samples coming under the excellent category as per SAR classification were about 98% for the PRM and about 96% uniformly for the SWM, NEM, and POM seasons, respectively. About 2% of the PRM samples and 4% of the samples from other seasons fall within the good category. It implies that groundwater was fit for irrigation purposes in most parts of the study area. The emergence of salinity due to rock-water interaction, leaching from geologic materials, and anthropogenic sources could increase the concentration of Na^+ in irrigation water and affect soil permeability (Venkatramanan et al., 2015).

Based on this classification, during the PRM, about 42% of samples were in the categories of excellent and good, about 12% were under permissible, and about 2% were in the doubtful and unsuitable categories, respectively. In the SWM season, of the 50 samples considered, about 38% came under the excellent category, 44 under the good category, 12 under the permissible limits, and six under the doubtful category. During the NEM season, about 8% of samples were under the excellent category, 56 were under the good category, 30 were under the permissible, and the rest six were under the doubtful category. In the POM season, about 10% of the samples were under the excellent category, 58 were under the good category, 20 were under the permissible, and the rest 12 were under the doubtful category (Table 12.6). Vetrimurugan and Elango (2015) have reported a high value of Na (%) from the coastal parts of Karaikal. In the NEM and POM seasons, the coastal sites had a high sodium percentage; these sampling points might have gained sodium from seawater due to saline intrusion. The sodium ions in groundwater are absorbed by clay particles, displacing Ca^{2+} and Mg^{2+} ions. This process of exchange of ions between Na^+ in groundwater with Ca^{2+} and Mg^{2+} in soil results in a reduction of permeability and leads to poor internal drainage in the soil (Vetrimurugan and Elango, 2015). A high sodium percentage in waters is caused by the dissolution of minerals from lithology, chemical fertilizers, and the extended residence period of waters (Qiyan and Baoping, 2002). As per the RSC values (in Table 12.6), there was no bad-category groundwater sample for the irrigation in the PRM. But the most bad-category samples (6%) were found during the NEM season, whereas good quality groundwater was observed during the SWM season. An equal number of samples (~90%) were observed in the good category during both the NEM and POM seasons. The samples falling into the bad category had a high concentration of carbonates and bicarbonates. Irrigation with these waters will result in the precipitation of calcite in the soil, leading to soil dispersion and impaired nutrient uptake by plants (Krishnakumar et al., 2014). Based on the KR, groundwater samples had been classified as either suitable or unsuitable for irrigation purposes. During all the seasons, about 84%–96% of samples were under the suitable categories (Table 12.6). But about 12%–16% of groundwater samples were unsuitable during the NEM and

TABLE 12.6
Groundwater Classification for Irrigation Purpose in Cauvery and Vennar Sub-basins

Irrigation Indices		PRM		SWM		NEM		POM	
	Category	Samples	Percentage	Samples	Percentage	Samples	Percentage	Samples	Percentage
SAR									
0–10	Excellent	49	98	48	96	48	96	48	96
Oct–18	Good	1	2	2	4	2	4	2	4
18–26	Fair	0	0	0	0	0	0	0	0
Na%									
0–20	Excellent	21	42	19	38	4	8	5	10
20–40	Good	21	42	22	44	28	56	29	58
40–60	Permissible	6	12	6	12	15	30	10	20
60–82	Doubtful	1	2	3	6	3	6	6	12
>80	Unsuitable	1	2	0	0	0	0	0	0
MAR									
<50	Suitable	23	46	21	42	31	62	18	36
>50	Unsuitable	27	54	29	58	19	38	32	64
RSC									
>1.25	Good	50	100	47	94	45	90	45	90
1.25–2.5	Medium	0	0	0	0	1	2	2	4
>2.5	Bad	0	0	3	6	4	8	3	6
Kelly's ratio									
<1	Suitable	48	96	46	92	44	88	42	84
>1	Unsuitable	2	4	4	8	6	12	8	16

POM seasons. The samples falling into the unsuitable category had a high concentration of Na^+ due to the intense cation exchange process (El-Amier et al., 2021).

During the PRM, SWM, NEM, and POM seasons, about 46%, 42%, 62%, and 36% of the samples fall within the suitable categories. A MAR value of more than 50 was deemed unsuitable for irrigation. During the PRM, SWM, NEM, and POM, about 54%, 58%, 38%, and 64% of the samples fall within the unsuitable category, respectively, as presented in Table 12.6. If the concentration of Mg^{2+} is higher than calcium, it may have a negative impact on agricultural yield (Nagaraju et al., 2006). The MAR values suggest an enrichment of magnesium relative to calcium; this was due to the influence of seawater intrusion into the coastal aquifers (Sajil Kumar et al., 2013). The direct cation exchange process during seawater intrusion causes the absorption of calcium and the enrichment of sodium and magnesium (Bouderbala et al., 2016). In general, the groundwater in the study area was fit for irrigation as per the SAR, sodium percentage, RSC, and KR. But the MAR, a metric for determining the suitability of groundwater for irrigation, revealed that the water quality in some areas of the Cauvery and Vennar sub-basins was unsuitable for irrigation.

12.7 HYDROGEOCHEMICAL PROCESSES

The combined effect of both natural and anthropogenic factors controls the hydrogeochemical processes in an aquifer. These factors varied in space and time in the study area. The Gibbs plot and Chadha diagram had provided an insight into the various hydrogeochemical processes in the study area.

12.8 GIBBS PLOT

The reaction between groundwater and aquifer materials is a major process that controls groundwater geochemistry. It has a significant role in deciding the quality of groundwater. This technique aids in the comprehension of the genesis of water (Gibbs, 1970). Several researchers employ the Gibbs plot to elucidate the factors influencing hydrogeochemistry (Nageswara Rao et al., 2017; Brindha et al., 2017; Salifu et al., 2017). This diagram is based on the ratios of Na^++K^+: $Na^++K^++Ca^{2+}$ *vs* TDS, and Cl^-: Cl^-+HCO^{3-} *vs* TDS. It is useful in assessing the influence of rainfall, evaporation, and rock formations on the groundwater system. The Gibbs plot was generated for all four seasons, as shown in Figure 12.9. The results showed that the majority of the samples fall within the zones with rock dominance in the Cauvery sub-basin compared to the Vennar sub-basin in the delta and also suggested that percolating water interacts with rock chemistry. The chemical weathering occurred in rock-dominant zones, with the dissolution of rock-forming minerals influencing groundwater chemistry. Anthropogenic and marine factors contributed to the movement of samples from the rock-dominant zone to the evaporation-dominant zone, leading to an increase in Na^+, Cl^-, and TDS in groundwater (Srinivasamoorthy et al., 2008).

12.9 CHADHA DIAGRAM

The Figure provides details of the number of groundwater samples that fall into the four fields as per the Chadha diagrams are presented in Tables 12.7a and b. Spatial interpolation maps showing the seasonal evolution of geochemical processes based on the Chadha plots are shown in Figure 12.10. In the PRM, SWM, NEM, and POM seasons, about 92%, 50%, 32%, and 48% samples, respectively, fall in Field 6 (Ca–Mg–Cl type). In this group, the major water types were Ca-Mg-Cl, and Ca-Mg dominant with Cl^- or Cl^- dominant with Ca-Mg, showing reverse ion exchange. The samples falling into this category showed higher $Ca^{2+}+Mg^{2+}$ than Na^++K^+. Appelo and Postma (1999) suggested that the saline water ingression into fresh coastal aquifers results in $CaCl_2$ or Mg-Cl_2 type of water. In the PRM season, the scarcity of surface water in the Cauvery deltaic region compels the farmers to depend on groundwater for irrigation. This leads to the overutilization of groundwater

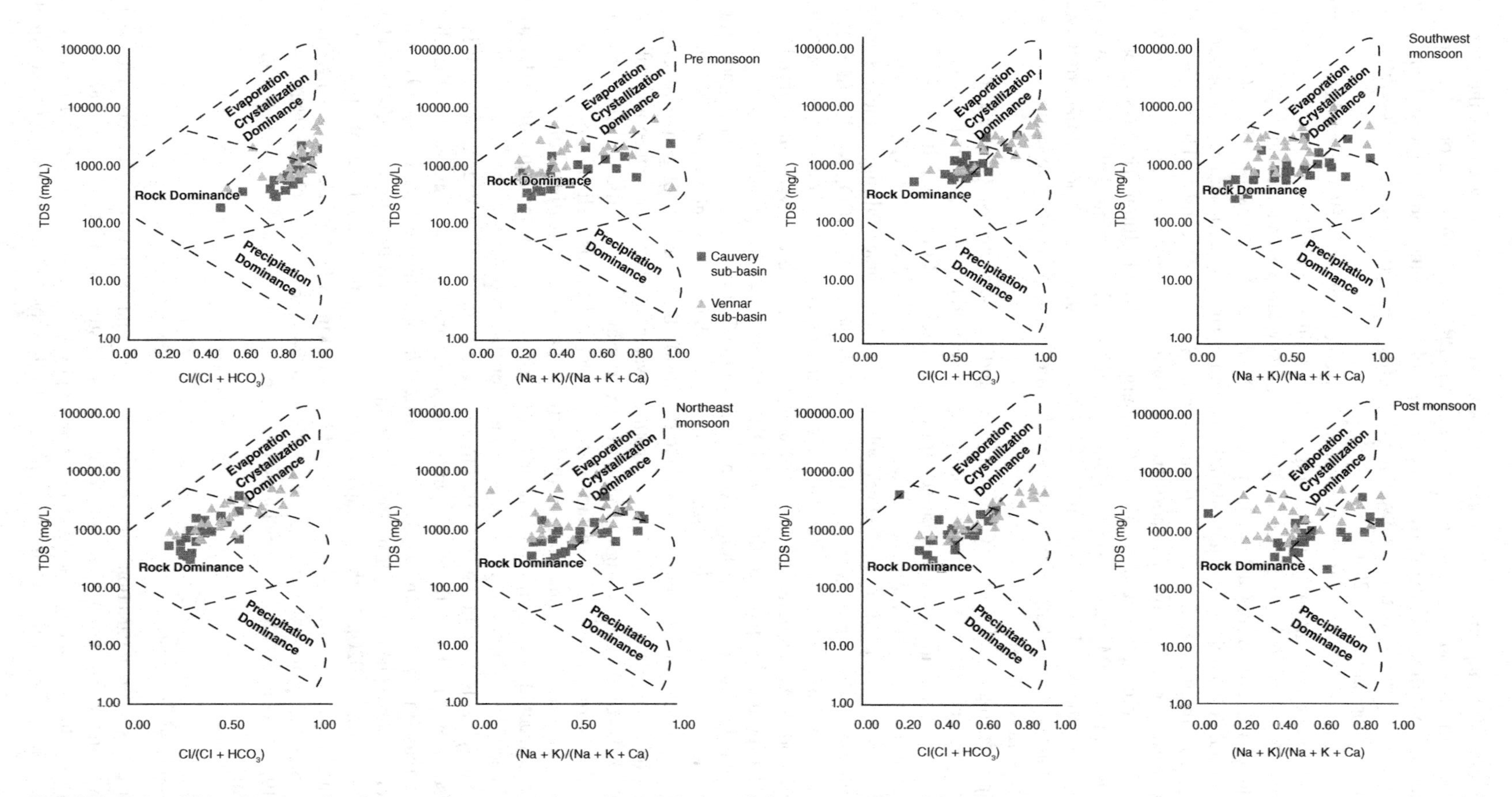

FIGURE 12.9 Gibbs diagrams for the groundwater samples collected during all four seasons from the Cauvery and Vennar sub-basins within the Cauvery delta, Southern India.

in the study area (Sivanappan, 2007). The overexploitation of groundwater causes the ingression of seawater into groundwater. Due to this process in the PRM and SWM seasons, the Ca-Mg-Cl type of water was prominent in the study area. Analysis of the groundwater in the northern part of Nagapattinam and Karaikal districts showed that Ca^{2+} - Mg^{2+} - Cl^- types were the dominant hydrochemical facies (Gopinath et al., 2016).

When surface water infiltrates, it brings along dissolved carbonate as HCO_3, causing temporary hardness. This process is known as recharging water. During the PRM, SWM, NEM, and POM seasons, about 2%, 40%, 46%, and 32% samples, respectively, fall within the category of recharging waters (Field 5). The shift from Ca-Mg-Cl type in the PRM and SWM seasons to Ca-HCO_3 type in the NEM was due to the 69% of annual rainfall received during this season. Excess surface water was released from the upstream storage to the delta from August to December. The presence of the Kollidam (Coleroon) river, a prominent northern distributary along the boundary of the study area, causes lateral recharge on the western part. The recharge from the Kollidam river is evident in the NEM and also in the POM season, as shown in Figure 12.10. The combined effect of rainfall and river water during the NEM season recharges the alluvium aquifer and improves the groundwater quality. Vetrimurugan and Elango (2015) reported that the Ca-Mg-Cl type is the second-most dominant water type in Karaikal.

The samples falling in Field 7 of the Chadha diagram indicate seawater. Na-Cl type, Na^+-dominant Cl type, or Cl-dominant Na type waters exhibited typical seawater mixing, primarily seen in coastal areas. Drinking and irrigation waters become salinized as a result of this water. In the PRM, SWM, NEM, and POM seasons, about 6%, 2%, 12%, and 10% of samples, respectively, had fallen in Field 7 (seawater), mostly observed in the Vennar sub-basin. The samples falling in Field 8 indicate base-ion exchange waters. The major types were observed, such as Na-HCO_3, Na-dominant HCO_3, or HCO_3-dominant Na-type waters with base-ion exchange process or RSC disposition. During the SWM, NEM, and POM, about 8%, 10%, and 10% of the samples, respectively, fall within Field 8.

In the delta, before flowing into the Bay of Bengal, the Cauvery splits into 36 major rivers/streams (Mohanakrishnan, 2011). Along the shoreline of the Cauvery delta, the diurnal tidal ranges are low—0.6 to 1.2 m, varying with the season. Because of the delta's gentle slope, the tidal influx and amount of inundation are relatively great (ICID, 2000). Deterioration of groundwater quality and salinization along the coastal area result from seawater intrusion via surface and subsurface channels (Trabelsi et al., 2012). High groundwater withdrawal from shallow aquifers results in the intrusion of salinity from these surface water bodies, and this is well evident from the high values of potassium in the groundwater. In the southern part of Nagapattinam district, deep tube wells were constructed in salt pans to pump brine water. Few salt pans and aquaculture farms had constructed channels from the sea to bring seawater for the production of salt/aquaculture farming. Many paddy fields were being converted to aquaculture farms. The Vedaranyam salt swamp region is also located in the southernmost part of the study area. It had been observed that the salt pan areas increased from 2.8% in 1978 to 10.43% in 2017, covering a spatial area of 19.88 km^2. In the same period, the aquaculture farms showed an increase from 0.19 (0.03%) to 7.74 km^2 (1.41%) (Sathiya Bama et al., 2020). In the Nagapattinam town area, the spatial extent of aquaculture farms had increased from 4.02 km^2 (5.60%) in 2006 to 9.80 km^2 (13.67%) in 2016. During the same period, the frame salt pan was increased from 0.76 km^2 (0.74%) to 1.53 km^2 (2.13%) (Nagamani and Suresh, 2019). These aquaculture farms and salt pans posed an adverse effect on the groundwater quality in this delta.

12.10 CONCLUSIONS

The spatial and seasonal variations in the quality of groundwater in the Cauvery and Vennar sub-basins within the Cauvery delta in Southeast India were studied with the help of 200 groundwater samples collected seasonally. The concentrations of physical parameters and major ions revealed that EC, Ca^{2+}, Mg^{2+}, Na^+, and TH are not within the allowable limits of drinking water quality

TABLE 12.7A
Groundwater Classification Based on the Chadha Diagram for the Entire Study Area

Type of Water		PRM		SWM		NEM		POM	
		No. of Samples	Percentage	No. of Samples	Percentage	No. of Samples	Percentage	No. of Samples	Percentage
Field 5	$Ca–HCO_3$ type of recharging waters	1	2	20	40	23	46	16	32
Field 6	Ca–Mg–Cl type of reverse ion-exchange waters	46	92	25	50	16	32	24	48
Field 7	Na–Cl type of end-member waters (seawater)	3	6	1	2	6	12	5	10
Field 8	$Na–HCO_3$ type of base ion-exchange waters	0	0	4	8	5	10	5	10

TABLE 12.7B
Classification of Groundwater Based on the Chadha Diagram in the Cauvery and Vennar Sub-basins

		PRM		SWM		NEM		POM	
Type of Water	Type of Water	Cauvery Sub-basin	Vennar Sub-basin	Cauvery Sub-basin	Vennar Sub-basin	Cauvery Sub-basin	Vennar Sub-basin	Cauvery Sub-basin	Vennar Sub-basin
Field 5	$Ca–HCO_3$ type of recharging waters	1	0	14	6	16	7	11	5
Field 6	Ca–Mg–Cl type of reverse ion-exchange waters	21	25	5	20	1	15	6	18
Field 7	Na–Cl type of end-member waters (seawater)	1	2	1	0	2	4	1	4
Field 8	$Na–HCO_3$ type of base ion-exchange waters	0	0	3	1	4	1	5	0
Total		23	27	23	27	23	27	23	27

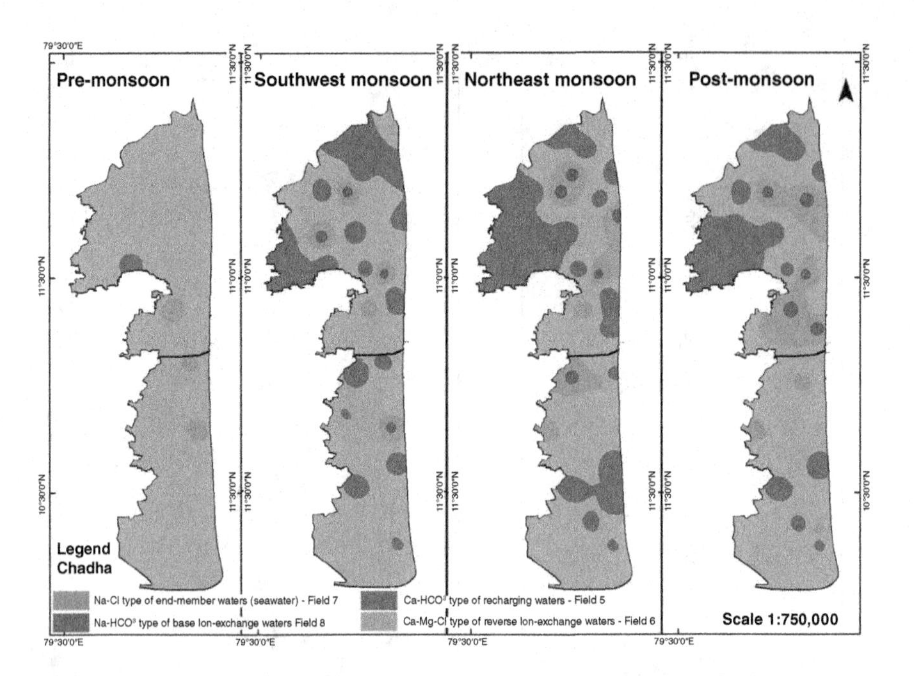

FIGURE 12.10 Spatial distribution maps showing seasonal evolution of geochemical processes in the different seasons based on the Chadha diagrams in both the Cauvery and Vennar sub-basins.

standards as specified by WHO and BIS in all the seasons. The WQI map clearly showed that the majority of the area has poor water for drinking purposes. The quality of groundwater in the Vennar sub-basin was much poorer when compared to the Cauvery sub-basin. The MAR shows that the groundwater quality for irrigation is not satisfactory in several parts within the delta. However, the SAR, Na (%), RSC, and KR do not indicate conspicuous irrigation water quality problems. The results based on Gibbs plots suggested that in most of the samples, the geochemistry is controlled by the dominance of rock-water interaction. The influence of anthropogenic activities and the sediment depositional environment in the Vennar sub-basin are reflected in the Gibbs plot. During the PRM and SWM seasons, the hydrogeochemical facies dominant water type is Ca-Mg-Cl, which usually occurs when the seawater intrudes into the fresh coastal aquifers, thereby indicating salinization in the study area. The shift in water type from Ca–Mg–Cl to Ca–HCO_3-type is an indicator of recharging, which highly improves the groundwater quality during only the rainy season. The results have also brought into light the variation in the hydrogeochemical processes of the Cauvery and Vennar sub-basins. Overall, the depositional environment of quaternary sediments, hydrogeological settings, anthropogenic activities, and the availability of surface water in the delta are the significant factors that control the hydrogeochemistry. The current research is expected to help in the water management of the thickly populated Cauvery delta, known as the "rice bowl" of Tamil Nadu state in Southern India.

ACKNOWLEDGEMENTS

Prof. V.M. Tiwari, Director of CSIR-NGRI, Hyderabad, has been permitted to publish this article (Ref. No. NGRI/Lib/2020/134). The work reported in this chapter has been carried out by the first author as part of the doctoral programme at the Water Institute of Karunya Institute of Technology and Sciences with the funding of the Ministry of Environment, Forestry and Climate Change, Government of India (GoI), New Delhi (Sanction Number F.No.13–16/2008-RE). The authors acknowledge all the support and facilities provided.

REFERENCES

Abbasi, T. and Abbasi, S.A. 2012. *Water Quality Indices*, 1st edn. Elsevier, Amsterdam.

APHA. 2012. *Standard Methods for the Examination of Water and Wastewater*, 22nd edn. American Public Health Association, Washington DC.

Appelo, C.A.J. and Postma, D. 1999. *Chemical Analysis of Groundwater, Geochemistry, Groundwater and Pollution*. Balkema, Rotterdam.

Arumugam, K. and Elangovan, K. 2009. Hydrochemical characteristics and groundwater quality assessment in Tirupur Region, Coimbatore District, Tamil Nadu, India. *Environmental Geology*, 58, 1509. https://doi.org/10.1007/s00254-008-1652-y

Asokan, R. and Hameed, P.S. 1992. Distribution of natural radio nuclide 40K in biotic and abiotic components of the Cauvery river system, Tiruchirapalli, India. *Journal of Biosciences*, 17, 491–497. https://doi.org/10.1007/BF02720104

BIS, 2012. Indian Standard: Drinking Water - Specification (Second Revision). New Delhi.

Bouderbala, A., Remini, B., Saaed, Hamoudi A., and Pulido-Bosch, A. 2016. Application of multivariate statistical techniques for characterization of groundwater quality in the coastal aquifer of nador, Tipaza (Algeria). *ActaGeophysica*, 64(3), 670–693. https://doi.org/10.1515/ACGEO-2016-0027

Brindha, K., Pavelic, P., Sotoukee, T., et al. 2017. Geochemical characteristics and groundwater quality in the vientiane plain, Laos. *Exposure and Health*, 9, 89–104. https://doi.org/10.1007/s12403-016-0224-8

Brown, C.J. and Schoonen, M.A.A. 2004. The origin of high sulfate concentrations in a coastal plain aquifer, Long Island, New York. *Applied Geochemistry*, 19, 343–358. https://doi.org/10.1016/S0883-2927(03)00154-9

Bu, H., Tan, X., Li, S., and Zhang, Q. 2010. Temporal and spatial variations of water quality in the Jinshui River of the South Qinling Mts., China. *Ecotoxicology and Environmental Safety*, 73, 907–913. https://doi.org/10.1016/j.ecoenv.2009.11.007

Burrough, P.A. and McDonnell, R.A. 1998. *Principles of GIS*. Oxford University Press, London, UK.

CGWB, 2020. Aquifer Mapping and Management of Ground Water Resources, Lower Cauvery Systems, Tamil Nadu, Chennai.

Chadha, D.K. 1999. A proposed new diagram for geochemical classification of natural waters and interpretation of chemical data. *Hydrogeology Journal*, 7, 431–439. https://doi.org/10.1007/s100400050216

Chandrasekar, N., Selvakumar, S. and Srinivas, Y., et al. 2014. Hydrogeochemical assessment of groundwater quality along the coastal aquifers of southern Tamil Nadu, India. *Environmental Earth Sciences*, 71, 4739–4750. https://doi.org/10.1007/s12665-013-2864-3

Das, S. 1991. Hydrogeological features of deltas and estuarine tracts of India. *Memoir of Geological Society of India*, 22, 183–225.

Davis, S., de Wiest, N., and Roger, J.M. 1966. *Hydrogeology*. John Wiley & Sons, New York.

Dhinagaran, V. 2008. District Groundwater Brochure Nagapattinam District Tamil Nadu, Chennai.

Drever, J.I. and Stillings, L.L. 1997. The role of organic acids in mineral weathering. *Colloids and Surfaces A: Physicochemical and Engineering Aspects*, 120, 167–181. https://doi.org/10.1016/S0927-7757(96)03720-X

El-Amier, Y.A., Kotb, W.K., Bonanomi, G., et al, 2021. Hydrochemical assessment of the irrigation water quality of the El-Salam Canal, Egypt. *Water*, 13, 2428. https://doi.org/10.3390/W13172428

Foster, S., Pulido-Bosch, A., Vallejos, Á., et al. 2018. Impact of irrigated agriculture on groundwater-recharge salinity: a major sustainability concern in semi-arid regions. *Hydrogeology Journal*, 26, 2781–2791. https://doi.org/10.1007/s10040-018-1830-2

Gibbs, R.J.J. 1970. Mechanisms controlling world water chemistry. *Science*, 170, 1088–1090. https://doi.org/10.1126/science.170.3962.1088

Gnanachandrasamy, G., Ramkumar, T., Venkatramanan, S., et al. 2015. Accessing groundwater quality in lower part of Nagapattinam district, Southern India: using hydrogeochemistry and GIS interpolation techniques. *Applied Water Science*, 5, 39–55. https://doi.org/10.1007/s13201-014-0172-z

Gopinath, S., Srinivasamoorthy, K., Saravanan, K., et al. 2015. Hydrogeochemical characteristics of coastal groundwater in Nagapattinam and Karaikal aquifers: implications for saline intrusion and agricultural suitability Nagapattinam and Karaikal aquifers: implications for saline intrusion. *Journal of Coastal Sciences*, 2, 1–11. https://doi.org/10.6084/m9.figshare.1512790

Gopinath, S., Srinivasamoorthy, K.,Vasanthavigar, M., et al 2016. Hydrochemical characteristics and salinity of groundwater in parts of Nagapattinam district of Tamil Nadu and the Union Territory of Puducherry, India. *Carbonates and Evaporites*, 33, 1–13. https://doi.org/10.1007/s13146-016-0300-y

Härdter, R., Rex, M., and Orlovius, K. 2004. Effects of different Mg fertilizer sources on the magnesium availability in soils. *Nutrient Cycling in Agroecosystems*, 70, 249–259. https://doi.org/10.1007/s10705-004-0408-7

Hosseinifard, S.J. and Mirzaei Aminiyan, M. 2015. Hydrochemical characterization of groundwater quality for drinking and agricultural purposes: a case study in rafsanjan plain, Iran. *Water Quality, Exposure and Health*, 7:531–544. https://doi.org/10.1007/s12403-015-0169-3

Hounslow, A. 1995. *Water Quality Data : Analysis and Interpretation*. Lewis Publishers, Boca Raton, FL.

ICID. 2000. 8th International Commission on Irrigation and Drainage (ICID), International Drainage Workshop. In: "Drainage in deltaic and tidal zones-Executive guidelines for designing drainage Channels in the Cauvery delta." New Delhi, pp. 303–453

Jacks, G., Warrier, C.U., and Shammas, M. 2008. Two coastal aquifers, in India and Oman management options. In: Paliwal, B.S. (ed.) *Global Groundwater Resources and Management*. Scientific Publishers (India), Jodhpur, International Geological Congress, Oslo, pp. 133–140

Jagadeshan, G. and Elango, L. 2015. Suitability of fluoride-contaminated groundwater for various purposes in a part of Vaniyar River Basin, Dharmapuri District, Tamil Nadu. *Water Quality, Exposure and Health*, 7:557–566. https://doi.org/10.1007/s12403-015-0172-8

Kelly, W.P. 1940. Permissible composition and concentration of irrigated waters. In: *Proceedings of the A.S.C.F*, 607.

Kelly, J., Thornton, I., and Simpson, P.R. 1996. Urban geochemistry: a study of the influence of anthropogenic activity on the heavy metal content of soils in traditionally industrial and non-industrial areas of Britain. *Applied Geochemistry*, 11, 363–370. https://doi.org/10.1016/0883-2927(95)00084-4

Kokkat, A. 2018. Hydrogeochemistry of Cauvery Delta with Special Reference to Water Quality Status. Coimbatore.

Kokkat, A., Palanichamy, J., and James, E. 2016. Spatial and temporal variation in groundwater quality and impact of sea water in the Cauvery Delta, South India. *International Journal of Earth Sciences and Engineering (IJEE)*, 9, 383–392.

Krishnakumar, P., Lakshumanan, C., Kishore, V.P., et al. 2014. Assessment of groundwater quality in and around Vedaraniyam, South India. *Environmental Earth Sciences*, 71, 2211–2225. https://doi.org/10.1007/s12665-013-2626-2

Kumar, M., Kumari, K., Ramanathan, A.L., and Saxena, R. 2007. A comparative evaluation of groundwater suitability for irrigation and drinking purposes in two intensively cultivated districts of Punjab, India. *Environmental Geology*, 53, 553–574. https://doi.org/10.1007/s00254-007-0672-3

Kumar, S.K., Rammohan, V., Sahayam, J.D., and Jeevanandam, M. 2009. Assessment of groundwater quality and hydrogeochemistry of Manimuktha River basin, Tamil Nadu, India. *Environmental Monitoring and Assessment*, 159, 341–351. https://doi.org/10.1007/s10661-008-0633-7

Kumari, S., Singh, A.K.,Verma, A.K., and Yaduvanshi, N.P.S. 2014. Assessment and spatial distribution of groundwater quality in industrial areas of Ghaziabad, India. *Environmental Monitoring and Assessment*, 186, 501–514. https://doi.org/10.1007/s10661-013-3393-y

Lopez, B., Baran, N., and Bourgine, B. 2015. An innovative procedure to assess multi-scale temporal trends in groundwater quality: example of the nitrate in the Seine–Normandy basin, France. *Journal of Hydrology*, 522, 1–10. https://doi.org/10.1016/J.JHYDROL.2014.12.002

Madramootoo, C.A. 2012. Sustainable groundwater use in agriculture. *Irrigation and Drainage*, 61, 26–33. https://doi.org/10.1002/ird.1658

Masoud, A.A., Meswara, E.A., el Bouraie, M.M., and Kamh, S.Z. 2018. Monitoring and assessment of the groundwater quality in wadi Al-Arish downstream area, North Sinai (Egypt). *Journal of African Earth Sciences*, 140, 225–240. https://doi.org/10.1016/J.JAFREARSCI.2018.01.016

Mohanakrishnan, A. 2011. A Few Novel and Interesting and Innovative Irrigation Structures Conceived, Designed and Executed in the plan project in Tamil Nadu, No.44, Pub. Irrigation Management Training Institute (IMTI), Thuvakudy, Trichy, Tamil Nadu

Mondal, N.C. and Singh, V.P. 2011. Hydrochemical analysis of salinization for a tannery belt in Southern India. *Journal of Hydrology*, 405, 235–247. https://doi.org/10.1016/j.jhydrol.2011.05.058

Mondal, N.C., Singh, V.P., Singh, V.S., and Saxena, V.K. 2010a. Determining the interaction between groundwater and saline water through groundwater major ions chemistry. *Journal of Hydrology*, 388, 100–111. https://doi.org/10.1016/j.jhydrol.2010.04.032

Mondal, N.C., Singh, V.S., Puranik, S.C., and Singh, V.P. 2010b. Trace element concentration in groundwater of Pesarlanka Island, Krishna Delta, India. *Environmental Monitoring and Assessment*, 163, 215–227. https://doi.org/10.1007/s10661-009-0828-6

Mondal, N.C., Singh, V.P., Singh, S., and Singh, V.S. 2011. Hydrochemical characteristic of coastal aquifer from Tuticorin, Tamil Nadu, India. *Environmental Monitoring and Assessment*, 175, 531–550. https://doi.org/10.1007/s10661-010-1549-6

Mondal, N.C., Tiwari, K.K., Sharma, K.C., and Ahmed, S. 2016. A diagnosis of groundwater quality from a semiarid region in Rajasthan, India. *Arabian Journal of Geosciences*, 9, 1–22. https://doi.org/10.1007/s12517-016-2619-z

Nagamani, K. and Suresh, Y. 2019. Evaluation of coastal aquaculture ponds using remote sensing and GIS. *Indian Journal of Geo Marine Sciences*, 48, 1205–1209.

Nagaraju, A., Suresh, S., Killham, K., and Hudson-Edwards, K. 2006. Hydrogeochemistry of waters of Mangampeta barite mining area, Cuddapah Basin, Andhra Pradesh, India. *Turkish Journal of Engineering and Environmental Sciences*, 30(4), 203–219.

Nageswara Rao, P.V., Appa Rao, S., and Subba Rao, N. 2017. Geochemical evolution of groundwater in the Western Delta region of River Godavari, Andhra Pradesh, India. *Applied Water Science*, 7, 813–822. https://doi.org/10.1007/s13201-015-0294-y

Qiyan, F. and Baoping, H. 2002. Hydrogeochemical simulation of water-rock interaction under water flood recovery in Renqiu Oilfield, Hebei Province, China. *Chinese Journal of Geochemistry*, 21, 156–162. https://doi.org/10.1007/BF02873773

Rahman, A., Mondal, N.C., and Tiwari, K.K. 2021. Anthropogenic nitrate in groundwater and its health risks in the view of background concentration in a semi arid area of Rajasthan, India. *Scientific Reports*, 11:9279. https://doi.org/10.1038/s41598-021-88600-1

Rainwater, F.H. and Thatcher, L.L. 1960. Methods for Collection and Analysis of Water Samples. Washington, DC.

Richards, L.A. 1954. *Diagnosis and Improvement of Saline and Alkali Soils*. U.S. Dept. of Agriculture, Washington, DC.

Sajil Kumar, P.J. 2016. Deciphering the groundwater–saline water interaction in a complex coastal aquifer in South India using statistical and hydrochemical mixing models. *Modeling Earth Systems and Environment*, 2, 1–11. https://doi.org/10.1007/s40808-016-0251-2

Sajil Kumar, P.J., Elango, L., and James, E.J. 2013. Assessment of hydrochemistry and groundwater quality in the coastal area of South Chennai, India. *Arabian Journal of Geosciences*, 7, 2641–2653. https://doi.org/10.1007/S12517-013-0940-3

Salifu, M., Yidana, S.M., and Anim-Gyampo, M., et al. 2017. Hydrogeochemical and isotopic studies of groundwater in the middle voltaian aquifers of the Gushegu district of the Northern region. *Applied Water Science*, 7, 1117–1129. https://doi.org/10.1007/s13201-015-0348-1

Sappa, G., Ergul, S., and Ferranti, F. 2014. Water quality assessment of carbonate aquifers in southern Latium region, Central Italy: a case study for irrigation and drinking purposes. *Applied Water Science*, 4, 115–128. https://doi.org/10.1007/s13201-013-0135-9

Sarath Prasanth, S.V., Magesh, N.S., and Jitheshlal, K.V., et al. 2012. Evaluation of groundwater quality and its suitability for drinking and agricultural use in the coastal stretch of Alappuzha District, Kerala, India. *Applied Water Science*, 2, 165–175. https://doi.org/10.1007/s13201-012-0042-5

Sathiya Bama, V.P., Rajakumari, S., and Ramesh, R. 2020. Coastal vulnerability assessment of Vedaranyam swamp coast based on land use and shoreline dynamics. *Natural Hazards*, 100, 829–842. https://doi.org/10.1007/S11069-019-03844-5

Satish, K., Amarender, B., Dhakate, R., et al. 2016. Assessment of groundwater quality for drinking and irrigation use in shallow hard rock aquifer of Pudunagaram, Palakkad District Kerala. *Applied Water Science*, 6, 149–167. https://doi.org/10.1007/s13201-014-0214-6

Sawyer, C.N. and McCarty, P.L. 1967. Chemistry for Sanitary Engineers. https://doi.org/10.3/JQUERY-UI.JS

Saxena, V.K., Singh, V.S., Mondal, N.C., and Jain, S.C. 2003. Use of hydrochemical parameters for the identification of fresh groundwater resources, Potharlanka Island, India. *Environmental Geology*, 44, 516–521. https://doi.org/10.1007/s00254-003-0807-0

Saxena, V.K., Mondal, N.C., and Singh, V.S. 2004. Identification of sea-water ingress using strontium and boron in Krishna Delta, India. *Current Science*, 86, 586–590. https://doi.org/10.2307/24107915

Schroeder, H.A. and Palmer, C.E. 1960. Relations between hardness of water and death rates from certain chronic and degenerative diseases in the United States. *Journal of Chronic Diseases*, 12, 586–591. https://doi.org/10.1016/0021-9681(60)90002-3

Selvam, S., Manimaran, G., Sivasubramanian. P., et al. 2014. GIS-based Evaluation of Water Quality Index of groundwater resources around Tuticorin coastal city, south India. *Environmental Earth Sciences*, 71, 2847–2867. https://doi.org/10.1007/s12665-013-2662-y

Sivanappan, R.K. 2007. Mapping and Study of Coastal Water Bodies in Nagapattinam District. NGO Co-ordination and Resource Centre, Nagapattinam.

Srinivasamoorthy, K., Chidambaram. S., Prasanna, M.V., et al. 2008. Identification of major sources controlling groundwater chemistry from a hard rock terrain — A case study from Mettur taluk, Salem district, Tamil Nadu, India. *Journal of Earth System Science*, 117:49–58. https://doi.org/10.1007/s12040-008-0012-3

Stumm, W. and Morgan, J.J. 1996. *Aquatic Chemistry : Chemical Equilibria and Rates in Natural Waters*. Wiley, New York.

Sukhija, B.S., Varma, V.N., Nagabhushanam, P., and Reddy, D.V. 1996. Differentiation of palaeomarine and modern seawater intruded salinities in coastal groundwaters (of Karaikal and Tanjavur, India) based on inorganic chemistry, organic biomarker fingerprints and radiocarbon dating. *Journal of Hydrology*, 174, 173–201. https://doi.org/10.1016/0022-1694(95)02712-2

UNDP. 1973. Groundwater Investigation in Tamil Nadu (Phase II) India, United Nations Development Program. New York, NY.

Vasanthavigar, M., Srinivasamoorthy. K., Vijayaragavan, K., et al. 2010. Application of water quality index for groundwater quality assessment: Thirumanimuttar sub-basin, Tamilnadu, India. *Environmental Monitoring and Assessment*, 171, 595–609. https://doi.org/10.1007/s10661-009-1302-1

Venkatramanan, S., Chung, S.Y., Ramkumar, T., et al. 2013. A multivariate statistical approaches on physico-chemical characteristics of ground water in and around Nagapattinam district, Cauvery deltaic region of Tamil Nadu, India. *Earth Sciences Research Journal*, 17, 97–103.

Venkatramanan, S., Chung, S.Y., Ramkumar, T., et al. 2015. Application of GIS and hydrogeochemistry of groundwater pollution status of Nagapattisnam district of Tamil Nadu, India. *Environmental Earth Sciences*, 73, 4429–4442. https://doi.org/10.1007/s12665-014-3728-1

Vetrimurugan, E. and Elango, L. 2015. Groundwater chemistry and quality in an intensively cultivated river delta. *Water Quality, Exposure and Health*, 7, 125–141. https://doi.org/10.1007/s12403-014-0133-7

Vetrimurugan, E., Elango, L., and Rajmohan, N. 2013. Sources of contaminants and groundwater quality in the coastal part of a river delta. *International Journal of Environmental Science and Technology*, 10, 473–486. https://doi.org/10.1007/s13762-012-0138-3

WHO. 2011. *Guidelines for Drinking-water Quality*, 4th edn. WHO Press, New York.

Wilcox, L.V. 1955. Classification and Use of Irrigation Waters, pp. 1–19. https://doi.org/USDA Circular No. 969.

Zeeshan, M. and Azeez, P.A. 2016. Hydro-chemical characterization and quality assessment of a Western Himalayan river, Munawar Tawi, flowing through Rajouri district, Jammu and Kashmir, India. *Environmental Monitoring and Assessment*, 188, 520. https://doi.org/10.1007/s10661-016-523-9

13 Groundwater Potential Zones Mapping based on the ANN and ML Models Using AHP and MIF techniques

Ahmed Elbeltagi, Renata Graf, Dimple, Jitendra Rajput, Ali Mokhtar, and Chaitanya B. Pande

13.1 INTRODUCTION

Groundwater (GW) is an important drinking water resource in many regions and countries, especially for areas with identified water deficits and the ongoing process of degradation of water resources. There is still a great need for information on the underground phase of water circulation in the catchment area, as well as documenting their resources and prognostic studies of changes in water circulation systems (Das et al., 2021). The problem of access to drinking water in the context of negative climate change and various forms of environmental degradation, as well as the process of globalization and conflicts on various grounds, including water, requires the implementation of integrated solutions promoting modern methods of monitoring and mapping of the GW (Melese and Belay, 2022; Srivastava et al., 2021, 2022; Khadke and Pattnaik, 2021). Strategies for sustainable management of GW resources and their protection include, among other things: recognizing the abundance of aquifers and the limits of their natural resistance, as well as assessing the potential of GW and its renewal capacity, and determining the nature and needs of resource users (Guru et al., 2017; Das, 2018). The GW potential (GWP) is shaped by the region's natural conditions, mainly the hydrogeological background dependent on geology, geomorphology, drainage and soil properties, climatic factors (precipitation), and economic activities carried out in the region. The differences in GW productivity and potential are due to the different natural conditions (Golkarian et al., 2018). Areas with significant and economically exploited GW resources are typified as potential groundwater zones (GWPZs) (Waikar and Nilawar, 2014; Owolabi et al., 2020). The identification of GWPZ is of particular importance in areas where the occurrence of GW is very limited due to unfavorable hydrogeological conditions (Nagarajan and Singh, 2009; Siddi Raju et al., 2016). In addition, assessing and mapping potential zones is an essential element for improving techniques for studying GW resources (Melese and Belay, 2022). Furthermore, there are many hydraulic and natural structures such as dams, tanks, lakes, ponds, etc. influencing the GW recharge potential, whose presence in GWPZ allows for managing the water resources more sustainably (Srivastava and Chinnasamy, 2021a; 2022). However, many countries and regions worldwide face a shortage of information and databases on GW and the aforementioned local structures from direct field research and hydrogeological exploration.

Remote sensing-based methods that use satellite data and GIS-integrated remote sensing techniques (Srinivasa Rao and Jugran, 2003; Das et al., 2019; Ettazarini and Jakani, 2020; Srivastava and Chinnasamy, 2021b) have great possibilities in the field of GWP identification. It is found that due to difficulties in GW monitoring, e.g., in rocky and hard-to-reach areas, these techniques are appropriate (Mohammed and Sayl, 2020). In addition, the geospatial technique is more cost-effective

DOI: 10.1201/9781003303237-13

in assessing GW resources than the traditional, field-based methods, which are based on drilling, geophysics, and stratigraphy, making them costly and time-consuming (Arkoprovo et al., 2012; Arulbalaji and Gurugnanam, 2016; Thapa et al., 2017). Remote sensing-based methods, integrated with GIS, enable the acquisition, processing, and interpretation of thematic data as input information necessary for the mapping of GWP (Rahmati et al., 2014; Helaly, 2017), while the statistical approaches used in the models allow for the classification of factors and the determination of their impact on the development of GWP (Çelik, 2019; Saravanan et al., 2021). GIS-based models can use a huge amount of spatial data (thematic layers) for maps of the GWP, and their choice is not limited (Singh et al., 2018; Ghosh, 2021). It is emphasized that remote sensing-based methods integrated with GIS reduce the ambiguity of hydrogeological data in various aspects of their interpretation (Melese and Belay, 2022). Validation of GWPZs, defined through remote sensing-GIS methods with appropriate classifiers, is most often carried out by verifying designated zones with field data on the yield of exploration boreholes (Senthilkumar et al., 2019). Remote sensing for determining GWPZs was already used in the 1970s (Karanth and Seshu babu, 1978; Moore, 1978) and was developed on a larger scale later (Sree Devi et al., 2001; Sankar, 2002; Srinivasa Rao and Jugran, 2003). Currently, remote sensing GIS-based methods integrated with the statistical approach are often used to identify GWPZs (Arabameri et al., 2019; Ahmad et al., 2020; Jaafarzadeh et al., 2021; Saravanan et al., 2021). These methods enable the development of models based on indexes and a quantitative approach (Daneshfar and Zeinivand, 2015; Abijith et al., 2020; Gyeltshen et al., 2020). Among remote sensing-based methods, the multi-criteria decision approach (MCDA) is very popular (Singh et al., 2018; Benjmel et al., 2020; Hagos and Andualem, 2021; Yıldırım, 2021), and especially the analytical hierarchical process (AHP) (Patra et al., 2018; Arulbalaji et al., 2019; Gebru et al., 2020; Kumar et al., 2020; Aykut, 2021; Ghosh, 2021). The MCDA is one of the most effective techniques for solving problems in mapping GWPZs in areas with different hydrogeological conditions. The AHP is considered a simple and effective method of assessing GWP (Çelik, 2019; Melese and Belay, 2022). The AHP model provides solutions to complex decision problems and was first introduced by Saaty (1980). In terms of identifying GWPZs, the decision-making analysis involves the selection of criteria for the assessment of potential zones, which requires good knowledge of the physical-geographical characteristics of the catchment area and careful weighing of the criteria (assessment of the importance of criteria) according to expert knowledge (Lentswe and Molwalefhe, 2020). The weights assigned to individual criteria according to AHP influence the result of the decision analysis. The AHP-MCDA assessment technique is often used to integrate the thematic maps (the so-called factor maps). Mohammed and Sayl (2020) compared several methods and models for identifying GWPZs that have evolved over the past few years. As a result of this assessment, it was found that most research on the identification of GWPZs uses GIS in integration with multi-criteria analysis and hydrological models, while the most important criteria for selecting the appropriate locations of GWPZs were: geology, slope, soil, and land use (or land cover). However, the thematic layers used to identify GWPZs differ depending on the region and the adopted research methods, and the choice of qualitative layers is free (Çelik, 2019). Modeling GWPZs was also carried out using the fuzzy-AHP approach, which is one of the decision-making methods (Mallick et al., 2019; Rajasekhar et al., 2019; Shao et al., 2020). Fuzzy-AHP is classified as one of the AHP methods based on fuzzy logic theory (Putra et al., 2018). This method works similar to the AHP approach, and in the case of GWPZ identification, fuzzy-AHP was employed to calculate the weight of the factors. The remote sensing-GIS approaches, in integration with the multi-influence factor (MIF) (Nasir et al., 2018; Das, 2019; Siddi Raju, 2019; Zghibi et al., 2020; Das et al., 2021; Khan et al., 2021) and frequency ratio model (FRM) (Al-Abadi, 2017; Jothibasu and Anbazhagan, 2017; Das, 2019; Arshad et al., 2020; Muavhi et al., 2021) are often used in the identification of GWPZs. In the MIF model, the influence of each major and secondary parameter on the GWP is determined by assigning appropriate weights to them. The assessment of the GWP is based on assessing the impact weight of a given factor by distinguishing factors favoring and limiting the potential. In contrast, the selection of criteria for assessing the potential and their number are different (Daneshfar and Zeinivand, 2015). The FRM allows for obtaining eventuality (probabilistic) correlation among dependent and independent variables, including multi-classified layers (Manap et al., 2014; Daneshfar and Zeinivand, 2015).

According to Taheri et al. (2019), the MIF and AHP methods give good results when used to identify GWPZs. Comparing both methods with the selection of factors showed a slightly better accuracy of MIF over AHP. The integration of different influencing factors in delineating the GWPZ using MIF and FRM was presented, among others, by Daneshfar and Zeinivand (2015) and Das et al. (2018). Jhariya et al. (2021) assessed GWPZs using MIF and MCDA, as well as electrical resistivity survey techniques, highlighting the great benefits of integrating research approaches in identifying GWP. Moreover, the weighted overlay method (WOM) and the weighted aggregation method (WAM) are also used in the identification of GWPZs (Senthilkumar et al., 2019; Ifediegwu, 2022). The weighted overlay tool uses one of the most common approaches to overlay analysis to solve multi-criteria problems such as location selection and suitability modeling of, e.g., different layers and base maps for identifying GWPZs (Chaudhari and Lal, 2018). In the GWPZs typing procedure, the WOM integrates layers of thematic maps useful for their assessment and finally for the mapping of GWPZs (Daneshfar and Zeinivand, 2015). Ultimately, the maps of GWP are prepared by assigning appropriate weightage to different thematic maps (or thematic layers) and applying them using the WOM. In this way, for each parameter obtained from thematic maps, based on their significance about the choice of location and weight, a place in the ranking of the impact on the identification of GWP is assigned. A novel approach to the weighted aggregation and determining the weights in an aggregation procedure constitutes the WAM. The weighted aggregation system consists of two components: the weighted transformation and the aggregation operator, which are induced by a common generator function (Dombi and Jónás, 2022). In WAM, the output can be taken as the result of a weighted aggregation procedure (Calvo et al., 2004), which is considered when the importance of a single criterion needs to be aggregated. The GWP is assessed by aggregating a selected group of criteria adapted to the hydrogeological region. In WAM, as extended aggregation functions, the quasi-arithmetic means and weighted quasi-arithmetic means are considered, as are the ordered weighted averaging operator and the quasi-ordered weighted averaging function (Zargini, 2020). These elements are included in the decision-making process. An innovative approach to determining the GWP was proposed by Al-Abadi et al. (2017), who applied entropy information theory and the linear WAM. In this study, entropy theory was used to assess the importance of GW conditioning factors in determining their potential. The GWP index was determined using a linear weighted aggregation of all the factors.

Advanced methods have great potential in the field of assessment and management of GW resources, including artificial neural networks (ANNs) (Lee et al., 2012, 2017; Tamiru and Wagari, 2021), machine learning models (ML) (Naghibi et al., 2017; Lee et al., 2017; Miraki et al., 2019), and hybrid intelligence approach (Miriaki et al., 2019). Extended ML methods allow not only GWP modeling but also their prediction (Ettazarini and Jakani, 2020).

ANN is one of the most famous algorithms for regression, classification, and GWP pattern recognition (Lee et al., 2017). Research confirms that ANN shows promising results in predicting GWP (Tamiru and Wagari, 2021). The ANNs are classified as nonlinear simulation methods that focus on the information processing algorithm. They simulate and calculate complex relationships between catchment parameters and define a multidimensional information space adapted to subsequent calculations (Lv et al., 2020). The ANN models, through training processes based on the existing situation, can generalize and predict the output data from the sample input data (Chan and Chan, 2020). By building an algorithm for data generation processes, ANN solves the non-linear nature of GW resources (Barbetta et al., 2016; Tsakiri et al., 2018). A novelty in the use of ANNs in the study of GWP is the integration of the field-based data with remote sensing data, based on which thematic maps are generated, which are the basis for the assessment of GWP. Tamiru and Wagari (2021) showed that the ANN model is more efficient in GWPZ identification compared to GIS techniques, which was confirmed by its greater accuracy and higher probability of delimiting zones in areas with data deficiency. In determining and mapping GWP, the adaptive neuro-fuzzy inference system (ANFIS) was also used, a type of ANN based on the Takagi-Sugeno fuzzy inference system (Razavi-Termeh et al., 2019; Chen et al., 2019a).

Unlike statistical models, ML based on artificial intelligence and data mining take non-linear relationships. Their use in the identification of GWPZs and their mapping is associated with finding the most economical and efficient procedure, which brings benefits to the identification of GW resources. The main benefit of ML is its ability to handle high-dimensional data and map those that have complex classification problems, such as GWP data. According to Kamali Maskooni et al. (2020), ML and statistical models have been implemented to identify GWP with highly satisfactory results. These authors used advanced ML algorithms to assess GWP using remote sensing-derived data. ML techniques in different spatial locations were used by: Hussein et al. (2020) in GW forecasting, Pourghasemi et al. (2020) in mapping the GW recharge potential zones, and Prased et al. (2020) in GWP mapping. Due to the extensive use of ML for the analysis of complex structures and stochastic data, they offer more reliable results in the exploration of GW resources and mapping their potential (Lee et al., 2020; Pourghasemi et al., 2020; Prasad et al., 2020). The GWP map is developed using various GIS-based ML algorithms such as support vector machines (SVMs), decision trees (DTs), random forests (RFs), and naive Bayes classifiers (NBCs), which are predictive neural networks (PNNs) (Sahoo et al., 2017; Khosravi et al., 2018; Kenda et al., 2020).

The SVM can efficiently handle the nonlinear relationship between predictors and responses in a multivariate space related to the determination of GWP (Lee et al., 2018). This algorithm effectively predicts and maps the GWPZs (Naghibi et al., 2017; Panahi et al., 2020). Guzman et al. (2015) and Lee et al. (2018) used to compare and predict GWP. Among others, ANN and SVM obtained significantly different performance results of both models in forecasting GW levels. Furthermore, the DT is a decision support method based on a hierarchical model that simplifies decision rules by carefully dividing independent factors into homogeneous zones (Lee and Lee, 2015). In the identification of GWPZs, the locations (e.g., wells) with a specific GWP are separated based on the spatial interaction between the well locations and each GW conditioning factor. The RF algorithm can solve and manage many problems related to large datasets. It does not require prior data to scale and transform the datasets (Elmahdy et al., 2020). For regression and classification of training datasets, a DT is developed to extract the output data by class and obtain a dependent variable (Sachdeva and Kumar, 2020). The RF algorithm is built from many DT and composites, making it possible to determine the relationship between factors influencing GW occurrence and shaping its potential (Kim et al., 2018). The NBC is a widely accepted ML classification algorithm developed based on the Bayes rule and operated on a variable-independent assumption (Soria et al., 2011). This algorithm is straightforward for modeling estimation schemes that do not require complex iteration parameters.

In contrast, the PNN estimates potentiality or susceptibility using classified sample data (Azimi et al., 2018). In the case of GWPZs, the potential indicator criterion is applied, and its classification is adopted. The PNN was developed based on probability density algorithms, which corresponds to the definition of GWPZ proposed by Khosravi et al. (2018). In some cases (for the training of algorithms), the GWP is defined as the probability of a well being successful. According to Dou et al. (2018) and Golkarian et al. (2018), the PNN models solve the clustering problems generated during the implementation of Bayesian decision theory. The research confirms that the PNN, RF, and DT models are very useful in mapping GWPZs. These models could be implemented for various spatial potential probability models (local and regional scale), which should be used to manage and protect GW resources (Dou et al., 2018). This is also confirmed by the studies of Naghibi et al. (2017), who applied SVM, RF, and genetic algorithm-optimized RF models in GWP mapping (Panahi et al., 2020). They developed the maps of GWP using a ML algorithm and a deep learning algorithm (DLA), specifically the SVR and convolutional neural network (CNN) functions. ML and statistical models have been implemented to produce the maps of GWP with highly satisfactory results (Kamali et al., 2020; Avand and Moradi, 2021), but bagger-boosting models usually show better results (Sachdeva and Kumar, 2020).

The research so far indicates the need to develop innovative forecasting systems to provide a more accurate estimation of the GWP for rational resource management, especially in regions with high water deficits. In terms of the use of advanced methods in the GWPZ mapping, the hybrid intelligence approach provides good performance compared to single models (Arabameri et al., 2020). For this reason, attempts are now being made to integrate techniques and develop ensemble intelligence techniques - hybrid models that are increasingly used in various spatial locations in GW forecasting (Miraki et al., 2019; Pham et al., 2019; Arabameri et al., 2021). Among the advanced algorithms, the multi-learning algorithm finds its use in the mapping of GWP. This algorithm is an ensemble super learner and a subfield of ML (Lee et al., 2022). In multi-task learning, multiple learning tasks are solved simultaneously while exploiting commonalities and differences across tasks. Lee et al. (2022) used a super learner that combines multiple different learning algorithms, which allowed generating maps of GWP.

Hybrid ML models that combine the base model with optimization algorithms or team techniques are more reliable in GWP mapping. The mapping of GWPZs using a novel hybrid intelligence approach was carried out by Miraki et al. (2019). The proposed innovative classifier methods, namely, the RF classifier based on the random subspace ensemble random forest (RS) (RS-RF). The modeling results showed that the new RS-RF hybrid model had a very high predictive ability for GWP mapping and showed the best performance among reference models such as RF, NB, and logistic regression (LR). Al-Fugara et al. (2020) used the combination of SVM and genetic algorithms (GA) to construct a hybrid model in GWP mapping, and Nguyen et al. (2020) developed an ensemble model combining ANN with the real AdaBoost ensemble technique (ANN-RAB). Arabameri et al. (2021) modeled the GWP with a novel GIS-based ML ensemble technique, which is a combination of RS with the multilayer perception (MLP), naïve Bayes tree (NB Tree), and classification and regression tree (CART) algorithms. The results of these studies showed that all models performed well for GWP mapping, and the MLP-RS hybrid model achieved high validation scores. Moreover, it was confirmed that among the GW conditioning factors (GWCFs), slope, elevation, height above nearest drainage, and terrain ruggedness index are the most important predictors of GW presence (Arabameri et al., 2021). For example, Tien Bui et al. (2019) developed the AB-AD tree by integrating an alternating decision tree classifier and an adaptive boosting ensemble model.

The ensemble - boosting and bagging-based ML for prediction of GWP were also used, such as: boosted generalized additive model (Gam Boost), adaptive boosting classification trees (AdaBoost), bagged classification, and regression trees (Bagged CART). Mosavi et al. (2021) showed that bagging models (i.e., RF and Bagged CART) had higher performance in identification of GWP than boosting models (i.e., AdaBoost and Gam Boost), and the best performance was achieved in the study group by the RF model. The hybrid computational intelligence approach was used to GWP mapping by Rizeei et al. (2019). These authors developed the ensemble multi-adoptive boosting logistic regression technique (MABLR), which shows excellent results in GWP mapping, mainly in terms of reduction of bias and variance due to insufficient sample size, oversimplification, and sensitivity to outliers. In a GWP analysis, Chen et al. (2018) used novel ensemble weights-of-evidence with LR and functional tree models, while Pradhan et al. (2019) used a novel hybrid integration approach of bagging-based Fisher's linear discriminant function (LDA). The model proposed by Chen et al. (2019b) obtained high accuracy of GWP mapping and used decision algorithms based on trees, adaptive boosting or bagging, RS, dagging, and rotation forest.

The ensemble learning models improve ML's computing ability and facilitate multi-dimensional analysis of huge datasets (Chen et al., 2018; Prasad et al., 2020). It is indicated that ensemble ML gives more accurate results in the assessment and mapping of GWPZs (Nguyen et al., 2020; Arabameri et al., 2021; Jaafarzadeh et al., 2021; Mosavi et al., 2021). Research is underway on their application and comparing their performance in the field of GWP mapping. The results of the identification and mapping of GWP can be useful for water resource managers to make the right decisions about the optimal use of GW resources for future planning, especially in critical areas with high water resource deficiency or confirmed water deficits.

GW is an essential component of nature that exists in the earth's gaps and fills the soil's void under the water table. GW has been established as a critical natural resource, serving as a primary water source in all areas of the globe. GW availability is determined by geological, morphological, biological, and atmospheric factors (Mukherjee, 1996; Oh et al., 2011; Das et al., 2018). Numerous strategies for assessing GWP take these aspects into account. The efficiency of the procedures varies; some are more effective, precise, time-saving, and cost-efficient, while others are time-consuming and expensive (Ahmadi et al., 2021).

GW is dynamic and highly changeable from location to location. The presence of GW in a particular region is not coincidental; it is the result of the interaction of numerous factors, including natural geography, climate, hydrology, geology, ecology, underlying soil layers, fracture density, surface slope, and land use (Manap et al., 2013; Mehrjardi-Taghizadesh et al., 2021). In particular because of their interdependence, selecting a single element to represent the recharge process reduces the credibility of the estimations for a given area. Four assumptions guided the selection of parameters that most significantly affect aquifer recharge: GWP rises as GW recharge rises (precipitation capable of infiltrating), with increased soil and rock permeability (geological and lithological units), with increased geological structures, and with flat slopes (Mehrjardi-Taghizadesh et al., 2021). To efficiently manage GW resources, it is critical to point out locations with a high potential for GW storage. The most effective strategy for assisting administrators in adopting management plans is to map the GWP (Nguyen et al., 2020; Termeh-Razavi et al., 2019). GWP utilizes both traditional and novel technologies, including GIS and RS.

Despite their accuracy in estimating the potential of GW, indigenous methods are impeding its usage in many regions of the world due to time and expense constraints, as well as a lack of data and methods for extraction (Nguyen et al., 2020). GIS-based approaches enable the storage, manipulation, and analysis of data in various formats and sizes, resulting in the generation of themed maps (Oh et al., 2011; Oilonomidis et al., 2015; Razandi et al., 2015; Band et al., 2020). Over the last decade, statistical, ML, and hybrid models combined with GISs have been employed to develop more accurate models for assessing GWP (Al-Fugara et al., 2020). Additionally, statistical models such as the frequency ratio (FR) (Ozdemir, 2011), LR (Park et al., 2017), the weight of evidence (WOE) (Madani et al., 2015), and the evidentiary belief function (EBF) (Khishtinat et al., 2019)) were employed to map GWP. Recently, academics have grown increasingly interested in GIS-based GWP mapping using ML algorithms. The selection of acceptable elements affecting GWP and adequate models for determining its prospective regions is one of the most important criteria for determining GWP. Because these models are good at predicting nonlinear relationships, various variables affect the GWP (Razavi-Termeh et al., 2020).

Recent research has focused on developing novel hybrid models that increase the accuracy of GWP mapping by optimizing the base model using different algorithms (Kordestani et al., 2018) or group tactics (Chen et al., 2019; Banadkooki et al., 2020; Prasad et al., 2020). Although the Bayesian network (BayesNet) model has been effectively employed in hazard modeling in the past (Lee et al., 2020; Zhang et al., 2020), one of its drawbacks is its inability to find the ideal network configuration. Appropriate site selection for good fields in newly planned urban centers and industrial estates adjacent to any significant wetland is a powerful pollution prevention tool that ensures the environmental soundness of any development program. As a result, it is critical to identify various GWPZ inside and next to the wetland environment.

Numerous criteria affecting the area's GW hydrology, such as geology, hydro-geomorphology, drainage density, aquifer thickness, and depth to the water table, may be integrated on a GIS platform with appropriate weights to classify an area into different GWPZs (Al-Hadithi et al., 2003; Sikdar and Bhattacharya, 2003; Sikdar et al., 2004; Sreedevi et al., 2005). The product will assist legislators, developers, and policymakers in developing strategies for many development initiatives and supplying water for domestic, agricultural, and industrial reasons. The present book chapter briefly describes various important techniques and methods for GWP estimation worldwide and their superiority over other techniques.

13.2 METHODS FOR THE IDENTIFICATION OF GROUNDWATER POTENTIAL ZONES

The identification of GWPZs is carried out on the basis of various methods, the most common of which are remote sensing-based methods or advanced methods. The choice of methods is related to many factors that relate to, inter alia, the scale of the study (regional, local), and the research area, including the physical and geographical features of the region, which translates into the specificity of GW occurrence conditions and their recognition. Another factor is the functioning or lack of a GW monitoring system in the area covered by the study and access to measurement data that comes from direct, systematic observations, or from irregularly conducted field surveys. Information on GW occurrence and circulation conditions in relation to geological, geomorphological, and hydrographic factors, as well as climate features, is supplemented by thematic maps and GIS databases prepared for a given area. Furthermore, investigating the presence of surface water storage structures in a given watershed is also considered one of the probable sources from which GW may be recharging. Thus, identification of such structures during watershed delineation can be imperative while understanding the GW dynamics (Srivastava and Chinnasamy, 2023a, 2023b). At various spatial scales, most often in the absence of access to measurement and observation data and thematic data, it is possible to estimate GWPZs using remote sensing - based methods that use satellite data and integrated GIS techniques. Among these methods, different statistical approaches are used to assess the determinants of GWP and to classify the zones of GWP. Applications in this regard have, among others, AHPs, MIFs, WOMs, FRMs, and WAMs, which are detailed in the next section. Remote sensing, compiled with GIS and statistical approaches, plays a significant role in the research and identification of GWP.

In the prediction and forecasting of GWPZs, mainly innovative forecasting systems are used, which are based on advanced methods, including: ANN, ML, and the hybrid intelligence approach. The advanced methods, such as nonlinear simulation and computation methods, allow for the determination of complex relationships between parameters (predictors) that affect the GWP size and spatial distribution of zones in a multi-dimensional space system. ANN and MLD solve the problems of regression, classification, and pattern recognition of GWP. Advanced techniques, such as the hybrid intelligence approach, which works more efficiently than single models, are more and more often used in GW research, including the multi-learning algorithm, which has become very popular recently in the mapping of GWP.

13.2.1 Remote Sensing-Based Methods

It is possible to evaluate vast volumes of geospatial data and designate GWP using GIS and RS investigations. Satellite data can be used to determine geological structures, geomorphic landforms, and their hydrologic characteristics (Rita et al., 2007). Using remote sensing data from an aeroplane or satellite is now permissible (Todd, 1980).

13.2.2 Analytical Hierarchical Process (AHP)

The most widely exercised and well-known GIS-based method for identifying GWPZs is multi-criteria decision analysis utilizing the AHP. According to the model, influencing parameters can be entered based on subject matter experts' opinions and knowledge (Kumar and Krishna, 2018). The effect and relevance of each element are described using a pairwise matrix (Saaty, 1990).

TABLE 13.1
Saaty's 1–9 Scale for Pair-Wise Comparison

Importance Rank	Definition
1	Equal weightage
II	Equal to moderate weightage
III	Moderate weightage
IV	Moderate +
V	Strong weightage
XI	Strong+
XII	Very strong or demonstrated weightage
XIII	Very to the extreme weightage
IX	Extreme weightage

Source: Saaty (1980).

TABLE 13.2
Random Inconsistency Index

n	1	II	III	IV	V	XI	XII	XIII	IX	X	XI	XII
RI	0.00	0.00	0.58	0.90	1.12	1.24	1.32	1.41	1.45	1.49	1.51	1.48

Source: Saaty (1980).

Once the weights are established, the consistency ratio of the matrix is computed using the equation proposed by (Saaty, 1990) (Tables 13.1 and 13.2).

$$CR = \frac{CI}{RI} \tag{13.1}$$

The consistency ratio is CR, the consistency index is CI, and the random index is RI (Table 13.2) (Saaty, 1990). The CI is calculated as follows:

$$CI = \frac{\lambda_{max} - n}{n - 1} \tag{13.2}$$

where λ_{max} is the principal eigenvalue of the matrix. The CR obtained must be less than 0.1, according to (Malczewski, 1999). If it is more than 0.1, the pairwise comparison matrix should be recalculated by changing the values of the factors (Saaty, 1977).

13.3 MULTI INFLUENCE FACTOR (MIF)

This empirical method for exploring GWPZ uses RS and a geographic information system. The MIF consisted of many parameters influencing the GWPZs like land use/land cover, slope, drainage density, geology, precipitation, and soil.

13.3.1 The MIF Approach Steps

1. The identification of characteristics affecting GWP through a review of the literature. The GWP of a given place changes according to various influencing characteristics that determine how GW aquifers are recharged.

2. The processing of parameters to assure consistency and the assignment of scores and weighting. The analysis of influencing parameters and the assignment of scores to subclasses within the parameter map. Subclasses with a significant impact on GW recharge (*A*) can be assigned a weight of 2, while those with a less impact (*B*) can be assigned a weight of 1. Subclasses that have no impact on GW recharge receive a weight of 0. To calculate the relative effect, the cumulative score ($A+B$) of both major (*A*) and minor (*B*) impacts is used. The weight assigned to each influencing parameter is determined using the following formula:

$$(A+B)/\sum(A+B)\times 100 \tag{13.3}$$

where *A* is a major influencing factor and *B* is a minor (less) influencing factor in the study.
3. The next step is to categorize all theme layers using the spatial analysis extension in ArcMap. The graphic will show the numerical effect of subclasses of contributing factors on GW recharge potential.
4. To classify the output layer into excellent, very good, good, and very poor GWPZ, the fourth phase integrates all influencing variables layers. This strategy can explore a large area with a difficult landscape (Magesh et al., 2012).

13.3.2 Weighted Overlay Method (WOM)

This method uses the Survey of India topo sheet at 1:50 K scale and IRS P6 LISS IV MX satellite data. GIS and RS are used to create drainage density, contour, and stream length maps. It is then used to create slope, aspect, contour, and accumulation maps. It is used for georeferencing and geometric rectification. You can utilize DEM data to create land use/cover classification and lineament maps. Thus, all thematic maps are analyzed in overlay, and weights and ranks are assigned to each theme layer (Waikar and Nilawar, 2014; Lakshmi and Reddy, 2018). Use the following equation to find a GWPZ:

$$\mathrm{GWPI} = \mathrm{IP}_{1w}\mathrm{IP}_{1r} + \mathrm{IP}_{2w}\mathrm{IP}_{2r} + \cdots + \mathrm{IP}_{nw}\mathrm{IP}_{nr} \tag{13.4}$$

where GWPI is the groundwater potential index, IP is the groundwater influncing parameters, *w* is the weight, *r* is the rank, and *n* is the *n*th number of parameters that vary according to the area.

13.3.3 Frequency Ratio Model (FRM)

A spatial database incorporating GW factors was built and implemented in this manner. In GIS software, the input layers are in vector format. Aspect maps, slope maps, and contour maps, among other things, are calculated using DEM data. Finally, all of these variables, such as lineament maps, slope maps, geology maps, and land use maps, are transformed into raster grids. Finally, weights are assigned to each thematic map, and the ranks for mapping prospective zones are evaluated (Manap et al., 2014).

13.3.4 Weighted Aggregation Method (WAM)

Weights applied to variables modify their values, which are subsequently aggregated into the final result. In other words, these weighted aggregation functions are composites of aggregation and weighing transformations. In this study, we will examine weighted aggregation functions. The approaches used to determine the weight of criteria or qualities aim to gain information about their importance. The results are then given as weights. Our study examines whether the weights can be used as 'descriptors' of the decision-makers expectations (expectations in short). In practice, a method's weighted version is often determined intuitively. An aggregation operator can be used to construct weighted aggregation functions (Dambi and Jonas, 2022).

13.3.5 Advanced Methods

As statistical models do not take nonlinear interactions into account, ML models based on artificial intelligence (AI) have been developed. The conditions required to improve GW capacity have been determined using ML techniques based on data mining. The CART; (Naghibi et al., 2019), RF; (Rahmati et al., 2016), multivariate adaptive regression splines (MARS), decision stumps, alternating decision trees (ADTree), ANNs, and SVMs were all tested for GWPZ mapping. Each's purpose is to identify the best cost-effective and efficient approach (Arabameri et al., 2021).

13.3.6 Artificial Neural Networks (ANN)

One type of computational mechanism that may acquire, represent, and compute a mapping from one multivariate information space to another is the ANN (Garrett, 1994). Many ANN algorithms are available; however, back-propagation training is the most common. The backpropagation method is trained on a set of input and output instances. After training, the network is utilized to classify the full dataset (Paola and Schowengerdt, 1995). A neural network comprises nodes, each representing a simple processing element (Atkinson and Tatnall, 1997). An ANN requires less formal statistical knowledge. Also, it can implicitly detect nonlinear correlations between independent and dependent variables and predictor variables. So it can be trained using various methods. But it's a "black box" with obvious limits in determining causal linkages. Also, ANN modeling requires more processing resources, and ANN development is empirical. Many methodological difficulties remain (Lee et al., 2017).

Three-layered network with acyclic linkages. The ANN's input-output relationship is as follows (Khashei and Bijari, 2010):

$$y_t = w_0 + \sum_{j=1}^{q} w_j \cdot g\left(w_{0j} + \sum_{i=1}^{p} w_{ij} \cdot y_{t-i} \right) + z_i \tag{13.5}$$

where y_t is the output, y_{t-i} is input, w_{ij} (0, 1, 2....,p, j=0,1,2....q) and w_j (j=0,1,2...,q) are the model parameter, p is the number of input nodes, and q is the number of hidden nodes.

13.4 MACHINE LEARNING MODELS (ML)

Recently, ML has been developed and used in various real-world circumstances, including GWPM (potential groundwater mapping). A recent study by Pal et al. (2020) used ML approaches including RF, Radial Basis, and ANN to analyze GWP in eastern India and Bangladesh. Naghibi et al. (2016) used BRT, CART, and RF models to map GWP in Iran's Koohrang Watershed. ML methods and their hybrid and ensemble models are promising to produce reliable GWP maps.

13.5 HYBRID INTELLIGENCE APPROACH

Recently, hybrid ensemble ML models combining a base model with optimization methods or ensemble techniques have been developed to improve GWP mapping reliability. Miraki et al. (2019) developed an ensemble model (RS-RF) combining RF and the RS ensemble technique to analyze GWP in the Qorveh-Dehgolan plain, Kurdistan province, Iran. Al-Fugara et al. (2020) used a hybrid SVM and GA model to map GWP in Jerash and Ajloun, Jordan. Naghibi et al. (2017) optimized

Naive Bayes using Adaboost, Bagging, and generalized additive. A recent study proposed using the whale optimization algorithm to optimize an ANN model for potential GW mapping and exhibited the hybrid model's improved predictive ability. The new RS-RF hybrid model outperformed existing benchmark models (LR, RF, and NB) regarding GWP mapping (Miraki et al., 2018). Researchers have yet to consent on a model for gauging GWP. This has led to an increase in the use of ensemble approaches in geohazard susceptibility mapping. The ANFIS, stepwise weighted assessment ratio analysis (SWARA) technique (Dehnavi et al., 2015), and EBF-fuzzy logic (Bui and al., 2015), and ANFIS combined (Chen et al., 2017). AND, OR, GAMMA 0.75, 0.8, 0.85, and GAMMA 0.9 are six fuzzy feature selection hybrid models (Mallick et al., 2021).

13.6 CONCLUSIONS AND RECOMMENDATIONS

The overall result illustrates that integrating RS and GIS techniques provides a powerful tool for analyzing GW resources and proposing an appropriate exploration plan for GW recharge in a studied area. Compared to conventional GW investigation techniques, remote sensing data that provides precise spatial information can be used cost-efficiently. Satty's AHP is one approach for determining the weightage for the GW study. The multi-parameter approach, carried out using GIS and an AHP technique, was cost-effective and stress-free work. Validation and comparison of the results may be varied from a method-to-method selection, which needs to be accounted for during assessment. This chapter would promote the efficient development and protection of GW supplies. Assessing GWP would help decision-makers manage GW and locate wells based on demand. Aquifer management can also be made more sustainable by augmenting available water resources.

REFERENCES

Abd Manap, M., Sulaiman, W.N.A., Ramli, M.F., Pradhan, B., Surip, N. 2013. A knowledge-driven GIS modeling technique for groundwater potential mapping at the upper langat basin, malaysia. *Arabian Journal of Geosciences*, 6. 1621–1637.

Abijith, D., Saravanan, S., Singh, L., Jennifer, J.J., Saranya, T., Parthasarathy, K.S.S. 2020. GIS-based multi-criteria analysis for identification of potential groundwater recharge zones-a case study from Ponnaniyaru watershed, Tamil Nadu, India. *Hydro-Research*, 3, 1–14.

Adhikary, P.P., Shit, P.K., Santra, P., Bhunia, G.S., Tiwari, A.K., Chaudhary, B.S. (eds.) Geostatistics and geospatial technologies for groundwater resources in India. *Springer Hydrogeology*. Springer, Cham. https://doi.org/10.1007/978-3-030-62397-5_10

Ahmad, I., Dar, M.A., Andualem, T.G., Teka, A.H., 2020. GIS-based multi-criteria evaluation of groundwater potential of the Beshilo River basin, Ethiopia. *Journal of African Earth Sciences*, 164, 103–747.

Ahmadi, H., Kaya, O.A., Babadagi, E., Savas, T., Pekkan, E. 2021. GIS-based groundwater potentiality mapping using AHP and FR models in central antalya, Turkey. *Environmental Sciences Proceedings*, 5, 11.

Al-Abadi, A.M. 2017. Modeling of groundwater productivity in northeastern Wasit Governorate, Iraq using frequency ratio and Shannon's entropy models. *Applied Water Science*, 7, 699–716. https:// doi.org/10.1007/s13201-015-0283-1

Al-Abadi, A.M., Shahid, S. 2016. Spatial mapping of artesian zone at Iraqi southern desert using a GIS-based random forest machine learning model. *Modeling Earth Systems and Environment*, 2. 1–17.

Al-Fugara, A.K., Ahmadlou, M., Al-Shabeeb, A.R., Al-Ayyash, S., Al-Amoush, H., Al-Adamat, R. 2020. Spatial mapping of groundwater springs potentiality using grid search-based and genetic algorithm-based support vector regression. *Geocarto International*, 2020, 1–20.

Al-Fugara, A.K. Pourghasemi, H.R., Al-Shabeeb, A.R., Habib, M., Al-Adamat, R., Al-Amoush, H., Collins, A.L. 2020. A com-parison of machine learning models for the mapping of groundwater spring potential. *Environmental Earth Sciences*, 79, 1–19.

Al-Fugara, A., Ahmadlou, M., Al-Shabeeb, A.R., Al-Ayyash S., Al-Amoush, H., Al-Adamat, R. 2022. Spatial mapping of groundwater springs potentiality using grid search-based and genetic algorithm-based support vector regression. *Geocarto International*, 37(1), 284–303.

Al-Hadithi, M., Singhal, D.C., Israil, M., et al. 2003. Evaluation of groundwater resources potential in Ratmau Pathri Rao Watershed Haridwar District, Uttaranchal, India, using geoelectrical, remote sensing and GIS techniques. In: V. P. Singh & R. N. Yadava (Eds.), *Ground Water Pollution*. New Delhi, Allied Publishers, pp. 9–17.

Arabameri, A., Jagabandhu, R., Saha, S., Blaschke, T., Ghorbanzadeh, O., Bui, D.T. 2019. Application of probabilistic and machine learning models for groundwater potentiality mapping in damghan sedimentary plain, Iran. *Remote Sensing*, 11, 3015.

Arabameri, A., Lee, S., Tiefenbacher, J.P., Thao, P., Ngo, T. 2020. Novel ensemble of MCDM-artificial intelligence techniques for groundwater-potential mapping in arid and semi-arid regions (Iran). *Remote Sensing*, 12(3), 1–28. https://doi.org/10.3390/rs120 30490

Arabameri, A., Pal, S.C., Rezaie, F., Nalivan, O.A., Chowdhuri, I., Saha, A., Lee, S., Moayedi, H. 2021. Modeling groundwater potential using novel GIS-based machine-learning ensemble techniques. *Journal of Hydrology Regional Studies*, 36, 100848. https://doi.org/10.1016/j.ejrh.2021.100848

Arkoprovo, B., Adarsa, J., Prakash, H.S. 2012. Delineation of groundwater potential zones using satellite remote sensing and geographic information system techniques: a case study from Ganjam district, Orissa, India. *Research Journal of Recent Sciences*, 1(9), 59–66.

Arshad, A., Zhang, Z., Zhang, W., Dilawar, A. 2020. Mapping favorable groundwater potential recharge zones using a GIS-based analytical hierarchical process and probability frequency ratio model: a case study from an agro-urban region of Pakistan. *Journal of Geoscience Frontiers*, 5, 1805–1819.

Arulbalaji, P., Gurugnanam, B. 2016. An Integrated Study to Assess the Groundwater Potential Zone Using Geospatial Tool in Salem District, South India. *Journal of Hydrogeology and Hydrologic Engineering*, 5, 2.

Arulbalaji, P., Padmalal, D., Sreelash, K. 2019. GIS and AHP techniques based delineation of groundwater potential zones: a case study from southern Western Ghats, India. *Scientific Reports*, 9(1), 1–17.

Atkinson, P.M., Tatnall, A.R.L. 1997. Neural networks in remote sensing. *The International Journal of Remote Sensing*, 18, 699–709.

Avand, M., Moradi, H. 2021. Spatial modeling of flood probability using geo- environmental variables and machine learning models, case study: Tajan Watershed, Iran. *Advances in Space Research*, 67(10), 3169–3186.

Avand, M., Janizadeh, S., Tien Bui, D., Pham, V.H., Ngo, P.T.T., Nhu, V.H. 2020. A tree-based intelligence ensemble approach for spatial prediction of potential groundwater. *International Journal of Digital Earth*, 13, 1–22.

Aykut, T. 2021. Determination of groundwater potential zones using geographical information systems (GIS) and analytic hierarchy process (AHP) between Edirne-Kalkansogut (northwestern Turkey). *Groundwater for Sustainable Development*, 12, 100–545.

Azimi, S. Azhdary, M.Z., Hashemi, M.S.A. 2018. Prediction of annual drinking water quality reduction based on Groundwater Resource Index using the artificial neural network and fuzzy clustering. *Journal of Contaminant Hydrology*, 220, 6–17. https://doi.org/10.1016/j.jconhyd.2018.10.010501

Banadkooki, F.B., Ehteram, M., Ahmed, A.N., Teo, F.Y., Fai, C.M., Afan, H.A., Sapitang, M., El-Shafie. 2020. A enhancement of groundwater-level prediction using an integrated machine learning model optimized by whale algorithm. *Natural Resources Research*, 29. 3233–3252.

Band, S.S., Janizadeh, S., Chandra Pal, S., Saha, A., Chakrabortty, R., Melesse, A.M., Mosavi, A. 2020. Flash flood susceptibility modeling using new approaches of hybrid and ensemble tree-based machine learning algorithms. *Remote Sensing*, 12, 3568.

Band, S.S., Janizadeh, S., Chandra Pal, S., Saha, A., Chakrabortty, R., Shokri, M., Mosavi, A. 2020. Novel ensemble approach of deep learning neural network (DLNN) model and particle swarm optimization (PSO) algorithm for prediction of gully erosion susceptibility. *Sensors*, 20, 5609.

Barbetta, S., Coccia, G., Moramarco, T., Todini, E., 2016. Case study: a real-time flood forecasting system with predictive uncertainty estimation for the Godavari River, India. *Water*, 8(10), 463.

Benjmel, K., Amraoui, F., Boutaleb, S., Ouchchen, M., Tahiri, A., Touab, A., 2020. Mapping of groundwater potential zones in crystalline terrain using remote sensing, GIS techniques, and multicriteria data analysis (case of the Ighrem region, Western Anti-Atlas, Morocco). *Water*, 12, 471.

Bui, D.T., Pradhan, B., Revhaug, I., Nguyen, D.B., Pham, H.V., Bui, Q.N. 2015. A novel hybrid evidential belief function-based fuzzy logic model in spatial prediction of rainfall-induced shallow landslides in the Lang Son city area (Vietnam). *Geomatics, Natural Hazards and Risk*, 6, 243–271.

Calvo, T., Mesiar, R., Yager, R.R., 2004. Quantitative weights and aggregation. *IEEE Transactions on Fuzzy Systems*, 12(1), 62–69.

Çelik, R., 2019. Evaluation of groundwater potential by gis-based multicriteria decision making as a spatial prediction tool: case study in the Tigris River Batman-Hasankeyf Sub-Basin, Turkey. *Water*, 11, 2630.

Chan, V.K.H., Chan, C.W., 2020. Towards explicit representation of an artificial neural network model: comparison of two artificial neural network rule extraction approaches. *Petroleum*, 6(4), 329–339.

Chaudhari, R., Lal, D., 2018. Weighted overlay analysis for delineation of ground water potential zone: a case study of Pirangut river basin. *International Journal of Remote Sensing & Geoscience (IJRSG)*, 7, 1.

Chen, W., Li, H., Hou, E., Wang, S., Wang, G., Peng, T. 2018. GIS-based groundwater potential analysis using novel ensemble weights-of-evidence with logistic regression and functional tree models. *Science of the Total Environment*, 634, 853–867.

Chen, W., Pourghasemi, H.R., Panahi, M., Kornejady, A., Wang, J., Xie, X., Cao, S. 2017. Spatial prediction of landslide susceptibility using an adaptive neuro-fuzzy inference system combined with frequency ratio, generalized additive model, and support vector machine techniques. *Geomorphology*, 297, 69–85.

Chen, W., Tsangaratos, P., Ilia, I., Duan, Z., Chen, X. 2019. Groundwater spring potential mapping using population-based evolutionary algorithms and data mining methods. *Science of the Total Environment*, 684, 31–49.

Chen, W., Panahi, M., Khosravi, K., Pourghasemi, H.R., Rezaie, F., Parvinnezhad, D. 2019a. Spatial prediction of groundwater potentiality using ANFIS ensembled with teaching-learning-based and biogeography-based optimization. *Journal of Hydrology*, 572, 435–448.

Chen, W., Pradhan, B., Li, S., Shahabi, H., Rizeei, H.M., Hou, E., Wang, S. 2019b. Novel hybrid integration approach of bagging-based fisher's linear discriminant function for groundwater potential analysis. *Natural* Resources Research, 28(4), 1239–1258.

Choubin, B., Rahmati, O., Soleimani, F., Alilou, H., Moradi, E., Alamdari, N. 2019. Regional groundwater potential analysis using classification and regression trees. In: *Spatial Modeling in Gis and R for Earth and Environmental Sciences*; Elsevier: Amsterdam, The Netherlands. pp. 485–498.

Daneshfar, M., Zeinivand H., 2015. Application of frequency ratio, weights of evidence and multi influencing factors models for groundwater potential mapping using GIS. *Journal of Applied Hydrology*, 2, 45–61.

Das, S. 2019. Comparison among influencing factor, frequency ratio, and analytical hierarchy process techniques for groundwater potential zonation in Vaitarna basin, Maharashtra, India. *Groundwater for Sustainable Development*, 8, 617–629.

Das, S., Pardeshi, S.D. 2018. Integration of different influencing factors in GIS to delineate groundwater potential areas using IF and FR techniques: a study of Pravara basin, Maharashtra, India. *Applied Water Science*, 8, 197.

Das, B., Pal, S.C., Malik, S., Chakrabortty, R. 2019. Modeling groundwater potential zones of Puruliya district, West Bengal, India using remote sensing and GIS techniques. *Geology, Ecology, and Landscapes*, 3(3), 223–237.

Das, N., Mondal, P., Sutradhar, S., Ghosh, R. 2021. Identification of groundwater potential zones using multi-influencing factors (MIF) technique: a geospatial study on Purba Bardhaman District of India. In: *Geostatistics and Geospatial Technologies for Groundwater Resources in India*. Springer, Cham. pp. 193–213.

Dehnavi, A., Aghdam, I.N., Pradhan, B., Morshed Varzandeh, M.H. 2015. A new hybrid model using step-wise weight assessment ratio analysis (SWARA) technique and adaptive neuro-fuzzy inference system (ANFIS) for regional landslide hazard assessment in Iran. *Catena*, 135, 122–148.

Dombi, J., Jónás, T. 2022. Weighted aggregation systems and an expectation level-based weighting and scoring procedure. *European Journal of Operational Research*, 299(2), 580–588.

Dou, J., Yamagishi, H., Zhu, Z., Yunus, A.P., Chen, C.W. 2018. TXT-tool 1.081-6.1 A Comparative Study of the Binary Logistic Regression (BLR) and Artificial Neural Network (ANN) Models for GIS-Based Spatial Predicting Landslides at a Regional Scale. Landslide Dynamics: ISDR-ICL Landslide Interactive Teaching Tools. Springer, Chem.

Elmahdy, S.I., Mohamed, M.M., Ali, T.A., Abdalla, J.E.D., Abouleish, M., 2020. Land subsidence and sinkholes susceptibility mapping and analysis using random forest and frequency ratio models in Al Ain, UAE. *Geocarto International*, 37, 1–17.

Ettazarini, S., Jakani M.E., 2020. Mapping of groundwater potentiality in fractured aquifers using remote sensing and GIS techniques: the case of Tafraoute region, Morocco. *Environmental Earth Sciences*, 79, 1–13.

Garrett, J. 1994. Where and why artificial neural networks are applicable in civil engineering. *Journal of Computing in Civil Engineering*, 8, 129–130.

Gebru, H., Gebreyohannes, T., Hagos. E., 2020. Identification of groundwater potential zones using analytical hierarchy process (AHP) and GIS-remote sensing integration, the case of Golina River Basin, Northern Ethiopia. *International Journal of Advanced Remote Sensing and GIS*, 9(1), 3289–3311.

Ghosh, B., 2021. Spatial mapping of groundwater potential using data-driven evidential belief function, knowledge-based analytic hierarchy process and an ensemble approach. *Environmental Earth Sciences*, 80, 625.

Golkarian, A., Naghibi, S.A., Kalantar, B., Pradhan, B. 2018. Groundwater potential mapping using C5, random forest, and multivariate adaptive regression spline models in GIS. *Environmental Monitoring and Assessment*, 190(3), 1–16.

Guru, B., Seshan, K., Bera, S. 2017. Frequency ratio model for groundwater potential mapping and its sustainable management in cold desert, India. *Journal of King Saud University*, 29, 333–347.

Guzman, S.M., Paz, J.O., Tagert, M.L.M., Mercer, A. 2015. Artificial neural networks and support vector machines: contrast study for groundwater level prediction. In: Paper Presented at the ASABE Annual International Meeting.

Gyeltshen, S., Tran, T.V., Teja Gunda, G.K., Kannaujiya, S., Chatterjee, R.S., Champatiray, P.K. 2020. Groundwater potential zones using a combination of geospatial technology and geophysical approach: case study in Dehradun, India. *Hydrological Sciences Journal*, 65(2), 169–182.

Hagos, Y.G., Andualem T.G. 2021. Geospatial and multi-criteria decision approach of groundwater potential zone identification in Cuma sub-basin, Southern Ethiopia. *Heliyon*, 7(9), e07963.

Helaly, A.S. 2017. Assessment of groundwater potentiality using geophysical techniques in Wadi Allaqi basin, Eastern Desert, Egypt -case study. NRIAG *Journal of* Astronomy *and* Geophysics, 6(2), 408–421.

Hussein, E.A., Thron, C., Ghaziasgar, M., Bagula, A., Vaccari, M. 2020. Groundwater prediction using machine-learning tools. *Algorithms*, 13(11), 300.

Ifediegwu, S.I. 2022. Assessment of groundwater potential zones using GIS and AHP techniques: a case study of the Lafia district, Nasarawa State, Nigeria. *Applied Water Science*,12, 10.

Jaafarzadeh, M.S., Tahmasebipour, N., Haghizadeh, A., Pourghasemi, H.R., Rouhani, H. 2021. Groundwater recharge potential zonation using an ensemble of machine learning and bivariate statistical models. *Scientific Reports*, 11(1), 1–18.

Jhariya, D.C., Khan, R., Mondal, K.C., Tarun Kumar, T., Indhulekha, K. Singh, V.K. 2021. Assessment of groundwater potential zone using GIS-based multi-influencing factor (MIF), multi-criteria decision analysis (MCDA) and electrical resistivity survey techniques in Raipur city, Chhattisgarh, India AQUA-Water Infrastructure. *Ecosystems and Society*, 70(3), 375–400.

Jothibasu, A., Anbazhagan, S. 2017. Spatial mapping of groundwater potential in Ponnaiyar River basin using probabilistic-based frequency ratio model. *Modeling Earth Systems and Environment*, 3, 33.

József, D., Tamás, J. 2022. Weighted aggregation systems and an expectation level-based weighting and scoring procedure. *European Journal of Operational Research*, 299(1), 580–588.

Kamali Maskooni, E., Naghibi, S.A., Hashemi, H., Berndtsson, R. 2020. Application of advanced machine learning algorithms to assess groundwater potential using remote sensing-derived data. *Remote Sensing*, 12(17), 2742.

Kanungo, D.P., Arora, M.K., Sarkar, S., Gupta, R.P. 2006. A comparative study of conventional, ANN black box, fuzzy and combined neural and fuzzy weighting procedures for landslide susceptibility zonation in Darjeeling Himalayas. *Engineering Geology*, 85, 347–366.

Karanth, K.R., Seshu babu, K. 1978. Identification of major lineaments on satellite imagery and on aerial photographs for delineation for possible potential groundwater zones in Penukonda and Dharmavaram taluks of Anantapur district. In: Proceeding of Joint Indo-US Workshop on Remote Sensing of Water Resources (NRSA, Hyderabad), Indian Society of Remote Sensing (ISRS) National Natural Resources Management System (NNRMS), Ahmedabad, India. pp. 188–197.

Kenda, K., Peternelj, J., Mellios, N., Kofinas, D., Čerin M., Rožanec, J., 2020. Usage of statistical modeling techniques in surface and groundwater level prediction. *Journal of Water Supply: Research and Technology-Aqua*, 69(3), 248–265.

Khan, U., Faheem, H., Jiang, Z. et al. 2021. Integrating a gis-based multi-influence factors model with hydro-geophysical exploration for groundwater potential and hydrogeological assessment: a case study in the Karak Watershed, Northern Pakistan. *Water*, 13(9), 1255.

Khashei, M., Bijari, M. 2010. An artificial neural network (p,d,q) model for timeseries forecasting. *Expert Systems With Applications,* 37, 479–489.

Khosravi, K., Panahi, M., Bui, D.T. 2018 Spatial prediction of groundwater spring potential mapping based on an adaptive neuro-fuzzy inference system and metaheuristic optimization. *Hydrology and Earth System Sciences*, 22, 4771–4792.

Khosravi, K., Sartaj, M., Tsai, F.T.-C., Singh, V.P., Kazakis, N., Melesse, A.M., Prakash, I., Bui, D.T., Pham, B.T. 2018. A comparison study of DRASTIC methods with various objective methods for groundwater vulnerability assessment. *Science of the Total Environment*, 642, 1032–1049.

Kim, J.C., Lee, S., Jung, H.S., Lee, S. 2018. Landslide susceptibility mapping using random forest and boosted tree models in Pyeong-Chang, Korea. *Geocarto International*, 33(9), 1000–1015.

Kordestani, M.D., Naghibi, S.A., Hashemi, H., Ahmadi, K., Kalantar, B., Pradhan, B. 2018. Groundwater potential mapping using a novel data-mining ensemble model. *Hydrogeology Journal*, 27, 211–224.

Kumar, A., Krishna, A.P. 2018. Assessment of groundwater potential zones in coal mining impacted hard-rock terrain of India by integrating geospatial and analytic hierarchy process (AHP) approach. *Geocarto International*, 33, 105–129.

Kumar, V.A., Mondal, N.C., Ahmed, S. 2020. Identification of groundwater potential zones using RS, GIS and AHP techniques: a case study in a part of Deccan volcanic province (DVP), Maharashtra, India. *Journal of the Indian Society of Remote Sensing*, 48(3), 497–511.

Lakshmi, S.V., Reddy, Y.V.K. 2018. Identification of groundwater potential zones using gis and remote sensing. *International Journal of Pure and Applied Mathematics*, 119(17), 3195–3210.

Lee, S., Lee, Ch-W. 2015. Application of decision-tree model to groundwater productivity-potential mapping. *Sustainability*, 7, 13416–13432.

Lee, S., Song, K., Kim, Y., Park, I. 2012. Regional groundwater productivity potential mapping using a geographic information system (GIS) based neural network model. *Hydrogeology Journal*, 20, 1511–1527.

Lee, S., Hong, S.M., Jung, H.S. 2017. GIS-based groundwater potential mapping using artificial neural network and support vector machine models: the case of Boryeong city in Korea. *Geocarto International*, 33, 847–861. https://doi.org/10.1080/10106049.2017.1303091.

Lee, S., Hong, S., Jung, H. 2017. GIS-based groundwater potential mapping using artifcial neural network and support vector machine models: the case of Boryeong city in Korea neural network and support vector machine models. *Geocarto International*, 604, 1–15.

Lee, S., Hong, S.M., Jung, H.S. 2018. GIS-based groundwater potential mapping using artificial neural network and support vector machine models: the case of Boryeong City in Korea. *Geocarto International*, 33(8), 847–861.

Lee, S., Hyun, Y., Lee, S., Lee, M. 2020. Groundwater potential mapping using remote sensing and GIS-based machine learning techniques. *Remote Sensing Journal*, 12, 1200.

Lee, S., Lee, M.-J., Jung, H.-S., Lee, S. 2020. Landslide susceptibility mapping using naïve bayes and bayesian network models in Umyeonsan, Korea. *Geocarto International*, 35, 1665–1679.

Lee, S., Kaown, D., Koh, E-H., Lee, H-L., Ko, K-S., Lee, K-K. 2022. Advanced utilization of multi-learning algorithm: ensemble super learner to map groundwater potential for potable mineral water. *Geocarto International*, 37, 9897–9916. https://doi.org/10.1080/10106049.2022.2025921.

Lentswe, G.B., Molwalefhe, L. 2020. Delineation of potential groundwater recharge zones using analytic hierarchy process-guided GIS in the semi-arid Motloutse watershed, eastern Botswana. *Journal of Hydrology: Regional Studies*, 28, 100674

Lv, Z., Zuo, J., Rodriguez, D. 2020. Predicting of runoff using an optimized SWAT-ANN: a case study. *Journal of Hydrology: Regional Studies*, 29, 100688.

Madani, A., Niyazi, B. 2015. Groundwater potential mapping using remote sensing techniques and weights of evidence GIS model: a case study from Wadi Yalamlam basin, Makkah Province, Western Saudi Arabia. *Environmental Earth Sciences*, 74, 5129–5142.

Magesh, N.S., Chandrasekar, N., Soundranayagam, J.P. 2012. Delineation of groundwater potential zones in Theni district, Tamil Nadu, using remote sensing, GIS and MIF techniques. *Geoscience Frontiers*, 3(2), 189–196.

Malczewski, J. 1999. *GIS and Multicriteria Decision Analysis*. John Wiley & Sons, Hoboken, NJ, USA.

Mallick, J., Khan, R.A., Ahmed, M, Alqadhi, S.D., Alsubih, M., Falqi, I., Hasan, M.A. 2019. Modeling groundwater potential zone in a semi-arid region of Aseer using Fuzzy-AHP and geo-information techniques. *Water*, 11, 2656.

Mallick, J., Talukdar, S., Kahla, N.B., Ahmed, M., Alsubih, M., Almesfer, M.K., Islam, A.R.M.T. 2021. A novel hybrid model for developing groundwater potentiality model using high resolution digital elevation model (DEM) derived factors. *Water*, 13, 2632. https://doi.org/10.3390/w13192632

Manap, M.A., Nampak, H., Pradhan, B., Lee, S., Sulaiman, W.N.A., Ramli, M.F., 2014. Application of probabilistic-based frequency ratio model in groundwater potential mapping using remote sensing data and GIS. Arabian *Journal of Geosciences*, 7(2), 711–724. https://doi.org/10.1007/s12517-012-0795-z

Mandal, I., Pal, S. 2020. Modelling human health vulnerability using different machine learning algorithms in stone quarrying and crushing areas of Dwarka river Basin, Eastern India. *Advances in Space Research*, 66(6), 1351–1371.

Melese, T., Belay, T. 2022. Groundwater potential zone mapping using analytical hierarchy process and GIS in Muga Watershed, Abay Basin, Ethiopia. *Global Challenges*, 6, 1–13.

Miraki, S., Zanganeh, S.H., Chapi, K., et al. 2019. Mapping groundwater potential using a novel hybrid intelligence approach. *Water Resources Management*, 33, 281–302. https://doi.org/10.1007/s11269-018-2102-6

Mohammed, A., Sayl, K.N. 2020. Determination of Groundwater Potential Zone in Arid and Semi-Arid Regions: A review," 2020 13th International Conference on Developments in eSystems Engineering (DeSE), pp. 76–81, https://doi.org/0.1109/DeSE51703.2020.9450782

Moore, G.K. 1978. The role of remote sensing in groundwater exploration. In: Proc. Joint Indo-US Workshop on Remote Sensing of Water Resources, pp. 22–40. National Remote Sensing Agency (NRSA), Hyderabad, India.

Mosavi, A., Sajedi Hosseini, F., Choubin, B., et al. 2021. Ensemble boosting and bagging based machine learning models for groundwater potential prediction. *Water Resources Management*, 35, 23–37.

Muavhi, N., Thamaga K.H., Mutoti, M.I. 2021. Mapping groundwater potential zones using relative frequency ratio, analytic hierarchy process and their hybrid models: case of Nzhelele-Makhado area in South Africa. *Geocarto International*, 37, 6311–6330. https://doi.org/10.1080/10106049.2021.1936212

Mukherjee, S. 1996. Targeting saline aquifer by remote sensing and geophysical methods in a part of Hamirpur-Kanpur, India. *Hydrogeology Journal*, 9, 53–64.

Nagarajan, M., Singh S. 2009. Assessment of groundwater potential zones using GIS technique. *Journal of the Indian Society of Remote Sensing*, 37 (1), 69–77.

Naghibi, S.A., Pourghasemi, H.R. 2015. A comparative assessment between three machine learning models and their performance comparison by bivariate and multivariate statistical methods in groundwater potential mapping. *Water Resources Management*, 29, 5217–5236.

Naghibi, S.A., Pourghasemi, H.R., Dixon, B. 2016. GIS-based groundwater potential mapping using boosted regression tree, classification and regression tree, and random forest machine learning models in Iran. *Environmental Monitoring and Assessment*, 188, 44.

Naghibi, S.A., Ahmadi, K., Daneshi, A. 2017. Application of support vector machine, random forest, and genetic algorithm optimized random forest models in groundwater potential mapping. *Water Resources Management*, 31, 2761–2775.

Naghibi, S.A., Pourghasemi, H.R., Abbaspour, K.A. 2017. Comparison between ten advanced and soft computing models for groundwater potential assessment in Iran using R and GIS. *Theoretical and Applied Climatology*, 131, 967–984.

Naghibi, S.A., Moghaddam, D.D., Kalantar, B., Pradhan, B. and Kisi, O. 2017. A comparative assessment of GIS-based data mining models and a novel ensemble model in groundwater well potential mapping. *Journal of Hydrology*, 548, 471–483.

Naghibi, S.A., Dolatkordestani, M., Rezaei, A., Amouzegari, P., Heravi, M.T., Kalantar, B., Pradhan, B. 2019. Application of rotation forest with decision trees as base classifier and a novel ensemble model in spatial modeling of groundwater potential. *Environmental Monitoring and Assessment*, 191, 248.

Nasir, M.J., Khan, S., Zahid, H., Khan, A. 2018. Delineation of groundwater potential zones using GIS and multi influence factor (MIF) techniques: a study of district Swat, Khyber Pakhtunkhwa Pakistan. *Environmental Earth Sciences*, 77(10), 367.

Nguyen, P.T., Ha, D.H., Avand, M., Jaafari, A., Nguyen, H.D., Al-Ansari, N., Van Phong, T., Sharma, R., Kumar, R., Van Le, H. et al. 2020. Soft computing ensemble models based on logistic regression for groundwater potential mapping. *Applied Sciences*, 10, 2469.

Oh, H.J., Kim, Y.S., Choi, J.K., Park, E., Lee, S. 2011. GIS mapping of regional probabilistic groundwater potential in the area of Pohang City, Korea. *Journal of Hydrology*, 399, 158–172.

Oikonomidis, D., Dimogianni, S., Kazakis, N., Voudouris, K. 2015. A GIS/Remote Sensing-based methodology for groundwater potentiality assessment in Tirnavos area, Greece. *Journal of Hydrology*, 525, 197–208.

Owolabi, S.T., Madi, K., Kalumba, A.M. et al. 2020. A groundwater potential zone mapping approach for semi-arid environments using remote sensing (RS), geographic information system (GIS), and analytical hierarchical process (AHP) techniques: a case study of Buffalo catchment, Eastern Cape, South Africa. *Arabian Journal of Geosciences*, 13, 1184. https://doi.org/10.1007/s12517-020-06166-0

Ozdemir, A. 2011. GIS-based groundwater spring potential mapping in the Sultan Mountains (Konya, Turkey) using frequency ratio, weights of evidence and logistic regression methods and their comparison. *Journal of Hydrology*, 411, 290–308.

Pal, S., Kundu, S., Mahato, S. 2020. Groundwater potential zones for sustainable management plans in a river basin of India and Bangladesh. *Journal of Cleaner Production*, 257, 120311.

Panahi, M., Sadhasivam, N., Pourghasemi, H.R., Rezaie, F., Lee, S., 2020. Spatial prediction of groundwater potential mapping based on convolutional neural network (CNN) and support vector regression (SVR). *Journal of Hydrology*, 588, 125033. https://doi.org/10.1016/j.jhydrol.2020.125033

Paola, J.D., Schowengerdt, R.A. 1995. A review and analysis of backpropagation neural networks for classification of remotely-sensed multi-spectral imagery. *International Journal of Remote Sensing*, 16, 3033–3058.

Park, S., Hamm, S.-Y., Jeon, H.-T., Kim, J. 2017. Evaluation of logistic regression and multivariate adaptive regression spline models for groundwater potential mapping using R and GIS. *Sustainability*, 9, 1157.

Patra, S., Mishra, P., Mahapatra, S.C. 2018. Delineation of groundwater potential zone for sustainable development: a case study from Ganga alluvial plain covering Hooghly district of India using remote sensing, geographic information system and analytic hierarchy process. *Journal of Cleaner Production*, 172, 2485–2502. https://doi.org/10. 1016/j.jclepro.2017.11.161

Peng, L., Niu, R., Huang, B., Wu, X., Zhao, Y., Ye, R. 2014. Landslide susceptibility mapping based on rough set theory and support vector machines: a case of the Three Gorges area, China. *Geomorphology*, 204, 287–301.

Pham, B.T., Jaafari, A., Prakash, I., Singh, S.K., Quoc, N.K., Bui, D.T. 2019. Hybrid computational intelligence models for groundwater potential mapping. *Catena*, 182, 104101.

Pourghasemi, H.R., Sadhasivam, N., Yousefi, S., Tavangar, S., Nazarlou, H.G., Santosh, M. 2020. Using machine learning algorithms to map the groundwater recharge potential zones. *Journal of Environmental Management*, 265, 110525.

Pradhan, B., Li, S., Shahabi, H., Rizeei, H.M., Hou, E., Wang, S. 2019. Novel hybrid integration approach of bagging-based fisher's linear discriminant function for groundwater potential analysis. *Natural Resources Research*, 28(4), 1239–1258.

Prasad, P., Loveson, V.J., Kotha, M., Yadav, R. 2020. Application of machine learning techniques in groundwater potential mapping along the west coast of India. *GIS Science and Remote Sensing*, 57(6), 735–752.

Putra, M.S.D., Fauziah, S.A., Gunaryati, A. 2018. Fuzzy analytical hierarchy process method to determine the quality of gemstones. *Advances in Fuzzy Systems*, 6, 9094380. https://doi.org/10.1155/2018/9094380

Rahmati, O., Samani, A.N., Mahdavi, M., Pourghasemi, H.R., Zeinivand, H. 2014. Groundwater potential mapping at Kurdistan region of Iran using analytic hierarchy process and GIS. *Arabian Journal of Geosciences*, 8, 7059–7071. https://doi.org/10.1007/s12517-014-1668-4

Rahmati, O., Pourghasemi, H.R., Melesse, A.M. 2016. Application of GIS-based data driven random forest and maximum entropy models for groundwater potential mapping: a case study at Mehran Region, Iran. *Catena*, 137, 360–372.

Rajasekhar, M., Raju, G.S., Sreenivasulu, Y., Raju, R.S. 2019. Delineation of groundwater potential zones in semi-arid region of Jilledubanderu river basin, Anantapur District, Andhra Pradesh, India using fuzzy logic, AHP and integrated fuzzy-AHP approaches. *Hydro Research*, 2, 97–108.

Razandi, Y., Pourghasemi, H.R., Neisani, N.S., Rahmati, O. 2015. Application of analytical hierarchy process, frequency ratio, and certainty factor models for groundwater potential mapping using GIS. *Earth Science Informatics*, 8, 867–883.

Razavi-Termeh, S.V., Sadeghi-Niaraki, A., Choi, S.-M. 2019. Groundwater potential mapping using an integrated ensemble of three bivariate statistical models with random forest and logistic model tree models. *Water*, 11, 1596.

Razavi-Termeh, S.V., Khosravi, K., Sadeghi-Niaraki, A., Choi, S.-M., Singh, V.P. 2020. Improving groundwater potential map-ping using metaheuristic approaches. *Hydrological Sciences Journal*, 65, 2729–2749.

Rita, D., Jiong, Y., Zuo, Z., Melanie, P.G., Adrian, R.K., Steven, P.G., Robin, R. 2007. SR proteins function in coupling RNAP II transcription to Pre-mRNA splicing. *Molecular Cell*, 26, 867–881.

Rizeei, H.M., Pradhan, B., Saharkhiz, M.A., Lee, S. 2019. Groundwater aquifer potential modeling using an ensemble multi-adoptive boosting logistic regression technique. *Journal of Hydrology*, 579, 124172.

Saaty, T.L. 1977. A scaling method for priorities in hierarchical structures. *Journal of Mathematical Psychology*, 15, 234–281. https://doi.org/10.1016/0022-2496(77)90033-5

Saaty, T.L. 1980. *The Analytic Hierarchy Process*. McGraw-Hill, New York.

Saaty, T. 1990. *Decision Making for Leaders: The Analytic Hierarchy Process for Decisions in a Complex World*. RWS Publications, Pittsburgh, PA, USA.

Sachdeva, S., Kumar, B. 2020. Comparison of gradient boosted decision trees and random forest for groundwater potential mapping in Dholpur (Rajasthan), India. *Stochastic Environmental Research and Risk Assessment*, 28, 1–20.

Sahoo, S., Russo, T.A., Elliott, J., Foster, I. 2017. Machine learning algorithms for modeling groundwater level changes in agricultural regions of the US. *Water Resources Research*, 53(5), 3878–3895.

Sankar, K. 2002. Evaluation of groundwater potential zones using remote sensing data in upper Vaigai river basin, Tamilnadu, India. *Journal of the Indian Society of Remote Sensing*, 30(3), 119–130.

Saravanan, S., Saranya, T., Abijith, D., Jacinth, J.J., Singh, L. 2021. Delineation of groundwater potential zones for Arkavathi sub-watershed, Karnataka, India using remote sensing and GIS. *Environmental Challenges*, 5, 100380.

Senthilkumar, M., Gnanasundar, D., Arumugam, R. 2019. Identifying groundwater recharge zones using remote sensing & GIS techniques in Amaravathi aquifer system, Tamil Nadu, South India. *Sustainable Environment Research*, 29(1), 15.

Shao, Z., Huq, M.E., Cai, B., Altan, O., Li, Y. 2020. Integrated remote sensing and GIS approach using Fuzzy-AHP to delineate and identify groundwater potential zones in semi-arid Shanxi Province. *Environmental Modelling & Software*, 134, 104868.

Siddi Raju, R., Sudarsana Raju, G., Rajasekhar, M. 2019. Identification of groundwater potential zones in Mandavi River basin, Andhra Pradesh, India using remote sensing, GIS and MIF techniques. *Hydro Research*, 2, 1–11.

Sikdar, P.K., Bhattacharya, P. 2003. Groundwater risk analysis and development plan of Calcutta. In S.P. Das Gupta (ed.), *Environmental Issues for the 21st Century* (pp. 83–119). New Delhi: Mittal Publishers.

Sikdar, P.K., Chakraborty, S., Adhya, E., Paul, P.K., et al. 2004. Landuse/landcover changes and groundwater potential zoning in and around Raniganj Coal Mining Area, Bardhaman district. W.B.-A GIS and remote sensing approach. *Journal of Spatial Hydrology*, 4(2), 1–24.

Singh, L.K., Jha, M.K., Chaudary, V.M. 2018. Assessing the accuracy of GIS-based multi-criteria decision analysis approaches for mapping groundwater potential. *Ecological Indicators*, 91, 24–37. https://doi.org/10.1016/j.ecolind.2018.03.070

Soria, D., Garibaldi, J.M., Ambrogi, F., Biganzoli, E.M., Ellis, I.O. 2011. A non-parametric version of the naive Bayes classifier. *Knowledge-Based System*, 24, 775–784.

Sree devi, P.D., Srinivasulu, S., Raju, K.K. 2001. Hydrogeomorphological and groundwater prospects of the Pageru river basin by using remote sensing data. *Environmental Geology*, 40, 1088–1094.

Sreedevi, P.D., Subrahmanyam, K., Ahmed, S., et al. 2005. Integrated approach for delineating potential zones to explore for groundwater in the Pageru River basin, Cuddapah District, Andhra Pradesh, India. *Hydrogeology Journal*, 13(3), 534–543.

Srinivasa Rao, Y., Jugran, D.K. 2003. Delineation of groundwater potential zones and zones of groundwater quality suitable for domestic purposes using remote sensing and GIS. *Hydrological Sciences Journal*, 48(5), 821–833. https://doi.org/10.1623/hysj.48.5.821.51452

Srivastava, A., Chinnasamy, P. 2021a. Developing village-level water management plans against extreme climatic events in Maharashtra (India)-a case study approach. In: Vaseashta, A., Maftei, C. (eds.) *Water Safety, Security and Sustainability. Advanced Sciences and Technologies for Security Applications*. Springer, Cham. https://doi.org/10.1007/978-3-030-76008-3_27

Srivastava, A., Chinnasamy, P. 2021b. Investigating impact of land-use and land cover changes on hydro-ecological balance using GIS: insights from IIT Bombay, India. *SN Applied Sciences*, 3, 343. https://doi.org/10.1007/s42452-021-04328-7

Srivastava, A., Chinnasamy, P. 2022. Assessing groundwater depletion in Southern India as a function of urbanization and change in hydrology: a threat to tank irrigation in Madurai City. In: Kolathayar, S., Mondal, A., Chian, S.C. (eds.) *Climate Change and Water Security. Lecture Notes in Civil Engineering*, vol 178. Springer, Singapore. https://doi.org/10.1007/978-981-16-5501-2_24

Srivastava, A., Chinnasamy, P. 2023a. Tank cascade system in Southern India as a traditional surface water infrastructure: a review. In: Chigullapalli, S., Susha Lekshmi, S.U., Deshpande, A.P. (eds.) *Rural Technology Development and Delivery. Design Science and Innovation*. Springer, Singapore. https://doi.org/10.1007/978-981-19-2312-8_15

Srivastava, A., Chinnasamy, P. 2023b. Understanding Declining storage capacity of tank cascade system of Madurai: potential for better water management for rural, peri-urban, and urban catchments. In: Chigullapalli, S., Susha Lekshmi, S.U., Deshpande, A.P. (eds.) *Rural Technology Development and Delivery. Design Science and Innovation*. Springer, Singapore. https://doi.org/10.1007/978-981-19-2312-8_14

Srivastava, A., Khadke, L., Chinnasamy, P. 2021. Web application tool for assessing groundwater sustainability—a case study in rural-Maharashtra, India. In: Vaseashta, A., Maftei, C. (eds.) *Water Safety, Security and Sustainability. Advanced Sciences and Technologies for Security Applications*. Springer, Cham. https://doi.org/10.1007/978-3-030-76008-3_28

Srivastava, A., Khadke, L., Chinnasamy, P. 2022. Developing a web application-based water budget calculator: attaining water security in rural-Nashik, India. In: Kolathayar, S., Mondal, A., Chian, S.C. (eds.) *Climate Change and Water Security. Lecture Notes in Civil Engineering*, vol 178. Springer, Singapore. https://doi.org/10.1007/978-981-16-5501-2_37

Taghizadeh-Mehrjardi, R., Schmidt, K., Toomanian, N., Heung, B., Behrens, T., Mosavi, A., Band, S.S., Amirian-Chakan, A., Fathabadi, A., Scholten, T.2021. Improving the spatial prediction of soil salinity in arid regions using wavelet transformation and support vector regression models. *Geoderma*, 383, 114793.

Taheri, K., Missimer, T.M., Taheri, M., Moayedi, H., Pour, F.M. 2019. Critical zone assessments of an alluvial aquifer system using the multi-infuencing factor (MIF) and analytical hierarchy process (AHP) models in Western Iran. *Natural Resources Research*, 29, 1–29.

Tamiru, H., Wagari, M., 2021. Comparison of ANN model and GIS tools for delineation of groundwater potential zones, Fincha Catchment, Abay Basin, Ethiopia. *Geocarto International*, 37, 6736-6754. https://doi.org/10.1080/10106 049.2021.1946171

Thapa, R., Gupta, S., Guin, S. et al. 2017. Assessment of groundwater potential zones using multi-influencing factor (MIF) and GIS: a case study from Birbhum district, West Bengal. *Applied Water Science*, 7, 4117–4131. https://doi.org/10.1007/s13201-017-0571-z

Todd, D.K. 1980. *Groundwater Hydrology*, (2nd edn). Wiley, New York.

Tsakiri, K., Marsellos, A., Kapetanakis, S. 2018. Artificial neural network and multiple linear regression for flood prediction in Mohawk River, New York. *Water*, 10(9), 1158.

Waikar, M.L., Nilawar, A.P. 2014. Identification of Groundwater Potential Zone using Remote Sensing and GIS Technique. *International Journal of Innovative Researchin Science, Engineering and Technology*, 3(5), 12163–12174.

Yıldırım, Ü. 2021. Identification of groundwater potential zones using GIS and multi-criteria decision-making techniques: a case study Upper Coruh River Basin (NE Turkey). *ISPRS International Journal of Geo-Information*, 10(6), 396.

Zabihi, M., Pourghasemi, H.R., Pourtaghi, Z.S., Behzadfar, M. 2016. GIS-based multivariate adaptive regression spline and random forest models for groundwater potential mapping in Iran. *Environmental Earth Sciences*, 75, 1–19.

Zargini, B. 2020. Multicriteria decision making problems using variable weights of criteria based on alternative preferences. *American Scientific Research Journal for Engineering, Technology, and Sciences (ASRJETS)*, 74(1), 1–14.

Zghibi, A., Mirchi A., Msaddek, M.H., Merzougui, A., Zouhri, L., Taupin, J.D., Chekirbane, A., Chenini, I., Tarhouni, J. 2020. Using analytical hierarchy process and multi-influencing factors to map groundwater recharge zones in a semi-arid Mediterranean coastal aquifer. *Water*, 12, 2525.

Zhang, S., Li, C., Zhang, L., Peng, M., Zhan, L., Xu, Q. 2020. Quantification of human vulnerability to earthquake-induced landslides using Bayesian network. *Engineering Geology*, 265, 105436.

14 Irrigation Water Requirement and Estimation of ETo Based on the Remote Sensing, GIS, and Other Technologies

Ahmed Elbeltagi, Wessam El-Ssawy, Amal Mohamed, Sobhy M. Mahmoud, Ali Mokhtar, and Chaitanya B. Pande

14.1 INTRODUCTION

Drought, stricter laws governing the use of water in agronomic production, and high challenges for water resources with non-agricultural uses such as domestic, industry, and the environment are all contributing to irrigation water shortages in many parts of the world, particularly in arid and semi-arid areas (Cherif et al., 2015). The hydrological cycle is thought to be incomplete without evapotranspiration (ET). In addition, (Nouri et al., 2014). ET is an essential criterion in water resource management for the water scarcity problem. As a result, water scarcity fears can be blamed for the recent high annual ET rate. To increase the efficiency of water supply systems, reliable research of ET factors is essential (Mohammadian et al., 2017). Water scarcity is anticipated to worsen as humans compete for resources, and global warming results from growing greenhouse gas (GHG) levels in the atmosphere.

In the future, many global climate models predict a more intense and varied hydrologic cycle, with altered precipitation patterns and a higher frequency of severe droughts (Field et al., 2014). Humans are concerned about a projected fall in crop yield as demand for agricultural products rises in the emerging scenario. It is critical at this stage to establish water management systems that are more water-efficient than previously used ones. To maximize the return from limited water (rainfall and irrigation) available for agricultural usage, farm counselors and producers require whole-system-based quantitative learning and knowledge on crop water needs and accessible water supplies (Hsiao et al., 2007). Throughout a watershed, water evaporates from a variety of surfaces. Precipitation absorbed by plant surfaces and evaporating back into the sky is an interception. This amount of precipitation does not reach the surface of the soil. Evaporation occurs when water bodies in a watershed and soil surfaces are exposed to the sun. Transpiration is the process through which plant roots absorb water from the ground and evaporate it through the plant leaves. Interception, transpiration, and evaporation from soils and water bodies make up ET, the total amount of water that evaporates from a watershed. The cycle repeats as evaporated water is lost to the atmosphere for a brief moment before returning to the earth's surface as precipitation in another location. Rainfall that falls on a watershed but does not evaporate might run over the soil surface or permeate into the soil. The moisture condition, water capacity, and pores within the soil matrix all influence the fate of penetrated water. Evaporation and transpiration are combined in the process of ET. Evaporation occurs on many water surfaces, including lakes, rivers, soils, and plants.

Removing plant tissues and their vapor into the atmosphere is known as transpiration (Jensen and Allen, 2016). The process of transpiration entails the transfer of water from the soil to the

DOI: 10.1201/9781003303237-14

plant leaves, followed by evaporation into the atmosphere through stomatal pores in the leaves. Because it permits important nutrients and minerals to be transferred through the plant's many components, transpiration is critical for plant growth and development. Solar radiation, air temperature and humidity, and wind speed are all factors that influence evaporation and transpiration processes (according to Jensen and Allen, 2016). Estimating crop (ETc) is important for making real-time irrigation scheduling decisions, comparing unconventional irrigation choices, and evaluating the financial impact of various cropping systems in various growing environments, including the potential impact of projected climate change scenarios. ETc is well associated with agricultural biomass and yield output, allowing crop yield estimation (Power et al., 2011; Cammarano et al., 2012).

14.2 POTENTIAL AND REFERENCES EVAPOTRANSPIRATION

It is necessary to clarify the concepts of potential and reference crop evapotranspiration (ETp and ETo), which have been confused for decades. One typical example is that Hargreaves and Samani (1982) used "potential", while Hargreavers and Samoni (1985) used the term "reference crop" (Allen et al., 1998; Peng et al., 2017; Mardikis et al., 2005; Dinpashoh et al., 2011).

14.3 POTENTIAL EVAPOTRANSPIRATION

Thornthwaite (1948) coined the phrase "potential evapotranspiration" after researching rainfall and water use in numerous areas in the United States. It represents the greatest amount of evaporation that might occur under perfect conditions. ETp expresses the evaporative requirement of the atmosphere, or the quantity of water that may be transported to the atmosphere from water or land. It is described as "the movement of water from the earth back to the atmosphere, represented by the combined evaporation from the soil surface and transpiration from plants, the reversal of rainfall" (Thornthwaite, 1948). It was described by Anon (1956) as defined it as "the rate of water vapor loss from a short grass canopy under the following conditions: growing on a large area, completely covering the soil".

14.4 REFERENCE CROP EVAPOTRANSPIRATION

The United Nations' Food and Agriculture Organization (FAO) established ETo. "The rate of evaporation from a wide surface of 8–15 cm tall, green grass cover of uniform height, actively developing, totally shading the ground, and not short of water", according to ETo. Grass and alfalfa were chosen as the small and high-location crops, respectively (Doorenbos and Pruitt, 1977; Beard, 1985; Snyder et al., 1987). That would be similar to how water evaporates off a large, uniformly heightening green grass cover that is actively developing (Allen et al., 1998). In FAO Irrigation and Drainage Paper No.56, the FAO formally adopted the ETo term (Allen et al., 1998). Another "alfalfa", defined as a tall crop with an estimated height of 0.5 m, was listed as the reference crop by the American Society of Civil Engineers (ASCE). Estimating ETo assists agronomists and farmers in managing water requirement difficulties and providing vital irrigation plan information. Because one of the main purposes is to distinguish the ET of plants from a broad territory, the definition of ETo is clearer and more explicit than that of ETp (Walter et al., 2000). Using unique and optimal plant constraints, estimating the amount of water evaporating from crop and vegetation surfaces is straightforward. When evaporative demand exceeds precipitation, irrigation is used in some regions to supplement natural precipitation and reduce crop production losses due to water shortages. Evaporation and transpiration absorb a considerable portion of the irrigation water supplied to agricultural lands. Plant growth and development, as well as grain output and quality, may be harmed during a growing season.

14.5 EFFECTIVE PRECIPITATION (EP) DEFINITION

The quantity of rain sprayed and held in the soil is known as effective precipitation (EP). During dry periods, less than 5 mm of daily rain is not considered effective because it is created by evaporation from the surface before it soaks into the ground. EP can be added to the soil and made available to crops. The severity and frequency of atmospheric droughts are increasing due to global warming and climate change. Long-term meteorological drought frequently results in a lack of soil moisture and a reduction in groundwater levels, limiting vegetation growth and posing a major threat to the environment (Chen et al., 2021; Li et al., 2020; Chang et al., 2016). As a result of decreasing precipitation (P) and the effects of climate change on water storage, ET, and evaporation rates, supplemental irrigation is utilized in some restoration and revegetation initiatives (Gorguner et al., 2020). Due to the shortage of water resources in these places, irrigation programs must be carefully planned; also, excessive irrigation may temporarily help vegetation survival, but it will increase soil salinity and negatively harm plant diversity in the long run (El-Keblawy et al., 2016). The majority of rainwater is used in agriculture to grow crops. However, some of the rain was lost. Water captured by live or dry plants, which is lost through evaporation during crop growth, which is lost owing to evaporation from the soil surface, and which contributes to leaching and infiltration, are all examples of these losses. Meteorologists cannot analyze or evaluate the EP problem based on tables of precipitation frequency, quantity, and intensity or on atmospheric physical phenomena (Rahman et al., 2008; Manalo, 1976). Many fields and sub-disciplines collide in this task. Soil types, cropping patterns, and socioeconomic and management considerations, for example, all have a direct impact on the amount of effective and ineffective rainfall in agriculture (Rahman et al., 2008; FAO, 1975).

14.6 MEASUREMENTS OF EFFECTIVE RAINFALL

Various approaches have been offered in the past to estimate rainfall successfully, including direct measurement, empirical methods, and soil water balancing methods; soil water balancing methods produce the best results (Patwardhan et al., 1990). The CROPWAT 8.0 model used four empirical equations to estimate EP for irrigation planning and design, according to these techniques (Kuo et al., 2001; Smith, 1991). These methods include the USDA-SC method for calculating EP, a constant percentage of precipitation, reliable precipitation, an empirical formula, and the USDA-SC method for calculating EP. For each approach used in predicting EP in CROPWAT 8.0, the input data is the same for each plot, but the outcome is different. Smith (1991) and Kuo et al. provide a full discussion of each of these experimental procedures (2001). To avoid water shortages that could lead to poorer yields, Bokke and Shoro (2020) proposed using the USDA-SC water scarcity region method and the water sufficient region approved rain method in determining EP. Simultaneously, enough drainage should be given to the plan to avoid water logging, which reduces production.

14.7 NET IRRIGATION WATER REQUIREMENT (NIWR)

The amount of water that will be provided to the crop by irrigation for maximum crop growth, taking into account losses incurred during irrigation water transportation and use, is known as the irrigation water requirement (IWR). IWR is changing due to climate change, which includes rising temperatures, variable precipitation, and increased atmospheric carbon dioxide (CO_2) concentrations (Wang et al., 2014). The following equation (USDA, 1993) represents the net amount of irrigation water required to satisfy a crop's water requirement.

$$\text{NIWR} = \text{ETc} - \text{Pe} - \text{GW} - \Delta\text{SW} \quad (14.1)$$

where:

NIWR = net irrigation water requirement

ETc = crop evapotranspiration
Pe = effective precipitation
GW = groundwater contribution
ΔSW = soil water depleted

14.8 DATASETS SOURCES TO ESTIMATE REFERENCE EVAPOTRANSPIRATION (ETo)

A greater understanding of water use accuracy is required since water shortages are becoming more frequent and competition for available water resources is growing. ET and consumer consumption are now interchangeable terms. For irrigation, energy, water transmission, flood control, municipal and industrial water, and wastewater reuse systems, as well as the economics of multipurpose water projects, ET is required for the design and operation of water resource projects (Rodrigues and Braga, 2021; Jensen and Allen, 2016). A good estimation of crop water requirements and ETc is necessary for agricultural water management. The strategy developed by Allen et al. (1998), based on the combined ETo and crop modulus, is a widely accepted method for determining ETc at the field level. As a result, ETo became an essential factor in predicting irrigation requirements. It also has various applications, including irrigation system design and scheduling (Rodrigues and Braga, 2021; Doorenbos and Pruitt, 1977).

Numerous empirical, semi-empirical, and physical formulas for calculating ETo are dependent on weather variables such as maximum and minimum temperatures, wind speed, relative humidity, and solar radiation (Rs). The FAO-56 Penman-Monteith (PM) equation has been proposed as a standard method for predicting ETo when complete weather databases are available (Rodrigues and Braga, 2021; Jensen and Allen, 2016). This approach has been extensively documented worldwide in numerous regions and climatic conditions (Allen et al., 2006; Allen, 2005). The International Water Management Institute (IWMI 2000) recently made a global climatic dataset with relatively high geographical resolution available. These indicators, which represent average conditions for the previous 30 years on a monthly basis, are provided. The 10-minute spatial resolution of Arc (about 16 km at the equator). The dataset was produced using observations collected over the previous 30 years from around 56,000 locations worldwide. With readings of humidity, sunshine, and wind speed dispersed throughout the grid, these were mostly temperature stations. The IWMI dataset is a great resource for information since the geographical resolution is substantially better than other global datasets used in climate change studies, and the range of variance in climatic conditions is very wide (Peter and Allen, 2002). To estimate the ETo, experimental, artificial, and physical intelligence models have been improved (Shiri and Landeras, 2012). A hypothetical turf with a height of 0.12 m, a surface resistance of 70 s/m, and a surface albedo of 0.23 are used to calculate ETo, which is the sum of evaporation and transpiration (Allen et al., 2012). The accepted technique for calculating ETo is now the Penman-Monteith (PM) equation. However, this technique requires many meteorological variables, many of which are frequently unavailable (Yasin et al., 2012). The PM equation is often used to predict agricultural water needs and setup and operate irrigation systems since it has a strong physical basis (Allen et al., 1998). The IWR is computed during the season using equations (Döll and Siebert, 2002; Shen et al., 2013) to calculate losses owing to irrigation inefficiency:

$$\text{IWR} = \text{ETc} - \text{Pe} / \text{Ic} \tag{14.2}$$

$$\text{ETc} = \text{Kc} * \text{ETo} \tag{14.3}$$

where ETo is the reference evapotranspiration and IWR is the crop-specific irrigation water demand (mm/day) (mm/day). The crop-specific evapotranspiration (ETc), which is measured in millimeters/day, is used to calculate the quantity of water needed for yield. The yield coefficient, or Kc, indicates key factors in irrigation scheduling and water distribution. Effective rainfall is measured in millimeters per day, while irrigation efficiency is measured in terms of water losses during conversion (Shen et al., 2013).

14.9 CALCULATION OF REFERENCE EVAPORATION

The ETo can be calculated from meteorological variables, and it can be used in various areas (Allen et al., 1998). For a hypothetical reference crop, daily ETo is identified according to the FAO-56 PM method (Allen et al., 1998).

14.10 APPROACHES USED FOR ESTIMATING THE REFERENCE EVAPOTRANSPIRATION (ETo) AND CROP WATER REQUIREMENTS (CWR)

Farmers use various irrigation techniques to deliver water to the soil, which serves as the plant's reservoir for growth and development. Depending on the climate, management practices, and crop type, plant absorption and evaporation from unplanted soil surfaces and water loss from the plant itself (transpiration) reduce the amount of water accessible in the soil and plant. Both of these processes are referred to as ET. Farmers must therefore make up for the water used in such procedures to maintain the plant's health and ensure a respectable crop. According to the crop type, growth stage, environment in which the plant lives, and agricultural management techniques employed during the season, various methods for determining how much water should be delivered to the plant have been developed. An accurate and precise assessment of ETc is essential for optimal crop development, maximum output, and minimal water use, especially in countries with limited water resources. This article will cover various approaches for calculating crop water needs and ETo. Reference evaporation is the amount of water lost through ET from a reference surface with no water shortage. Grass serves as the reference surface. It is 12 cm tall, with an albedo of 0.23 and a fixed surface resistance of 70 s/m (Mokhtar et al., 2020a, 2020b).

14.11 TRADITIONAL METHODS FOR ESTIMATING (ETo)

14.11.1 Pan Evaporation

Although no standardized pan is used, estimating evaporation with pans has a long history because it is easy and inexpensive. It can be used to calculate ETo for irrigation scheduling. It is advised as the standard instrument class by the World Food Organization. The FAO uses an evaporation pan to calculate ETo that is 1.21 m in diameter, 25.5 cm deep, and mounted to a wood platform with an open frame (Stanhill, 1976).

$$\text{ETo} = \text{Kpan} \times \text{Epan} \tag{14.4}$$

where Epan is pan evaporation, mm/day and Kpan is a pan evaporation coefficient.

Researchers created the following easy method for converting Epan to ETo for dry climate conditions Snyder et al. (2005), written as follows:

$$[\![\text{ET}]\!]\_(\text{o}) = 10 \sin \sin\left\{0.5\pi * \text{E\_pan} * [0.79 - 0.0035F\ )2 + 0.0622F)]/19.2\right\} \tag{14.5}$$

Equation (14.3) computes for fetch variances by first modifying the Epan rates to the predicted values for 100 m of grass fetch. In order to avoid the necessity for RH and u data, which are often unavailable, the approach requires calibration in more humid or windier locations (Snyder et al., 2005).

14.11.2 Radiation Methods

Empirical radiation approaches for calculating potential ET that included an energy component were developed by adding a solar radiation parameter. The most popular of these techniques was the Jensen-Haise alfalfa reference radiation approach (Jensen and Haise, 1963). The key factors were solar radiation, mean air temperature, and two constants. The constants were computed using elevation and saturation vapor pressures for the mean highest and lowest temperatures during the hottest month of the year.

14.11.3 Hargreaves and Samani Equation

The Hargreaves and Samani (1985) method is an equation that estimates global solar radiation at the surface, Rs, using air temperature measurements. When net solar radiation data is missing or of uncertain accuracy, this equation can be used to approximate ET values (Allen et al., 1998)

$$[\![\mathrm{ET}]\!]\_(\mathrm{o}) = 0.0023\big((T\_\mathrm{max}+\ T\_\mathrm{min})/2+17.8\big)\surd(T\_\mathrm{max}-\ T\_\mathrm{min}\ )\ R\_a \qquad (14.6)$$

where Ra is the extra-terrestrial solar radiation, T_{max} and T_{min} are the maximum and minimum air temperatures, respectively, and KRS is an empirical adjustment coefficient that depends on the site location. KRS takes a value of 0.16°C −0.5 for interior regions and 0.19°C −0.5 for coastal regions. KRS is fitted to Rs/Ra versus (T_{max}–T_{min}) and usually increases with increasing temperature.

The Hargreaves grass-related radiation framework focuses on solar radiation, which means air temperature (Hargreaves and Samani, 1982, 1985). In addition to solar radiation from extraterrestrial radiation, Hargreaves and Samani (1982, 1985) suggested measuring the difference between the average maximum and average lowest monthly temperatures. In this situation, the Hargreaves approach becomes a temperature-based strategy (Jensen et al., 1990). These radiation approaches have shown to be accurate at forecasting ET when adjusted for local variables. Because it takes little meteorological data and has a high consistency with observed values, Allen et al. (1998) support the Hargreaves and Samani (1985) model as a method to estimate ETo when solar radiation, wind speed, and/or relative humidity measurements are absent (Sabziparvar and Tabari, 2010; Tabari, 2010).

14.12 TEMPERATURE-BASED METHODS

14.12.1 Blaney-Criddle Equation

Because of the low data requirements, the Doorenbos and Pruitt (1977) equation, often known as the Blaney-Criddle equation, is the most widely used. The Blaney-Criddle approach, which was based on a consumptive consumption coefficient, average air temperature, and the percentage of seasonal daylight hours that occurred during the computation period, was the most commonly utilized in semiarid environments. The great performance could be attributed to the fact that it was designed for humid settings with little advective influence (Irmak et al., 2003; Ali and Shui, 2009). The Blaney-Criddle equation, according to George et al. (2002), delivers the best results in humid environments. The equation's inadequacy in dry conditions is also demonstrated.

14.12.2 Penman-Monteith Equation

Penman (1948, 1956) created the first combination equation by combining elements for the energy needed to continue evaporation and a technique for vapor evacuation. The PM equation is a Penman combination equation that considers surface resistance and aerodynamics (Jensen et al., 1990). The standard ET model is based on an idealized clipped grass crop with the following characteristics: 0.12 m height, 70 s/m surface resistance, and 0.23 albedo. At a height of 2 m, the average air temperature is represented by the letter T. This model, which incorporates assumptions about clipped grass as the crop, is a condensed version of the original PM equation.

$$[\![\mathrm{ET}]\!]\_(\mathrm{o}) = \left(0.408\Delta((R\_n - G)) + \gamma 900 / (T\_\mathrm{mean} + 273) U\_2\ (e\_s - e\_a\ )\right) / \left(\Delta + \gamma(\ 1 + 0.34 U\_2)\right) \tag{14.7}$$

14.12.3 Crop Evapotranspiration

A crop's ET rate under ideal circumstances (without limits on water quality, pests, or insufficient soil fertility) is known as ETc (Allen et al.,1998). Accurate prediction of ETc is essential for accurately forecasting agricultural water requirements, as it forms the basis for assessing water availability, crop water balance, and crop water requirements (Pereira et al., 1999). Crop water demand is the quantity of water needed by plants to make up for water lost through evaporation so they can grow healthily, whereas soil water balance is the volume of water added, taken out, or retained in the soil over time (Allen et al., 1998).

It is usual practise to use a dimensionless reduction coefficient, particularly a crop coefficient (Kc), to describe the overall impacts of many types of resistance, such as surface resistance, stomatal resistance, and diffusion resistance (Wang et al., 2018). The crop coefficient, as opposed to a reference crop, reveals the physiological and physical characteristics of the crop being studied (Ding et al., 2015). The single crop coefficient technique takes into account variations in soil water evaporation and crop transpiration rates between the crop and the grass reference surface. The Kc combines the effects of the four main characteristics that set a crop apart from the reference grass; it is essentially the ETc/ETo ratio (Allen et al., 2000). The single-coefficient method's equation is as follows:

$$\mathrm{ETc} = \mathrm{Kc} \times \mathrm{ETo} \tag{14.8}$$

Two crop coefficients are applied in the dual crop coefficient technique. The first is the base crop coefficient (Kcb), which describes plant transpiration, and the second is the soil water evaporation coefficient (Ke), which describes evaporation from the soil surface (Allen et al., 1998). The following is a description of the dual-crop coefficient relationship:

$$\mathrm{ETc} = (\mathrm{Kcb} + \mathrm{Ke}) \times \mathrm{ETo} \tag{14.9}$$

The various methods for determining ET include:

14.13 LYSIMETER EXPERIMENTS

A lysimeter is a container that isolates soil as nearly as possible from its surroundings while nevertheless reproducing nearby soil, crop density, and crop height. It generates a controlled soil-water environment that allows for precise water use and nutrient movement measurements within a set bottom boundary condition. (Liu et al., 2002; Dugas and Bland, 1989). The standards for precise and reliable ETc assessments, according to Allen et al. (2011), are: (1) a lower limit of 50 m green fetch around the lysimeter location; (2) close crop length conditions throughout the field; and (3) fraction of ground cover (fc), leaf area index (LAI), water management, and soil conditions that are similar inside and outside of the lysimeters.

Allen et al. (2011) list the benefits of lysimeters for measuring ET. Compared to micrometeorological approaches, fetch is less in demand. The system can also be virtually entirely investigated, and automation can be fully automated. Errors in managing vegetation or water may undermine accuracy since large chunks of soil constitute a small sample compared to a field area.

14.14 EDDY COVARIANCE

Eddy covariance (EC) sensors are being utilized increasingly frequently in ET monitoring due to their simplicity of use, decreased sensor costs, and capacity to simultaneously measure H, E, and CO_2 fluxes. Eddy correlation is based on the statistical correlation (correlation) of vertical vapor or sensible heat fluxes inside turbulent eddies' upper and lower legs. This necessitates fast measurements of *T*, w, e, or q, frequently at a frequency of 5–20 Hz (5–20 times/second). Tanner (1988) and Tanner et al. (1988) presented early instances of eddy instrumentation (1993).

High-frequency sensors, a basic knowledge of turbulence physics, and a plethora of "corrections" are all required for the EC approach to be accurate. The EC method needs a long fetch to create an equilibrium boundary layer (EBL) deeper than the sensor height (frequently broken) and elevated above the canopy to reduce roughness sublayer distortions. Allen et al. (2011) highlighted the benefits and drawbacks of the EC method as follows:

The capacity to measure ET over potential and non-potential surfaces, continuous, non-destructive direct sampling of the turbulent boundary layer, and the systems' current level of automation are all benefits. Over medium-sized areas (50–200 m), fluxes are evaluated. Disadvantages: the energy balance closure error (Rn G= _ E +H) can be 10%–30%, and a large fetch, generally 50–100 times the instrument's height above the zero plane displacement height, is required. Electronics, turbulence theory, and biology experts are needed.

14.15 REMOTE SENSING ENERGY BALANCE

E and ET were calculated based on the satellite data, particularly in large areas, using energy balance (Bastiaanssen et al., 1998a,b, 2005; Moran, 2000; Kustas et al., 2003; Tasumi et al., 2007; Allen et al., 2007a,b; Irmak et al., 2011). Users should keep in mind that satellite-based ET data are essentially retrievals, or best approximations, of an aerodynamic and radiative process as seen from space (Kamble and Irmak, 2009).

Allen et al. (2011) summarized the advantages and disadvantages of remote sensing-based energy balance (RSEB) models:

Advantages: the technique has a number of benefits, including the ability to sample and integrate data across a large range of places and the fact that it is frequently more affordable than point measurements.

Disadvantages: since images are only taken periodically for a certain location, such as every 16 days for a Landsat satellite, many satellite systems, especially those with high spatial resolution, have time gaps between estimates of ET.

14.15.1 Satellite-Based ET Using Vegetation Indices

Satellite-based or ground-based energy balancing techniques require a sizable time investment and specialized skill sets. On the other hand, the energy balance products can calibrate more straightforward methods that estimate crop coefficients using common vegetative indices (VI), such as Kc or ETrF (Tasumi et al., 2005; Tasumi and Allen, 2007; Singh and Irmak, 2009). Estimating Kc from VI is feasible due to the strong correlation between vegetation quantity and transpiration. Leaf area and transpiration increase as plant cover increases (Glenn et al., 2007). It is challenging to use VI-based approaches to determine soil surface evaporation after precipitation events or lower ET associated with a soil-water scarcity because these activities are not accurately reflected in the VI. The near infrared band (0.7–1.3 m) and the red band (0.6–0.7 m), which are two shortwave bands frequently detected by satellites, are used to create the NDVI (normalized difference vegetation index), a widely used VI. The following are the advantages and disadvantages of VI-based Kc and Kcb estimate, as outlined by Allen et al. (2011):

Advantages: quick evaluations may be done by mid-level specialists; huge areas can be covered; and spatial resolution is excellent, especially when aerial photography is used. Disadvantages:

calculating evaporation (from soil) is less trustworthy than estimating transpiration due to the lack of a direct link with vegetation quantity, and because the reference ET computation requires excellent weather data, quality estimates of the reference ET are required to transform Kc into ET.

14.16 CROP EVAPOTRANSPIRATION (ETC) AND CROP WATER REQUIREMENTS (CWR)

14.16.1 Crop Evapotranspiration (ETc)

"The amount of water lost by ET from disease-free, well-fertilized crops produced in vast fields under ideal soil water conditions and reaching full production under the given climatic conditions" is what ETc stands for. To calculate the crop coefficient and determine the ETc, the ET0 must be calculated (Kc). The ETc can be expressed as follows:

$$ETc = ETo * Kc \tag{14.10}$$

In contrast to a reference surface, which has a uniform appearance and is a full ground cover, the Kc combines the characteristics of a typical field crop that fluctuate over time. These features include canopy resistance, crop height, aerodynamic properties (roughness), albedo (reflectance), and soil evaporation from cultivated fields. Kc levels vary from crop to crop over the growing season and are influenced by the environment and irrigation techniques (Figure 14.1). The base crop coefficient (Kcb), which takes into account crop transpiration, and the evaporation coefficient (Ke),

which refers to soil evaporation, are the two components of the single and dual crop coefficients.

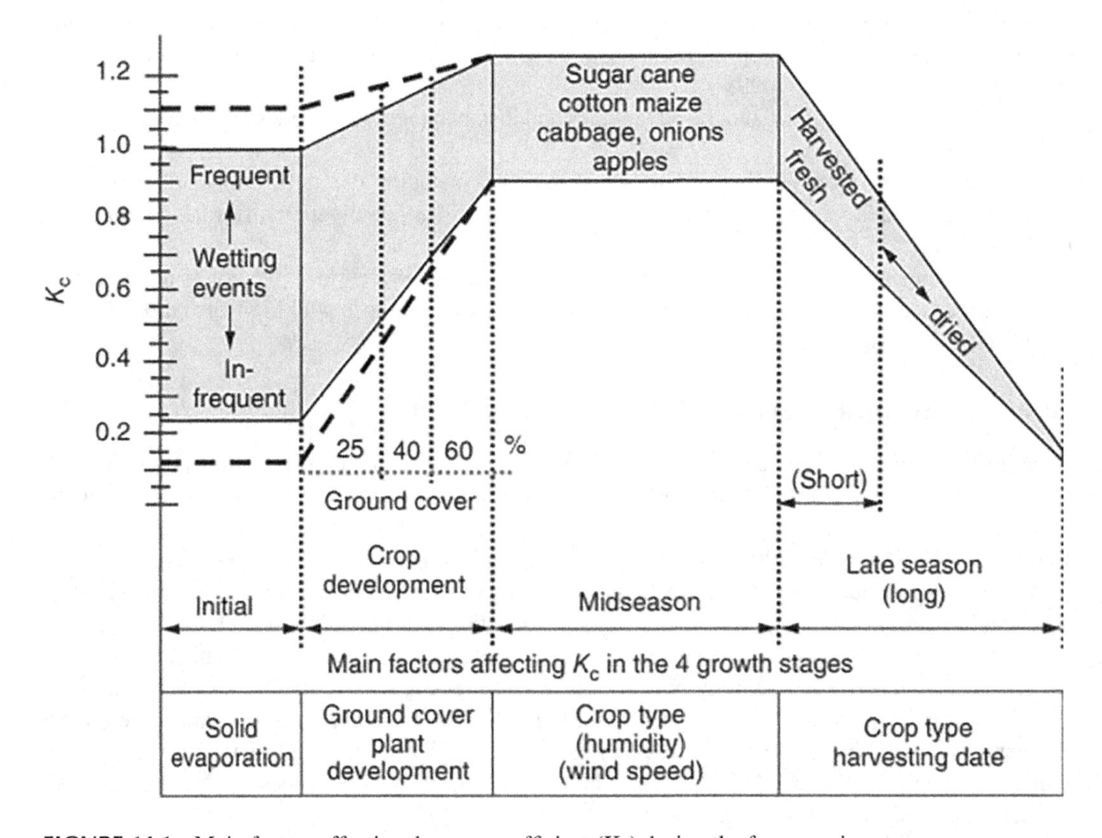

FIGURE 14.1 Main factors affecting the crop coefficient (Kc) during the four growing stages.

Source: Adapted from (Allen et al., 1998).

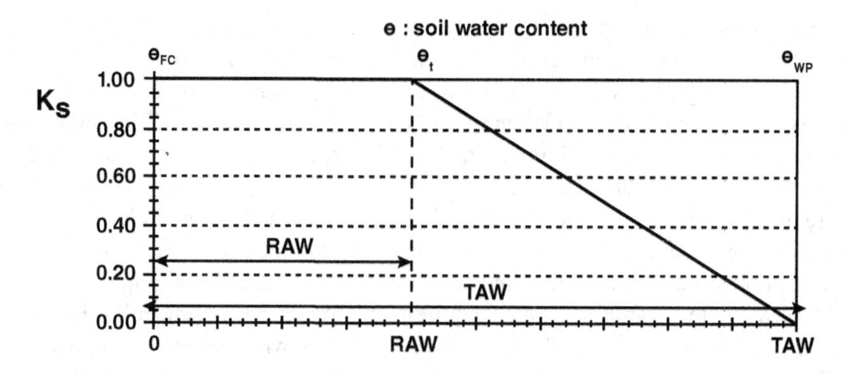

FIGURE 14.2 Water stress coefficient (Ks) adapted from the FAO paper no.56.

A list of Kc values for different crop development stages is supplied, along with the planting months and lengths of crop development stages for different crops, geographies, and climates. However, these statistics must only be used if none of the other possibilities are available (Allen et al., 1998).

The Kc in non-standard conditions refers to the ET from crops cultivated under non-standard management and environmental settings (ETc adj). When growing crops in fields, actual ETc may differ due to pests, illnesses, and less-than-ideal conditions. This might result in slower plant growth, lower plant density, and less evapotranspiration than ETc (Figure 14.2)

$$\text{ETactual} = \text{ETc, adj} = \text{Ks} * \text{Kc} * \text{ET0} \tag{14.11}$$

where Ks is a stress coefficient that ranges from 0 to 1, it happens for the plants and from a water stress point of view, it is a water depletion below the threshold that affects negatively on the yield, which is called readily available water (RAW), as shown in figure Ks is calculated by the following formula:

$$\text{KS} = (\text{TAW} - \text{Dr}) / (\text{TAW} - \text{RAW}) \tag{14.12}$$

where: Dr, root zone soil water depletion; TAW, total available water and RAW, readily available water – a fraction of TAW and threshold for water stress (Figure 14.2).

14.16.2 Crop Water Requirements (CWR)

CWR refers to the volume of water needed during a specific period to compensate for ET losses from a cropped field. It may be said that CWR is utilized to manage drives in determining irrigation water requirements, on-farm irrigation scheduling, and water distribution schedules.

$$\text{CWR} = \text{ETc} = \text{NIR} + \text{Peff} \tag{14.13}$$

where NIR is the net irrigation requirement and Peff is effective rainfall. The NIR is estimated as a difference between ETc and EP:

$$\text{NIR} = \text{ETc} - \text{Peff} \tag{14.14}$$

ET was estimated using a variety of empirical techniques, including the PM (Allen et al., 1998), Priestley-Taylor, Hargreaves, and Thornthwaite methods from the FAO, as well as additional techniques. However, reliable ET0 measurement at the regional scale is challenging due to the small number of meteorological stations. Additionally, in contemporary weather station networks, these requirements are not always met, affecting the model's uncertainty and ETo estimates (Cruz-Blanco et al., 2014). New strategies based on weather forecasting systems and remote sensing techniques have been developed to get around this restriction (Ramrez-Cuesta et al., 2017). Additionally, the coverage of weather station data is limited, and it occasionally delivers inaccurate information for fields nearby the station. Last but not least, few of these weather stations are frequently visited, and vast areas are dispersed far from well-maintained weather stations (Voogt, 2006; Collins, 2011). As a result, land surface conditions and the status of water resources can be monitored using satellite remote sensing photos at different spatial and temporal resolutions. Images from remote sensing can also be used to calculate agricultural water requirements for irrigation schedules. Using data from Landsat-7 ETM+, Landsat-5 TM, and TERRA/AQUA MODIS, which have spatial resolutions of 60, 120, and 1000m, respectively, daily ET was computed based on prior studies (Cristóbal et al., 2011). Additionally, ET can be modeled at large scales using GIS climate-based approaches like the global crop water model (GCWM) (Siebert and Döll, 2010), Lund-Potsdam-Jena managed Land (LPJmL), and GIS-based Environmental Policy Integrated Climate (GEPIC) (Liu et al., 2007, Liu and AJB). But it has been found that radiometric data from remote sensing and GIS is a good method for predicting ET at both local and global sizes (Cristóbal et al., 2005; Mu et al., 2007). On the other hand, Mishra et al. (2005) constructed paddy crop coefficients using remote sensing data; nevertheless, the impounded water backdrop in paddy fields may have affected crop reflectance and significantly influenced the results. On the other hand, a wide range of crop water requirements were forecasted using remote sensing vegetation indices (Ray and Dadhwal, 2001; Gontia and Tiwari, 2004). For instance, (Jayanthi et al., 2007) computed the potato's ET using crop coefficients based on basal and canopy reflectance. Additionally, the suggested approaches might have contributed to the development of a useful methodology for choosing the right number of weather stations to include in a network of weather stations as well as the appropriateness of their placements (Ramrez-Cuesta et al., 2017). On the other hand, industrialized nations without access to weather information must use remote sensing and GIS. Therefore, by determining the best interpolation techniques.

14.17 ARTIFICIAL INTELLIGENCE METHODS FOR ESTIMATING ET0

Despite the stringent requirements of the empirical equations for ETo, they provide accurate values in addition to being valid in various climatic conditions, which are the primary indicators in the empirical equations (Ferreira and da Cunha, 2020; Pereira et al., 2015). One of the most important criteria for irrigation water scheduling is quick and accurate methods that require the fewest meteorological indications, given the limited number of climatic factors available for computing ETo (Fan et al., 2021). The standard equation for estimating ETo is the PM equation, authorized by the United Nations Agriculture and Food Organization and accessible while greatly preserving the accuracy of the estimated ETo values in the same range (Fan et al., 2019). The accepted formula for calculating ETo is the PM equation, developed by the United Nations Agriculture and Food Organization. The PM equation must be calculated using a variety of daily climate variables, including maximum and minimum temperatures, wind speed, net solar radiation, and relative humidity. However, approaches that require less climatic data must be available due to the limited climatic characteristics present in the study area while still greatly maintaining the accuracy of the computed ETo values in the same range (Fan et al., 2019). The PM equation is challenging to implement, especially in developing countries that face several challenges including a lack of monitoring stations and their uneven distribution across the nation, where the focus is on the spread of irrigated agricultural areas, as well as the inability to accurately measure all the climatic criteria necessary to apply the PM equation (Fan et al., 2019). Support vector machines (SVMs), a recent computer software model, has shown impressive accuracy in calculating and projecting ETo (Fan et al., 2019). The ability to establish relationships between the inputs and outputs used in ETo estimation, which are primarily meteorological data, has led to an increase in the use of machine learning programs in recent years, leading to increased accuracy and power when using machine learning programs in ETo modeling (Ferreira and da Cunha, 2020; Kumar et al., 2011). Under local conditions and across stations, both random forest (RF) and generalized regression neural network (GRNN) were fairly accurate at forecasting ETo, but RF performed better than GRNN (Feng et al., 2017). Former ETo values are used as input data in machine learning models, which is a frequent technique for predicting ETo values (Ashrafzadeh et al., 2020; Landeras et al., 2009). Additionally, the performance of machine learning models and the precision of the forecasting ETo are considerably improved by using lagged ETo values in addition to the historical daily meteorological data used to estimate ETo via the PM equation. With the help of so-called deep learning models, machine learning programmes also offer the capability to forecast ETo values for upcoming agricultural seasons. These predictions are limited to knowing the anticipated changes in temperature and precipitation due to global warming, from which good irrigation water planning and management for the following years can be made (Ferreira and da Cunha, 2020). When the prediction time is prolonged from 1 to 30 days, the provided data's accuracy decreases (Ferreira and da Cunha, 2020).

14.18 CONCLUSIONS

Based on the physics of evaporation and ET, irrigation water needs are a complete method for determining the amount of water required for agricultural irrigation. Irrigation is an important part of water resource management and the development of agricultural policies. Climate change and increased population density are putting a strain on water security in desert areas. Traditional methods for predicting agricultural water requirements require significant computational power, money, and time. Using remote sensing, GIS, and other techniques such as physical models and artificial methodologies, this book section aims to improve all elements of water need estimation and ETo estimation. This chapter focuses on ETo and IWR applications with limited meteorological datasets, particularly in arid and semi-arid locations. ETo and IWR applications are also examined, demonstrating the value of ETo in agricultural water management and future planning. Brief introductions are also interpreted, as is their flexibility. Furthermore, we demonstrated how the methods used to estimate ETo and IWR would assist decision-makers and developers in obtaining higher performance with less computational cost while also ensuring agricultural sustainability.

REFERENCES

Ali, M.H., Shui, L.T. 2009. Potential evapotranspiration model for Muda irrigation project. Malaysia. *Water Resources Management*, 23, 57–69.

Allen, R., Pereira, L.S., Raes, D. 1998. Crop evapotranspiration-Guidelines for computing crop evapotranspiration-FAO Irrigation and Drainage Paper 56. FAO - Food and Agriculture Organization of the United Nations Rome, pp. 1–15.

Allen, R.G., Clemmens, A.J., Burt, C.M., Solomon, K., O'Halloran, T. 2005. Prediction accuracy for project wide evapotranspiration using crop coefficients and reference evapotranspiration. *Journal* of *Irrigation* and *Drainage Engineering*, 131, 24–36.

Allen, R.G., Pruitt, W.O., Wright, J.L., Howell, T.A., Ventura, F., Snyder, R., Itenfisu, D., Steduto, P., Berengena, J., Yrisarry, J.B. 2006. A recommendation on standardized surface resistance for hourly calculation of reference ETo by the FAO56 Penman Monteith method. *Agricultural Water Management*, 81, 1–22.

Allen, R.G., Tasumi, M., Trezza, R. 2007a. Satellite-based energy balance for mapping evapotranspiration with internalized calibration (METRIC)—Model. *Journal* of *Irrigation* and *Drainage Engineering*, 133(4), 380–394.

Allen, R.G., Tasumi, M., Morse, A., Trezza, R., Wright, J.L., Bastiaanssen, W., Kramber, W., Lorite, I., Robison, C. W. 2007b. Satellite-based energy balance for mapping evapotranspiration with internalized calibration (METRIC)—Applications. *Journal* of *Irrigation* and *Drainage Engineering*, 133(4), 395–406.

Allen, R.G., Pereira, L.S., Howell, T.A., Jensen, M.E. 2011. Evapotranspiration information reporting: I. Factors governing measurement accuracy. *Agricultural Water Management*, 98(2011), 899–920. https://doi.org/10.1016/j.agwat.2010.12.015.

Anon, J. 1956. Proceeding of the informal meeting on physics in agriculture. *Netherlands Journal of Agricultural science*, 4, 162.

Ashrafzadeh, A., Kişi, O., Aghelpour, P., Biazar, S.M., Masouleh, M.A. 2020. Comparative study of time series models, support vector machines, and GMDH in forecasting long-term evapotranspiration rates in northern Iran. *Journal of Irrigation and Drainage Engineering*, 146, 04020010.

Bastiaanssen, W.G.M., Menenti, M., Feddes, R.A., Holtslag, A.A.M. 1998a. The surface energy balance algorithm for land (SEBAL): part 1 formulation. *Journal of Hydrology*, 212–213, 198–212.

Bastiaanssen, W.G.M., Pelgrum, H., Wang, J., Ma, Y., Moreno, J., Roerink, G.J., van der Wal, T. 1998b. The surface energy balance algorithm for land (SEBAL): part 2 validation. *Journal of Hydrology*, 212–213, 213–229.

Bastiaanssen, W.G.M., Noordman, E.J.M., Pelgrum, H., Davids, G., Thoreson, B.P., Allen, R.G. 2005. SEBAL model with remotely sensed data to improve water resources management under actual field conditions. *Journal* of *Irrigation* and *Drainage Engineering*, 131(1), 85–93.

Beard, J. 1985. An assessment of water use by turfgrasses, Turfgrass water conservation. *University of California Division of Agriculture and Natural Resources, UCANR Publications,* 21, 45–60.

Blaney, H.F., Criddle, W.D. 1962. Determining consumptive use and irrigation water requirements. In: *USDA Technical Bulletin 1275*, US Department of Agriculture, Beltsvill, MD.

Bokke, A.S. Shoro, K.E. 2020. Impact of effective rainfall on net irrigation water requirement: the case of Ethiopia. *Water Science 2020*, 34(1), 155–163. https://doi.org/10.1080/11104929.2020.1749780

Brouwer, C., Heibloem, Y. 1986. *Irrigation Water Needs [Irrigation Water Management Training Manual no. 3]*. Food and Agriculture Organization of the United Nations, Rome, Italy.

Cammarano, D., Grace, P., Payero, J.O., Basso, B., Wilkens, P. 2012. Agronomic and economic evaluation of irrigation strategies on cotton lint yield in Australia. *Crop & Pasture Science*, 63, 647–655.

Chang, J., Guo, A., Wang, Y., Ha, Y., Zhang, R., Xue, L.U., Tu, Z. 2019. Reservoir operations to mitigate drought effects with a hedging policy triggered by the drought prevention limiting water level. *Water Resources Research*, 55(2), 904–922.

Chen, S., Huang, Y., Wang, G. 2021. Detecting drought-induced GPP spatiotemporal variabilities with sun-induced chlorophyll fluorescence during the 2009/2010 droughts in China. *Ecological Indicators*, 121, 107092. https://doi.org/10.1016/j. ecolind.2020.107092.

Cherif, I., Jarlan, L., Fieuzal, R., Rodriguez, J.C. 2015. Improving remotely sensed actual evapotranspiration estimation with raster meteorological data. *International Journal* of *Remote* Sensing, 1161, 4606e4620. https://doi.org/10.1080/01431161.2015.1084439.

Collins, J.M. 2011. Temperature variability over Africa. *Journal of climate*, 24, 3649–3666.

Cristóbal, J., Pons, X., Ninyerola, M. 2005. Modelling actual evapotranspiration in Catalonia (Spain) by means of remote sensing and geographical information systems. *Göttinger Geographische Abhandlungen*, 113, 144–150.

Cristóbal, J., Poyatos, R., Ninyerola, M., Llorens, P., Pons, X. 2011. Combining remote sensing and GIS climate modelling to estimate daily forest evapotranspiration in a Mediterranean mountain area. *Hydrology and Earth System Sciences*, 15, 1563–1575.

Cruz-Blanco, M., Gavilán, P., Santos, C., Lorite, I.J. 2014. Assessment of reference evapotranspiration using remote sensing and forecasting tools under semi-arid conditions. *International Journal of Applied Earth Observation and Geoinformation*, 33, 280–289.

Ding, R., Tong, L., Li, F., Zhang, Y., Hao, X., Kang, S. 2015. Variations of crop coefficient and its influencing factors in an arid advective cropland of northwest China. *Hydrological Processes*, 29, 239–249.

Dinpashoh, Y., Jhajharia, D., Fakheri-Fard, A., Singh, V.P., Kahya, E. 2011. Trends in reference crop evapotranspiration over Iran. *Journal of Hydrology*, 399, 422–433.

Döll, P., Siebert, S. 2002. Globalmodeling of irrigation water requirements. *Water Resources Research*, 38, 1037. https://doi.org/10.1029/2001WR000355.

Doorenbos, J., Pruitt, W. 1977. Guidelines for predicting crop water requirements, Irrig. Drain. Paper No. 24. FAO, Rome, Italy.

Dugas, W., Bland, W. 1989. The accuracy of evaporation measurements from small lysimeters. *Agricultural and Forest Meteorology*, 46, 119–129.

El-Keblawy, A. 2016. Impact of fencing and irrigation on species composition and diversity of desert plant communities in the United Arab Emirates. *Land Degradation & Development*, 28, 1354–1362.

Fan, J., Ma, X., Wu, L., Zhang, F., Yu, X., Zeng, W. 2019. Light gradient boosting machine: an efficient soft computing model for estimating daily reference evapotranspiration with local and external meteorological data. *Agricultural Water Management*, 225, 105758.

Fan, J., Zheng, J., Wu, L., Zhang, F. 2021. Estimation of daily maize transpiration using support vector machines, extreme gradient boosting, artificial and deep neural networks models. *Agricultural Water Management*, 245, 106547.

FAO, 1975. Production végétale et protection des plantes. Surveillance agrométéorologique pour la prévision des récoltes, N°117.

Fares, A., Awal, R., Fares, S., Johnson, A., Valenzuela, H. 2016. Irrigation water requirements for seed corn and coffee under potential climate change scenarios. *Journal* of *Water* and *Climate Change*, 7, 39–51.

Feng, Y., Cui, N., Gong, D., Zhang, Q., Zhao, L. 2017. Evaluation of random forests and generalized regression neural networks for daily reference evapotranspiration modelling. *Agricultural Water Management*, 193, 163–173.

Ferreira, L.B., Da Cunha, F.F. 2020. Multi-step ahead forecasting of daily reference evapotranspiration using deep learning. *Computers and Electronics in Agriculture*, 178, 105728.

Field, C.B., Barros, V.R., Dokken, D.J., Mach, K.J., Mastrandrea, M.D., Bilir, T.E.,Chatterjee, M., Ebi, K.L., Estrada, Y.O., Genova, R.C., Girma, B., Kissel, E.S., Levy,A.N., MacCracken, S., Mastrandrea, P.R., White, L.L. (Eds.), 2014. Climate Change 2014: Impacts, Adaptation, and Vulnerability. Contribution of Working GroupII to the Fifth Assessment Report of the Intergovernmental Panel on Climate Change. Cambridge University Press, Cambridge.

Gardner, F.P., Pearce, R.B., Mitchell, R.L. 1985. Physiology of crop plants. Iowa State University Press, Ames, USA.

George, B.A., Reddy, B., Raghuwanshi, N., Wallender, W.W. 2002. Decision support system for estimating reference evapotranspiration. *Journal* of *Irrigation* and *Drainage Engineering*, 128, 1–10.

Glenn, E.P., Huete, A., Nagler, P., Hirschoock, K., Brown, P. 2007. Integrating remote sensing and ground methods to estimate evapotranspiration. *Critical* Reviews in *Plant Sciences*, 26, 139–168.

Gontia, N, Tiwari, K. Crop evapotranspiration estimation using temporally distributed normalized difference vegetation index. International conference on emerging technologies in agricultural and food engineering (ETAE-2004) held at IIT Kharagpur, India, 2004. pp. 14–17.

Gorguner, M., Kavvas, M.L. 2020. Modeling impacts of future climate change on reservoir storages and irrigation water demands in a mediterranean basin. *Science of the Total Environment*, 748, 141246.

Hargreaves, G.H., Samani, Z.A. 1982. Estimating potential evapotranspiration. *Journal* of *Irrigation* and *Drai nage Engineering Division*, 108, 225–230.

Hargreaves, G.H., Samani, Z.A. 1985. Reference crop evapotranspiration from temperature. *Applied Engineering in Agriculture*, 1, 96–99.

Hsiao, T.C., Steduto, P., Fereres, E. 2007. A systematic and quantitative approach to improve water use efficiency in agriculture. *Irrigation Science*, 25, 209–231.

Irmak, S., Allen, R., Whitty, E. 2003. Daily grass and alfalfa-reference evapotranspiration estimates and alfalfa-to-grass evapotranspiration ratios in Florida. *Journal* of *Irrigation* and *Drainage Engineering*, 129, 360–370.

IWMI, International Water Management Institute, 2000. World Water and Climate Atlas. http://www.iwmi.org

Jayanthi, H., Neale, C.M., Wright, J.L. 2007. Development and validation of canopy reflectance-based crop coefficient for potato. *Agricultural Water Management*, 88, 235–246.

Jensen, M.E., Haise, H.R. 1963. Estimating evapotranspiration from solar radiation. Proceedings of the American Society of Civil Engineers. *Journal* of *Irrigation* and *Drainage Engineering Division*, 89, 15–41.

Jensen, M.E., Allen, R.G. 2016. *Evaporation, Evapotranspiration, and Irrigation Water Requirements. American Society* of *Civil Engineers,* Reston, VA.

Kamble, B., Irmak, A. 2009. Combining remote sensing measurements and model estimates through hybrid data assimilation scheme to predict hydrological fluxes. In: Proceedings of the IEEE International Geosciences and Remote Sensing Symposium, July 6–11, 2008. Boston, MA, USA.

Katerji, N., Rana, G. 2011. Crop reference evapotranspiration: a discussion of the concept, analysis of the process and validation. *Water Resources Management* 25(6), 1581–1600.

Kumar, M., Raghuwanshi, N., Singh, R. 2011. Artificial neural networks approach in evapotranspiration modeling: a review. *Irrigation Science*, 29, 11–25.

Kuo, S.F., Lin, B.J., Shieh, H.J. (2001). CROPWAT model to evaluate cropwater requirement in Taiwan. International Commission on Irrigation and Drainage 1st Asian Regional Conference, Seoul, S.Korea.

Kustas, W.P., Norman, J.M., Anderson, M.C., French, A.N. 2003. Estimating subpixel surface temperatures and energy fluxes from the vegetation index-radiometric temperature relationship. *Remote Sensing of Environment,* 85, 429–440.

Landeras, G., Ortiz-Barredo, A., López, J.J. 2009. Forecasting weekly evapotranspiration with ARIMA and artificial neural network models. *Journal of Irrigation and Drainage Engineering*, 135, 323–334.
Li, J., Zhang, S., Huang, L., Zhang, T., Feng, P. 2020. Drought prediction models driven by meteorological and remote sensing data in Guanzhong Area, China. *Hydrology* Research, 51, 942–958.
Liu, C., Zhang, X., Zhang, Y. 2002. Determination of daily evaporation and evapotranspiration of winter wheat and maize by large-scale weighing lysimeter andmicro-lysimeter. *Agricultural and Forest Meteorology*, 111, 109–120.
Liu, J., Ajb, Z.H.Y. 2009. Global consumptive water use for crop production: the importance of green water and virtual water. *Water Resources Research*, 45, 5.
Liu, J., Williams, J.R., Zehnder, A.J., Yang, H. 2007. GEPIC–modelling wheat yield and crop water productivity with high resolution on a global scale. *Agricultural Systems*, 94, 478–493.
Makkink, G.F. 1957. Testing the Penman formula by means of lysimeters. *Journal* of the *Institution* of *Water Engineers*, 11, 277–288.
Manalo, E.B. 1976. *Agro-climatic Survey of Bangladesh.* International Rice Research Institute, Manila, Philippine.
Mardikis, M., Kalivas, D., Kollias, V. 2005. Comparison of interpolation methods for the prediction of reference evapotranspiration—an application in Greece. *Water Resources Management*, 19, 251–278.
Mohammadian, M., Arfania, R., Sahour, H. 2017. Evaluation of SEBS algorithm for estimation of daily evapotranspiration using landsat-8 dataset in a semi-arid region of Central Iran. *Geology*,7, 335e347. https://doi.org/10.4236/ojg.2017.73023.
Mokhtar, A., He, H., Alsafadi, K., LI, Y., Zhao, H., Keo, S., Bai, C., Abuarab, M., Zhang, C., Elbagoury, K., Wang, J., He, Q. 2020a. Evapotranspiration as a response to climate variability and ecosystem changes in southwest, China. *Environmental Earth Sciences*, 79, 312.
Mokhtar, A., He, H., Zhao, H., Keo, S., Bai, C., Zhang, C., Ma, Y., Ibrahim, A., Li, Y., Li, F. 2020b. Risks to water resources and development of a management strategy in the river basins of the Hengduan Mountains, Southwest China. *Environmental Science: Water Research & Technology*, 6, 12.
Monteith, J.L. 1965. Evaporation and environment. *Symposia of the Society* for *Experimental Biology*, 19, 205–234.
Moran, S. 2000. Use of remote sensing for monitoring evaporation over managed watersheds. ASCE CD ROM, pp. 13.
Mu, Q., Heinsch, F.A., Zhao, M., Running, S.W. 2007. Development of a global evapotranspiration algorithm based on MODIS and global meteorology data. *Remote sensing of Environment*, 111, 519–536.
Nouri, H., Beecham, S., Anderson, S., Hassanli, A.M. 2014. Remote sensing techniques for predicting evapotranspiration from mixed vegetated surfaces. *Urban Water Journal*, 37e41. https://doi.org/10.5194/hessd-10-3897-2013 (February 2015).
Patwardhan, A.S., Nieber, J.L., Johns, E.L. 1990. Effective rainfall estimation methods. *Journal of Irrigation and Drainage Engineering*, 116(2), 182–193. https://doi.org/10.1061/(ASCE)0733–9437(1990)116:2(182).
Peng, L., Li, Y., Feng, H. 2017. The best alternative for estimating reference crop evapotranspiration in different sub-regions of mainland China. *Scientific Reports*, 7, 54–58.
Penman, H.L. 1948. Natural evaporation from open water, hare soil and grass. *Proceedings of the Royal Society of London*, 193, 120–145.
Penman, H.L. 1956. Evaporation: an introductory survey. *Netherlands Journal of Agricultural Science*, 4, 9–29.
Pereira, L.S., Perrier, A., Allen, R.G., Alves, I. 1999. Evapotranspiration: concepts and future trends. *Journal* o f *Irrigation* and *Drainage Engineering*, 125, 45–51.
Pereira, L.S., Paredes, P., Rodrigues, G.C., Neves, M. 2015. Modeling malt barley water use and evapotranspiration partitioning in two contrasting rainfall years. Assessing AquaCrop and SIMDualKc models. *Agricultural Water Management*, 159, 239–254.
Peter, D., Allen, R.G. 2002. Estimating reference evapotranspiration under inaccurate data conditions. *Irrigation and Drainage Systems*, 16: 33–45, 2002.
Power, B., Rodriguez, D., DeVoil, P., Harris, G., Payero, J.O. 2011. A multi-field bioeconomic model of irrigated grain-cotton farming systems. *Field Crops Research*, 124(2), 171–179.
Priestley, C., Taylor, R. 1972. On the assessment of surface heat flux and evaporation using large-scale parameters. *Monthly Weather Review*, 100, 81–92.
Rahman, M.M., Islam, M.O., Hasanuzzaman, M. 2008. Study of effective rainfall for irrigated agriculture in South-Eastern Part of Bangladesh. *World Journal of Agricultural Sciences*, 4(4), 453–457.
Ramírez-Cuesta, J.M., Cruz-Blanco, M., Santos, C., Lorite, I.J. 2017. Assessing reference evapotranspiration at regional scale based on remote sensing, weather forecast and GIS tools. *International Journal of Applied Earth Observation and Geoinformation*, 55, 32–42.

Ray, S., Dadhwal, V. 2001. Estimation of crop evapotranspiration of irrigation command area using remote sensing and GIS. *Agricultural Water Management*, 49, 239–249.

Rodrigues, G.C., Braga, R.P. 2021. Estimation of daily reference evapotranspiration from NASA POWER reanalysis products in a hot summer mediterranean climate. *Agronomy*, 11, 2077. https://doi.org/10.3390/agronomy11102077.

Sabziparvar, A.A., Tabari, H. 2010. Regional estimation of reference evapotranspiration in arid and semi-arid regions. *Journal* of *Irrigation* and *Drainage Engineering ASCE*, 136(10), 724–731. http://doi.org/10.1061/(ASCE)IR.1943-4774.0000242.

Sayama, T., Mcdonnell, J.J. 2009. A new time-space accounting scheme to predict stream water residence time and hydrograph source components at the watershed scale. *Water Resources Research*, 45, 9.

Shen, Y.J., Li, S., Chen, Y.N., Qi, Y.Q., Zhang, S.W. 2013. Estimation of regional irrigation water requirement and water supply risk in the arid region of Northwestern China 1989–2010. *Agricultural Water Management*, 128, 55–64.

Shiri, J., Kisi, O., Landeras, G. 2012. Daily reference evapotranspiration modeling by using genetic programming approach in the Basque Country (Northern Spain). *Journal of Hydrology*, 414(2012), 302–316.

Siebert, S., Döll, P. 2010. Quantifying blue and green virtual water contents in global crop production as well as potential production losses without irrigation. *Journal of Hydrology*, 384, 198–217.

Singh, R., Irmak, A. 2009. Estimation of crop coefficients using satellite remote sensing. *Journal* of *Irrigation* and *Drainage Engineering ASCE*, 135(5), 597–608.

Smith, M. 1991. *Manual and Guidelines*. FAO of UN, Rome, Italy.

Snyder, R.L., Lanini, B.J., Shaw, D.A., Pruitt, W.O. 1987. Using reference evapotranspiration (ETo) and crop coefficients to estimate crop evapotranspiration (ETc) for trees and vines. *Leaflet University California Natural Resources*, 1987, 12–27.

Snyder, R.L., Orang, M., Matyac, S., Grismer, M.E. 2005. Simplified estimation of reference evapotranspiration from pan evaporation data in California. *Journal* of *Irrigation* and *Drainage Engineering*, 131, 249–253.

Stanhill, G. 1976. The CIMO international evaporimeter comparisons (1st ed.), Secretariat of the WMO.

Tabari, H. 2010. Evaluation of reference crop evapotranspiration equations in various climates. *Water Resources Management*, 24, 2311–2337. http://doi.org/10.1007/s11269-009-9553-8.

Tanner, B.D. 1988. Use requirements for Bowen ratio and eddy correlation determination of evapotranspiration. In: Hay, D.R. (Ed.), Planning Now for Irrigation and Drainage in the 21st Century, Proc., Irrig. and Drain. Div, Conf. ASCE, Lincoln, NE, USA, pp. 605–616.

Tanner, B.D., Swiatek, E., Green, J.P. 1993. Density fluctuations and use of the krypton hygrometer in surface flux measurements. In: Allen, R.G., Neale, C.M.U. (Eds.), Management of Irrigation and Drainage Systems: Integrated Perspectives. Proceedings of National Conference on Irrigation and Drainage Engineering. ASCE, Park City, UT, July, pp. 945–952.

Tasumi, M., Allen, R.G. 2007. Satellite-based ET mapping to assess variation in ET with timing of crop development. *Agricultural Water Managemente*, 88 (1–3), 54–62.

Tasumi, M., Allen, R.G., Trezza, R., Wright, J.L. 2005. Satellite-based energy balance to assess within-population variance of crop coefficient curves. *Journal* of *Irrigation* and *Drainage Engineering*, 131(1), 94–109.

Thornthwaite, C.W. 1948. An approach toward a rational Classification of climate. *Geography* Revision , 38, 55–94.

Voogt, M. 2006. Meteolook, a physically based regional distribution model for measured meteorological variables. M. Sc. Thesis, University of Technology, Delft, The Netherlands.

Walter, I.A., Allen, R.G., Elliott, R., Jensen, M., Itenfisu, D., Mecham, B., Howell, T., Snyder, R., Brown, P., Echings, S. 2000. ASCE's standardized reference evapotranspiration equation. *Watershed Management and Operations Management*, 2000, 1–11.

Wang, J., Zhang, Y., Gong, S., Xu, D., Juan, S., Zhao, Y. 2018. Evapotranspiration, crop coefficient and yield for drip-irrigated winter wheat with straw mulching in North China Plain. *Field Crops Reseatch*, 217, 218–228.

Wang, W.G., Yu, Z.B., Zhang,W., Shao, Q.X., Zhang, Y.W., Luo, Y.F., Jiao, X.Y., Xu, J.Z. 2014. Responses of rice yield, irrigation water requirement and water use efficiency to climate change in China: historical simulation and future projections. *Agricultural Water Management*, 146, 249–261.

Yassin, M., Alazba, A.A., Mattar, M. 2012. Artificial neural networks versus gene expression programming for estimating reference evapotranspiration in arid climate. *Agricultural Water Management*, 163:110–124.

Zhang, B., AghaKouchak, A., Yang, Y., Wei, J., Wang, G. 2019a. A water-energy balance approach for multi-category drought assessment across globally diverse hydrological basins. *Agricultural and Forest Meteorology*, 264, 247–265.

15 Smart Irrigation Water Using Sensors and Internet of Things (IOT)

Ahmed Elbeltagi, Papri Mukherjee, Rituparna Saha, Amit Biswas, and Chaitanya B. Pande

15.1 INTRODUCTION

Water scarcity is a global issue in the present era. We all know that this water is a must requirement for life, maybe plants, animals, or humans. Agriculture, which is the primary source of food for animals and plants, consumes lot of water for its growth (Biswas et al., 2021). But, due to industrial revolution, urban and village improvements and developments over the years we have contaminated a large amount of fresh water. The sewage and industrial effluents have been disposed of directly into rivers, lakes, and other water bodies. These have decreased the amount of fresh water required to meet the future demand, i.e., the sustainability failed. Moreover, increased population density resulted in use of more fresh water, over-exploitation of ground water due to carelessness, and lack of awareness (Biswas et al., 2021). Besides, there is increased frequency and intensity of different extreme climatic events, which affected the water reservoirs and groundwater levels (Khadke and Pattnaik, 2021; Srivastava et al., 2021, 2022). All these have ultimately depleted usable water quality and quantity. So, water shortage has become a major issue in the present world. When the problem before was limited to lack of drinking water in naturally arid areas, it now has come up to drying up of humid regions and urban areas. The issue has taken its root even more in large cities. Rising water demands rooted in urbanization coincides with sealed surfaces and dwindling groundwater levels, resulting in water shortage during the dry period which in turn reduces plants and trees productivity, wilting, and obviously brings misery to the local dwellers. In fact, there existed a rich culture of water harvesting and management practices since time immemorial (e.g., the tank cascade system in South Asia) that gradually witnessed degradation, majorly attributed to reasons discussed before (Srivastava and Chinnasamy, 2022, 2023a, 2023b). Hence, critical discussion at the interface of water science, applications, and society is imperative in the 21st century.

Water demand in the agricultural field is very high. In case of irregular or inadequate rainfall, the crop water requirement is not properly met. At this point, irrigation comes into the picture. It is a process by which we provide water to the cultivation land in surplus to rainfall. Over the years, different types of irrigation system have been developed and are adopted (Biswas et al., 2021). It is generally done using canal systems, where the water is pumped into the fields after a defined and regular interval of time. In this conventional method, no feedback from the field on water level is involved. But it has been found that the efficiency has been less and the water demand by plants is varying with different factors like soil type, cropping pattern, temperature, humidity, etc. Also, the conventional irrigation practices resulted in either insufficiently or overly irrigated lands due to variable water-holding capacity of the land, water infiltration, and run-off water. In case of excess water in the lands, the water replaces the air-containing pores of soil, so the plant roots lack air. On the other hand, an area with under-irrigation suffers water stress (Jury and Vaux, 2007). The crops

DOI: 10.1201/9781003303237-15

that are highly sensitive to water content are affected due to these and result in poor yields. So, the water management is very essential in agriculture.

At this point, smart irrigation comes into view (Gutiérrez et al., 2014). It can save about 80% of the water. It is a time-saving, easy-to-use, and trouble-saving means for providing water to crops and plants. It is capable of saving a huge amount of water as it automatically waters plants based on their needs in the correct proportion (Darshana et al., 2015). It is efficient in both small and large scale. It reduces down the risk of the work as well as makes the task easier. Now, this smart irrigation is dependent totally on Internet of Things (IoT). IoT plays a huge role in the evolution of technology. It involves the network of physical objects that are embedded with sensors, software, and other technologies for the purpose of connecting and exchanging data with other devices and systems over the internet. Ideas evolved a lot starting from Industry 1.0. In the 21st century, currently, we are standing in Industry 4.0 which is also known as Industrial Internet of Things (IoT). It features an IP address for internet connectivity between the network of objects, and the communication that occurs between them and over Internet-enabled devices and systems (Pathan and Hate, 2016). IoT allows us to gather the non-available data, determine the water demand, and act accordingly, thus ensuring sustainable water use. Smart irrigation systems have become an internal of smart city, and it is globally accepted. It has already been proved beneficial to the citizens. It is commonly used in precision farming and crop irrigation (Kagalkar, 2017).

The focus of the smart irrigation system is mainly on energy and water conservation and automatic functioning according to the requirement or the user demand (Ashwini, 2018). In contrast to the traditional method, it involves feedback mechanism. It is done with the help of sensors like moisture, temperature, and humidity sensors (Krishna and Priyanka, 2014). An irrigation control based on soil moisture uses tensiometric and volumetric methods. These methods are simple to use but are dependent on soil water characteristics, which differ with the soil type (Nemali and Van Iersel, 2006). Maintenance should be given proper importance in this irrigation system because any fault in the sensors or the setup will mislead the readings and can result in watering at incorrect time and amount (Gutiérrez et al., 2014). This will ultimately affect the crop yield and also cause usable water loss. The system can be made crop-specific. Proper scheduling of irrigation will be a critical matter to handle for efficient crop water requirement management (Jaguey et al., 2015). It needs utmost care when there is water scarcity. The frequency, duration, and amount are the three important parameters in this system. Evapo-transpiration, thermal imaging, capacitive methods, neutron scattering method, and gypsum blocks are some relevant technologies used for moisture sensing. Among this, the capacitive method is instantaneous, is costly, and has to be calibrated often with changing temperature and soil type. Neutron scattering method serves the purpose best with the most accurate results but has disadvantages like radiation hazards, calibration difficulty, and costlier.

15.2 BENEFITS OF SMART IRRIGATION SYSTEM AND DATASETS

Water conservation is one of the biggest benefits of a smart irrigation system. Generally speaking, traditional watering techniques can waste up to 50% of the water they utilize owing to irrigation, evaporation, and over-watering inefficiencies. Smart irrigation systems employ sensors to collect data in real-time or overtime to change watering schedules and inform watering practices.

15.3 SMART IRRIGATION SYSTEM COMPONENTS

15.4 ARDUINO UNO

The Arduino Uno is an open-source microcontroller board created by Arduino.cc that is based on the Microchip ATmega328P microprocessor. A variety of expansion boards (shields) and other circuits can be interfaced with the board's sets of digital and analogue input/output (I/O) pins.

FIGURE 15.1 Arduino Uno.

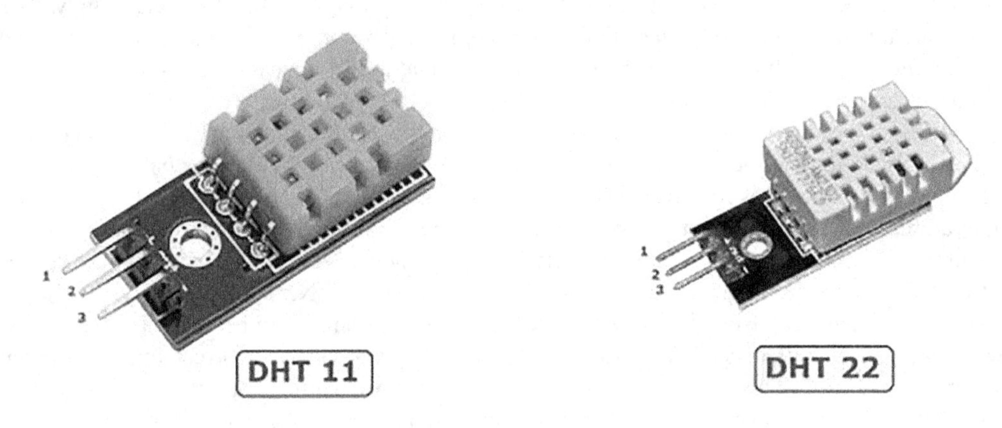

FIGURE 15.2 DHTll temperature and humidity sensor.

The board contains six analogue I/O pins and 14 digital I/O pins, six of which can be used for PWM output. It can be programmed using the Arduino IDE (Integrated Development Environment) via a type B USB connector (Arduino et al., 2018) (Figure 15.1).

15.5 DHT11 TEMPERATURE AND HUMIDITY SENSOR

The thermistor and capacitive humidity sensor are the two components that make up the DHT sensors. A straightforward chip that converts analog data to digital data and outputs a digital signal with the temperature and humidity is also included. Any microcontroller can easily read the digital signal (Figure 15.2).

15.6 CONNECTER WIRE

The wires from a timer to a valve are connected with ire connectors. For subterranean connections in a valve box, wire connectors saturated with silicone, gel, or grease form a waterproof connection. For each valve, you will need two wire connectors (Figure 15.3).

15.7 DIFFERENT SENSORS USED IN SMART IRRIGATION SYSTEM

Sensors are the core part of smart irrigation system. It helps in improving the efficiency. Some of the most common sensors used in irrigation are soil moisture sensor, rain sensor, and temperature sensor (Krishna and Priyanka, 2014). Apart from helping in smart irrigation system, these sensors can

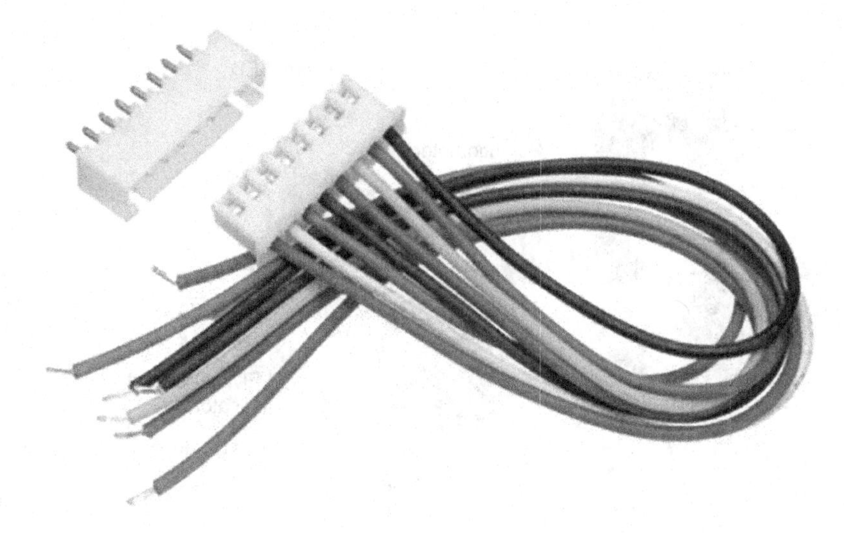

FIGURE 15.3 Temperature and humidity sensor.

also be used in some of the existing systems, where it can be easily installed and produced almost similar results. These are more affordable in case there is a compatible irrigation timer already installed on the site.

15.7.1 Soil Moisture Sensor

It is a low-cost electronic sensor, which is used to detect the moisture of the soil. It measures the volumetric content of water inside the soil. Soil moisture sensor consists of mainly two parts: sensing probes and sensor module (Krishna and Priyanka, 2014). Probes allow the current to pass through the soil and then it gets the resistance value according to the moisture value in soil, while on the other hand, sensor module reads data from the sensor probes and processes the data and converts it into a digital/analog output. They can provide both types of output: digital output (DO) and analog output (AO).

Probes consist of two nickel-coated copper tracks with two header pins. These pins are internally connected to the two copper tracks and are used to connect the sensing probes to sensor module through two jumper wires. One pin of the sensor module provides a +5 V current to one probe, and another pin of the sensor module receives the return current from the other probe. Under dry soil conditions, it provides high resistance and less conductivity, and in case of more water in the soil, resistance will decrease and conductivity will increase.

The module part is having a number of important parts. Variable resistor is applied to set the sensitivity of the moisture sensor. An onboard LED is present, which indicates that the sensor power supply is ON or OFF. Output LED indicates the presence of moisture in the soil, and finally, LM393 is used as a voltage comparator in this moisture sensor module.

As for the working principle, sensing probe is connected to the sensor module circuit using the jumper wire and enters the probes into the dry soil. Then, the sensor is connected to the 5 V power supply. The threshold voltage is set in dry soil condition by rotating the potentiometer knob for setting the sensor sensitivity. When there is more water in the soil, then probe's conductivity will increase and resistance will decrease. So, if a low amount of voltage from the sensing probe is given, then the LM393 Comparator IC compares this voltage with the threshold voltage. In this condition, this input voltage is less than the threshold voltage, so the soil sensor output goes low (0). When there is less water in the soil, then probes will have low conductivity and high resistance. So, a high amount of voltage from the sensing probe is given. Then the LM393 Comparator IC compares

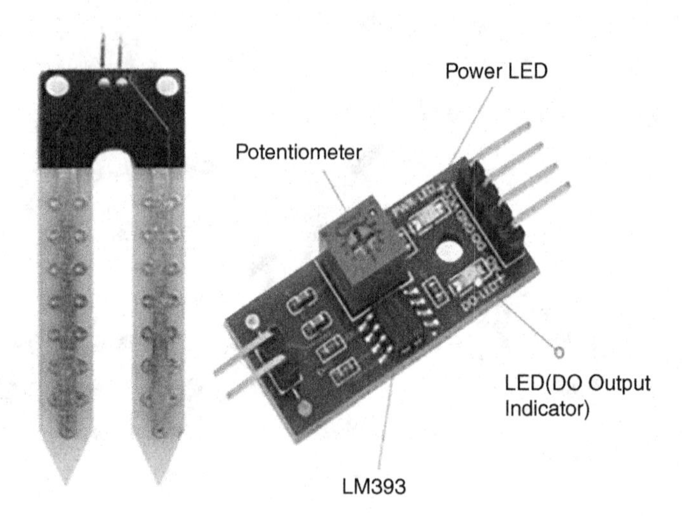

FIGURE 15.4 Soil moisture sensor probe with sensor module.

this voltage with the threshold voltage. In this condition, this input voltage is greater than the threshold voltage, so the sensor output goes high (1) (Figure 15.4).

15.7.2 Rain Sensor

Rain sensor is used to observe the water drops or rainfall. It works like a switch and has two parts: sensing pad and sensor module. Whenever rain falls on the surface of a sensing pad, then the sensor module reads the data from the sensor pad to process and convert it into an analog or digital output.

Sensing pad includes two series copper tracks coated with nickel. It has two header pins that are connected internally to the copper tracks of the pad, which helps to connect the sensing pad with the rain sensor module with the help of two jumper wires. Here, the rain sensor module's one pin provides a +5 V power supply toward one path of the sensing pad, whereas the other pin gets the return power from another path of the pad. Under dry situations, this pad gives huge resistance as well as less conductive. Once water falls on the surface of the sensor pad, then its resistance will be reduced and conductivity will be enhanced.

The sensor module includes some essential components like variable resistor, LM393 IC, output LED, and power LED. Variable resistor fixes the rain sensor's sensitivity. Power LED indicates the power supply of the sensor is ON/OFF. Output LED detects and informs about rain drop. Lastly, LM393 IC is used as a voltage comparator.

The working principle includes first connecting the sensing pad to the sensor module using a jumper wire. Next, both the pins of rain sensor modules like GND and VCC are connected to a 5 V power supply pin. After that, the threshold voltage is fixed at the non-inverting terminal of the LM393 IC in the dry state of the pad. If the volume of raindrops on the surface of the pad increases, then its conductivity increases and resistance decreases. After that from the pad, a less amount of voltage can be provided to the inverting input terminal of the LM393 IC. Then this IC evaluates this voltage through the threshold voltage. In this state, the input voltage is low as compared to the threshold voltage; as a result, the output of the rain sensor goes low. When no rain falls on the surface of the pad, then it has high resistance and less conductivity. After that, the high voltage will be assigned across the pad. Thus, the high voltage from the pad can be provided to the inverting input of the IC. IC evaluates this voltage by using the threshold voltage. So, in this state, this input voltage is higher as compared to the threshold voltage. As a result, the output of the sensor module goes high.

15.7.3 Temperature Sensor

This sensor gives temperature measurement in a readable form through an electrical signal. The working of a temperature meter depends upon the voltage across the diode. The temperature change is directly proportional to the diode's resistance. The cooler the temperature, lesser will be the resistance, and vice versa. The resistance across the diode is measured and converted into readable units of temperature (Fahrenheit, Celsius, Centigrade, etc.), and displayed in numeric form over readout units. These sensors play an important role in monitoring the irrigation system.

There are two main types of temperature sensors: contact-type temperature sensors, which measure the degree of hotness or coolness in an object by being in direct contact with it, and non-contact-type temperature sensors, which are not in direct contact with the object, rather they measure the degree of hotness or coolness through the radiation emitted by the heat source.

15.8 ALGORITHMS USED FOR SMART IRRIGATION

Machine learning (ML) can be described as the scientific study of algorithms and statistical models used by computer systems to perform a specific task without using explicit instructions, relying on patterns and inference instead. It is a subset of artificial intelligence (AI). Machine learning algorithms help to prepare a mathematical model based on sample data (training data), in order to make predictions or decisions without being explicitly programmed to perform the task. They have a wide range of applications, such as email filtering and computer vision, where it is difficult or infeasible to develop a conventional algorithm for effectively performing the task. ML helps in predictions using computers as it involves computational statistics. The study of mathematical optimization delivers methods, theory, and application domains to the field of machine learning. Data mining is a field of study within machine learning which focuses on exploratory data analysis through unsupervised learning. Machine learning is often referred to as predictive analytics in business problems.

ML can be effectively used in smart irrigation system. It is having different techniques. Algorithm based on a combination of supervised and unsupervised machine learning techniques can be developed using support vector regression (SVR) and *k*-means clustering for the estimation of difference/change in the factors (e.g., soil moisture) affecting yield on site. This method can give good accuracy and less mean squared error (MSE) in the prediction of the soil moisture in the upcoming days. To perform that, it will take help of the data from the sensors. The data can about air temperature, air relative humidity, soil temperature, radiation, and soil moisture difference. The SVR model will be trained according to the received data.

Suppose if we predict the soil moisture difference using trained SVR model, then the same predicted value will be given as an input for *k*-means clustering. This will further improve the accuracy of soil moisture difference, and MSE will reduce even more. After receiving the final prediction, it can be used in the development of smart irrigation scheduling algorithm. It will effectively increase the utilization of the precipitation information for enhancing the irrigation. For visualizing the data of upcoming days on which irrigation is dependent, responsive web portal can be developed. Support vector regression (SVR) has evolved from support vector machine (SVM), in which the dependent variable is numerical instead of categorical.

SVR aims at reducing the error by determining the hyper-plane and minimizing the range between the predicted and observed values. It is a non-parametric technique and allows the creation of non-linear models. SVR performs better performance prediction than other algorithms like linear regression and KNN due to improved optimization strategies for a broad set of variables. It is also flexible in dealing with geometry, transmission, data generalization, and an additional functionality of kernel. This additional functionality enhances the model capacity for predictions by considering the quality of features. This method utilizes kernel functions to generate the model. Some of the frequently used kernel functions are polynomial, linear, radial basis, and sigmodal. *k*-means clustering takes a straightforward and simple methodology to group a given information set into a definite number of clusters. The objective is to find k centroids, one for each bunch. First, it divides n number of the objects into k non-empty subgroups/cluster and then finds the cluster centroids (mean point) of every subgroup/cluster. Then, it calculates the distances from every point to the centroids and allocates each object to a specific cluster, where the distance is minimum from the centroid. The process iterates to re-assign the points and identify the centroid of the new clusters.

Data mining algorithms are used to take decisions on drip irrigation system. Naïve Bayes algorithm is common for the drip irrigation system. It is a classification-based machine learning algorithm that helps to classify the data based on the conditional probability values using computation fast algorithm for classification problems. This algorithm is good for real-time prediction, multi-class prediction, recommendation system, text classification, and sentiment analysis use cases. Naïve Bayes algorithm can be built using Gaussian, multinomial, and Bernoulli distribution, and is scalable and easy to implement for a large dataset. Thus, we will use the algorithms in system for controlling water quantity as per the requirement and condition.

15.9 APPLICATIONS OF SMART IRRIGATION IN AGRICULTURAL WATER SAVING

The world had abundant supply of water resources for maintaining life. But with the rise in population and over-exploitation of water, it is now lacking that abundance. The water availability is in stress; if it continues, we will face severe water scarcity. So, it is high time that we conserve water in every way possible. As already discussed, traditional irrigation lacks monitoring, and up to 50% of the water used can be wasted due to inefficiencies in irrigation, evaporation, and over-watering. Smart irrigation will use real-time or historical data to inform watering routines and modify watering schedules to improve efficiency, and thus save water from being wasted at a lower operating expense. The system uses microcontroller, which increases the system life with a lower power consumption (Kumar et al., 2017).

Two important aspects of smart irrigation include control and delivery types. Control type is the one where the way the irrigation is controlled and delivery type is when the type of water delivery systems used. Also, the control for smart irrigation systems is mainly of two types, weather-based and soil-based, each of them varies in its technical method of sensing and supplying information.

Weather-based smart irrigation systems use local weather information of the area, drawn from reliable weather sources, sensors, or historical data to support informed decisions about watering schedule. This system is also called evapotranspiration system, owing to the water loss through evaporation from the land and transpiration from plants. Analytical assessments will be performed on factors like local temperature, humidity, insolation, and wind. After that, water schedule will be decided based on the output.

Soil-based smart irrigation systems use local soil moisture data drawn from sensors in the ground to support informed decisions about watering schedules (Kumar et al., 2017). Users manage irrigation on demand with the help of these systems, i.e., if a land is dried then the water is supplied, while if the land is wet and is saturated, then the water supply is stopped. To serve this purpose, we can set the threshold level for the sensors to act accordingly. Thus, eventually controlling these two points reduces the amount of water used by linking it to the moisture level needed in the soil for a particular crop.

Coming to the delivery, they are of four types: surface, sprinklers, trickle, and sub-surface methods. Surface irrigation is the most traditional one, which distributes water through irrigation ditches and over the soil surface withy the help of gravity. It has four phases: first phase is called advance phase, where water is supplied to the top end of the fields, which naturally follows to the field length and gradually runs off to the pond. Second is the storage phase. The period of time is between the end of the advance phase and the shut-off phase. Third is the depletion phase, which is a short period of time when the length of the field is submerged. The fourth and last is the recession phase, where the time period is while the waterfront goes toward the downstream end of the field. Sprinklers allow application of water under high pressure with the help of a pump. It releases water through a small diameter nozzle placed in the pipes. Water is distributed through a system of pipes, sprayed into air, and irrigates in most of the soil type due to a wide range of discharge capacity. It can be fixed or mobile. Trickle irrigation spreads water very locally to the ground surface. It supplies water directly to the plants or roots system. The system needs extensive tubing to ensure that all of the plants in a garden are reached by the irrigation, but it results in less wastage of water. It is having several other advantages like reduced nutrient loss in the soil, decreased leaching in water table, local watercourses, and reduces root diseases as water accumulation around the roots are prevented. Subsurface methods are buried next to the plant's root zone and apply water below the ground. Delivery is done directly to the infiltrative surface of the soil using specially manufactured polyethylene tubing with built-in turbulent flow emitters. This method requires little field preparation and labor, and causes minimum evaporation loss and surface waste. The irrigation water should be of good quality to prevent the soil from excess salinity, and the flow rate should be low to prevent waterlogging of the field. Trickle systems and subsurface methods are generally most effective in saving water. But, in practice, the sub-surface method has limited use. This is because its efficiency is dependent on a large number of factors like permeable soil in the root zone to permit free and rapid movement of water laterally and vertically, relatively impervious layer at 2–3 m in the substratum to prevent deep percolation of water, etc. Considering the popularity of these smart irrigation systems, it is essential that the stakeholders (such as farmers, etc.) must be introduced with such advanced systems so as to reduce the technology gap and applications. This will demand decentralized and people-centric approaches while implementing smart irrigation technologies. Hence, in order to enhance the user-friendly dimensions, it is important that the target villages, concerned stakeholders, and their development strategies (that may influence land use land cover) must be understood carefully before devising and designing irrigation technologies (Srivastava and Chinnasamy, 2021a, 2021b).

15.10 CONCLUSIONS AND RECOMMENDATIONS

Smart irrigation water using sensors and internet of things can be effectively used for precision farming and crop improvement under dry land conditions by maintaining the irrigation water as per crop needs. Applications of smart irrigation in agricultural water saving are also discussed and showed the importance of this system. Brief introductions of this system based on IoT and sensors with their adaptability are also interpreted. Furthermore, we illustrated how the intelligence method tool will help the decision makers, and developers, in achieving agricultural water sustainability. We will suggest that this irrigation can be applied at village levels or plot level by farmers and NGOs.

REFERENCES

Arduino UNO for beginners - Projects, 2017. Programming and Parts, makerspaces.com. 7 February 2017. Retrieved 4 February 2018.

Ashwini, B.V., 2018. A study on smart irrigation system using IoT for surveillance of crop-field. *International Journal of Engineering and Technology*, 7(4.5), 370–371.

Biswas A., D. R. Mailapalli, and N. S. Raghuwanshi, 2021. Modelling the effect of changing transplanting date on consumptive water footprints for paddy under system of rice intensification. *Journal of the Science of Food and Agriculture*, 101, 13.

Darshana, S., T. Sangavi, S. Mohan, A. Soundharya, and S. Desikan, 2015. Smart irrigation system. *IOSR Journal of Electronics and Communication Engineering*, 10(3), 32–33.

Gutiérrez, J., J. F. Villa-Medina, A. NietoGaribay, and M. A. Porta- Gándara, 2014. Automated irrigation system using a wireless sensor network and GPRS module. *IEEE Transactions On Instrumentation and Measurement*, 63(1), 166–167.

Jaguey, J.G., J. F. Villa-Medina, A. Lopez Guzman, and M. A. Porta Gandara, 2015. Smartphone irrigation sensor. *IEEE Sensors Journal*, 15(9), 5122–5127.

Jury, W. A. and H. J.Vaux, 2007. The emerging global water crisis: managing scarcity and conflict between water users. *Advances in* Agronomy, 95, 1–76.

Kagalkar, A., 2017. Smart irrigation system. *International Journal of Engineering Research & Technology*, 6(05), 982–983.

Khadke, L. and S. Pattnaik, 2021. Impact of initial conditions and cloud parameterization on the heavy rainfall event of Kerala (2018). *Modeling Earth Systems and Environment*, 7, 2809–2822. https://doi.org/10.1007/s40808-020-01073-5

Krishna, B.V. and K. Priyanka, 2014. Soil moisture sensor design for crop management system by using cellular communication. *International Journal of Advanced Research in Electrical, Electronics and Instrumentation Engineering*, 03(10), 12408–12414.

Kumar, B. D., P. Srivastava, R. Agrawal, and V. Tiwari, 2017. Micro controller based automatic plant irrigation system. *International Research Journal of Engineering and Technology*, 04(05), 1436–1437.

Nemali, K. S. and M. V. Van Iersel, 2006. An automated system for con-trolling drought stress and irrigation in potted plants. *Scientia Horticulturae*, 110(3), 292–297.

Pathan, S. A. and S. G. Hate, 2016. Automated irrigation system using wireless sensor network. *International Journal of Engineering Research & Technology*, 5(06), 6–9.

Srivastava, A. and P. Chinnasamy, 2021a. Developing village-level water management plans against extreme climatic events in Maharashtra (India)—a case study approach. In: Vaseashta A., Maftei C. (eds.) *Water Safety, Security and Sustainability. Advanced Sciences and Technologies for Security Applications.* Springer, Cham. https://doi.org/10.1007/978-3-030-76008-3_27

Srivastava, A. and P. Chinnasamy, 2021b. Investigating impact of land-use and land cover changes on hydro-ecological balance using GIS: insights from IIT Bombay, India. *SN Applied Sciences*, 3, 343. https://doi.org/10.1007/s42452-021-04328-7

Srivastava, A. and P. Chinnasamy, 2022. Assessing groundwater depletion in Southern India as a function of urbanization and change in hydrology: a threat to tank irrigation in Madurai City. In: Kolathayar S., Mondal A., Chian S.C. (eds.) *Climate Change and Water Security. Lecture Notes in Civil Engineering*, vol 178. Springer, Singapore. https://doi.org/10.1007/978-981-16-5501-2_24

Srivastava, A. and P. Chinnasamy, 2023a. Tank cascade system in Southern India as a traditional surface water infrastructure: a review. In: Chigullapalli S., SushaLekshmi S.U., Deshpande A.P. (eds.) *Rural Technology Development and Delivery. Design Science and Innovation.* Springer, Singapore. https://doi.org/10.1007/978-981-19-2312-8_15

Srivastava, A. and P. Chinnasamy, 2023b. Understanding declining storage capacity of tank cascade system of Madurai: potential for better water management for rural, peri-urban, and urban catchments. In: Chigullapalli S., SushaLekshmi S.U., Deshpande A.P. (eds.) *Rural Technology Development and Delivery. Design Science and Innovation.* Springer, Singapore. https://doi.org/10.1007/978-981-19-2312-8_14

Srivastava, A., L. Khadke, and P. Chinnasamy, 2021. Web application tool for assessing groundwater sustainability—a case study in rural-Maharashtra, India. In: Vaseashta A., Maftei C. (eds.) *Water Safety, Security and Sustainability. Advanced Sciences and Technologies for Security Applications.* Springer, Cham. https://doi.org/10.1007/978-3-030-76008-3_28

Srivastava, A., L. Khadke, and P. Chinnasamy, 2022. Developing a web application-based water budget calculator: attaining water security in rural-Nashik, India. In: Kolathayar S., Mondal A., Chian S.C. (eds.) *Climate Change and Water Security. Lecture Notes in Civil Engineering*, vol 178. Springer, Singapore. https://doi.org/10.1007/978-981-16-5501-2_37

16 Analysis of Climate Variability and Change Impact on Rainfall Trend Pattern in Nigeria

A Case Study of Calabar River Basin, South-eastern Nigeria

Abali Temple Probyne

16.1 INTRODUCTION

The coastal environment has captivated the interest of both scholars and policymakers due to its ever-changing climate and land use pattern. Coastal resources are under severe strain as a result of the world's ever-increasing human population and demand for agriculture (Cunningham et al., 2005; Khadke and Pattnaik, 2021). Similarly, the coastal belt has the potential to expand residential, industrial, agricultural, recreational, and commercial land uses, making it a hotspot for population growth, rapid land-use transformation, loosing traditional wisdom of conservation and protection, and consequently, the pollution (Oyegun, 1993; Abali, 2016; Srivastava and Chinnasamy, 2021a, b). Coping with the effects of increasing sea levels as a result of global warming is one of the most pressing concerns confronting coastal residents. Despite a trend toward increasingly conservative projections of sea level rise this century, the rate and magnitude of the changes to be expected still differ (Goudie, 2004). However, the Intergovernmental Panel on Climate Change (IPCC)'s third assessment report of her working group revealed that sea level would rise by 0.09–0.88 m between 1990 and 2100. There would be intrusion of salt water into rivers, estuaries, and groundwater. This will in no doubt affect tidal range, oceanic currents, upwelling patterns, salinity levels, biological processes, runoff, and terrain erosion patterns and tidal flooding as a result of global warming and rising sea levels. Furthermore, cliffs will be more vulnerable to falls, landslides, and other mass movements, exacerbating problems where loose or weak materials are already deteriorating rapidly as a result of increasing rates of erosion. Nonetheless, owing to the characteristics of the coast, such as slope, wave environment, tidal regime, and erosion susceptibility, the effects of rising sea levels will vary. According to estimates, about half of the world's population dwells in coastal lowlands, subsiding river deltas, and river floodplains. Climate change will wreak havoc on these densely populated places. Whether natural systems, such as coastal marshes and coral reefs, can successfully adapt to changing conditions is dependent on the rate of sea-level change, rather than the total quantity. Increasing the frequency of extreme events is more difficult for human and environmental systems to adapt than slowly changing average climatic conditions. Higher sea temperatures, on the other hand, have yet to be determined whether they would increase the number of tropical storms and their severity, spreading their effects more poleward, or whether there are greater temperature differences between the land and the water may intensify monsoons and affect their timings. The availability of resources and the monetary value of the land under danger will affect human responses to rising sea levels. These will have negative impacts on salt-marshes, mangroves, coral reefs, lagoons, and the ice-covered Arctic shores. These are likely to receive less attention due to high beachfront property values, motivating economic investment to prevent rising sea levels in cities, while there is

DOI: 10.1201/9781003303237-16

likelihood of less attention on negative impacts on saltmarshes, mangroves, coral reefs, lagoons, and ice-covered Arctic shores. The decision-making process connected to coastal erosion and floods is challenging due to financial constraints and a variety of factors such as physical, social, economic, legal, political, and aesthetic. The buildup of carbon dioxide, methane, nitrous oxide, and CFCs as greenhouse gases in the atmosphere would intensify the greenhouse effect, culminating in global warming brought on by human activities such as deforestation, urbanization, and industrialization. Nigeria, in general, and Calabar, in particular, had experienced significant climate variability and change, evidenced by the late onset and early cessation of the rains (NIMET, 2007; Abali, 2021). In Nigeria, Cross River State is home to more than 40% of the country's remaining tropical high forests (THFs) (Balogun, 1994). Ogbonaya (1996) estimated the rate of forest loss in Nigeria to be about 4,000 km^2 a year. However, if this rate continues unabated, Nigeria will have no natural forest left in less than 14 years. However, 5% of the forest now remains, and out of this, 4% is in Cross River Region (NEST, 1991). In Nigeria, Cross River State presently has over 40% of the remaining tropical high forests (THFs) of the entire nation (Balogun, 1994). The forest resource base includes the mangrove swamps and tropical rainforest in the south, the central, and the derived Guinea savanna toward the north (Ogar, 2006). Thus, Cross River has become the most forested State in Nigeria with at least 75% of its population inhabiting rural communities. In Nigeria, Ogbonaya (1996) estimated the rate of forest loss to be about 4,000 km^2 a year. The researcher also envisaged that if this rate continues unabated, Nigeria will have no natural forest left in less than 14 years. However, 5% of the forest now remains, and of this, 4% is in Cross River Region (NEST, 1991). At present, 34% of Cross River State is forest (Dunn, 1994). In the forest reserves of the Calabar river catchment, the trend has manifested in the level of exploitation for agriculture, fuel energy, construction, urbanization, etc.

Due to the lack of funding for the Forestry Department, growing population and immigration, and the establishment of plantations, Cross River State has lost roughly 19% of its tropical high forests during the last few decades (Akintoye, 2002). There are around 2,000 communities in the rural areas of the state, which house 70% of the population. Cross Riverians' major occupations include farming, hunting, and the extraction and collection of timber and non-timber forest products. At the moment, forests encompass 34% of Cross River State (Dunn, 1994). These forests are made up of forest reserves, community forests, and woods in Cross River National Park. The level of exploitation for agriculture, fuel production, construction, urbanization, and other reasons in the forest reserves of the Calabar river basin has demonstrated this pattern (Table 16.1). As a result of this trend, climate change poses a concern in the area.

For instance, even the Cross River National Park (CRNP) with a forested land of about 4,000 km^2 was established in 1994, the depletion of the rainforests within and outside its delineation boundaries through commercial logging is increasing rapidly. Akintoye (2002) asserted that this precious vegetation has gradually become the commercial logging targets of multi-nationals logging and wood processing firms like Hanseatic Nigeria Limited (German), Kisari Investment Company Limited (Belgian), and most recently, Wempco Agro-Forestry Company Limited (Hong Kong/Chinese). Egbai et al. (2011) are of the view that there is no correlation between apparent lack of vegetation cover, sandy nature of soil, high temperature regimes, and the location of Calabar as a city. In their opinion, the exposure of the soil to rainwash cannot be attributed to the geographical location of the city. Climate change and fluctuation pose a serious danger to a region's global socio-economic development. It has an impact on food supply, accessibility, and consumption, as well as human health and livelihood. These effects are both immediate and long term, as a result of more frequent and more intense extreme weather occurrences as a consequence of global temperature and precipitation patterns changing. In a world where the climate is changing, livelihoods and lives are being impacted on a daily basis; everything from clothing to recreation to the types of crops grown and animals raised is influenced by the weather and climate of a location. Consequently, regular monitoring and evaluation of the changing climate are required to ensure a sustainable future.

TABLE 16.1
Exploitation of Trees in the Calabar River Basin's Forest Reserves

	Number by Year				
Location	**1988**	**1990**	**2000**	**2001**	**2003**
Awi – Akampa	1,176	159	1,113	1,375	667
Oban	295	1,139	447	721	547
Odukpani	110	41	76	70	67
Total	**1,581**	**1,339**	**1,636**	**2,166**	**1,281**

Source: Cross River State Forestry Commission, Calabar (1988, 1990, 2000, 2001, and 2003).

To facilitate the attainment of the goal of this study, the following research questions were put forward:

i. What is the climate variability and change pattern in the study area overtime?
ii. What is the impact of climate variability and change on rainfall trend pattern?
iii. How does the impact of climate variability and change on rainfall trend pattern affect the livelihood of the inhabitants in the study area?
iv. What are the adaptive and mitigation measures to climate variability and change in the study area?

16.2 AIM AND OBJECTIVES

The aim of the study is to determine the impact of climate variability and change on rainfall trend pattern in the study area. Specific objectives include to determine

i. the climate variability and change pattern in the study area overtime;
ii. the impact of climate variability and change on rainfall trend pattern;
iii. the influence of climate variability and change and rainfall trend pattern on the livelihood of the inhabitants in the study area;
iv. The adaptive and mitigation measures to climate variability and change in the study area.

16.3 STUDY AREA

The Calabar River Basin, with an estimated area of 460 km^2, is located between the latitudes of 4°45′N and 5°10′ N and longitudes 8°05′ and 8°45′ E in south-eastern Nigeria. It is a low-land (plain) with coastal plain sands from the Benin formation. The average annual temperature is around 27°C throughout the year, with a total rainfall of around 2,900 mm. The relative humidity is expected to be in the 90% range. It lies within the hydrological borders of the Calabar river system, with tropical rainforest and lateritic grained sand as vegetation and soil, respectively. There are over 399,761 inhabitants (2010 Projected Population Census). Among the people's main occupations are farming, hunting, wood extraction and gathering, and non-timber forest product gathering.

16.4 LITERATURE REVIEW

Globally, there are different climatic conditions prevailing and experienced in many regions of the globe with attendant significant variations occurring at any given time. This is as a result of the energy of the sun emitted as heat radiations caused by anthropogenic activities that produce greenhouse gases. Some of the radiations (near infrared rays) are absorbed and retained by the earth's surface.

Ochuko (2009) observed that the temperature of the earth's surface is determined by the energy balance between the heat energy reaching the earth's surface and the heat energy that is radiated back to space. Carbon dioxide, methane, chlorofluorocarbons, ozone, and water vapor concentrations have reached historic levels due to fossil fuel-dependent industrialization and man's degenerative lifestyle based on over-exploitation of resources like coal, oil, and gases that take generations to replenish (Dhameja, 2007). These greenhouse gases operate like the glass of a greenhouse at lower altitudes in the atmosphere. They are transparent to near-infrared (short wavelength) rays but opaque to heat radiated by the heated ground (longer wavelength heat rays) and trap them, similar to glass. Greenhouse gases contribute to the heat already existing on the earth's surface by preventing sun rays from escaping into space. The greenhouse effect is the outcome of this, and it causes an increase in temperature. This is referred to as global warming on a global scale. The following severe environmental problems are posed by the greenhouse effect: The melting of ice masses in the Arctic and Antarctic regions, as well as rising sea levels, will result in the submergence of many low-lying coastal areas. Flooding of coastal areas will cause massive soil erosion and siltation, as well as contamination of water and the spread of water-borne diseases. In temperate regions, summers will be longer and hotter, while winters will be shorter and warmer. Due to rising carbon dioxide concentrations, plant growth and yield will increase, resulting in depletion of nutrients from the soil and ecosystem disruption caused by increased rainfall (9%–10%), altered crop patterns, and adverse weather on effects on the environment's flora and fauna. Climate change evidence is overwhelming: sea levels are rising, glaciers are melting, precipitation patterns are shifting, and the world is warming (Adedeji et al., 2014). A rapid cooling of the earth's climate occurred 65 million years ago. Seventy-five percent of existing plant and animal species are considered to have perished as a result of this. Strong warming trend that began around 1,880 and has been slowly increasing world temperatures. It has also been reported that the 20th century was the warmest in 600 years, with the 1990s being the hottest decade of the century. According to Okebukola (1997), the total amount of carbon dioxide in the atmosphere remained nearly constant until the 20th century, when the burning of fossil fuels such as coal, fuel oil, gasoline, kerosene, diesel, and natural gas began to release large amounts of carbon dioxide into the atmosphere. The United Nations Intergovernmental Panel on Climate Change (IPCC), which comprises physicists, geochemists, geographers, oceanographers, botanists, limnologists, and paleoclimatologists, was established in 1988 by a group of 2,000 scientists from around the world to study global warming and make recommendations. In its third assessment report in 2001, the IPCC released a projected a probable temperature increase of at least 1.4°C (2.5°F) by 2,100. Under its "worst case" scenario, the increase would be 5.8°C (10.4°F). Climate change is caused by a number of factors, including deforestation, ozone layer depletion, increased carbon dioxide, and greenhouse gas emission into the atmosphere. Humans cut down trees for timber or to clear land for farming or construction. This has the potential to both liberate carbon stored in trees and lower the number of trees available to absorb CO_2. Carbon dioxide levels had climbed to a record high of 379 parts per million (ppm) in 2007, according to the IPCC, and are increasing at a rate of 1.9 ppm per year. Carbon dioxide levels are expected to reach 970 parts per million by 2,100 under a higher emission scenario, more than triple the pre-industrial levels (Manstrandrea and Schneider, 2009). The negative effects of such a trend in carbon dioxide concentrations, particularly on agricultural systems, are exceedingly worrying and deadly.

The African continent is under severe climate stress and is extremely vulnerable to climate change's effects. The expected rise in sea level will drown Nigeria's, Mozambique's, Kenya's, Gambia's, and Egypt's low-lying coastal areas (Oyegun, 2007). It is important to note that Africa is already experiencing severe food shortages. Many factors contribute and compound the impacts of current climate variability in Africa and will have negative effects on the continent's ability to cope with climate change. Poverty, illiteracy, and a lack of skills are among them, as are weak institutions, limited infrastructure, a lack of technology and information, low levels of primary education and healthcare, poor resource access, and armed conflicts (UNFCCC, 2007). Additional hazards include over-exploitation of land resources, such as forests, population growth, desertification, and land degradation (UNDP, 2006). Dust and sand storms in the Sahara and Sahel have a negative impact on agriculture, infrastructure, and health. Africa's climate is likely to become more unpredictable, with more frequent and severe extreme weather events, and risk to health and life. This includes increasing risk of drought and flooding in new areas (Christensen et al., 2007) and inundation due to sea-level rise in the continent's coastal areas. Malaria, tuberculosis, and diarrhea are among the climate-sensitive diseases that afflict Africa (Guernier et al., 2004; McMichael et al., 2006).

Carbon is absorbed through the growth of forest and non-forest trees as well as the abandonment of managed areas. Carbon emissions from biomass harvesting and the conversion of forests and savannas to agricultural lands are estimated to be 112.23 $TgCO_2$ in the same study (30.61 $TgCO_2$-C). This resulted in a net CO_2 emission of 75.54 Tg (20.6 Tg CO_2-C) of 75.54 Tg CO_2 (FME, 2003). Nigeria's gas flaring accounted for 58.1 million tons of CO_2 or 50.4% of total CO_2 emissions from the energy sector, in 1994. The sector's consumption of liquid and gaseous fuels resulted in CO_2 emissions of 51.3 and 5.4 million tons, respectively (FME, 2003).

It is projected that the sea level will rise by 10–30 cm by the year 2025; by 44–71 cm by 2075, and 66–110 cm by 2100. Against the background of the induced subsidence along the coast of Nigeria, these values would be much higher (Oyegun, 2007). Fubara and Alabo (2007) observed that as much as 20 km strip of the Niger Delta inland from the coast would be submerged under water in the next 50 years at the current rates of increase in sea level. This would amount to loss of over 25,000 km^2 from the entire coastline of the country with all coastal settlements in the current geography of the nation being lost to the Atlantic Ocean.

16.5 METHODOLOGY

The primary sources of data employed for this research are the climatic (rainfall) data documentation of the Nigerian Meteorological Agency (NIMET), Calabar Station between 1971 and 2014 (Table 16.2), and the monthly annual temperature and rainfall documentation of the Nigerian Meteorological Agency (NIMET), Calabar station (Table 16.3). The data analysis was carried out using rainfall and temperature documentation of Nigerian Meteorological Agency (NIMET), Calabar station. The climatic (rainfall) variables (Table 16.1) were used to directly plot line graphs to show variability and change over time (Figures 16.2–16.14).

16.6 RESULTS

The 43 years' trend (1971–2014) of rainfall and temperature in Calabar as recorded by NIMET was compared against NIMET's historical meteorological data maps between 1941–1970 and 1971–2000 (Figures 16.1–16.14). The computation was predicated on meteorological parameters that indicate status of the climate such as temperature, rainfall, and extreme weather events like flooding, drought, dust storm, and heat wave. The assessment was also based on scientific and credible data from the Agency's archive that spans many decades.

The mean annual temperature is 27°C with two periods of high temperature recorded (double maxima) in the months of July and September. The annual range of temperature is a characteristic of equatorial climate. The difference between the temperature of the hottest month and the coldest month is as low as 5.0°C. The mean monthly rainfall of 242 mm and mean monthly annual rainfall is over 2,900 mm. The average annual relative humidity is about 90%, and the evaporation rate is above 1.7 mm (Table 16.3).

From the graphs, one can deduce that there is significant variability of rainfall in Calabar. Different researchers classified the areas into three temperature zones: hottest areas – over 27°C, moderately low areas (24°C–27°C), and cool areas – less than 24°C. Calabar falls under the hottest areas with mean annual temperature of 27°C. The region is accompanied by anthropogenic activities that produce greenhouse gases (mostly carbon dioxide) in the environment, and this constitutes abnormal global warming in the area. Unless high temperatures are accompanied by high humidity, the weather is not as comfortable as it would appear to be. It is much more uncomfortable under

TABLE 16.2
Climatic (Rainfall) Data for Calabar (1971–2014)

Year	Jan. (mm)	Feb. (mm)	Mar. (mm)	Apr. (mm)	May (mm)	Jun. (mm)	Jul. (mm)	Aug. (mm)	Sept. (mm)	Oct. (mm)	Nov. (mm)	Dec. (mm)	Mean (mm)	Annual (mm)	Max. (mm)	Min. (mm)
1971	Trace	84.0	78.8	220.4	291.2	374.3	661.4	325.4	322.5	404.0	103.2	86.4	**268.3**	**3,219.9**	**661.4**	**78.8**
1972	47.0	171.1	108.1	179.9	147.4	180.8	510.8	495.8	487.3	289.7	128.1	45.7	**249.5**	**2,994.2**	**510.8**	**45.7**
1973	30.3	76.2	159.5	181.9	210.2	124.0	262.9	373.3	329.7	235.3	58.1	68.1	**189.0**	**2,268.2**	**373.3**	**58.1**
1974	9.8	16.2	49.7	203.4	222.7	318.6	269.0	440.8	588.4	337.6	87.9	12.5	**231.5**	**2,778.3**	**588.4**	**12.5**
1975	0.0	20.7	136.0	227.1	218.1	186.1	552.6	270.9	469.0	248.6	335.9	10.2	**243.2**	**2,918.4**	**552.6**	**10.2**
1976	35.7	176.9	210.0	188.5	208.4	52.5	194.9	528.8	284.1	358.6	418.8	4.1	**238.7**	**2,864.3**	**528.8**	**4.1**
1977	5.0	5.6	92.8	82.1	253.1	597.6	340.6	556.2	235.7	354.9	76.1	47.0	**240.2**	**2,881.9**	**597.6**	**5.6**
1978	70.1	46.5	146.8	296.6	211.7	488.2	648.2	304.0	721.2	308.6	123.4	6.4	**300.2**	**3,601.7**	**721.2**	**6.4**
1979	0.0	107.3	87.0	264.3	188.6	599.3	353.9	364.2	374.7	366.4	105.4	5.7	**256.1**	**3,072.9**	**599.3**	**5.7**
1980	40.0	5.5	126.5	163.0	316.0	620.3	364.1	728.7	500.5	489.1	198.1	1.7	**319.4**	**3,832.9**	**728.7**	**1.7**
1981	10.5	TRACE	159.0	163.5	360.7	355.8	519.7	400.4	333.9	325.4	67.6	10.3	**269.63**	**2,965.93**	**519.7**	**10.3**
1982	57.7	64.0	159.2	190.1	351.3	396.0	505.7	381.7	476.8	131.8	73.4	27.9	**250.7**	**3,008.6**	**505.7**	**27.9**
1983	0.0	13.5	15.9	141.5	300.0	504.6	291.4	341.4	292.4	296.4	114.1	67.0	**216.2**	**2,594.4**	**504.6**	**13.5**
1984	0.0	24.8	187.8	143.2	312.0	437.3	274.5	121.2	414.5	310.4	273.4	TRACE	**249.91**	**2,749.01**	**437.3**	**24.8**
1985	16.8	0.0	153.6	250.0	520.5	456.6	370.1	326.3	403.2	313.3	148.9	4.5	**267.90**	**3,214.9**	**520.5**	**0.0**
1986	0.0	72.1	283.8	118.5	274.3	152.8	533.8	264.1	404.3	414.6	92.2	0.0	**237.32**	**2,847.8**	**533.8**	**0.0**
1987	24.3	37.4	194.2	258.8	337.6	375.9	350.8	493.8	329.5	399.2	112.3	94.8	**271.3**	**3,255.6**	**493.8**	**37.4**
1988	51.2	31.1	156.1	278.6	224.1	507.9	367.0	147.3	538.3	233.7	131.6	56.3	**242.90**	**2,914.90**	**538.3**	**31.1**
1989	0.0	0.0	121.5	341.8	225.6	540.5	626.5	286.5	279.0	218.3	95.9	TRACE	**273.6**	**3,009.2**	**626.5**	**0.0**
1990	20.8	Trace	5.4	53.6	341.0	473.6	702.7	225.7	336.5	310.0	203.3	56.5	**270.83**	**2,979.13**	**702.7**	**5.4**
1991	Trace	11.0	88.3	259.3	258.4	479.0	439.6	505.6	169.8	359.0	72.2	19.7	**242.0**	**2,903.9**	**505.6**	**11.0**
1992	2.8	0.5	271.5	206.3	161.8	337.7	455.1	400.2	481.4	408.5	120.7	0.1	**258.5**	**3,102.3**	**481.4**	**0.1**

(Continued)

TABLE 16.2 (*Continued*)
Climatic (Rainfall) Data for Calabar (1971–2014)

Year	Jan. (mm)	Feb. (mm)	Mar. (mm)	Apr. (mm)	May (mm)	Jun. (mm)	Jul. (mm)	Aug. (mm)	Sept. (mm)	Oct. (mm)	Nov. (mm)	Dec. (mm)	Mean (mm)	Annual (mm)	Max. (mm)	Min. (mm)
1993	64.3	39.9	177.7	161.2	230.1	281.0	272.4	479.1	420.5	242.6	119.7	22.8	**222.5**	**2,669.5**	**479.1**	**22.8**
1994	40.7	0.0	167.3	328.2	255.3	249.2	609.5	424.5	290.0	265.2	229.7	0.0	**256.3**	**3,075.2**	**609.5**	**0.0**
1995	Trace	21.1	366.1	248.3	208.6	375.2	632.4	467.4	696.8	496.1	168.0	69.7	**340.9**	**4,090.6**	**696.8**	**21.1**
1996	2.4	162.2	161.5	314.5	299.5	435.3	401.1	425.1	615.2	357.4	40.5	0.6	**292.1**	**3,505.0**	**615.2**	**0.6**
1997	61.0	0.0	139.4	228.1	328.2	633.1	796.6	492.6	211.2	319.1	212.3	68.2	**311.8**	**3,740.5**	**796.6**	**0.0**
1998	25.4	6.0	174.0	149.1	211.6	504.5	255.2	353.7	365.0	437.1	396.3	33.6	**262.4**	**3,148.5**	**504.5**	**6.0**
1999	86.0	49.8	223.0	311.4	180.0	270.3	349.9	494.5	368.3	463.7	207.3	0.3	**265.3**	**3,183.8**	**494.5**	**0.3**
2000	66.0	0.0	951.9	166.4	217.0	250.6	597.9	392.0	577.6	232.9	153.6	57.5	**327.0**	**3,924.4**	**951.9**	**0.0**
2001	0.0	11.6	151.7	371.8	419.4	390.5	268.5	457.0	455.7	381.0	217.1	5.7	**284.5**	**3,414.5**	**457**	**5.7**
2002	0.0	13.5	154.6	383.3	301.3	344.6	274.1	623.5	284.3	285.8	126.0	6.8	**254.3**	**3,052.1**	**623.5**	**6.8**
2003	26.7	103.2	226.6	283.0	315.3	202.2	327.4	398.6	399.2	224.1	148.5	2.9	**239.2**	**2,870.2**	**399.2**	**2.9**
2004	9.9	19.9	73.5	278.4	270.2	308.0	303.5	391.9	335.5	196.4	168.3	0.6	**213.3**	**2,559.5**	**391.9**	**0.6**
2005	33.8	35.5	295.7	299.9	263.9	615.6	828.2	634.4	230.4	279.8	182.3	71.5	**339.7**	**4,076.9**	**828.2**	**35.5**
2006	84.7	57.1	323.0	166.1	430.8	227.7	484.0	273.4	536.0	175.3	134.4	0.1	**255.3**	**3,063.2**	**536.0**	**0.1**
2007	0.0	51.1	181.0	265.9	384.2	583.5	492.7	415.5	561.7	197.4	262.1	33.1	**311.7**	**3,739.9**	**583.5**	**33.1**
2008	15.1	1.0	108.1	216.9	386.8	437.0	597.7	509.2	217.9	315.0	105.1	77.1	**270.2**	**3,242.0**	**597.7**	**1.0**
2009	89.7	38.5	87.5	150.5	308.9	218.4	577.4	507.3	273.9	148.1	126.9	0.0	**221.6**	**2,659.0**	**577.4**	**0.0**
2010	31.8	88.2	63.6	130.4	306.5	611.3	384.0	406.7	451.3	269.6	272.1	56.2	**276.4**	**3,316.3**	**611.3**	**56.2**
2011	Trace	153.4	123.1	208.8	340.9	388.6	468.6	573.7	251.8	519.9	235.2	43.8	**300.7**	**3,608.5**	**573.7**	**43.8**
2012	32.6	376.7	36.0	99.9	439.4	398.8	637.1	861.3	619.4	410.4	126.5	30.6	**366.9**	**4,403.0**	**861.3**	**30.6**
2013	141.0	83.7	231.4	286.9	466.9	458.6	477.0	411.1	340.4	306.4	220.9	81.1	**292.1**	**3,505.4**	**477.0**	**81.1**
2014	248.0	61.6	366.2	245.0	332.2	220.0	714.4	410.3	501.5	255.0	136.8	136.8	**302.3**	**3,627.8**	**714.4**	**61.6**

Unit of measurement = millimeter (mm); trace = rain not measurable.

Source: Nigeria Meteorological Agency (NIMET).

TABLE 16.3
Monthly Annual Temperature and Rainfall for the Study Area

Month	Temperature (°C)	Rainfall (mm)
January	23.70	32.30
February	28.70	26.70
March	28.40	165.80
April	27.90	199.10
May	27.30	287.60
June	26.30	404.40
July	25.40	456.10
August	25.20	407.20
September	25.80	404.60
October	26.10	343.30
November	26.90	145.50
December	27.10	28.10
	Summary	
Maximum	28.70	456.10
Minimum	23.70	26.70
Mean	26.60	242.00
Total	**318.80**	**2,900.70**

Source: Adapted from Nigeria Meteorological Agency (NIMET), Calabar Station.

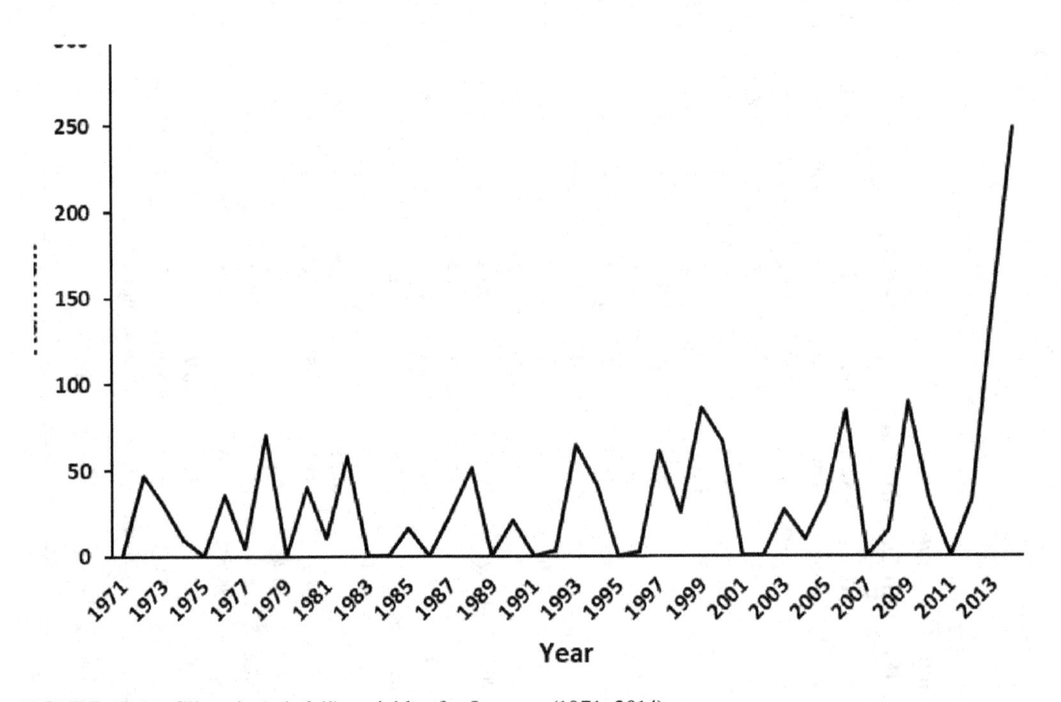

FIGURE 16.1 Climatic (rainfall) variables for January (1971–2014).

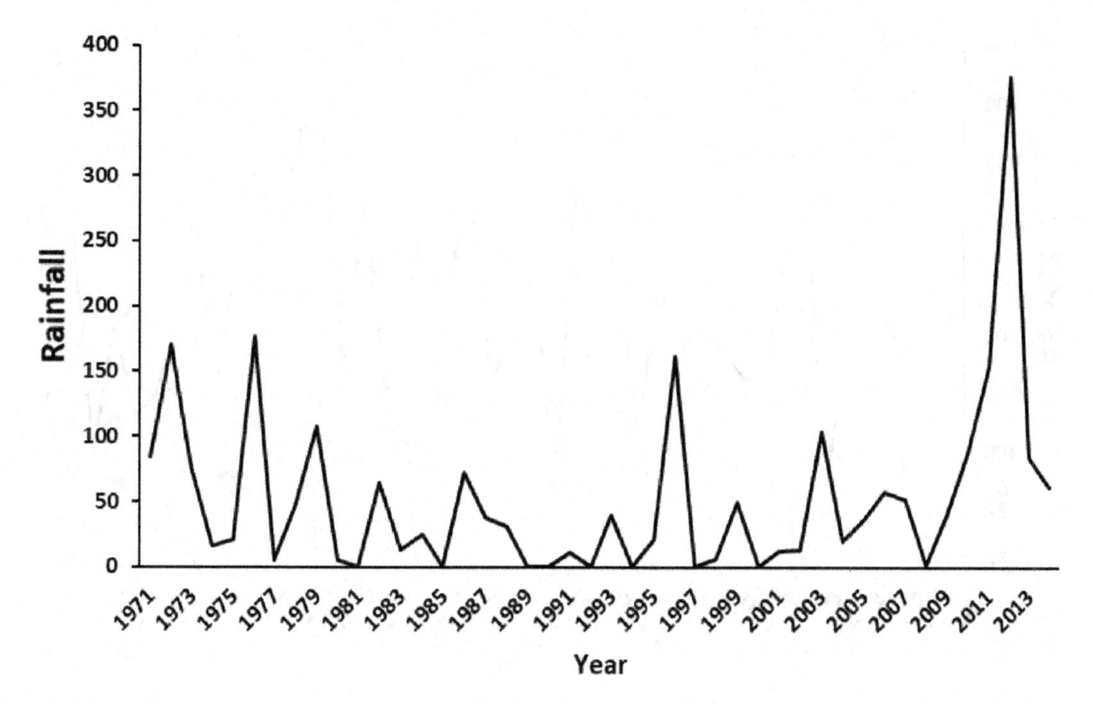

FIGURE 16.2 Climatic (rainfall) variables for February (1971–2014).

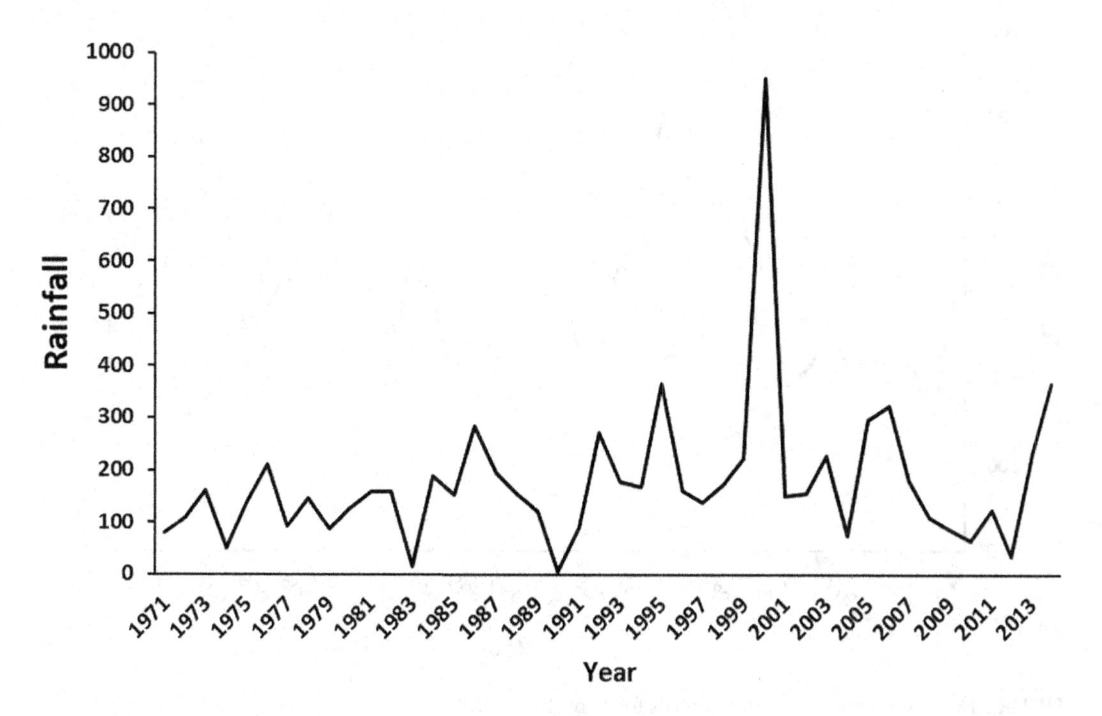

FIGURE 16.3 Climatic (rainfall) variables for March (1971–2014).

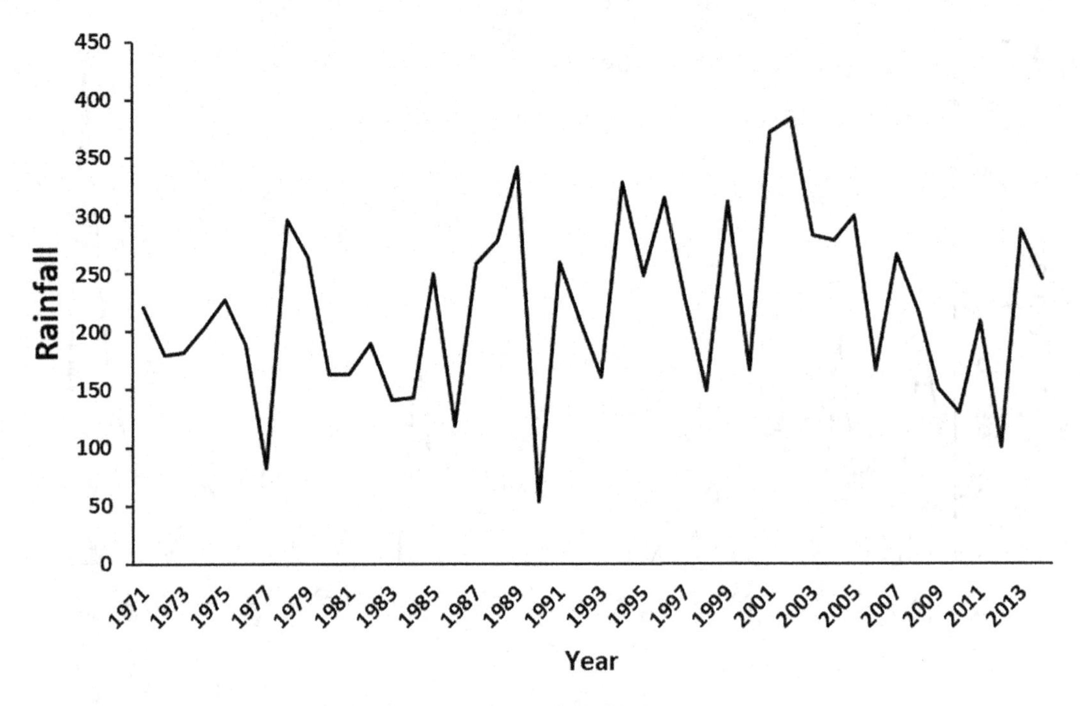

FIGURE 16.4 Climatic (rainfall) variables for April (1971–2014).

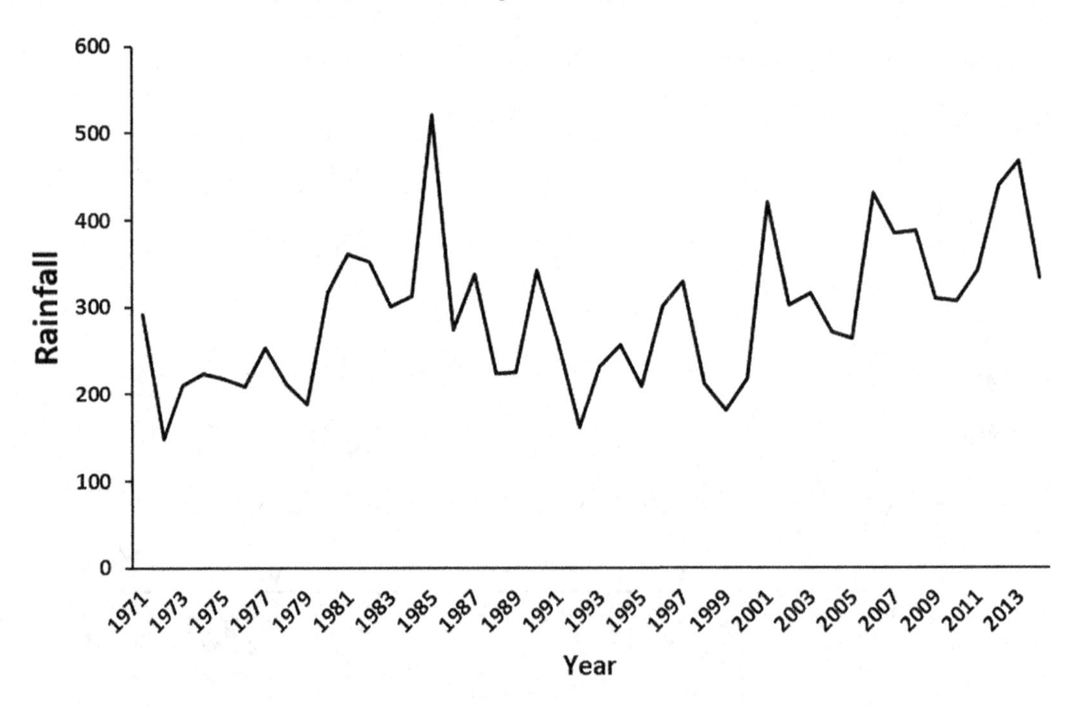

FIGURE 16.5 Climatic (rainfall) variables for May (1971–2014).

the sticky heat of Calabar where the air is always charged with water vapor. The high temperatures with high humidity are more oppressive than hot but dry conditions (Iloeje, 1979). The above trend depicts significant climate change in Calabar, evidenced by late onset and early cessations. The late onset and early cessation of rains have revealed the contraction of the length of the rainy season. This has impacted negatively on farming practices in the region, which has also changed

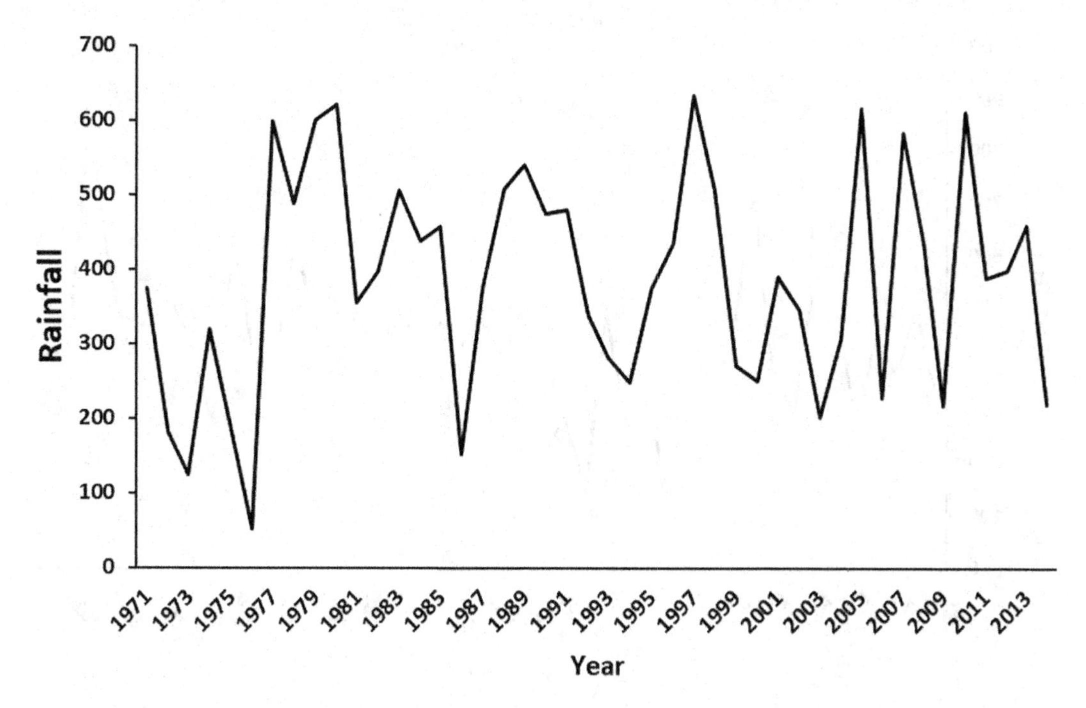

FIGURE 16.6 Climatic (rainfall) variables for June (1971–2014).

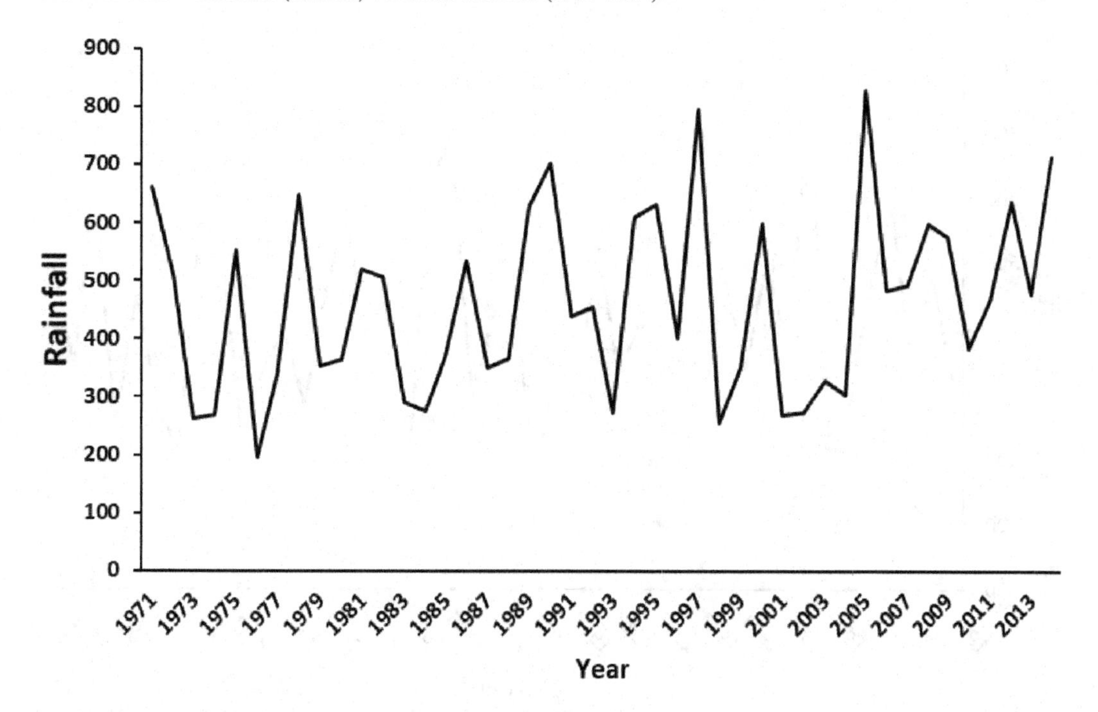

FIGURE 16.7 Climatic (rainfall) variables for July (1971–2014).

in line with this observed pattern. There is also evidence of significant changes in known weather patterns in the region. For example, the little dry season, then commonly known as August Break, has become less significant in the region. Similarly, analysis has revealed that the environment has become warmer as temperatures have risen considerably and Harmattan dust haze has also become more pronounced in recent years.

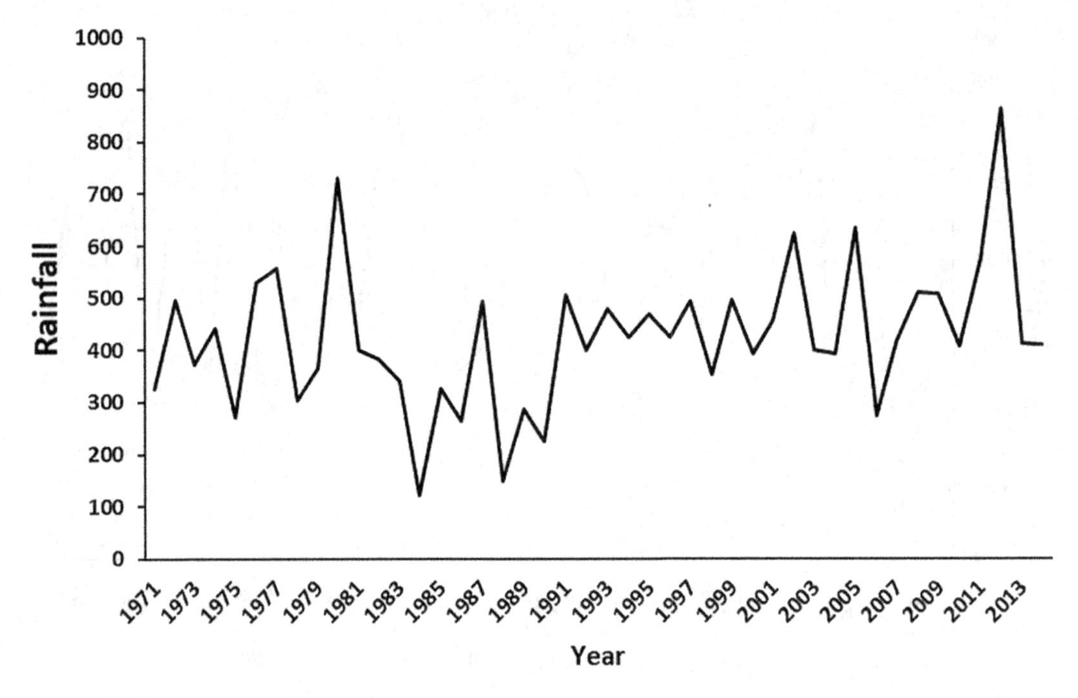

FIGURE 16.8 Climatic (rainfall) variables for August (1971–2014).

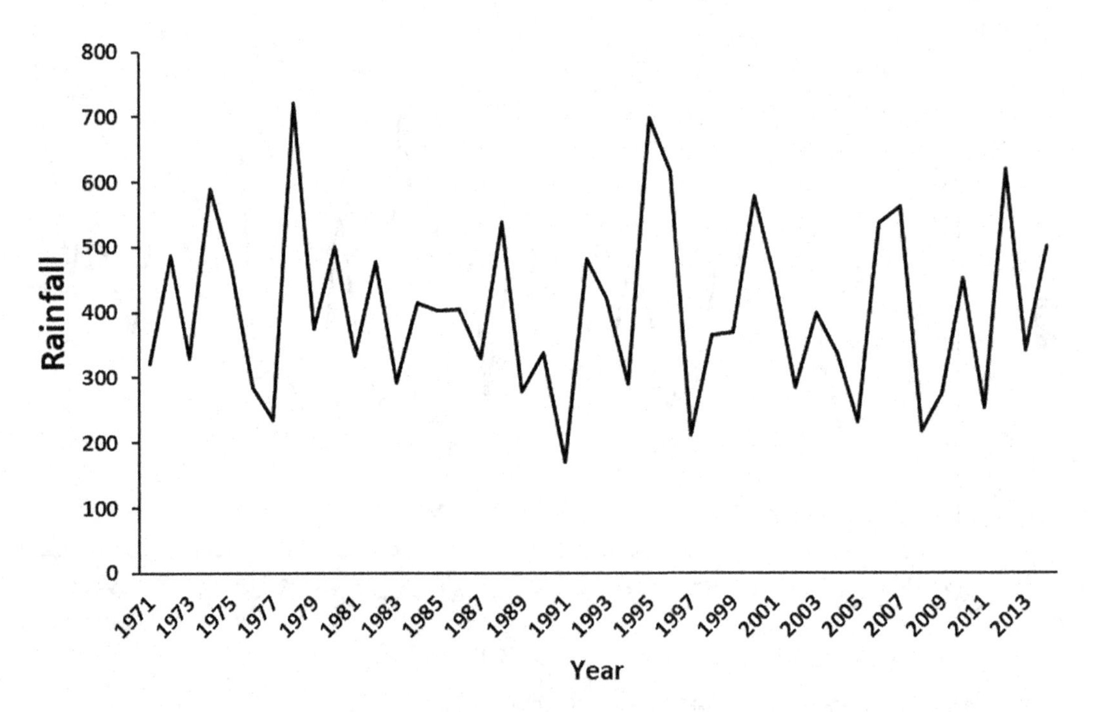

FIGURE 16.9 Climatic (rainfall) variables for September (1971–2014).

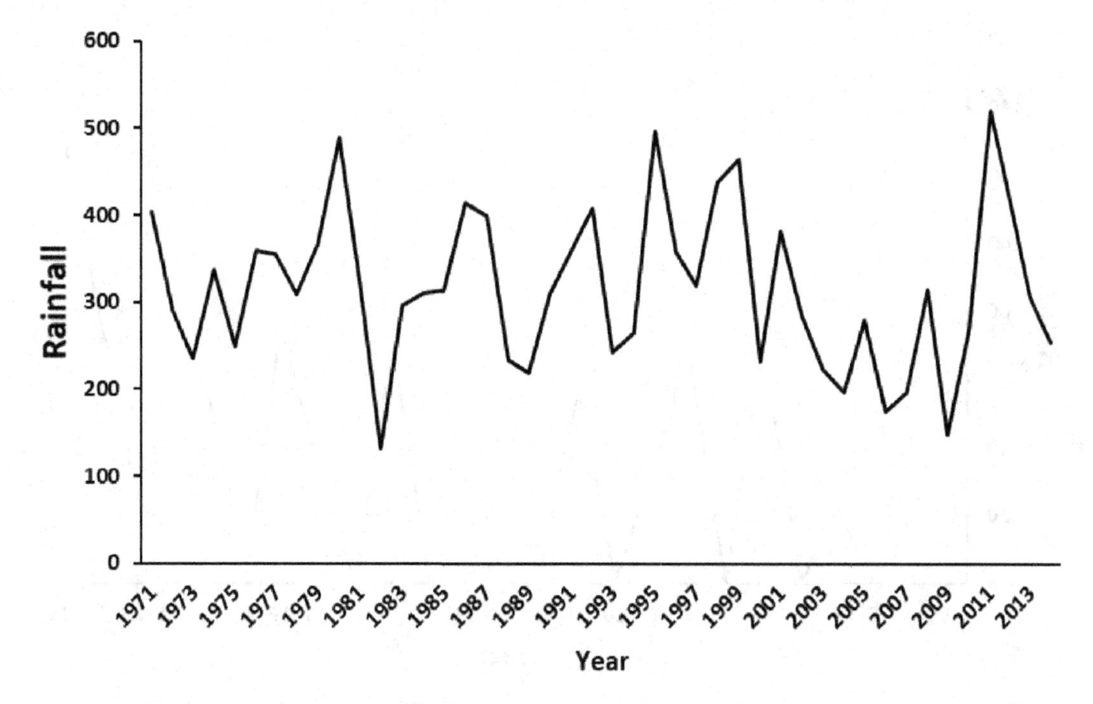

FIGURE 16.10 Climatic (rainfall) variables for October (1971–2014).

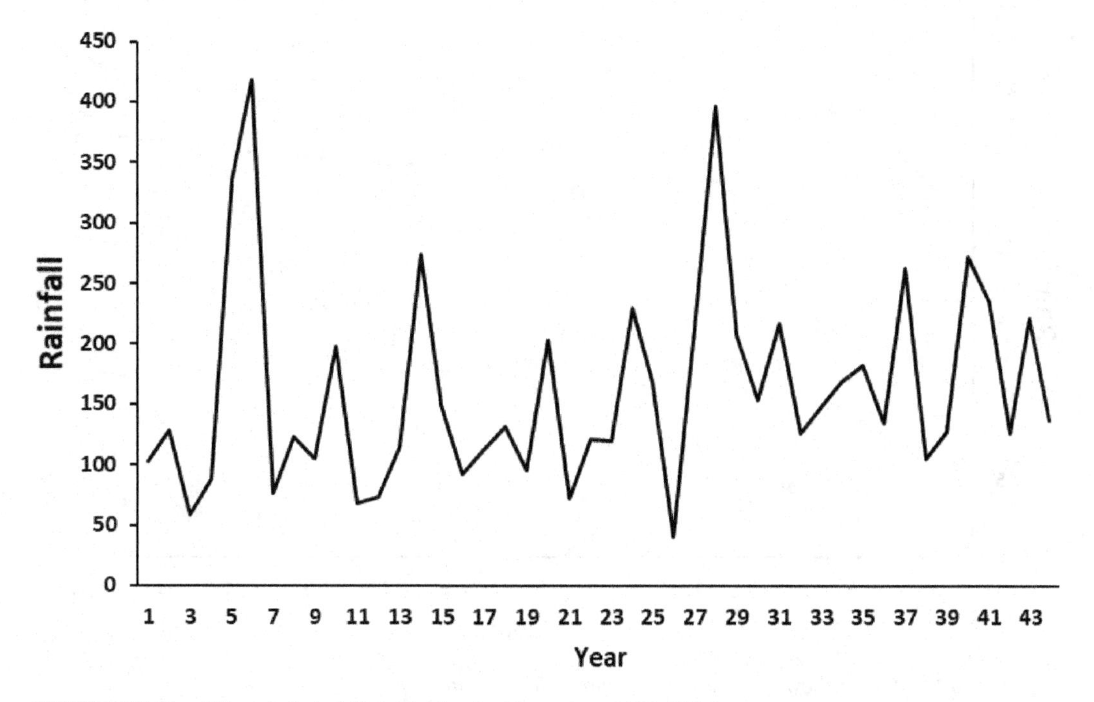

FIGURE 16.11 Climatic (rainfall) variables for November (1971–2014)

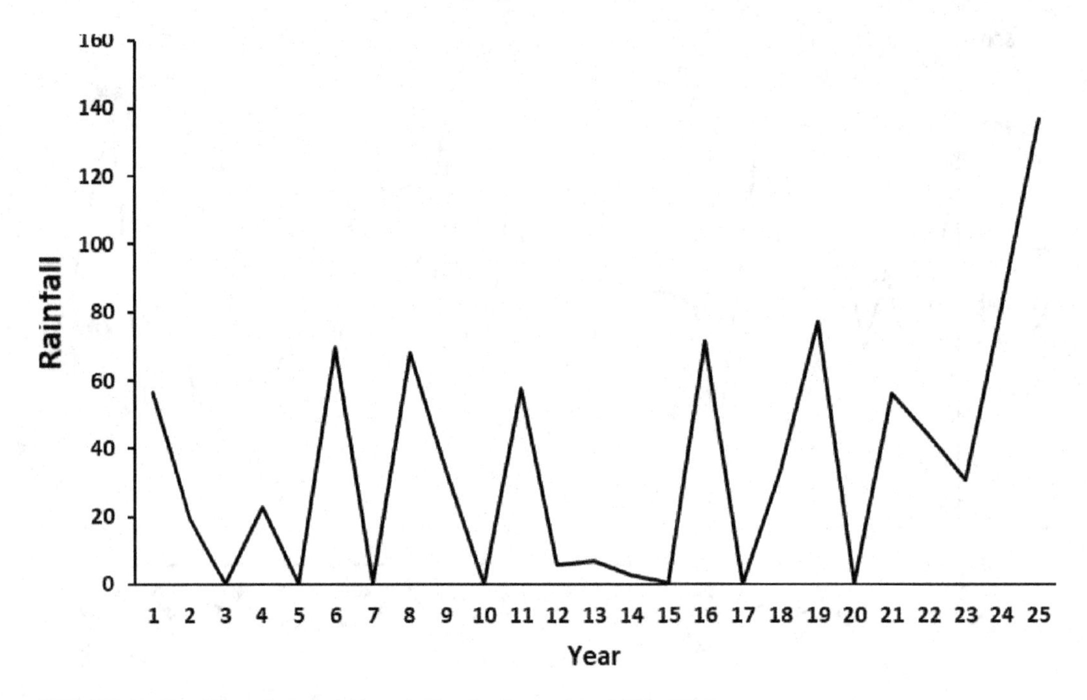

FIGURE 16.12 Climatic (rainfall) variables for December (1971–2014).

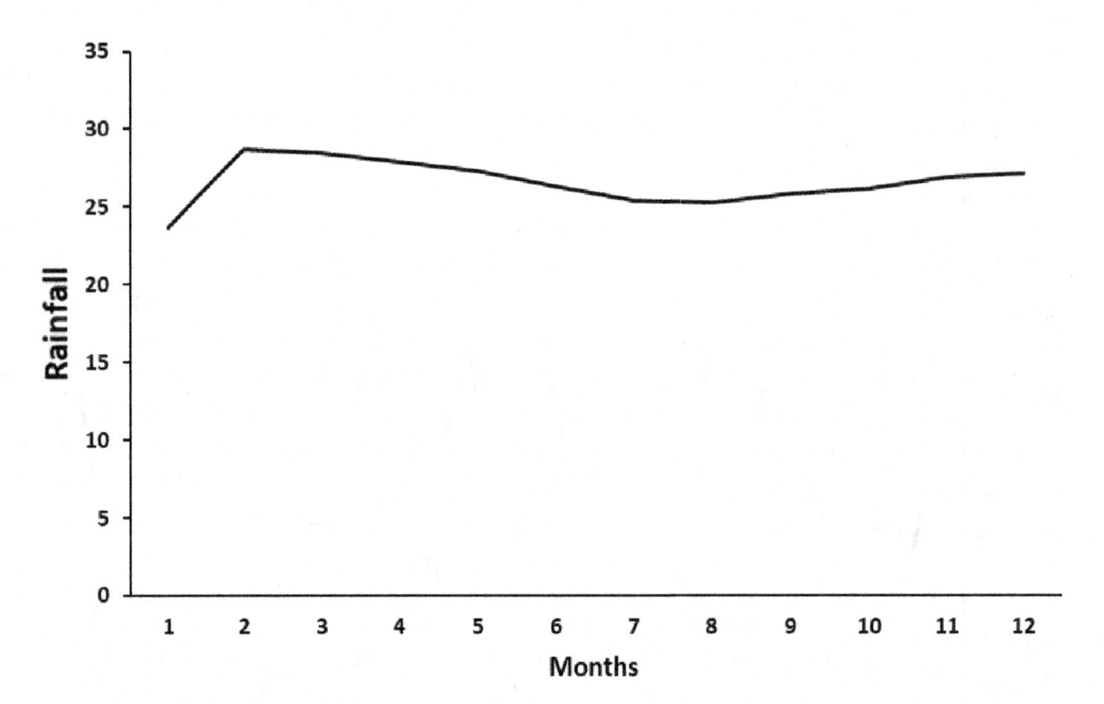

FIGURE 16.13 Graph showing average temperature of study area.

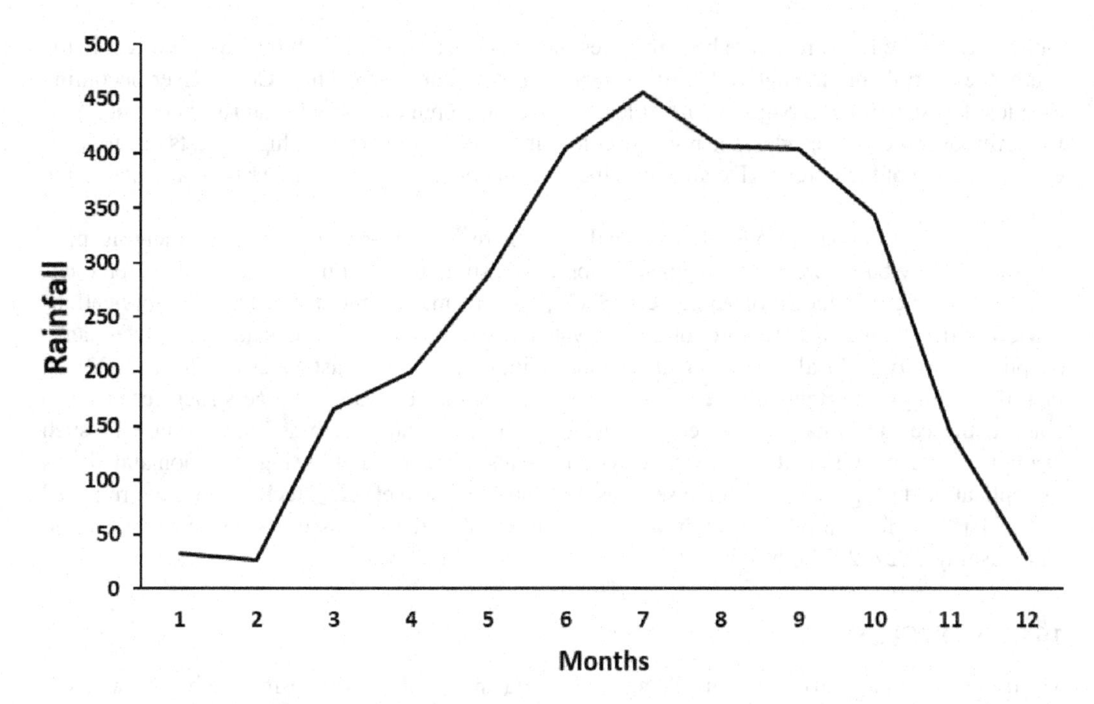

FIGURE 16.14 Graph showing average rainfall of study area.

Below shows the trend or evidence of climate change between 1941–1970 and 1971–2000, based on NIMET's historical meteorological data of Nigeria. During the middle of the 20th century (1941–1970), Calabar had normal or above normal onset dates of rainy season. As the century progressed (1971–2000), most parts of the country, including Calabar till date, have early cessation.

16.7 DISCUSSIONS AND RECOMMENDATIONS

This study conducted analysis of climatic data of the Nigerian Meteorological Agency (NIMET), Calabar Station, that span 43 years (1971–2014) of the study area and compared it against NIMET's historical meteorological maps of Nigeria between 1941–1970 and 1971–2000, to estimate climate variability and change. The comparison revealed a historical sequential rise in temperature, evidenced by late onset and early cessation of the rains. The late onset and early cessation of rains have necessitated the contraction of the length of the rainy season. This has impacted negatively on farming practices in the region. Furthermore, there is also evidence of significant changes in known weather patterns in the region. For example, the little dry season, then commonly known as August Break, has become less significant in the region. Similarly, the analysis has revealed that the environment has become warmer as temperatures have risen considerably and Harmattan dust haze has also become more pronounced in recent years. The evaluation of climate change pattern of Calabar river basin was necessitated owing to its distinct socio-economic benefit to Nigeria. The region plays host to Africa's foremost leisure resort (Tinapa), one of world's largest rubber plantations (Pamol), the National Integrated Power Project (NIPP), Nigerian Police Training School, etc. In Nigeria, Cross River State has over 40% of the remaining tropical high forests (THFs) of the entire nation. The forest resource base includes the mangrove swamps and tropical rainforest in the south, the central, and the derived Guinea savanna toward the north. Thus, Cross River becoming the most forested State in Nigeria with at least 75% of its population inhabiting rural communities. However, over the last decades, the region has lost about 19% of its tropical high forests due to inadequate funding of the Forestry Department, increase in population and immigration, and plantation establishment.

Considering the conceptions of water balance, there is a strong need of implementing public-participatory-based water management approaches so as to minimize impacts of extreme climatic events such as floods, droughts, etc. Such practices may demand user-friendly applications in view of developing surface water or groundwater budgeting tools. This can additionally enhance adaptive capacity in local masses against global climate issues (Srivastava et al., 2021, 2022). In fact, the findings on establishing cascading networks of surface water storage structures so as to control surface runoff have remained promising in similar regions such as South Asia. One such example includes traditional tank cascade systems that stores rainwater during monsoon and allows its application during non-monsoonal seasons. Similar green and efficient technologies are required to be identified for the present study area so as to combat the discussed issues (Srivastava and Chinnasamy, 2022, 2023a, b).

16.8 CONCLUSION

The trends of climate variability and change in Nigeria and Calabar river basin have been evaluated. Climate variability and change in Nigeria and Calabar result from human activities like deforestation and burning of fuel (in machines and vehicles) into the atmosphere that contributes to these phenomena. It has been observed that there is a consistent and persistent increase in temperature as well as variability in rainfall pattern in the region resulting in late onset and early cessation of the rains, warmer environment, and pronounced Harmattan dust haze. Recommendations are made on the control of greenhouse effect to prevent future dangers. This work in no doubt is designed to add to the pool of existing knowledge in the field of global warming and climate change. To ameliorate the effects of greenhouse gases, there is need for improved energy efficiency, shift to renewable resources/cleaner source of energy (solar and wind), reduced deforestation and planting of trees. There should be absolute compliance with international organizations' action plans that fight global warming, e.g., the Montreal Protocol of 1987, the 1979 convention on long-range trans-boundary air pollution, the Kyoto Protocol of 1997, the clean air act (1990), UNCED (1992), USA (2008) congress, etc.

REFERENCES

Abali, T. P. (2016). Land cover Dynamics and Sediment Loss from Land use Types in Calabar River Catchment, Cross River State. Ph.D Thesis, University of Port Harcourt.

Abali, T. P (2021). Climate change pattern in Calabar, Cross River State, Nigeria. In: Chukwu-Okeah, G. O., Amangabara, G. T., & Elenwo, E. I. (Eds.), *Man, Environment and Sustainable Development: Essays in Honour of Professor Charles Uwadiae Oyegun, University of Port Harcourt, Nigeria* (pp. 132–179). Amajov & Co., Port Harcourt, Nigeria.

Adedeji, O., Reuben, O., Olatoye, O. (2014). Global climate change. *Journal of Geoscience and Environment Protection*, 2, 114–122.

Akintoye, O. A. (2002). Impact of Logging on Non-logged Species and Effects on Rural Socio-economic Development in Ikom LGA. M.Sc Thesis, University of Calabar.

Balogun, P. (1994). The Impact of Social and Economic Issues on Forestry in Cross River State, Nigeria, Cross River Forestry Department, Calabar.

Christensen, J.H., Hewitson, B., Busuioc, A., Chen, A., Gao, X., Held, I., Jones, R., Kolli, R. K., Kwon, W.-T., Laprise, R., Magaña Rueda, V., Mearns, L., Menéndez, C. G., Räisänen, J., Rinke, A., Sarr, A., Whetton, P. (2007). Regional climate projections. In: Solomon, S., Qin, D., Manning, M., Chen, Z., Marquis, M.,

Averyt, K. B., Tignor, M., Miller, H. L. (eds.) *Climate Change: The Physical Science Basis. Contribution of Working Group I to the Fourth Assessment Report of the Intergovernmental Panel on Climate Change.* Cambridge University Press, Cambridge.

Cross River Basin Development Authority, CRBDA. (1982). Inventory of natural site conditions, soil slopes, hydrology, land use and vegetation throughout the area of operation of the authority. Progress Report, 4.

Cross River State Forestry Commission. (1988, 1990, 2000, 2001 and 2003). Annual Report. Forestry Commission, Calabar.

Cunningham, W. P., Cunningham, M. A., Siago, B. (2005). *Environmental Science: A Global Concern.* McGraw – Hill, New York (Chapter 12).

Dhameja, S. K. (2007). *Environmental Studies.* S. K. Kataria & Sons, Delhi.

Dunn, J. (1994). *An Overview of Agroforestry in Cross River State.* Cross River Forestry Project (ODA Assisted), Calabar.

Egbai, O. O., Ndik, E. J. (2011). Influence of soil textural properties and land use cover type on erosion of a characteristic ultisols in Betem, Southern Nigeria. *Journal of Soil and Environmental Management* (in press).

Federal Ministry of Environment (FME). (2003). Nigeria's First National Communication under the United Nations Framework Convention on Climate Change. The Federal Ministry of Environment, Abuja. Landmark Publications, Lagos, Nigeria.

Fubara, M. J., Alabo, D. (2007). Sustainable flood hazard mitigation in the Niger Delta In: Oyegun, C. U. (eds.). Climate change and Nigeria's coastal resources. An Inaugural Lecture Series No 56, University of Port Harcourt.

Goudie, A. S. (2004). *Encyclopedia of Geomorphology*, Routledge, New York.

Guernier, V., Hochberg, M. E., Guegan, J. F. (2004). Ecology drives the worldwide distribution of human diseases. *PLOS Biology*, 2(6), 740–746.

Intergovernmental Panel on Climate Change (IPCC). (2001). Climate change 2001: Third Assessment Report, Impacts, Adaptation and Vulnerability of Climate Change. In: McCarthy, J. J., et al. (eds.) Cambridge University Press, Cambridge UK.

Khadke, L., Pattnaik, S. (2021). Impact of initial conditions and cloud parameterization on the heavy rainfall event of Kerala (2018). *Modeling Earth Systems and Environment*, 7, 2809–2822. https://doi.org/10.1007/s40808-020-01073-5

Manstrandrea, M. D., Schneider, S. H. (2009). Global Warming. Microsoft® Encarta® Online Encyclopedia.

McMichael, A. J., Campbell-Lendrum, D., Kovats, R. S., Edwards, S., Wilkinson, P., Edmonds, N., Nicholls, N., Hales, S., Tanser, F. C., Le Sueur, D., Schlesinger, M., Andronova, N. (2004). Climate Change in Comparative Quantification of Health Risks: Global and Regional Burden of Disease due to Selected Major Risk Factors. In: Ezzati, M., Lopez, A. D., Rodgers, A., Murray, C. J. L. (eds.) Ch. 20, pp. 1543–1649. World Health Organization, Geneva.

Nigeria Environmental Study/Action Team, NEST (1991). *Nigeria Threatened Environment: A NEST Publication.* INTEC Printers Limited, Ibadan.

Nigerian Meteorological Station NIMET (2007). Nigeria Climate Review Bulletin, In: Akenge, N. U (2008). *The Environment Outreach: Nigeria's Environment and Development*, 1(1), 32–39.

Ochuko, E. E. (2009). Climate change environmental implications in global warming/green housing effects in Warri, Nigeria. *International Journal of environmental Science* 5(3), 145–156.

Ogar, T. O. (2006). Forest Distribution from Logging Operations in Etung Local Government Area, Cross River State. M.Sc Thesis, University of Port Harcourt.

Ogbonaya, O. (1996). Forest policy and strategy: An NGO's vision, In: Obot, E., Baker, J. (eds.): Essential Partnership: The Forest and the People: Proceedings of Workshop on rainforest of South Eastern Nigeria and Cameroun. Obudu Cattle Ranch and Resort. 20th–24th October, pp. 156–160.

Okebukola, P. A. O. (1997). Things Worth Knowing About Global Warming and Ozone Layer Depletion: Onitsha In: Peter Okebukola, P., Ben, A. (eds.) Strategies For Environmental Education, Science Teachers Association of Nigeria (STAN). Environmental Education Series, African-Feb Ltd, Onitsha.

Oyegun, C. U. (1993). Land degradation and the coastal environment of Nigeria. *Catena*, 20, 215–225.

Oyegun, C. U. (2007). Climate change and Nigeria's coastal resources. An Inaugural Lecture Series No 56, University of Port Harcourt.

Srivastava A., Chinnasamy P. (2021a). Developing village-level water management plans against extreme climatic events in Maharashtra (India)—a case study approach. In: Vaseashta A., Maftei C. (eds) *Water Safety, Security and Sustainability.* Advanced Sciences and Technologies for Security Applications. Springer, Cham. https://doi.org/10.1007/978-3-030-76008-3_27

Srivastava, A., Chinnasamy, P. (2021b). Investigating impact of land-use and land cover changes on hydro-ecological balance using GIS: insights from IIT Bombay, India. *SN Applied Science*, 3, 343. https://doi.org/10.1007/s42452-021-04328-7

Srivastava A., Chinnasamy P. (2022). Assessing groundwater depletion in Southern India as a function of urbanization and change in hydrology: a threat to tank irrigation in Madurai City. In: Kolathayar, S., Mondal, A., Chian, S.C. (eds.) *Climate Change and Water Security. Lecture Notes in Civil Engineering*, vol 178. Springer, Singapore. https://doi.org/10.1007/978-981-16-5501-2_24

Srivastava, A., Chinnasamy, P. (2023a). Tank cascade system in Southern India as a traditional surface water infrastructure: a review. In: Chigullapalli, S., Susha Lekshmi, S.U., Deshpande, A.P. (eds.) *Rural Technology Development and Delivery. Design Science and Innovation.* Springer, Singapore. https://doi.org/10.1007/978-981-19-2312-8_15

Srivastava, A., Chinnasamy, P. (2023b). Understanding declining storage capacity of tank cascade system of Madurai: potential for better water management for rural, peri-urban, and urban catchments. In: Chigullapalli, S., Susha Lekshmi, S. U., Deshpande, A. P. (eds.) *Rural Technology Development and Delivery. Design Science and Innovation.* Springer, Singapore. https://doi.org/10.1007/978-981-19-2312-8_14

Srivastava, A., Khadke, L., Chinnasamy, P. (2021). Web application tool for assessing groundwater sustainability—a case study in Rural-Maharashtra, India. In: Vaseashta, A., Maftei, C. (eds.) *Water Safety, Security and Sustainability. Advanced Sciences and Technologies for Security Applications.* Springer, Cham. https://doi.org/10.1007/978-3-030-76008-3_28

Srivastava A., Khadke L., Chinnasamy P. (2022). Developing a web application-based water budget calculator: attaining water security in rural-Nashik, India. In: Kolathayar, S., Mondal, A., Chian, S.C. (eds.) *Climate Change and Water Security. Lecture Notes in Civil Engineering*, vol 178. Springer, Singapore. https://doi.org/10.1007/978-981-16-5501-2_37

United Nations Development Programme, UNDP (1995). Studies on erosion, flood and landslide in Abia, Akwa Ibom, Cross River and Imo States. *Baseline Report*, 3, 10–15.

United Nations Development Programme, UNDP (2006). Human Development Report 2006. Beyond Scarcity: Power, poverty and the global water crisis, United Nations Development Programme. http://hdr.undp.org/hdr2006/report.cfm

United Nation's Framework Convention on Climate Change, UNFCCC (2007). Climate Change: Impacts, Vulnerabilities and Adaptation in Developing Countries. UNFCCC Secretariat. Bonn, Germany.

17 Assessment of Hydrochemistry and Water Quality Index for Groundwater Quality – A Case Study of Maharashtra

K. R. Aher, S. M. Deshpande, M. L. Dhumal, R. O. Yenkie, and A. M. Varade

17.1 INTRODUCTION

Groundwater, being a primary source of water supply in arid and semi-arid regions, it is extensively used for drinking, irrigational, and industrial purposes as well. According to world water usability statistics, 65% of groundwater is used for drinking purposes, 20% for agricultural and livestock purposes, and 15% for the industrial and mining sectors (Saied et al., 2018), and approximately one-third of the world's population relies on groundwater for drinking purposes (UNEP, 1999; Adimalla and Venkatayogi, 2018; Adimalla et al., 2020). Due to factors such as geology, climate, hydrochemical conditions, and natural/anthropogenic conditions, the occurrence and movement of groundwater are highly uncertain (Das, 2008; Khare and Varade, 2018; Aher et al., 2019). Groundwater measures have become indispensably significant, which should be embraced in semi-arid areas to fulfill increasing demand and deteriorating water quality (Srivastava et al., 2021, 2022). The groundwater hydrochemistry is constrained by hydrogeology, evaporation and precipitation, and anthropogenic activities, which in turn control the concentrations of major and minor ions in groundwater (Singh et al., 2011; Rajesh et al., 2012). Surface water and groundwater contamination is characterized as the creatively actuated debasement of natural water quality. Contamination can hinder water use and make perils to general wellbeing through toxicity or the spread of disease. Most pollution originated with the disposal of wastewater following the use of water for a wide variety of purposes. Accordingly, many sources and causes can alter groundwater quality, ranging from the septic tank to irrigated agriculture. Conversely, with surface water pollution, sub-surface pollution is hard to recognize, considerably more challenging to control, and may endure for quite a long time. With the improving acknowledgment of the significance of underground water assets, endeavors are expanding to forestall, diminish, and dispose of groundwater contamination. Factors such as the composition of recharge water, soil-rock-water interaction, the residence time of groundwater, and the reactions that ensue within the aquifers define the groundwater quality of the region (Freeze and Cherry, 1979; Fetter, 1994; Appelo and Postma, 2005; Batabyal, 2018). Similarly, the quality of the water also gets affected by various anthropogenic activities. In rural areas, the population is widely dependent on groundwater for drinking and irrigation purposes, due to which water quality has turned into a vital concern. Furthermore, surface water storage structures such as lakes, ponds, cascading tanks, and irrigation structures also play a vital role in maintaining the quality and quantity (health) of the groundwater (Batabyal and Chakraborty, 2015), such that their wide-spread applications can be seen in the drylands of Maharashtra and South India (Srivastava

DOI: 10.1201/9781003303237-17

and Chinnasamy, 2021, 2022, 2023a, b). The hydrogeochemical properties of water are an essential factor determining its use for domestic, irrigation, and industrial purposes. The interaction of water with lithologic units through which it flows significantly controls the chemistry and quality of water (Liu et al., 2003; Subramani et al., 2009; Şener et al., 2017).

Anthropogenic pollutants carried from the surface into the aquifer as a result of leaks, spills, and haphazard chemical application at the surface have an impact on the quality of groundwater. Furthermore, irrigation returns recharge shallow aquifers with water from deeper aquifers, and seawater intrusion in coastal aquifers owing to overexploitation are both factors that degrade groundwater quality (Boniol, 1996; Kazakis et al., 2017). The worsening of groundwater quality has also resulted in a reduction in the quantities that may be exploited. Groundwater quality degradation is caused by two major factors: geogenic processes and anthropogenic activities. Excessive fertilizer use, for example, has resulted in nitrate contamination of groundwater considerably over water quality requirements in agricultural areas (Machiwal et al., 2011, 2018; Paradis et al., 2016). Groundwater quality must be assessed in order to determine suitability and design more sustainable groundwater management to fulfil present and future drinking water demands. The examination of the physicochemical qualities of the soil and their effects on crop yield is required for the assessment of water quality requirements. Numerous studies have found that urban development and agricultural activities have an impact on groundwater quality, either directly or indirectly. Regular fertilizer treatments on cropped fields and a process that accumulates nutrients in groundwater are the most common causes of agricultural contamination of groundwater. Nitrate is the most common contaminant in groundwater, and it is produced by both urban and agricultural activities (Wen et al., 2005; Srinivasa Rao et al., 1997; Zhu et al., 2009). Over-use of manure and fertilizers can contaminate groundwater with NO3 and Cl (Subbarao et al., 1995; Anku et al., 2008; Vasanthavigar et al., 2012).

For human consumption, the suitability of water sources is defined as the water quality index (WQI) and considered the best way to define water quality. It is observed that the use of individual water quality variables to explain the water quality for the common public is not easily understandable (Akoteyon et al., 2011; Bharti and Katyal, 2011; Lateef, 2011). The WQI technique makes use of water quality data and aids in the modification of regulations developed by various environmental monitoring agencies. The method simplifies and logically expresses the huge data by reducing the bulk of it to a single number (Stambuck-Giljanovic, 1999; Stigter et al., 2006; Reza and Singh, 2010; Saeedi et al., 2010; BabaeiSemiromi et al., 2011; Tyagi et al., 2013; Aher and Gaikwad, 2017; Aher et al., 2020; Deshpande et al., 2020; Dhumal, 2021). Therefore, WQI is one of the most popular methods for presenting water quality information to decision-makers. Presently, various workers involved in the field of hydrogeology, hydrology, environmental sciences, and researchers employ the WQI as a tool to evaluate the quality of drinking water (Chakraborty et al., 2007; Ramakrishnaiah et al., 2009; Vasanthavigar et al., 2010; Aly et al., 2014; Chung et al., 2014; Adimalla and Qian, 2019; Adimalla and Taloor, 2020; Deshpande et al., 2021). Considering the efficacy of WQI, the present work aims to evaluate groundwater quality and suitability by involving the computation of the WQI (Horton, 1965). The Yeola block under investigation represents a rural area where most of the population depends on groundwater, the primary source for drinking and irrigation purposes. The groundwater is being exploited through Dugand Borewells.

17.2 STUDY AREA SETUP

The Yeola block is situated in the western parts of Maharashtra, located between the latitudes 19°55′:20°15′ N and longitudes 74°15′:74°45′E and covers 1,064.47 km^2 area of the Yeola block, Nashik district (Figure 17.1). The drainages in the study are ephemeral and follow the general slope direction. During summer, the maximum temperature recorded is ~42.5°C, and the minimum temperature of ~5°C is experienced during winter. Relative humidity ranges from 43% to 62%. Rainfall

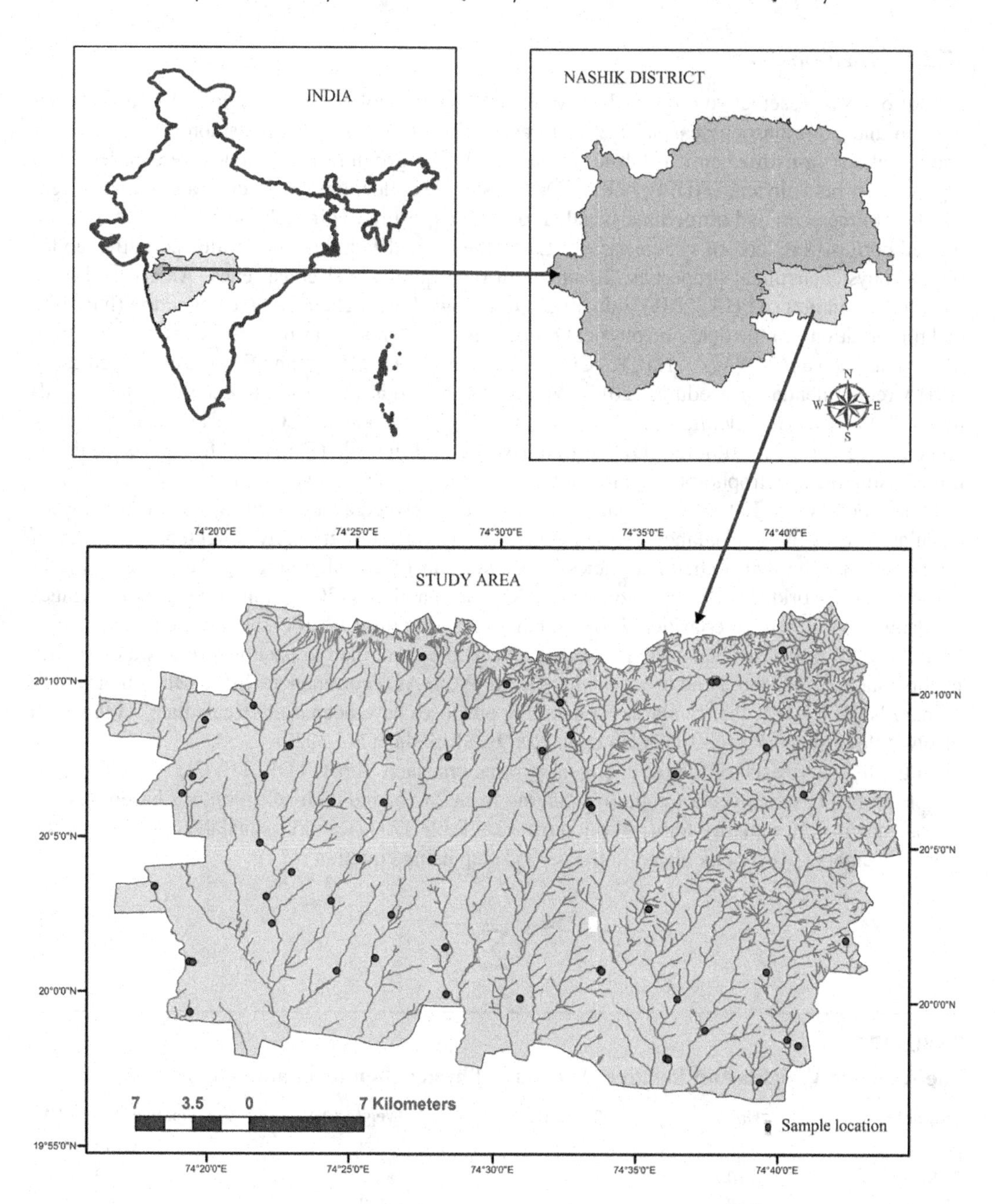

FIGURE 17.1 Location map of the study area.

is the primary water source, and the average annual rainfall in the study area is ~610.5 mm. The common crops in the area include bajra, cotton, wheat, sugarcane, etc. (CGWB, 2014). The Yeola block area is covered by the basaltic lava flows of the lower and upper Ratangarh Formation exposed in the ascending order of megacryst flows representing the Western Deccan Volcanic Province. Groundwater in alluvium occurs both under semi-confined and confined conditions. Significant rainfall was recorded during the Southeast monsoon, which prevailed from June to October. The rainfall, infiltration, and seepage of water from surface water bodies are responsible for the groundwater variation (Aher et al., 2019).

17.2.1 Methodology

A total of 55 representative dug and borewell samples were collected in the post-monsoon season of 2016 and pre-monsoon season of 2017 to ascertain the chemical composition of groundwater and variations in hydrochemical fabric (Figure 17.1). The groundwater samples were collected in high-density polyethylene (HDPE) bottles. During sample collection, the groundwater samples were filled up to the brim and immediately sealed to avoid exposure to air. Subsequently, all the sealed, packed bottles were labeled systematically. Later, the labeled samples were evaluated in the lab for several physicochemical properties. To maintain data quality and consistency, American Public Health Association (APHA, 2015) guidelines are followed throughout sample collection (handling and preservation). In the field, the physical properties of water, such as pH and EC, were measured. Major ions such as Ca^{++}, Mg^{++}, Na^{+}, K^{+}, CO_3^{-}, HCO_3^{-}, SO_4^{-}, Cl^{-}, NO_3^{-}, and F^{-} were examined using APHA-recommended procedures (2015). Volumetric methods were used to determine total hardness (TH) as $CaCO_3$, calcium (Ca^{++}), magnesium (Mg^{++}), carbonate (CO_3^{-}), bicarbonate (HCO_3^{-}), and chloride (Cl^{-}). The sulfate (SO_4^{-}), nitrate (NO_3^{-}), and fluoride (F^{-}) concentrations were determined using the spectrophotometric method. The flame photometer was used to determine sodium (Na^{+}) and potassium (K^{+}) among the analyzed ions. The verified accuracy of the chemical analysis, calculated from the ion balance method, falls within the internationally accepted error bars of around ±10%. The hydro-chemical facies (piper diagram) were plotted using AquaChem. v.2014 software. The World Health Organization (WHO, 2004) and BIS (2012) standard guidelines values for drinking water were considered to compare the suitability of groundwater for drinking purposes. The WQI is described as an index that reflects the combined impact of several water quality factors that are considered and used to calculate the WQI. As recommended by the Bureau of Indian Standards (BIS, 2012), the standards for drinking purposes have been used to calculate WQI, which involves three steps (Ramakrishnalah et al., 2009; Adimalla et al., 2020).

In the first step, each of the ten chemical parameters, such as pH, TDS, TH, Ca, Mg, HCO_3, Cl, SO_4, F, and NO_3, was given a weight (w_i) ranging from 2 to 5, and their selection was based on their importance in water quality for drinking purposes (Table 17.1) (Ramakrishnalah et al., 2009). The relative weights (W_i) are calculated in the second step using equation (17.1).

$$W_i = \frac{w_i}{\sum_{i=1}^{n} w_i} \tag{17.1}$$

TABLE 17.1
The Assigned Weights and Relative Weights of Physicochemical Parameters

Parameters	Units	BIS (2012)	Weight (w_i)	Relative Weight (W_i)
pH	–	6.5–8.5	3	0.073
TDS	mg/L	500	5	0.122
TH	mg/L	200	3	0.073
Ca^{++}	mg/L	200	3	0.073
Mg^{++}	mg/L	30	3	0.073
HCO_3^-	mg/L	120	3	0.073
Cl^-	mg/L	250	4	0.098
SO_4^-	mg/L	200	3	0.073
F^-	mg/L	1.5	5	0.122
NO_3^-	mg/L	45	5	0.122
Na^+	mg/L	200	2	0.049
K^+	mg/L	12	2	0.049
Sum			ΣW_i=41	ΣW_i=1.00

where (W_i) is the relative weight, (w_i) is the weight of each parameter, and (n) is the number of parameters, as shown in Table 17.1. In the third step, each individual parameter's quality rating scale (Q_i) is calculated by dividing its concentration in each groundwater sample by BIS (2012) drinking water quality standards, then multiplying by 100 using equation (17.2).

$$Q_i = \left(C_i / S_i\right) \times 100 \tag{17.2}$$

where Q_i is the quality rating, C_i is the concentration of each chemical parameter in each water sample in milligrams per liter (mg/L), and S_i is the Indian drinking water (BIS, 2012) guidelines for each chemical parameter. Eventuality, water quality sub-index (SIi) for each chemical parameter was computed by equation (17.3), and the WQI was determined by equation (17.4).

$$SI_i = W_i \times Q_i \tag{17.3}$$

$$\text{WQI} = \sum SI_{i-n} \tag{17.4}$$

where

SI_i is the sub-index of the ith parameter,

Q_i is the rating based on the concentration of ith parameter, and n is the total number of parameters.

17.3 RESULTS AND DISCUSSION

Table 17.2 shows the fundamental descriptive statistics of groundwater chemistry. To determine the acceptability of the groundwater samples taken for drinking purposes, the percent compliance of the physicochemical data concerning BIS (2012) and WHO (2004) for suitability for drinking water is reported in Table 17.3.

TABLE 17.2
Descriptive Statistics of the Physicochemical Parameters

		Pre-monsoon				Post-monsoon			
Parameter	Units	Minimum	Maximum	Mean	Standard Deviation	Minimum	Maximum	Mean	Standard Deviation
pH	–	7.07	8.22	7.68	0.23	6.90	8.75	7.32	0.35
EC	μS/Cm	112	4,058	811.05	603.88	166	3,370	911.16	515.23
TDS	mg/L	73	2,638	527.24	392.53	108	2,191	592.38	334.94
TH	mg/L	60	1,352	314.55	225.27	120	1,000	327.71	159.20
Ca^{++}	mg/L	11	306	60.96	43.29	16	155	47.49	29.36
Mg^{++}	mg/L	8	300	38.85	45.44	7	197	49.65	35.50
Na^{+}	mg/L		703	89.27	103.94	9	398	82.80	65.76
K^{+}	mg/L	0.10	2.30	0.35	0.43	0.20	80	6.74	12.99
CO_3^{-}	mg/L	0	0.02	0	0	0	88	2.76	13.17
HCO_3^{-}	mg/L	64	992	243.49	139.75	68	508	269.82	98.03
Cl^{-}	mg/L	16	1106	134.04	170.90	20	849	117.87	120.00
SO_4^{-}	mg/L	8	210	57.05	34.38	12	280	77.53	44.39
F^{-}	mg/L	0.01	1.98	0.44	0.38	0.05	1.68	0.38	0.25
NO_3^{-}	mg/L	1.00	262	52.91	54.11	1	162	47.40	43.06

TABLE 17.3
Compliance of Physicochemical Data Respecting Drinking Water Standards

		Percent Compliance			Percent Compliance	
Parameter	BIS limit (2012)	Pre-Monsoon	Post-Monsoon	WHO Limit (2004)	pre-Monsoon	Post-Monsoon
pH	6.5–8.5	100	98	6.5–8.5	100	98
EC	–		–	1,500	89	93
TDS	500–2,000	98	98	500–1,500	98	98
TH	300–600	93	96	100–500	91	87
F^-	1–1.5	98	98	1–1.5	98	98
Cl^-	200–1,000	98	100	200–600	96	98
SO_4^-	200–400	98	98	200–400	98	98
NO_3^-	45	49	51	45	49	51
Na^+	–	–	–	200	89	91
K^+	–	–	–	20	100	100
Ca^{++}	75–200	98	100	75–200	98	100
Mg^{++}	30–100	95	95	30–150	96	100

17.4 PHYSICOCHEMICAL PARAMETERS OF GROUNDWATER

The results of the physicochemical analysis were analyzed using descriptive statistics. Accordingly, the statistical parameters were calculated (Table 17.2). The pH ranged between 7.07–8.22 and 6.90–8.75, with an average of 7.68 and 7.31 during pre- and post-monsoon seasons. It indicated the neutral to alkaline nature of the groundwater during both seasons. The bulk of the samples show their slightly alkaline nature. However, 2% of groundwater exceeded the suggested limits of BIS (2012) (Tables 17.2 and 17.3).

Electrical conductivity (EC) denotes the ability of a substance to transmit electric current through it. The dissolved salts in water are found in their ionic forms, and they make the water a good conductor of electric current and thus helpful in assessing the water quality. EC values range from 112–4,058 to 166–3,370 µS/cm, with an average of 811.05 and 911.16 µS/cm during pre- and post-monsoon, respectively. The maximum limit of EC in drinking water is prescribed as 1,500 µS/cm as per WHO standard (2004). A total of 11% and 7% samples of the study area exceed the permissible limit during pre- and post-monsoon seasons, respectively (Tables 17.2 and 17.3). TDS expresses the total dissolved solids in water, indicating water salinity and water utility for human consumption. The TDS values vary from 73 to 2,638 mg/L and 108 to 2,191 mg/L with a mean of 527.24 mg/L and 592.38 mg/L during pre- and post-monsoon, respectively. Almost 82% of groundwater samples were found within the acceptable limit (500 mg/L). On the other hand, 98% of illustrations fell under the permissible limit (BIS, 2012) and were noted as effectively suitable for drinking purposes (Tables 17.2 and 17.3). TH represents the sum of Ca_2O and Mg_2O dissolved in groundwater and is expressed as $CaCO_3$. The TH values of the water samples ranged between 60–1,352 mg/L (mean 314.55 mg/L) and 120–1,000 mg/L (mean 327.71 mg/L) during pre- and post-monsoon, respectively (Table 17.2). The highest TH allowed limit for drinking is 600 mg/L as per BIS (2012). Results indicated only 4 and 2 groundwater samples above the maximum permissible limit (500 mg/L) in pre- and post-monsoon season, respectively. The high values of TH may be due to the leaching of calcium and magnesium bicarbonate through recharge.

Calcium indicates water hardness and is found in water as calcium ions. The calcium values ranged from 11 to 306 mg/L and 16 to 155 mg/L with a mean of 60.96 and 47.49 mg/L during pre- and post-monsoon seasons. Out of all samples, almost 98% fell below the permissible limit specified by BIS (2012) and WHO (2004) in both seasons (Table 17.3).

Olivine, augite, diopside, biotite, and hornblende, these ferromagnetic minerals contributes magnesium in the natural water (Singh, 2012; Aher, 2012; Aher and Deshpande, 2016). The concentration of Mg^{++} ranges from 8 to 300 mg/L (mean 38.85 mg/L) and 7 to 197 mg/L (mean 49.65) in pre- and post-monsoon seasons, respectively, clearly indicating the potable nature of groundwater (except three samples) in the area (BIS, 2012) (Table 17.3). The average concentrations of Na^{+} and K^{+} ions found in groundwater samples in the study area were observed in the range of 89.27 and 82.80 mg/L in pre- and post-monsoon seasons, respectively. All groundwater samples fall under the maximum permissible limits prescribed by WHO (2004) (Tables 17.2 and 17.3). The average values of CO_3^- were observed to be between 0 and 2.76 mg/L. In comparison, the value of HCO_3^- was between 64 and 992 mg/L (mean 243.49 mg/L) and 68 to 508 mg/L (mean 269.82 mg/L) during pre- and post-monsoon, respectively (Tables 17.2 and 17.3). Cl^- values vary from 16 to 1,106 mg/L and 20 to 849 mg/L, with an average of 134.03 and 117.87 mg/L during pre- and the post-monsoon seasons, respectively (Tables 17.2 and 17.3). For Indian conditions, the acceptable and permissible level in potable water is 250 and 1,000 mg/L (BIS, 2012). Except for one (sample no. 31) taken only during the pre-monsoon season, which exceeded the limit, all groundwater samples fell within the maximum allowable limits, making them most suitable for drinking. The SO_4^- values range between 8–210 mg/L and 12–280 mg/L, with a mean value of 57.08 and 77.53 mg/L, respectively, in pre- and post-monsoon seasons (Tables 17.2 and 17.3). As per BIS (2012) and WHO (2004), the acceptable and permissible limits of SO_4^- in potable water are 200 mg/L and 400 mg/L, and all the groundwater samples were observed within the maximum allowable limits (Tables 17.2 and 17.3), which are most suitable for drinking purposes. In the study area, F^- concentration in groundwater varies between 0.01–1.98 mg/L and 0.05–1.68 mg/L, with an average value of 0.44–0.38 mg/L during pre- and post-monsoon seasons, respectively. Most of the analyzed samples (98%) exhibited their suitability for drinking purposes.

Nitrate content fell in a broad range of 1–262 mg/L and 1–162 mg/L, with averages of 52.90 and 47.40 mg/L in pre- and post-monsoon seasons, one-to-one. The primary sources of nitrogen in the water, biological fixation, and precipitation, etc., movement were recognized by Berner and Berner (1987). In the present case, a high concentration of nitrates can cause methemoglobinemia. Therefore, groundwater samples having high nitrate concentrations are suggested to be avoided for drinking purposes (Freeze and Cherry, 1979; Reddy et al., 2014).

The major ion chemistry revealed that Na^{+} is the leading cation and HCO_3^- is the most dominant anion in both seasons. The array of an abundance of cations was recorded as $Na^{++>} Ca^{++>}Mg^{++>}K^{+}$ in the pre-monsoon and post-monsoon periods; comparatively, the order of anions was recorded as $HCO_{3,>}Cl^{-}{}^{->}SO^{4->}NO^{3-}$ for both seasons. The overall concentration pattern of the major ions may be ordered as $HCO_3{}^{->}Cl^{->}Na^{+>}Ca^{++>}SO^{4->}NO^{3->}Mg^{++>}F^{->}K^{+>}CO^{3}$ and $HCO_3{}^{->}Cl^{->}Na^{+>}SO^{4-}$ ${}^{>}Mg^{++>}Ca^{++>}NO^{3->}K^{+>}CO^{3->}F^{-}$ in pre- and post-monsoon seasons, respectively. In the majority of the cases, major cations and anions showed values within the permissible and safe limits concerning BIS (2012) and WHO (2004) standards, except for a few locations. Apart from agricultural activities, there have been no other substantial anthropogenic activities in the study area that could be linked to elevated nitrate concentrations.

17.5 WATER TYPES

Groundwater samples of the study area were plotted for two seasons on the Piper diagram to evaluate the water type using the hydrogeochemical software, AquaChem v.2014.1 (Piper 1944) (Figures 17.2 and 17.3). The study showed 45% and 47% in post-monsoon season samples belong to the $CaHCO_3$ type, followed by mixed CaMgCl type (pre-monsoon 38% and post-monsoon 33%), $CaNaHCO_3$ type (pre-monsoon 4% and post-monsoon 10%), NaCl type (pre-monsoon 11% and post-monsoon 6%), and CaCl type (pre-monsoon 2% and post-monsoon 4%). $CaHCO_3$ and $CaNaHCO_3$ have been formed from the recharge water and the dissolution of minerals along the flow path. Overall, the water types, mixed CaMgCl and NaCl, confirmed the wastewater's mixing with the $CaHCO_3$ water, followed by ion-exchange reactions.

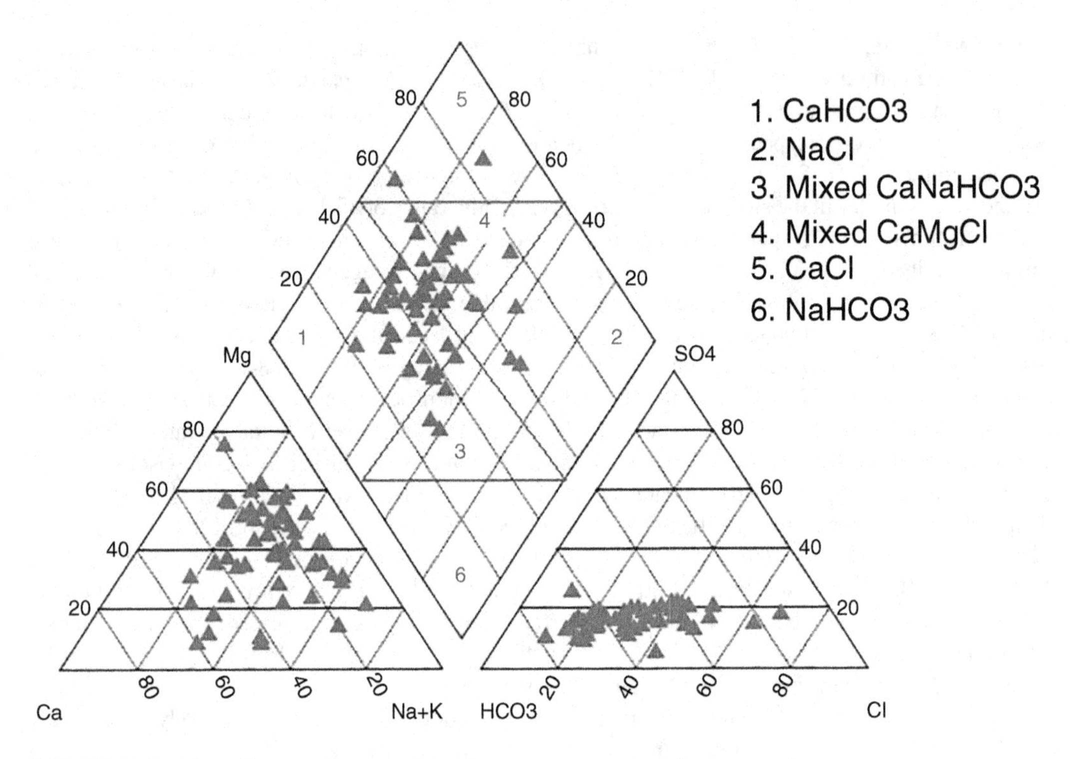

FIGURE 17.2 Piper diagram showing plots of post-monsoon groundwater samples.

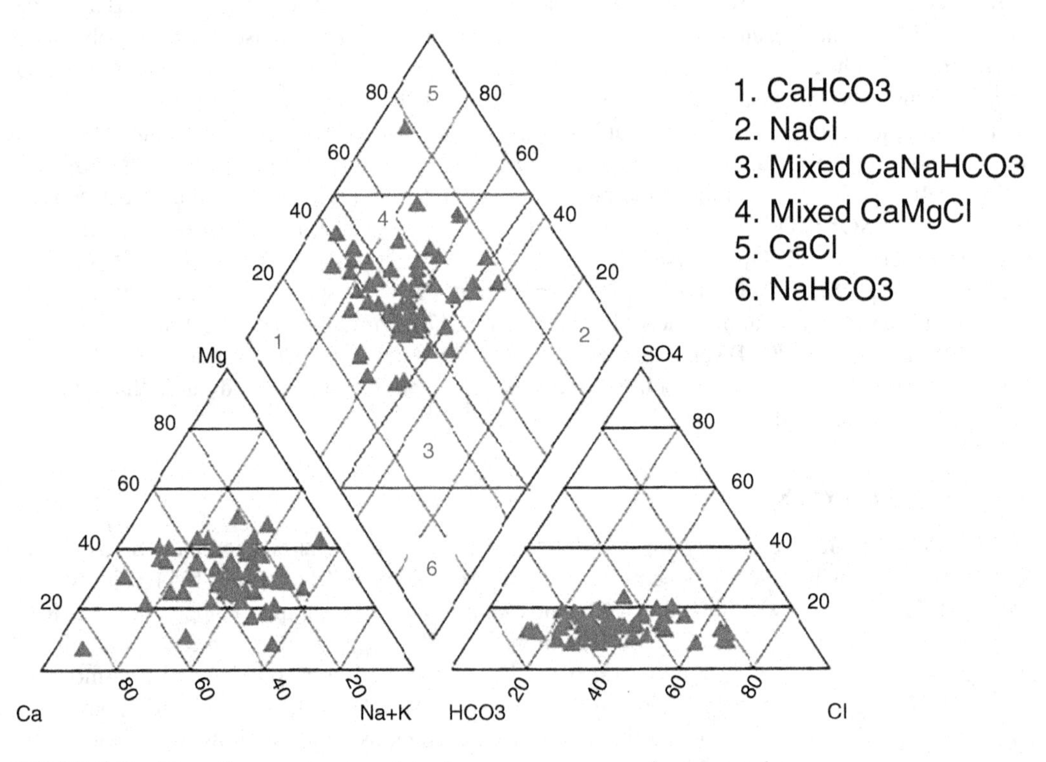

FIGURE 17.3 Piper diagram showing plots of pre-monsoon groundwater samples.

17.6 CORRELATION ANALYSIS

17.6.1 Correlation Coefficient

The investigation of the linear correlation method looked at the statistical link between the water quality metrics. A correlation coefficient is a tool used to assess how closely two variables are related and how strongly they are associated (Voudouris et al., 2000; Davis, 1986; El-Fakharany et al., 2017). A linear link between two variables can be measured by the value r, also known as the linear correlation coefficient. In honor of its creator, Karl Pearson, the linear correlation coefficient is occasionally referred to as the Pearson product-moment correlation coefficient. The mathematical formula for computing r is

$$r = \frac{n\sum xy - \left(\sum x\right)\left(\sum y\right)}{\sqrt{n\left(\sum x^2\right) - \left(\sum x\right)^2}\sqrt{n\left(\sum y^2\right) - \left(\sum y\right)^2}} \tag{17.5}$$

where x and y are the variables and n is the number of water parameters. The correlation coefficient ranges from −1 to +1. A correlation of +1 indicates a perfect positive relationship between two variables; as the value of one variable increases, the value of the other variable increases at the same rate. A correlation of −1 indicates that one variable changes inversely with relation to the other; as the value of one variable increases, the value of the other variable decreases at the same rate (Wang et al., 2007).

17.7 CORRELATION INVESTIGATION OF GROUNDWATER

Tables 17.4 and 17.5 provide the correlation matrices for the water quality characteristics of groundwater samples taken during the pre- and post-monsoon seasons, which were calculated using AquaChem v.2014 software. This study aimed to use a correlation matrix with 15 variables. Pre- and post-monsoon season parameter correlations have always followed the same pattern. There are noticeable connections between the various physicochemical characteristics, ranging from strong (r=>0.75) to moderate (r = 0.75−0.5). Electrical conductivity and TDS are positively correlated in groundwater samples (pre-monsoon r=1.00; post-monsoon r=1.00), demonstrating a perfect positive relationship between these two variables and demonstrating that conductivity rises as the concentration of all dissolved ions rises. While Na is significantly connected with total hardness, SO_4^- and Cl^-, the TDS shows a substantial positive correlation with Na^+, Cl^-, and SO_4^-. The strong correlation between the TDS and Na^+, Cl^-, and SO_4^- in the research area's groundwater reveals that sewage discharge, landfill waste sites, septic tanks, and domestic and industrial effluents percolate and mix there (Raja and Venkatesan, 2010). The correlation matrix and the WQI data were studied together in order to determine how groundwater characteristics affected the WQI values. Before and after the monsoon, the WQI was substantially linked with the EC, TDS, and TH parameters; during the post-monsoon, chloride and sulfate showed only a minor correlation with the WQI.

17.8 WATER QUALITY INDEX

In order to assess the water quality in urban, rural, and industrial areas, it is essential to use the WQI. The WQI values and types of groundwater samples (n=55) have been given (Table 17.6 and Figure 17.4). As a result, the WQI can be classified as excellent water type if the values are less than 150, good water type if the values are between 150 and 120, poor water type if the values are between 200 and 250, very poor water type if the values are between 250 and 300, and unfit for drinking if the values are above 300 (Ramakrishnalah et al., 2009). Tables 17.6 and 17.7 show the

TABLE 17.4
Correlation Coefficient Matrix of Physicochemical Parameters of Pre-monsoon Groundwater Samples

Variables	pH	EC	TDS	TH	Ca^{++}	Mg^{++}	Na^{+}	K^{+}	CO_3^-	HCO_3^-	Cl^-	SO_4^-	F^-	NO_3^-	WQI
pH	1	−0.463**	−0.463**	−0.395**	−0.200	−0.355**	−0.398**	−0.204	0.143	−0.355**	−0.375**	−0.492**	−0.449**	−0.753**	−0.593**
EC		1	1.000**	$\mathbf{0.929}^{**}$	0.384**	0.885**	$\mathbf{0.902}^{**}$	0.795**	−0.076	0.856**	$\mathbf{0.935}^{**}$	$\mathbf{0.929}^{**}$	0.304*	0.691**	$\mathbf{0.961}^{**}$
TDS			1	$\mathbf{0.929}^{**}$	0.384**	0.885**	$\mathbf{0.902}^{**}$	0.795**	−0.075	0.856**	$\mathbf{0.935}^{**}$	$\mathbf{0.929}^{**}$	0.303*	0.691**	$\mathbf{0.961}^{**}$
TH				1	0.545**	0.878**	0.774**	0.795**	−0.006	0.789**	$\mathbf{0.909}^{**}$	0.893**	0.227	0.623**	$\mathbf{0.926}^{**}$
Ca^{++}					1	0.077	0.080	0.251	0.040	0.176	0.327*	0.394**	0.069	0.340*	0.408**
Mg^{++}						1	0.874**	0.803**	−0.032	0.838**	0.893**	0.837**	0.232	0.545**	0.868**
Na^{+}							1	0.669**	−0.073	0.839**	0.921**	0.798**	0.253	0.624**	0.868**
K^{+}								1	0.009	0.672**	0.770**	0.700**	0.288*	0.349**	0.722**
CO_3^-									1	−0.045	−0.009	−0.085	0.005	−0.020	−0.032
HCO_3^-										1	0.754**	0.772**	0.344*	0.504**	0.823**
Cl^-											1	0.828**	0.168	0.611**	$\mathbf{0.900}^{**}$
SO_4^-												1	0.333*	0.738**	$\mathbf{0.937}^{**}$
F^-													1	0.329*	0.387**
NO_3^-														1	0.832**
WQI															1

**, Correlation is significant at the 0.01 level; *, Correlation is significant at the 0.05 level, bold, strong correlation (r=0.9).

TABLE 17.5

Correlation Coefficient Matrix of Physicochemical Parameters of Post-monsoon Groundwater Samples

Variables	pH	EC	TDS	TH	Ca^{++}	Mg^{++}	Na^{+}	K^{+}	CO_3^-	HCO_3^-	Cl^-	SO_4^-	F^-	NO_3^-	WQI
pH	1	0.018	0.017	−0.097	0.079	−0.142	0.330^{*}	−0.144	0.770^{**}	0.076	0.044	0.046	−0.117	−0.248	−0.102
EC		1	1.000^{**}	**0.923****	0.356^{**}	0.809^{**}	0.743^{**}	0.378^{**}	0.142	0.615^{**}	**0.919****	**0.919****	0.063	0.591^{**}	**0.938****
TDS			1	**0.922****	0.356^{**}	0.809^{**}	0.743^{**}	0.378^{**}	0.142	0.615^{**}	**0.919****	**0.919****	0.063	0.591^{**}	**0.938****
TH				1	0.360^{**}	0.888^{**}	0.502^{**}	0.327^{*}	0.033	0.639^{**}	0.810^{**}	0.845^{**}	0.010	0.610^{**}	**0.933****
Ca^{++}					1	−0.108	0.209	0.073	−0.090	0.210	0.341^{*}	0.297^{*}	0.036	0.124	0.286^{*}
Mg^{++}						1	0.434^{**}	0.313^{*}	0.080	0.577^{**}	0.697^{**}	0.756^{**}	−0.005	0.590^{**}	0.854^{**}
Na^{+}							1	0.049	0.334^{*}	0.517^{**}	0.787^{**}	0.759^{**}	0.052	0.348^{**}	0.632^{**}
K^{+}								1	−0.035	0.194	0.286^{*}	0.175	0.059	0.089	0.279^{*}
CO_3^-									1	−0.036	0.097	0.159	−0.020	−0.074	0.042
HCO_3^-										1	0.431^{**}	0.588^{**}	−0.086	0.270^{*}	0.601^{**}
Cl^-											1	0.853^{**}	0.045	0.464^{**}	0.825^{**}
SO_4^-												1	0.037	0.589^{**}	0.892^{**}
F^-													1	0.111	0.131
NO_3^-														1	0.806^{**}
WQI															1

**, Correlation is significant at the 0.01 level; *, Correlation is significant at the 0.05 level; bold, strong correlation (r=0.9).

various sample WQI values and water types, respectively. In the pre- and post-monsoon seasons, the WQI ranged widely from 20.07 to 402.95 and 30.45 to 286.69, with average values of 85.06 and 91.11, respectively. 96.4% of groundwater samples taken during the pre-monsoon period were excellent, 1.8% were good, and 1.8% were unfit for drinking. 92.7% of post-monsoon water samples were excellent for drinking, 5.5% were good, and 1.8% were very poor. Additionally, one sample or 1.8% of the total samples, was deemed unfit for consumption during the pre-monsoon seasons (Table 17.6).

It was observed that groundwater's dissolved ions, primarily Na^+, Mg^{++}, Cl^-, HCO_3^-, NO_3^-, and SO_4^- during the pre-monsoon period and Na^+, Mg^{++}, Cl^-, and $SO4^-$ during the post-monsoon period, had an impact on WQI values. Excessive nitrate concentrations in groundwater mostly drove high WQI values, but they were also influenced by high magnesium, alkalinity, sulfate, and chloride concentrations. The correlation coefficients of the WQI for chloride, sulfate, alkalinity, and magnesium show that these effects are present (Tables 17.6 and 17.7). Strong correlations exist between WQI readings in both seasons and sulfates and chlorides. The WQI readings (Table 17.7) demonstrate that, in some areas, groundwater quality declined after the monsoon (sample numbers 31, 49).

TABLE 17.6
The WQI Values, Type of Water, and Percentage of Sample

		No. of Samples		% of Samples	
WQI Values	**Water Quality**	**Pre-Monsoon**	**Post-Monsoon**	**Pre-Monsoon**	**Post-Monsoon**
<150	Excellent	53	51	96	93
150–200	Good	1	3	2	5
200–250	Poor	–	–	–	–
250–300	Very poor	–	1	–	2
>300	Unsuitable	1	–	2	–

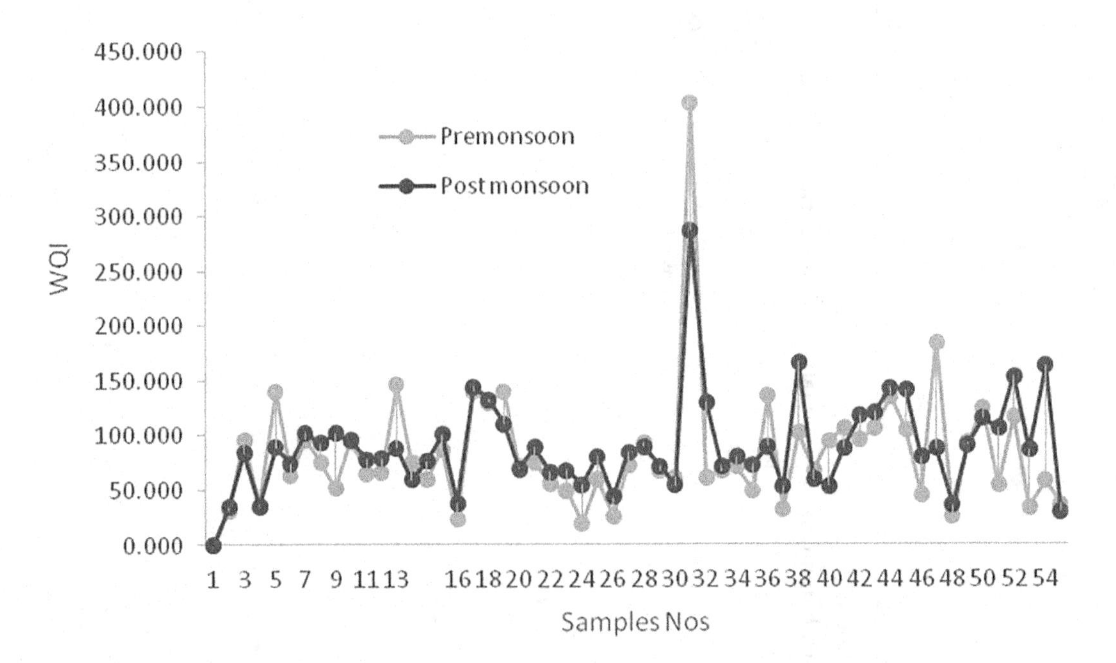

FIGURE 17.4 Plot of WQI values for pre- and post-monsoon seasons of the study area.

Table 17.7
Computation of Water Quality Index (WQI) for Individual Groundwater Samples

Sample No.	WQI Pre-Monsoon	Water Type	WQI Post-Monsoon	Water Type
1	31.249	Excellent	34.72	Excellent
2	95.707	Excellent	83.91	Excellent
3	36.821	Excellent	35.08	Excellent
4	140.164	Excellent	89.00	Excellent
5	63.710	Excellent	74.00	Excellent
6	96.507	Excellent	102.46	Excellent
7	75.204	Excellent	92.90	Excellent
8	51.239	Excellent	101.89	Excellent
9	95.242	Excellent	95.51	Excellent
10	65.220	Excellent	77.89	Excellent
11	66.422	Excellent	78.67	Excellent
12	146.212	Excellent	88.52	Excellent
13	75.732	Excellent	59.29	Excellent
14	59.074	Excellent	75.86	Excellent
15	87.034	Excellent	100.71	Excellent
16	23.099	Excellent	37.87	Excellent
17	140.458	Excellent	144.42	Excellent
18	130.113	Excellent	132.35	Excellent
19	140.190	Excellent	110.80	Excellent
20	71.221	Excellent	69.28	Excellent
21	75.335	Excellent	89.01	Excellent
22	56.193	Excellent	66.24	Excellent
23	49.842	Excellent	67.65	Excellent
24	20.077	Excellent	55.06	Excellent
25	59.306	Excellent	80.77	Excellent
26	26.536	Excellent	43.91	Excellent
27	72.605	Excellent	84.90	Excellent
28	92.796	Excellent	89.64	Excellent
29	68.023	Excellent	71.70	Excellent
30	61.156	Excellent	54.19	Excellent
31	402.957	Unsuitable	286.69	Very Poor
32	61.093	Excellent	129.89	Excellent
33	67.238	Excellent	71.69	Excellent
34	71.175	Excellent	80.00	Excellent
35	49.246	Excellent	73.02	Excellent
36	136.508	Excellent	89.55	Excellent
37	32.087	Excellent	53.03	Excellent
38	101.901	Excellent	166.15	Good
39	67.956	Excellent	59.04	Excellent
40	95.080	Excellent	53.24	Excellent
41	106.289	Excellent	88.77	Excellent
42	96.337	Excellent	117.34	Excellent
43	105.982	Excellent	121.11	Excellent
44	135.187	Excellent	142.83	Excellent
45	104.484	Excellent	140.72	Excellent
46	45.701	Excellent	80.33	Excellent
47	184.668	Good	87.84	Excellent

(Continued)

Table 17.7 (*Continued*)
Computation of Water Quality Index (WQI) for Individual Groundwater Samples

Sample No.	WQI Pre-Monsoon	Water Type	WQI Post-Monsoon	Water Type
48	25.723	Excellent	36.12	Excellent
49	92.828	Excellent	91.03	Excellent
50	123.986	Excellent	114.88	Excellent
51	54.433	Excellent	106.18	Excellent
52	116.283	Excellent	153.23	Good
53	34.291	Excellent	86.57	Excellent
54	57.807	Excellent	163.14	Good
55	36.638	Excellent	30.45	Excellent
Minimum	20.077		30.453	
Maximum	402.957		286.692	
Average	85.061		91.110	

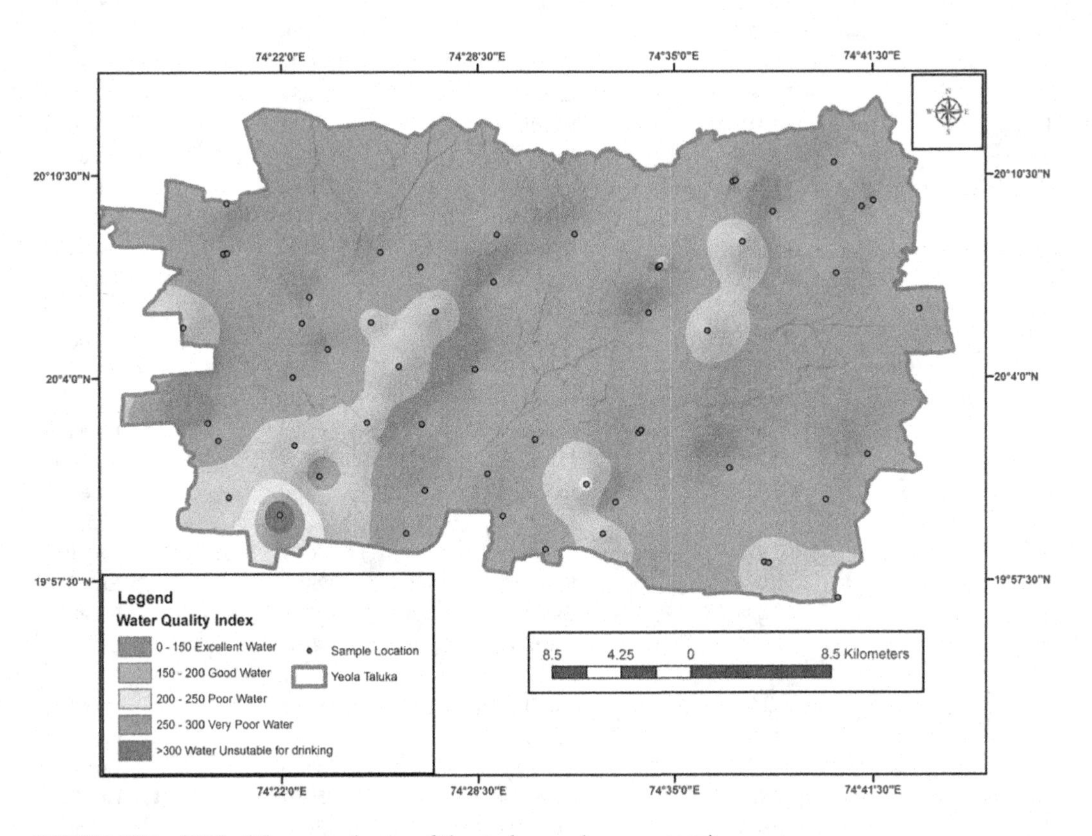

FIGURE 17.5 WQI of the groundwater of the study area (pre-monsoon).

This decline could be the result of some localized effects. Figures 17.5 and 17.6, respectively, show the spatial distribution of different water types during the pre- and post-monsoon seasons. The distribution of water types demonstrates that, in most instances, both seasons have excellent water types. As a result, it can be said that the WQI values in the groundwater of the study region range from 20.07 to 402.95, with a mean value of 85.06 in the pre-monsoon season. In comparison, it fluctuates between 30.45 and 286.69, with a mean value of 91.11, in the post-monsoon seasons. The WQI values were influenced by the following elements: Na^+, Mg^{++}, Cl^-, HCO_3^-, NO_3^-, and SO_4^- during

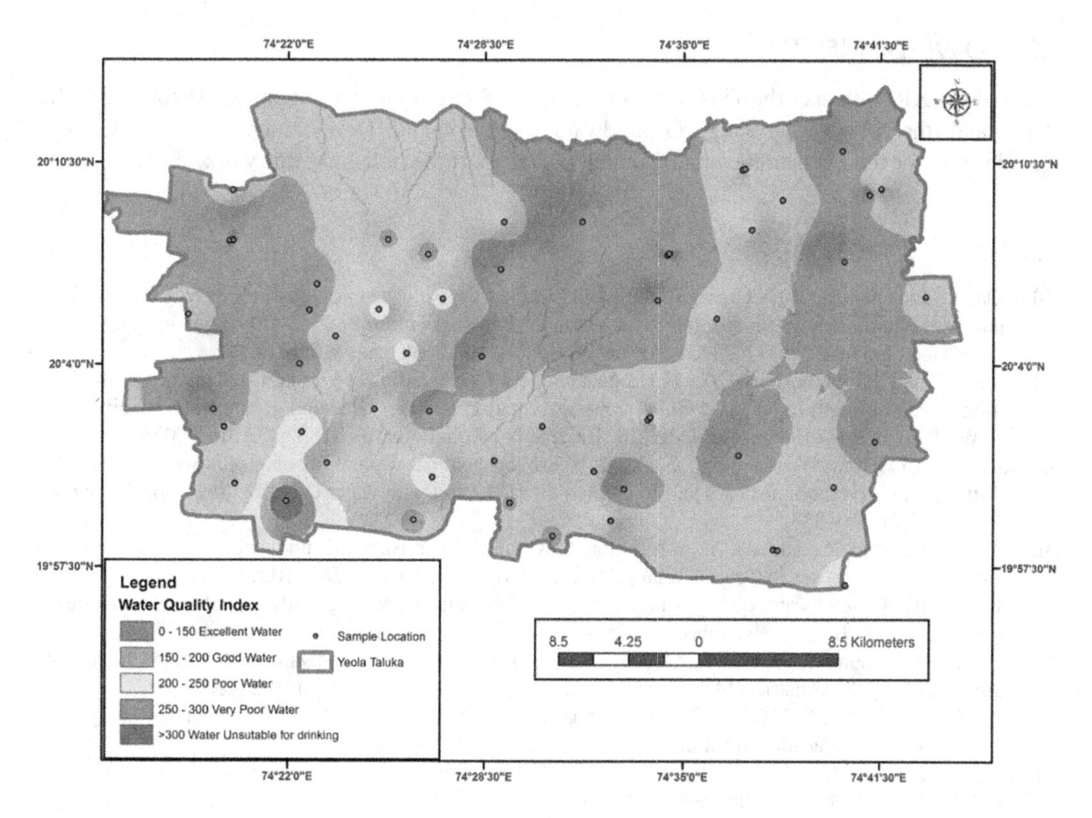

FIGURE 17.6 WQI of the groundwater of the study area (post-monsoon).

the pre-monsoon seasons (Figure 17.5) and Na^+, Mg^{++}, Cl^-, and SO_4^- during the post-monsoon seasons (Figure 17.6). High nitrate content in groundwater led to high WQI values; sulfates, magnesium, and chloride also significantly increased these values at some locations, and the quality of groundwater deteriorated.

17.9 CONCLUSIONS

In the Yeola block of Nashik District, Maharashtra, India, people are primarily dependent on groundwater, and it is the only water source for drinking needs. The hydrochemical study of the groundwater samples shows that the groundwater is neutral to alkaline in nature. The analysis result shows the predominance of major ions in the order of $HCO_3^{-}>Cl^{-}>Na^{+}>Ca^{++}>SO_4^{-}>NO_3^{-}>Mg^{++}>F^{-}>K^{+}>CO^{3-}$ and $HCO_3>Cl^{-}>Na^{+}>SO^{4-}>Mg^{++}>Ca^{++}>NO^{3-}>K^{+}>CO^{3-}>F^{-}$ in pre- and post-monsoon periods, respectively. The enhanced TDS and EC values are attributed to agricultural practices. In contrast, due to anthropogenic activities, higher values crossed the permissible limits of TH, magnesium, chloride, sulfate, and nitrates. About 51% and 49% of groundwater samples have exceeded the maximum permissible nitrate limits in pre- and post-monsoon seasons, respectively. Therefore, it is suggested that groundwater should not be used for drinking purposes with high nitrate concentrations. The WQI ranges from 20.07 to 402.95 and 30.45 to 286.69, with an average of 85.06 and 91.11 for pre- and post-monsoon seasons, respectively. During the pre-monsoon, 96.4% of groundwater samples were excellent, 1.8% were good, and 1.8% were unsuitable for drinking. In post-monsoon, 92.7% of water samples were excellent; 5.5% were good; and 1.8% were very poor. In addition, 1.8% of the sample, i.e., only one, is unsuitable for drinking purposes in pre-monsoon seasons. Preventive steps like water treatment, improvement of the local sanitation system, training, and public awareness programs are recommended to maintain water quality for its further use.

ACKNOWLEDGMENTS

The author (KRA) thanks the Director, Groundwater Surveys and Development Agency (GSDA), Pune, and the Deputy Director, Groundwater Surveys and Development Agency (GSDA), Aurangabad, for their constant encouragement and support in publishing this work.

REFERENCES

Adimalla, N., and Qian, H. 2019. Groundwater quality evaluation using water quality index (WQI) for drinking purposes and human health risk (HHR) assessment in an agricultural region of Nanganur, South India. Ecotoxicol. *Environ. Saf.* 176, 153–161. https://doi.org/10.1016/j.ecoenv.2019.03.066.

Adimalla, N., and Taloor, A. K. 2020a. Hydrogeochemical investigation of groundwater quality in the hard rock terrain of South India using Geographic Information System (GIS) and groundwater quality index (GWQI) techniques. *Environ. Process.* 10, 100288. https://doi.org/10.1016/j.gsd.2019.100288.

Adimalla, N., Li, P., and Venkatayogi, S. 2018. Hydrogeochemical evaluation of groundwater quality for drinking and irrigation purposes and integrated interpretation with water quality index studies. *Environ. Process.* 5(2), 363–383.

Adimalla, N., Dhakate, R., Kasarla, A., and Taloor, A. K. 2020. Appraisal of groundwater quality for drinking and irrigation purposes in Central Telangana, India. *Groundw. Sustain. Dev.* 10, 100334.

Aher, K. R. 2012. Ground water quality studies of Chikalthana area of Aurangabad, unpublished Ph.D thesis, submitted to Dr. Babasaheb Ambedkar Marathwada University, Aurangabad, pp. 20–49.

Aher, K. R., and Deshpande, S. M. 2016. Geochemistry and evaluation of groundwater pollution in Chikalthana area of Aurangabad District, Maharashtra, India. *J. Applied Geochem.* 18(2), 192–202.

Aher, K. R., and Gaikwad, S. G. 2017. Irrigation groundwater quality based on hydrochemical analysis of Nandgaon block, Nashik district in Maharashtra. *Int. J. Adv. Geosci.* 5(1), 1–5.

Aher, K. R., Deshpande, S. M., and Varade, A. M. 2019. Groundwater quality assessment atudies in Yeola Block of Nashik District, Maharashtra. *J. Geosci. Res.* 4(1), 11–22.

Aher, K. R., Gaikwad, S. G., and Salve, P. L. 2020. Assessing groundwater quality using water quality index in semiarid region of Aurangabad district, central India. *Int. J. Adv. Geosci.* 8(2), 249–258.

Akoteyon, I. S., Omotayo, A. O., Soladoye, O., and Olaoye, H. O. 2011. Determination of water quality index and suitability of urban river for municipal water supply in Lagos-Nigeria. *Europ. J. Scientific Res.* 54(2), 263–271.

Aly, A. A., Al-Omran, A. M., and Alharby, M. M. 2014. The water quality index and hydrochemical characterization of groundwater resources in HafarAlbatin, Saudi Arabia. *Arab. J. Geosci.* 8, 4177–4190. https://doi.org/10.1007/s12517-014-1463-2.

American Public Health Association. 2015. *Standard Methods for Examination of Water and Wastewater*, 21st ed. American Public Health Association, Washington, DC.

Anku, Y. S., Banoeng-Yakubo, B., Asiedu, D. K., and Yidana, S. M. 2008. Water quality analysis of groundwater in crystalline basement rocks, Northern Ghana Yidana. *Environ. Geol.* 58, 989–997. https://doi.org/10.1007/s00254-008-1578-4.

Appelo, C. A. J., and Postma, D. 2005. *Geochemistry, Groundwater, and Pollution*, 2nd ed. A.A. Balkema, Rotterdam, Netherlands.

Babaei Semiromi, F., Hassani, A. H., Torabian, A., Karbassi, A. R., and Hosseinzadeh Lotfi, F. 2011. Water quality index development using fuzzy logic: a case study of the Karoon river of Iran. *African J. Biotech.* 10(50), 10125–10133.

Batabyal, A. K. 2018. Hydrogeochemistry and quality of groundwater in a part of Damodar Valley, Eastern India: an integrated geochemical and statistical approach. *Stoch. Environ. Res. Risk Assess.* 32(8), 2351–2368.

Batabyal, A. K., and Chakraborty, S. 2015. Hydrogeochemistry and water quality index in the assessment of groundwater quality for drinking uses. *Water Environ. Res.* 87(7), 607–617.

Berner, E. K., and Berner, R. A. 1987. *The Global Water Cycle: Geochemistry and Environment.* Prentice Hall, Englewood Cliffs.

Bharti, N., and Katyal, D. 2011. Water quality indices used for surface water vulnerability assessment. *Int. J. Environ. Sci.* 2(1), 154–173.

Boniol, D. 1996. Summary of groundwater quality in the St. Johns River Water Management District. Special Publication SJ96-SP13.St. Johns River Water Management District, Palatka, Special Publication SJ96- SP13.

Bureau of Indian Standards. 2012. Indian Standard Drinking Water- Specification. 1st rev. Bureau of Indian Standards, New Dehli, India.

CGWB. 2014. Groundwater information Nashik district, Maharashtra, Central Groundwater Board, MOWR.1838/DBR, 1–21.

Chakraborty, S., Sikdar, P. K., and Paul, P. K. 2007. Hydrogeochemical Framework of Quaternary Aquifer of English Bazaar Block, MaldaDistrict, West Bengal. *Icfai J. Earth Sci.* 1, 61–74.

Chung, S. Y., Ramkumar, T., Venkatramanan, S., Kim, T. H, and Kim, D. S. 2014. Influence of hydrogeochemical processes and assessment of suitability for groundwater uses in Busan City, Korea. *Environ. Dev. Sustain.* 17, 423–441. https://doi.org/10.1007/s10668-014-9552-7.

Das, S., 2008. Hydrogeological research in India. *Geol. Soc. India.* 67, 589.

Davis, J. C., 1986. *Statistics and Data Analysis in Geology*, 2nd edn. John Wiley & Sons, New York, pp. 1–656.

Deshpande, S. M., Shinde, P. D., and Aher, K.R. 2020. Groundwater quality assessment for drinking purpose in Aurangabad urban city, Maharashtra, India. *J. Geosci. Res.* 5(2), 123–132.

Deshpande S. M., Bhagwat, U. S., and Aher, K. R. 2021. Assessment of groundwater quality in parts of Jalna district of Maharashtra, India using water quality index. *J. Water Engg. Mgt.* 2(2),40–47.

Dhumal, M. L. 2021. Mathematical computation of water quality index to determine surface water quality of manyad storage tank, Aurangabad district, Maharashtra, India. *Int. J. Phy. Res.* 9(1), 38–41.

El-Fakharany, M. A., Mansour, N. M., Yehia, M. M., and Monem, M. 2017. Evaluation of groundwater quality of the Quaternary aquifer through multivariate statistical techniques at the southeastern part of the Nile Delta, Egypt. *Sustain. Water Resour. Manag.* 3(1), 71–81.

Fantong, W. Y., Satake, H., Aka, F. T., Ayonghe, S. N., Asai, K., Mandal, A. K., and Ako, A. 2009. Hydrochemical and isotopic evidence of recharge, apparent age, and flow direction of groundwater in Mayo Tsanaga River Basin, Cameroon: Bearings on contamination. *Environ. Earth Sci.* 60(1), 107–120. https://doi.org/10.1007/s12665-009-0173-7.

Fetter, C. W. 1994. *Applied Hydrogeology*, 3rd ed. Macmillan College Publication, New York.

Freeze, R. A., and Cherry, J. A. 1979. *Groundwater Prentice-Hall International*. Englewood Cliffs, NJ.

Giridharan, L., Venugopal, T., and Jayaprakash, M. 2008. Evaluation of the seasonal variation on the geochemical parameters and quality assessment of the groundwater in the proximity of River Cooum, Chennai, India. Environ. Monit. Assess. 143, 161–178. https://doi.org/10.1007/s10661-007-9965-y.

Horton R. K. 1965. An index number system for rating water quality. *J. Water Pollut. Control Fed.* 37, 300–305

Islam, A. T., Shen, S., Haque, M. A., Bodrud-Doza, M., Maw, K. W., and Habib, M. A. 2018.Assessing groundwater quality and its sustainability in Joypurhat district of Bangladesh using GIS and multivariate statistical approaches. *Environ. Dev. Sustain*, 20(5), 1935–1959.

Kazakis, N., Mattas, C., Pavlou, A., Patrikaki, O., and Voudouris, K. 2017. Multivariate statistical analysis for the assessment of groundwater quality under different hydrogeological regimes. *Environ. Earth Sci.* 76(9), 349.

Khare, Y. D., and Varade, A. M. 2018. Approach to groundwater management towards sustainable development in India. *Italian J. Groundw.* 308, 29–36. https://doi.org/10.7343/as-2018–308.

Kumar, D., and Ahmed, S. 2003. Seasonal behaviour of spatial variability of groundwater level in a granitic aquifer in monsoon climate. *Curr. Sci.* 84(2), 188–196.

Kumar, M., Ramanathan, A. L., and RaoBhishm M. S. K. 2006. Identification and evaluation of hydrogeochemical processes in the groundwater environment of Delhi, India. *Environ. Geol.* 50, 1025–1039. https://doi.org/10.1007/s00254-006-0275-4.

Lateef K.H., 2011. Evaluation of groundwater quality for drinking purpose for Tikrit and Samarra cities using water quality index. *Eur. J. Sci. Res.* 58(4),472–481.

Liu C., Lin, K., and Kuo, Y. 2003. Application of factor analysis in the assessment of groundwater quality in a blackfoot disease area in Taiwa. *Sci. Total Environ.* 313, 77–89.

Machiwal, D., Jha, M. K., and Mal, B. C. 2011. GIS-based assessment and characterization of groundwater quality in a hard-rock hilly terrain of Western India. *Environ. Monit. Assess.* 174, 645–663.

Machiwal, D., Cloutier, V., Güler, C., and Kazakis, N. 2018. A review of GIS-integrated statistical techniques for groundwater quality evaluation and protection. *Environ. Earth Sci.* 77(19), 681.

Mohsen, J. 2007. Hydrochemical identification of groundwater resources and their changes under the impacts of human activity in the Chah basin in western Iran. *Environ. Monit. Assess.* 130, 347–364.

Paradis, D., Vigneault, H., Lefebvre, R., Savard, M. M., Ballard, J. M., and Qian, B. 2016. Groundwater nitrate concentration evolution under climate change and agricultural adaptation scenarios: Prince Edward Island, Canada. *Earth Syst. Dyn.* 7(1), 183–202.

Piper A. M., 1944. A graphic procedure in the geochemical interpretation of water analysis. Trans. Am. Geophys. Union. 25, 914–928.

Raja, G., and Venkatesan, P. 2010. Assessment of groundwater pollution and its impact in and around Punnam area of Karur District, Tamilnadu, India. E-Journal of Chemistry, 7.

Rajesh, R., Brindha, K., Murugan, R., and Elango, L. 2012. Influence of hydrogeochemical processes on temporal changes in groundwater quality in a part of Nalgonda district, Andhra Pradesh, India. *Environ. Earth Sci.* 65(4), 1203–1213. https://doi.org/ 10.1007/s12665-011-1368-2.

Ramakrishnaiah, C. R., Sadashivaiah, C., and Ranganna, G. 2009. Assessment of water quality index for the groundwater in TumkurTaluk, Karnataka State, India. *E-J. Chem.* 6(2), 523–530.

Reddy, C. S. L., Deshpande, S. M., Aher, K. R., and Humane, K. 2014. Impact of mining on groundwater quality in and around Mangampeta, Andhra Pradesh. *Gond. Geol. Magz.* 14, 177–185.

Reza, R., and Singh, G. 2010. Assessment of groundwater quality status by using water quality index method in Orissa, India. *World Appl. Sci. J.* 9(12), 1392–1397.

Saeedi, M., Abessi, O., Sharifi, F., and Meraji, H. 2010. Development of groundwater quality index. *Environ. Monit. Assess.* 163, 327–335.

Saeid, S., Chizari, M., Sadighi, H., and Bijani, M. 2018. Assessment of agricultural groundwater users in Iran: a cultural environmental bias. *Hydrogeol. J.* 26(1), 285–295. https://doi.org/10.1007/s10040-017-1634-9.

Şener, Ş., Şener, E., and Davraz, A. 2017. Evaluation of water quality using water quality index (WQI) method and GIS in Aksu River (SW-Turkey). *Sci. Total Environ.* 584, 131–144.

Singh, K., Hundal, H. S., and Singh, D. 2011. Geochemistry and assessment of hydrogeochemical processes in groundwater in the southern part of Bathinda district of Punjab, northwest India. *Environ. Earth Sci.* 64(7), 1823–1833. https://doi.org/ 10.1007/s12665-011-0989-9.

Singh, V. K., Bikundia, D. S., Sarswat, A., and Mohan, D. 2012. Groundwater quality assessment in the village of LutfullapurNawada, Loni, District Ghaziabad, Uttar Pradesh, India. *Environ. Monit. Assess.* 184, 4473–4488.

Srivastava, A., and Chinnasamy, P. 2021. Developing village-level water management plans against extreme climatic events in Maharashtra (India)—a case study approach. In: Vaseashta, A., Maftei, C. (eds) *Water Safety, Security and Sustainability. Advanced Sciences and Technologies for Security Applications.* Springer, Cham. https://doi.org/10.1007/978-3-030-76008-3_27.

Srivastava, A., and Chinnasamy, P. 2022. Assessing groundwater depletion in southern india as a function of urbanization and change in hydrology: a threat to tank irrigation in Madurai City. In: Kolathayar, S., Mondal, A., Chian, S. C. (eds) *Climate Change and Water Security. Lecture Notes in Civil Engineering*, vol 178. Springer, Singapore. https://doi.org/10.1007/978-981-16-5501-2_24.

Srivastava, A., and Chinnasamy, P. 2023a. Tank cascade system in southern india as a traditional surface water infrastructure: a review. In: Chigullapalli, S., SushaLekshmi, S. U., Deshpande, A. P. (eds) *Rural Technology Development and Delivery. Design Science and Innovation.* Springer, Singapore. https://doi.org/10.1007/978-981-19-2312-8_15.

Srivastava, A., and Chinnasamy, P. 2023b. Understanding declining storage capacity of tank cascade system of Madurai: potential for better water management for rural, peri-urban, and urban catchments. In: Chigullapalli, S., SushaLekshmi, S. U., Deshpande, A. P. (eds) *Rural Technology Development and Delivery. Design Science and Innovation.* Springer, Singapore. https://doi.org/10.1007/978-981-19-2312-8_14.

Srinivasa Rao, Y., Reddy, T. V. K., and Nayudu, P. T. 1997.Groundwater quality in the Niva River basin, Chitoor district, Andhra Pradesh, India. *Environ. Geol.* 32(1), 56–63.

Srivastava, A., Khadke, L., and Chinnasamy, P. 2021. Web application tool for assessing groundwater sustainability—a case study in rural-Maharashtra, India. In: Vaseashta, A., Maftei, C. (eds.) *Water Safety, Security and Sustainability. Advanced Sciences and Technologies for Security Applications.* Springer, Cham. https://doi.org/10.1007/978-3-030-76008-3_28.

Srivastava A., Khadke, L., and Chinnasamy, P. 2022. Developing a web application-based water budget calculator: attaining water security in rural-Nashik, India. In: Kolathayar, S., Mondal, A., Chian, S. C. (eds.) *Climate Change and Water Security. Lecture Notes in Civil Engineering*, vol 178. Springer, Singapore. https://doi.org/10.1007/978-981-16-5501-2_37.

Stambuk, G. N. 1999. Water Quality Evaluation by Index in Dalmatia. *Water Res.* 33(16), 3423–3440.

Stigter T. Y., Ribeiro, L., and Carvalho Dill, A. M. 2006. Application of a groundwater quality index as an assessment and communication tool in agro-environmental policies—two Portuguese case studies. *J. Hydrol.* 327, 578–591.

Subbarao, C., Subbarao, N. V., and Chandu, S. N. 1995.Characterisation of groundwater contamination using factor analysis. *Environ. Geol.* 28(4), 175–180. https://doi.org/10.1007/s002540050091.

Subramani, T., Rajmohan, N., and Elango, L. 2009. Groundwater geochemistry and identification of hydrogeochemical processes in a hard rock region, Southern India. *Environ. Monit. Assess.* 162(1–4), 123–137.

Tatawat, R. K., and Chandel, C. P. S. 2008. A hydrochemical profile for assessing the groundwater quality of Jaipur City. *Environ. Monit. Assess*, 143, 337–343. https://doi.org/10.1007/s10661-007-9936-3.

Tyagi, S., Sharma, B., Singh, P., and Dobhal, R. 2013. Water quality assessment in terms of water quality index. *Am. J. Water Res.* 1(3), 34–38.

UNEP, 1999. *Global Environment Outlook 2000*. Earthscan, UK.

Vasanthavigar, M., Srinivasamoorthy, K. Vijayaragavan, K., Rajiv Ganthi, R., Chidambaram, S., Anandhan, P., Manivannan, R., and Vasudevan, S. 2010. Application of water quality index for groundwater quality assessment: Thirumanimuttar Sub-Basin, Tamilnadu, India. *Environ. Monit. Assess.* 171, 595–609.

Vasanthavigar, M., Srinivasamoorthy, K., and Prasanna, M. V. 2012. Evaluation of groundwater suitability for domestic, irrigational, and industrial purposes: a case study from Thirumanimuttar river basin, Tamilnadu, India. *Environ. Monit. Assess.* 184(1), 405–420.

Voudouris, K., Panagopoulos, A., and Koumantakis, J. 2000. Multivariate statistical analysis in the assessment of hydrochemistry of the northern korinthia prefecture alluvial aquifer system (Peloponnese, Greece). *Nat. Res. Res.* 9(2), 135–146.

Wang, S. W., Liu, C. W., and Jang, C. S. 2007. Factors responsible for high arsenic concentration in two groundwater catchments in Taiwan. *Appl. Geochem.* 22, 460–476.

Wen, X., Wu, Y., Zhang, Y., and Liu, F. 2005. Hydrochemical characteristics and salinity of groundwater in the Ejina basin, Northwestern China. *Environ. Geol.* 48, 665–675. https://doi.org/10.1007/s00254-005-0001-7.

World Health Organization. 2004. *Guideline for Drinking Water Quality*, 2nd ed., vol 1. World Health Organization, Geneva, Switzerland.

Zhu, G., Su, Y., Huang, C., Feng, Q., and Liu, Z. 2009. Hydrogeochemical processes in the groundwater environment of Heihe River Basin, northwest China. *Environ. Earth Sci.* 60(1), 139–153. https://doi.org/10.1007/s12665-009-0175-5.

18 Assessment of Heavy Metal Contamination in Water and Sediments of Major Industrial Streams

Pankaj Bakshe and Ravin Jugade

18.1 INTRODUCTION

Nemours serious challenges that humanity has been suffering in the past couple of generations are associated with water quality and resources (Schwarzenbach et al., 2010). Streams and rivers perform a variety of human and ecological activities, including transportation of water, ecological tourism, aquaculture activities, habitation facilities, and impact mitigation (Hasan et al., 2021; Zhuang et al., 2018). The basic and essential parts of the fluvial environment are stream/river water and sediments, which not only provide nutrients for aquatic life but also act as latent sinks and consequent sources of contaminants, such as heavy metals (HMs) (Ali et al., 2020; Jiang et al., 2013; Kumar et al., 2021). HMs pose a significant environmental hazard to human life and aquatic ecosystems because of their non-biodegradability, tendency to undergo bioaccumulation, reasonable ecological stability, tenacity, and bio toxicity properties (Zhang et al., 2019). These non-biodegradable HMs may accrue and persist in the surface sediment for an extended period of time, and also mix with flowing water and migrate from one place to the other. In this way, it can damage the ecological environment and infiltrate the food chain, resulting in serious and long-lasting effects on the human body (Weber et al., 2020). Several researchers reported that geogenic and human-induced factors influence the three-dimensional distribution of HMs in water as well as the sediments in contact with it. The geogenic factors comprise rock-water survival, sediment loss, benthic agony, and flow regime changes all examples of geogenic factors, whereas human-induced factors consist of metropolitan wastewaters, sewage disposal (also called muck), industrial effluents, wastewater irrigation, and seeping of fertilizer and pesticides, fertilizers, and other chemicals (Islam et al., 2021; Fang et al., 2019). HMs are deposited in streams and rivers when they carry surface water through areas with a variety of human-induced inputs. For example, cascading networks of tanks (surface water-bearing structures) are one such network that are interconnected with streams and are either receiving their supply from a river or yielding surplus water to an adjoining river. Herein, these tanks may undergo phenomenal sediment deposition. Consequently, such networks are vulnerable to not only declining storage but simultaneously aggravating and accumulating sediment-carrying pollutants (as reported by Srivastava and Chinnasamy, 2021; 2022, 2023a, 2023b for the state of Maharashtra and South India). It may also alter the geographical distribution forms of HMs in superficial water and sediment (Kumar et al., 2021). Thus, evaluating the spatial distribution of HMs in water and sediments is important for the risk assessment and environmental conservation of the aquatic ecosystem.

In the present era, industrial effluent is a major concern among human-induced factors. Industries are being established to meet the requirements of the rising population in the country.

DOI: 10.1201/9781003303237-18

The establishment of industries on the one hand produces essential products, but at the same time, it also releases waste products in the form of solid, liquid, or gas, resulting in hazards, pollution, and energy losses. A large portion of domestic waste and industrial effluent is liquidated into the soil as well as water bodies, posing a substantial threat to public health and the proper functioning of biodiversity (Tariq et al., 2006). For instance, in India, 764 major industrial units are operational, including distilleries, chemical industries, food and dairy production, paper manufacturing, textile, and many more that generate wastewater of 501 MLD (Million Liter per Day) and this is the main source for polluting most of the rivers in India (ENVIS, 2018).

As a result, the main challenge for the scientific community is balancing of sustainability of the available resources, hand in hand with the need of the hour industrial development. The environment is a self-cleaning system if we allow it to regenerate itself. The toxicants find their way from the hydrosphere to the lithosphere and vice versa in a wonderful manner. This equilibrium can be best understood by analyzing both water bodies and sediments in contact with them. Heavy metal interactions with water and sediments are typically of the nature of adsorption–desorption cycles (Bradl, 2004; Wijesiri et al., 2019). Such interactions are significantly influenced by physicochemical properties like pH and salinity of water, and in the case of sediments, involve mineral content, specific surface area, and ion exchange capacity (Keshavarzifard et al., 2019). Hence, in order to accurately assess the aquatic environment in terms of heavy metal contamination, it is most relevant to examine the interaction between water and in-contact sediment. Therefore, in this paper, the correlation of heavy metal contamination in water bodies and their uptake by sediments or soil in contact with it has been studied in detail. Central India or more precisely a part of Maharashtra state of India is the present study site. Past studies have repeatedly reported the aggravating issues due to heavy metal contamination in water and sediments, such that decentralized solutions are being devised to combat the same (Wagh et al., 2018; Shree and Nishikant, 2019; Srivastava et al., 2021, 2022). The mobility of the HMs in sediment and the overlay water system is a core factor in determining the current situation of metal pollution. The HMs contamination sources may resume input if sediment continues to obtain metals from overlying water; conversely, the HMs contamination sources may reduce if sediment has a greater tendency to release metals into overlying water. Therefore, the main intentions of this study were to (1) quantify the heavy metal pollution strength in water and sediment of the study area, (2) compare the concentration of metals (Pb, Cu, Cr, Co, Ni, Zn, Cd, Hg, Fe, and Mn) to the international quality guidelines for water and sediment, and (3) explore the interactions of heavy metals comprised in sediment and the covering water.

18.2 MATERIALS AND METHODS

18.2.1 Study Area and Sample Collection

For the current study, the water samples were collected from the same industrial place as the stream sediment samples were collected in the Central India region named as district of Chandrapur, located on the eastern side of Maharashtra, with latitude ranging from 78°.48′ to 80°.48′.55′ East longitude and latitudes ranging from 18°.41′ to 20°–50′ North latitudes. The district is bordered on the western and eastern sides by the Wardha River and the Wainganga River, both of which are tributaries of the Godavari River. The district covers a total area of 11,443 km^2, of which 32.34 km^2 are used for industrial purposes. A total of 702 industries exist in this district, with 47 industries that can be classified as large-scale, 18 as medium-scale, and 637 as small-scale industries. Out of these industries, Ballarpur Industries Limited (BILT) and Chandrapur Super Thermal Power Plant (CSTP) are the two major industries in this district, and these were the sites from which stream sediment and water samples were collected Ballarpur Industries Ltd (BILT) has been in operating since 1953 adjacent to the Wardha River near Ballarshah town, nearly 16 km from Chandrapur. It manufactures 116,000 tons of paper of various textures and types. Along with this, it emits 55,000 m^3 of wastewater every day. There are two streams released from this industry, which eventually merge at the Wardha River. The streams were labeled B-A and B-B for convenience, as they travel 1.5 and 1 km from the industry to the river, respectively

The Chandrapur Super Thermal Power Plant (CSTP) is located 6 km on the northern side of Chandrapur, near Durgapur village. It occupies an area of approximately 11,266 hectares. This industry makes use of water from a dam erected over the Erai River. There are 9 units that produce 3,340 MW of thermal electricity. However, it produces wastewater in the range of 50–70 M^3 per MW. In this industrial area, there is one major stream that travels around 3.5 km from the industry and finally meets the Erai River. For clarity, this stream was again labeled as C-C. The study area and sample location are shown in Figure 18.1.

For the present study, 12, 9, and 32 water samples from streams B-A, B-B, and C-C were collected and labeled as A1 to A12, B1 to B9, and C1 to C32, respectively. Each sample was taken at a distance of around 100 m. All water samples were taken in duplicate at each location, 1,000 mL for general analysis and 250 mL for heavy metals analysis, at a depth of 0.2–0.3 m below the water surface (Taylor and Governor, 2015), using clean polyethylene bottles (washed with distilled water and rinsed with samples twice). For heavy metal analysis, the samples were filtered through 0.45 m millipore filters and acidified with 1% nitric acid to retain pH<2 except for Hg, which was maintained by adding 5% potassium chromate and 1% sulfuric acid (Hamlin, 1989). All water samples were preserved in cooler containers with ice packs before they were transported to the laboratory for storage at 4°C until further analysis.

18.2.2 Water Analysis

In the site location, physiochemical parameters such as temperature, pH, electrical conductivity, and dissolved oxygen were measured. In this case, a HANNA portable instrument HI 9829 multiparameter water quality meter was used. Whereas major cations and anions (including Na^+, K^+, Ca^{2+}, Mg^{2+}, CO_3^{2-}, HCO^{3-}, Cl^-, SO_4^{2-}, NO^{3-}, F^-, and PO_4^{3-}) were determined in accordance with the procedures described in the Standard Methods for the Examination of Water and Wastewater in the laboratory (Rodger and Bridgewater, 2017). Although, for the present study, these cations and anions are not used for any statistical analysis.

For heavy metal analysis, the nitric acid digestion method was adopted for the flame atomic absorption spectrophotometer. For this, 100 mL of acidified filtered water sample was measured in a 150 mL glass beaker and 5 mL of conc. HNO_3 was added. Then the beaker was transferred to a hot plate to maintain a reflux temperature of around 95°C that slowly evaporated the sample to the lowest volume possible (about 5–10 mL) before precipitation occurred. Once a light-colored, clear solution was obtained, its volume was made up to 25 mL (Rodger and Bridgewater, 2017). All water samples were treated similarly.

The concentration of heavy metals was quantitated by an atomic absorption spectrophotometer (Varian model AA240FS flame AAS) using different wavelengths, Pb (217.0 nm), Cu (324.8 nm), Cr (356.9 nm), Co (240.7 nm), Cd (228.8 nm), Mn (279.5 nm), Ni (232.0 nm), Zn (213.9), and Fe (248.3 nm), whereas the mercury analysis was carried out by the Direct Mercury Analyzer (DMA) of Milestone Inc., USA, by using a preserved water sample (with 5% potassium chromate and 1% sulfuric acid) without any further treatment at the wavelength 253.6 nm.

We have reported analysis and influencing parameters for sediment contaminants in the same area located in Chandrapur district (Bakshe and Jugade, 2021). This is an extension of the work reported therein and we report the correlation between water quality and sediment composition in this work.

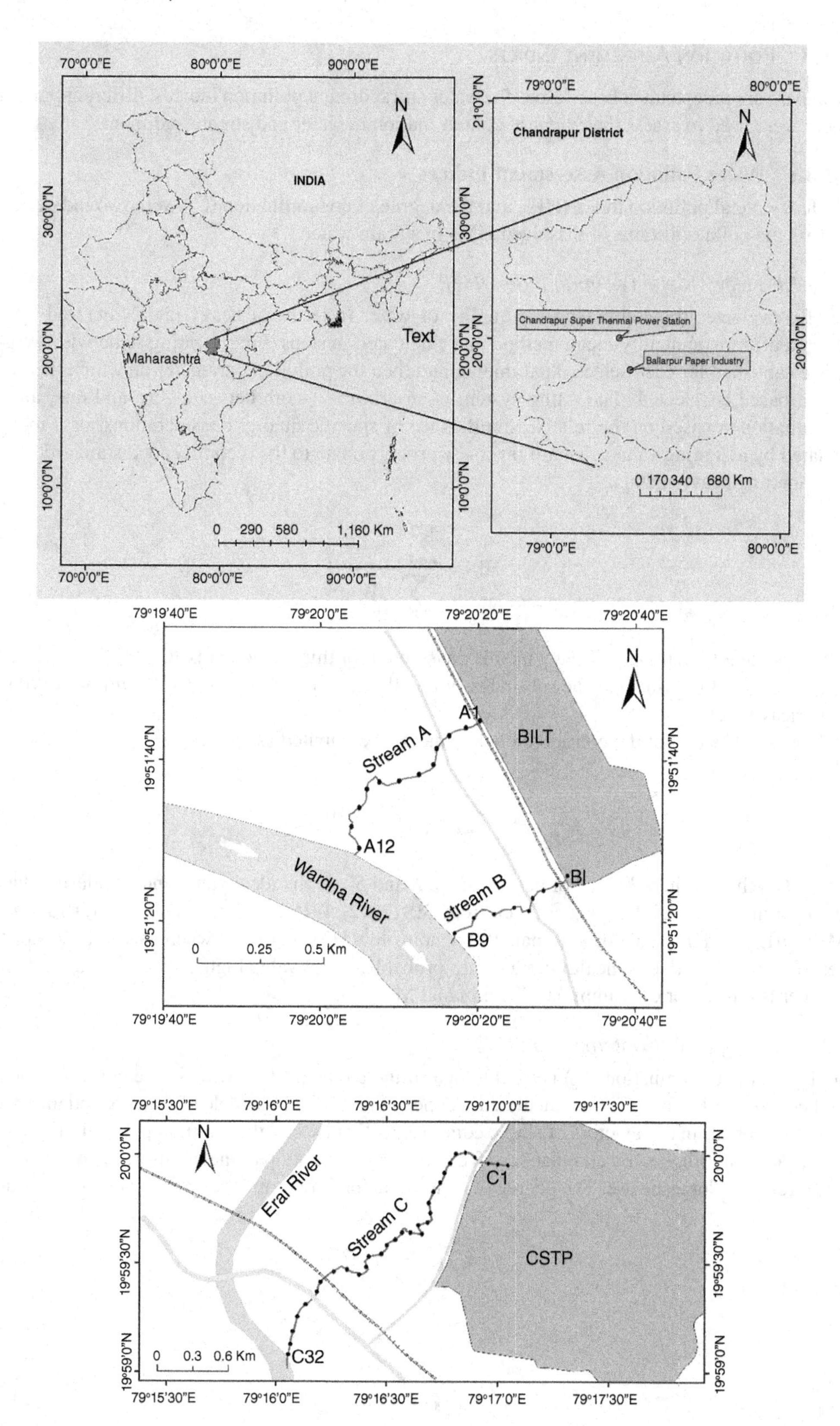

FIGURE 18.1 Locations of study areas with spatial distribution of streams and sampling locations.

18.2.3 Pollution Assessment Indices

Since there are independent base values for water and sediment pollution indices, different pollution indices were used to assess the extent of contamination in water and stream sediment.

18.2.3.1 Water Pollution Assessment Indices

The Heavy metal pollution index (HPI) and the degree of contamination (C_d) were two indices used to assess the pollution status of dissolved HMs in stream water.

18.2.3.1.1 Heavy Metal Pollution Index (HPI)

The HPI evaluates the comprehensive quality of water in terms of heavy metals derived by the weighted arithmetic quality mean method and calculated in steps. First, a rating scale with weightage is established for each selected parameter, and then the pollution parameter on which the index is to be based is selected. The rating system is an arbitrary score between zero and one, and its determination is based on the relative significance of specific quality considerations, or it may be evaluated by assigning values that are inversely proportionate to the recommended standard for the corresponding heavy metals.

$$\mathrm{HPI} = \frac{\sum_{i=1}^{n} w_i Q_i}{\sum_{i=1}^{n} W_i} \tag{18.1}$$

where n is the total number of heavy metals considered for this study that is 10 (Pb, Cu, Cr, Co, Ni, Zn, Cd, Hg, Fe, and Mn), Q_i is the sub-index of the ith heavy metal, and Wi is the unit weighting of the ith heavy metal.

The sub-index (Q_i) of the considered heavy metal is computed as

$$Q_i = \sum_{i=1}^{n} \frac{\{M_i - I_i\}}{S_i - I_i} \times 100 \tag{18.2}$$

where M_i is the monitored value of heavy metal, I_i and S_i are the ideal value and standard value of the ith parameter for drinking water taken from BIS (2012), WHO (2017), and USEPA (2018) for the HMs (µg/L). An HPI value of less than 100 denotes low HM pollution, whereas 100 indicates the threshold at which adverse health impacts are probable, and a value higher than 100 suggests that the water is unsafe for consumption (Sheykhi and Moore, 2012).

18.2.3.1.2 Degree of Contamination (C_d)

The degree of contamination (C_d) is used to determine the quality of water in the context of heavy metal contamination The C_d was calculated independently for each sample of water tested and it is a sum of the contamination factors of single components that exceed the maximum permissible limit. Hence, the C_d highlights the cumulative effects of several quality parameters that are considered to be hazardous to domestic use. The degree of contamination is determined by the following equation.

$$C_d = \sum_{i=1}^{n} Cf_i \tag{18.3}$$

Where,

$$Cf_i = \frac{M_i}{\mathrm{MAC}_i} - 1 \tag{18.4}$$

where Cf_i is the contamination factor for the ith heavy metal (HM) and MAC is the maximum admissible concentration of the ith HM. The following are the categories used to describe HM contamination based on *Cd*: <1 low; 1–3 moderate; >3 high contaminations of heavy metals in water (Backman et al., 1998).

18.2.3.2 Stream Sediment Pollution Indices

Whereas sediment contains different sizes of particles and organic matter, heavy metals strongly bind with sediment, resulting in greater values of heavy metals than water. So, here are two different indices used to assess the heavy metal contamination in stream sediment.

18.2.3.2.1 Modified Degree of Contamination (mC_d)

To illustrate an overall average value for a number of contaminants, the modified degree of contamination (mC_d) is the best tool to assess the quality of stream sediment with respect to heavy metals.

$$mC_d = \frac{\left(\sum_{i=1}^{n} C_f^i\right)}{n} \tag{18.5}$$

where n=number of analyzed elements, i=ith element, and C_f=Contamination factor, which is the ratio of monitored value and baseline value in sediment (mg/Kg). Abrahim and Parker, 2008, suggested the following gradations for the classification and description of the modified degree of contamination (mC_d) in sediments as shown in Table 18.1

18.2.3.2.2 Pollution Load Index

The extents of heavy metal contamination in sediment were evaluated by using the pollution load index (PLI) which was proposed by Tomlinson et al., 1980. PLI gives a simple, comparative means for evaluating a site or sediment quality. A value of zero denotes excellence, a value of one denotes exactly the background levels of contaminants existing, and values greater than one show progressive deterioration of the site and sediment quality. It is computed as:

$$\text{PLI} = \left(C_{f1} \times C_{f2} \times C_{f3} \times \cdots \times C_{fn}\right)^{1/n} \tag{18.6}$$

where n is the number of metals (n=8 in this study) and C_f is the contamination factor.

TABLE 18.1
Classification and Description of the Modified Degree of Contamination (mC_d)

Classification	Degree of Contamination
$mC_d<1.5$	Nil to very low
$1.5 \le mC_d<2$	Low
$2 \le mC_d<4$	Moderate
$4 \le mC_d<8$	High
$8 \le mC_d<16$	Very high
$16 \le mC_d<32$	Extremely high
$mC_d \ge 32$	Ultra high

18.2.4 Quality Guidelines for Water and Sediment

Quality guidelines have been formulated in response to a variety of environmental concerns and regulatory programs. Water and sediment quality guidelines have been proposed separately by either government bodies or independent researchers. For this study, the most widely accepted quality guidelines were used as a reference and baseline, which are as follows:

18.2.4.1 Water Quality Guidelines

Internationally WHO sets the guidelines for drinking water quality, similarly the guideline from the Environmental Protection Agency, (EPA) United States of America is also acknowledged globally. Because of geographical concerns, the quality guidelines from the Bureau of Indian Standards (BIS) are more appropriate in India. Most of the standards used in this study were from the BIS and WHO.

18.2.4.2 Sediment Quality Guidelines

Empirically based Sediment Quality Guidelines (SQGs), which are derived from field sediment concentrations and field and/or laboratory biological effects studies were used to evaluate the extent of heavy metal contamination in sediment as shown in Table 18.2

18.2.5 Statistical Analysis

The average values and standard deviations of the heavy metal concentrations in water were computed. A one-way ANOVA was used to compare heavy metal concentrations in sediment and overlying water, followed by post hoc Tukey testing. SPSS version 20 statistical software was used for the statistical analysis, and a significance level of $p<0.05$ was considered statistically substantial. The multivariate statistical method of principal components analysis (PCA) was employed to summarize the variation of a data set between samples on one side to a set of uncorrelated components on the other side. This method was used to evaluate the elemental level-based diversity of trace elements with physicochemical attributes (Astatkie et al., 2021)

TABLE 18.2
Important Types of SQGs with Description

Type of SQG	Acronym	Description	Reference
Lowest effect level	LEL	Clean to marginally polluted sediments. The sediment-dwelling organisms are generally not affected below this concentration	Persaud el al. (1993)
Threshold effect level	TEL	Below this concentration, a rare occurrence of adverse effects takes place	Smith et al. (1996)
Effect range—low	ERL	Rarely observed negative impact below this chemical concentration	Long and Morgan (1991)
Severe effect level	SEL	Highly polluted sediments show adverse effects on the majority of sediment-residence creatures	Persaud et al. (1993)
Probable effect level	PEL	Expected to have frequent adverse effects above these concentrations	Smith et al. (1996)
Effect range-median	ERM	Frequent occurrence of adverse effects over this concentration	Long and Morgan (1991)
Toxic effect threshold	TET	Heavy pollution load on sediments affects most of the organisms living in it	EC and MENV1Q (1992)

18.2.5.1 Regression Model

The relationship between heavy metals in sediment and overlying water was evaluated using a linear regression model. Linear regression evaluates relationships that are readily expressed by straight lines or may be expanded to several dimensions. It is especially useful for assessing and adjusting for confounding variables. The relationship's model is hypothesized and estimates of the parameter values are employed to make an estimated regression equation. If the model is acceptable, the estimated regression equation may be used to estimate the value of the dependent variable given values for the independent variables (Ideriah et al., 2012; Zaid, 2015) In this case, heavy metal concentrations in water are considered independent variables, whereas sediment concentrations are termed dependent variables. A linear regression equation is represented as follows:

$$y = mx + c \tag{18.7}$$

where y and x are the concentrations of metals in sediment and water respectively, m is the coefficient corresponding to the slope and c represents the intercept on the y axis.

18.2.5.2 Distribution Coefficient

The distribution coefficient is one of the most important metrics for determining the potential migration of a pollutant in the liquid phase when it comes into contact with sediment or suspended materials (Takata et al., 2014; Thanh-Nho et al., 2018). The distribution coefficient (K_d) is the ratio of contaminants in sediments to those in water, and it's calculated using the equation below (Boyer et al., 2018).

$$\text{Distribution Coefficient } (K_d) = \frac{\text{Metal in sediment } (\text{mg/Kg})}{\text{Metal in water} (\text{mg/L})} \tag{18.8}$$

According to Sedeño-Díaz et al. (2019), the Log K_d index provides valuable information about contaminant affinity in an aquatic media and it is categorized as Log (K_d)<3: Metals prevailing in the liquid phase (loosely bound to solid), 3<Log (K_d)<4: Metals released from the solid phase toward the liquid phase (moderately bound to solid) and Log (K_d)>5 corresponding to metals remaining in the solid phase (strongly bound to solid).

18.3 RESULTS AND DISCUSSION

18.3.1 Dispersal of Heavy Metals in Water

The concentrations of ten metals (Pb, Cu, Cr, Co, Ni, Zn, Cd, Hg, Fe, and Mn) in streams B-A, B-B, and C-C are compared to the concentrations of heavy metals in the corresponding stream sediment as shown in Figure. Also, the descriptive data of water analysis are related to water quality guidelines provided by the Bureau of Indian Standards which are shown in Table 18.3.

The mean concentration of Pb was found to be 0.086, 0.128, and 0.055 mg/L in the B-A, B-B, and C-C streams, respectively. In all three streams, it exceeds the BIS permitted limit. The highest concentrations of Pb were found to be 0.11, 0.182, and 0.071 mg/L in the A1, B1, and C1 sampling sites of the above streams respectively. This indicates that the Pb contamination in these streams was caused by industrial effluent. Among these three streams, B-B showed the most contamination of Pb. Similarly, the mean concentration of Cu was found to be 0.052, 0.068, and 0.059 mg/L in the B-A, B-B, and C-C streams. This heavy metal also exceeds the BIS-permitted limit in all three streams. The maximum concentrations of Cu were found to be 0.069, 0.077, and 0.059 mg/L in the A10, B9, and C32 sampling sites of the B-A, B-B, and C-C streams, respectively. Further, 7

TABLE 18.3
Descriptive Concentration (mg/L) of Heavy Metal in Water and Water Quality Guidelines

Metal→ Description↓	Pb	Cu	Cr	Co	Ni	Zn	Cd	Hg	Fe	Mn
					B-A					
Min	0.059	0.028	0.044	0.028	0.008	0.028	0.0014	0.00036	0.14	0.042
Max	0.11	0.069	0.057	0.052	0.02	0.081	0.0028	0.00067	0.358	0.087
Mean	**0.086**	**0.052**	**0.052**	**0.043**	**0.013**	**0.053**	**0.00208**	**0.00050**	**0.251**	**0.063**
Std. dev.	0.017	0.013	0.005	0.008	0.003	0.017	0.00052	0.00009	0.070	0.012
					B-B					
Min	0.092	0.056	0.038	0.022	0.011	0.026	0.001	0.00036	0.3	0.087
Max	0.182	0.077	0.056	0.047	0.018	0.061	0.0019	0.00062	0.421	0.172
Mean	**0.128**	**0.068**	**0.045**	**0.033**	**0.014**	**0.041**	**0.00137**	**0.00045**	**0.366**	**0.118**
Std. dev.	0.029	0.006	0.007	0.008	0.002	0.010	0.00034	0.00008	0.047	0.031
					C-C					
Min	0.048	0.035	0.032	0.022	0.01	0.061	0.002	0.00018	28.26	1.22
Max	0.071	0.109	0.071	0.052	0.026	0.47	0.0081	0.00029	42.36	2.8
Mean	**0.055**	**0.059**	**0.051**	**0.033**	**0.016**	**0.193**	**0.00434**	**0.00023**	**36.58**	**1.99**
Std. dev.	0.005	0.014	0.008	0.008	0.003	0.103	0.00178	0.00003	3.806	0.432
					Water Quality Guideline					
BIS	0.01	0.05	NA	NA	0.02	5	0.003	0.001	0.3	0.1
WHO	0.01	2	0.05	*	0.07	*	0.003	0.006	*	0.4

NA, not available; *, No health-based guideline is proposed.

out of 12, 9 out of 9, and 26 out of 32 sampling sites of B-A, B-B, and C-C streams, respectively, were found to exceed the BIS permitted limit. This implies that Cu pollution is primarily caused by industrial effluent, and the streams also serve as carriers of this metal in the river.

In contrast to the above, the Cr in the B-B stream was reported to be within the WHO permitted limit, with a mean concentration of 0.045 mg/L, whereas this metal was found to be above the WHO permitted limit (no BIS guideline for Cr) in the B-A and C-C streams, with mean concentrations of 0.052 mg/L and 0.051 mg/L, respectively. However, in the B-B stream, 0.056 mg/L and 0.52 mg/L of Cr were detected in samples B9 and B1, indicating that there was marginal pollution of Cr due to the mixing of river-contaminated water and industrial effluent in the respective samples. In the B-A and C-C streams, the Cr pollution is mainly due to industrial effluent, with the highest concentration of 0.057 mg/L in the A9 sample and 0.071 mg/L in the C1 samples, respectively. For Co, there are no guidelines from BIS or WHO, however, the mean concentration of this metal was found to be 0.043 mg/L, 0.033 mg/L, and 0.033 mg/L in the B-A, B-B, and C-C streams, respectively. The maximum concentrations were found in the B-A, B-B, and C-C streams, with 0.052 mg/L in A1, 0.047 mg/L in B2 & B9, and 0.052 mg/L in C32 samples, respectively. In these streams, there was no significant contribution of industrial effluent to Co contamination.

The mean concentration of Ni in the B-A, B-B, and C-C streams was 0.013 mg/L, 0.14 mg/L, and 0.16 mg/L, respectively, which is also under the BIS guideline limit. This metal was found at the highest concentrations in the A11, B9, and C31 samples, with 0.020 mg/L, 0.018 mg/L, and 0.026 mg/L in the B-A, B-B, and C-C streams, respectively. All of these sites were at the confluence of streams and rivers. This suggested that the high value of Ni in these sites was due to contaminated river water. The same observations were recorded for Zn as noted previously in Ni. Here also, all the concentrations of Zn were within the permissible limit of BIS in all these three streams. The mean concentration and maximum value of Zn in the B-A, B-B, and C-C streams were noted with 0.053–0.081 mg/L, 0.041–0.061 mg/L, and 0.19–0.47 mg/L respectively. The highest concentration of this metal was detected in the C32 sample of the C-C stream, which was caused by contaminated river water mixing with stream water; otherwise, there was no practical contamination of Zn in any of the three streams.

There is a lot of evidence that thermal power plants contribute to Cd pollution in the environment (Agrawal et al., 2010; Yaylali-Abanuz, 2011). This evidence remains true for Cd contamination in the C-C stream as well. The mean concentration for Cd was reported at 0.00434 mg/L, which exceeds the BIS permissible limit with a maximum concentration of 0.00810 mg/L in the C32 sample of the C-C stream. It was also found that a number of samples near the CSTP were polluted with Cd, leading to the conclusion that the Cd pollution in this stream was mostly caused by this industry. However, in the B-A and B-B streams, the mean and maximum concentrations of Cd were below the BIS guideline limit with 0.00208–0.0028 mg/L and 0.00137–0.0019 mg/L in A12 and B8 samples of respective streams. Likewise, all of the water samples had low levels of Hg, and no Hg pollution was found in any of the streams studied, and it was under BIS guidelines limits. In the B-A, B-B, and C-C streams, the mean concentration of Hg was found to be 0.00050, 0.00045, and 0.00023 mg/L and the range was 0.00036–0.00067 mg/L, 0.00036–0.00062 mg/L, and 0.00018–0.00029 mg/L, respectively.

Surprisingly, Fe concentrations in all CSTP -C stream water samples were found to be extremely high. It ranged from 28.26 to 42.36 mg/L with a mean value of 36.58 mg/L, which is 100 times greater than the BIS permissible guideline and that is also the most significant finding in this study. The high Fe concentration in this stream might be due to corrosion of the water supply system piping and the formation of an iron oxide scale on the inner walls of the boilers. Whereas, the B-B stream was also found to have a moderate level of Fe pollution, with concentrations ranging from 0.3 to 0.421 mg/L and an average value of 0.0366 mg/L. This might be due to pipe corrosion or the usage of iron-containing compounds such as ferrous sulfate and ferrous chloride etc. In contrast to the earlier, there was reasonable Fe contamination in the B-A stream, with a mean value of 0.0251 mg/L. The highest concentration of Fe, 0.358 mg/L, was detected in the A1 sample, indicating that BILT industrial effluent was also contributing to Fe pollution in rivers and streams.

Furthermore, there was no Mn contamination in the B-A stream, and the mean concentration was found to be 0.063 mg/L. It ranged from 0.042 to 0.087 mg/L, which is much below the BIS permissible level. In contrast, there was considerable Mn pollution in the B-B stream, with a mean value of 0.118 mg/L and a range of 0.087–0.172 mg/L, with the maximum concentration detected in the B1 sample, indicating that this heavy metal was also contaminated by the BILT industry. Similar to Fe contamination, all of the samples in the C-C stream exceeded the BIS permissible limit by more than 10–20 times, with a mean value of 1.99 mg/L observed. It ranged from 1.22 to 2.80 mg/L due to its leaching from coal fly ash (Prasad and Mondal, 2009)

18.3.2 Heavy Metals Correlation in Water and Sediment

The relationship between heavy metal concentrations in water and sediment was estimated using a regression model, the natural logarithm of the distribution coefficient, and statistical analysis for significant differences as follows.

18.3.2.1 Lead

The association between the concentrations of Pb in sediment and water is shown in Figure. The slope of equations in all three streams was positive and greater than unity, indicating a high concentration of Pb in water with a high concentration of Pb in sediment. Whereas in the B-A and B-B streams, the squared correlation coefficient was 0.8294 and 0.8634, respectively, denoting a large positive linear association and indicating that Pb in sediment increased as Pb in water increased, in the C-C stream it was 0.3848, showing a comparatively less positive linear association of Pb concentration in water and sediment. The log of distribution coefficient for this metal was less than 3 (2.7, 2.5, and 2.7 in B-A, B-B, and C-C streams, respectively), in all three streams, indicating that Pb was more prevalent in the liquid phase. In the B-A stream, the statistical analysis ($p < 0.05$) showed a statistically significant relationship between the concentrations of Pb in the water and sediment, whereas in the B-B and C-C streams, it ($p > 0.05$) showed no statistically significant relationship.

18.3.2.2 Copper

The association between the concentrations of Cu in sediment and water is shown in Figure. In all three streams, the slope of the equations was positive and larger than one, suggesting a large amount of Cu in the water with a large amount of Cu in the sediment. In the B-A, B-B, and C-C streams, the squared correlation coefficients were 0.2277, 0.3856, and 0.3796, showing a positive and moderate linear relationship between Cu content in water and sediment. The log of distribution coefficient for this metal was almost 3 (3.1, 3.3, and 3.0 in B-A, B-B, and C-C streams, respectively) in all three streams, showing that Cu was slightly bound to sediment and had a propensity to mix with water. Furthermore, similar statistical analysis results were obtained for Pb, where, in the B-A stream, the statistical analysis ($p < 0.05$) indicated a statistically significant relationship between the concentrations of Pb in the water and sediment, while in the B-B and C-C streams, it ($p > 0.05$) established no statistically significant relationship.

18.3.2.3 Chromium

The association between the concentrations of Cr in sediment and water is shown in Figure. In all three streams, the slope of the equations was positive and greater than unity, inferring that there was a high concentration of Cr in the water with a high concentration of Cr in the sediment. The squared correlation coefficients in the B-A, B-B, and C-C streams were 0.1557, 0.0921, and 0.3743, respectively, indicating a poor and positive linear association between Cr in the water and the sediment of the B-A and B-B streams, while in the C-C stream, there was a relatively strong positive linear association between Cr in the water and sediment. This also indicates that in the B-A and B-B streams, the Cr in sediment was not proportionally increased as the Cr in water increased, but in the C-C stream there was proportionality for Cr in the water and sediment. In all three streams, the log of distribution coefficient for this metal was greater than 3 (3.5, 3.5, and 3.3 in B-A, B-B, and C-C streams, respectively), indicating that Cr was poorly bound to sediment and had a strong tendency to mix with water. Furthermore, in the C-C stream the statistical analysis ($p < 0.05$) revealed a statistically significant relationship between Cr concentrations in the water and sediment, whereas in the B-A and B-B streams, the statistical analysis ($p > 0.05$) revealed no statistically significant relationship between water and sediment for this metal.

18.3.2.4 Cobalt

The interaction between Co concentrations in sediment and water is depicted in Figure. The slope of the equations was positive and greater than one in all three streams, revealing a high concentration of Co in water with a high concentration of Co in sediment. The squared correlation coefficient in the B-A and B-B streams was 0.4881 and 0.3840, respectively, indicating a moderate positive linear association and indicating that Co in sediment increased as Co in water increased, whereas it was 0.1632 in the C-C stream, indicating a less positive linear association of Co concentration in water and sediment indicating that the concentration of Co in sediment was not thoroughly dependent on the concentration of Co in water. In all three streams, the log of distribution coefficient for this metal was around 3 (3.2, 3.1, and 3.0 in B-A, B-B, and C-C streams, respectively), indicating that Co was partially bound to sediment and had a tendency to mix with water. Moreover, in the B-B and C-C streams, the statistical analysis ($p<0.05$) showed a statistically significant relationship between the concentrations of Co in the water and sediment, whereas in the B-A stream, it ($p > 0.05$) showed no statistically significant relationship.

18.3.2.5 Nickel

The relationship between Ni concentrations in sediment and water is depicted in Figure. The slope of the equations in all three streams was positive and greater than unity, implying a high concentration of Ni in the water with a high concentration of Ni in the sediment. The squared correlation coefficients for the B-A and B-B streams were 0.3732 and 0.5288, respectively, reflecting a significant positive linear association and indicating that Ni in sediment increased as Ni in water increased, whereas the squared correlation coefficient for the C-C stream was 0.1204, denoting a weak positive linear association of Ni concentration in water and sediment. The log of distribution coefficient for Ni was greater than 3 in all three streams (3.7, 3.5, and 3.6, respectively, in B-A, B-B, and C-C), indicating that this metal was weakly bound to sediment and had a significant propensity to mix with water. Furthermore, in the B-B stream, the statistical analysis ($p<0.05$) indicated a statistically significant relationship between the concentrations of Ni in the water and sediment, while in the B-A and C-C streams, it ($p>0.05$) developed no statistically significant relationship in between the concentrations of Ni in the water and sediment.

18.3.2.6 Zinc

The association between the concentrations of Zn in sediment and water is shown in Figure. The slope of the equations in all three streams was positive and greater than one, suggesting a high concentration of Zn in the water with a high concentration of Zn in the sediment. The squared correlation coefficients for the B-A and B-B streams were 0.0412 and 0.0067, indicating an extremely weak and positive linear association of Zn concentration in water and sediment, respectively, whereas the squared correlation coefficient for the C-C stream was 0.4778, indicating a significant positive linear association and indicating that Zn in sediment accelerated as Zn in water increased. The log of distribution coefficient for Zn in both BILT streams (i.e., B-A and B-B) was 3.5, indicating that this metal was loosely bound to sediment and had a high proclivity to mix with water. In the C-C stream, however, it was less than 3 (2.8), indicating that the Zn was little bound to sediment and had a very strong tendency to mix with water. Furthermore, the statistical analysis for the CSTP stream was $p=0.081$, which, while slightly greater than 0.05, reflects a considerable relationship between Zn concentrations in the water and sediment. While in the B-A and B-B streams, it was ($p>0.05$) denoting no statistically significant relationship between Zn concentrations in the water and sediment.

18.3.2.7 Cadmium

The association between Cd concentrations in sediment and water is depicted in Figure. The slope of the equations in all three streams was positive and above the unity, expressing a high concentration of Cd in the water with a high concentration of Cd in the sediment. The same regression model trends that are shown for Zn were also reported for Cd. The B-A and B-B streams had squared correlation coefficients of 0.0565 and 0.0955, indicating an overall weak and positive linear association of Cd concentration in water and sediment, respectively, while the C-C stream had a squared correlation coefficient of 0.2101, showing a considerable positive linear association and inferring that Cd in sediment increased as Cd in water increased. The log of distribution coefficient for this metal was below 3 in all three streams (2.4, 2.6, and 1.4, respectively, in the B-A, B-B, and C-C streams), suggesting that the Cd was very weakly bonded to sediment and had a high significant tendency to mix with water. Moreover, the statistical analysis for the CSTP stream was $p=0.064$, which, somewhat above 0.05, reflects a considerable relationship between Cd concentrations in the water and sediment. While in the B-A and B-B streams, it was ($p>0.05$) representing no statistically significant relationship between Cd concentrations in the water and sediment.

18.3.2.8 Mercury

The association between the concentrations of Hg in sediment and water is shown in Figure. In all three streams, the slope of the equations was positive and greater than unity, inferring that there was a high concentration of Hg in the water with a high concentration of Hg in the sediment. The squared correlation coefficient in the B-A and B-B streams was 0.4552 and 0.4300, respectively, indicating a significant positive linear association and revealing that Hg in sediment increased as Hg in water increased, whereas it was 0.0720 in the C-C stream, indicating a weak linear association of Hg concentration in water and sediment. The log of distribution coefficient for this metal was less than 3 in all three streams (2.8, 2.6, and 2.3 in the B-A, B-B, and C-C streams, respectively), indicating that the Hg was remarkably weakly attached to sediment and had a high significant propensity to mix with water. In the B-A and B-B streams, the statistical analysis ($p > 0.05$) developed no statistically significant relationship between the concentrations of Hg in the water and sediment while in the C-C stream, it ($p > 0.05$) developed a statistically significant relationship in between the concentrations of Hg in the water and sediment.

18.3.2.9 Iron

The relationship between Fe concentrations in sediment and water is represented in Figure. In all three streams, the slope of the equations was positive and above one, inferring that there was a high concentration of Fe in the water with a high concentration of Fe in the sediment. The squared correlation coefficients in the B-A, B-B, and C-C streams were 0.2213, 0.0471, and 0.1164, respectively, inferring a weak linear association between Fe in the water and sediment in the B-B stream, whereas a relatively strong positive linear association between Fe in the water and sediment of the B-A and C-C streams. The log of distribution coefficient for Fe in both BILT streams was greater than 5 (5.4 and 5.2 in the B-A and B-B, respectively), reflecting that this metal was strongly bound to sediment and had a high proclivity to attach with sediment, and thus the concentration of Fe in sediment was extremely high than in water. In contrast, the log of distribution coefficient for Fe in the C-C stream (3.2) was greater than 3, implying that this metal was loosely bound to sediment and had a significant propensity to mix with water, resulting in a high concentration of Fe in water, which was a surprising result of this study. Furthermore, in the B-A and B-B streams, the statistical analysis ($p < 0.05$) established a statistically significant relationship between the concentrations of Fe in the water and sediment, whereas, in the C-C stream it ($p > 0.05$) established no statistically significant relationship between the concentrations of Fe in the water and sediment.

18.3.2.10 Manganese

The association between the concentrations of Mn in sediment and water is shown in Figure. In all three streams, the slope of the equations was positive and greater than one, denoting that there was a high concentration of Mn in the water with a high concentration of Mn in the sediment. The squared correlation coefficient in the B-B streams was 0.2738, indicating a moderate linear association and suggesting that Mn in sediment increased as Mn in water increased, whereas it was 0.0182 and 0.0678 in the B-A and C-C streams, denoting a less linear association of Mn concentration in water and sediment and suggesting that Mn in sediment was not exclusively dependent on Mn in water. The log of the distribution coefficient for Mn in both BILT streams was greater than 4 (4.8 and 4.4 in the B-A and B-B, respectively), revealing that this metal was tightly bound to sediment and had a high tendency to attach with sediment, and thus the concentration of Mn in sediment was much higher than in water. In contrast, the log of distribution coefficient for Mn in the C-C stream (3.2) was greater than 3, showing that this metal was loosely bound to sediment and had a consistent proclivity to mix with water, leading to a higher Mn concentration in water. Moreover, only the B-B stream had a statistically significant relationship between the concentrations of Mn in the water and sediment according to the statistical analysis ($p < 0.05$), whereas the B-A and C-C streams had no statistically significant relationship between the concentrations of Mn in the water and sediment as per the statistical analysis ($p > 0.05$).

18.3.3 Assessment of Water Pollution

In order to assess water pollution, only six heavy metals (Pb, Cu, Ni, Zn, Cd, and Hg) were considered to compute the Heavy metal pollution index (HPI) and degree of contamination, while the remaining four metals (Cr, Co, Fe, and Mn) were not used due to non-availability of ideal values. According to HPI, all three streams, B-A, BIL-B, and C-C, were critically polluted with respect to these six heavy metals, with HPI values of 174, 257, and 160, respectively. It was also observed that the main influential heavy metal in the HPI was Pb, which contributed the highest index value with 130, 201, and 94 due to high contamination of this metal in the streams, followed by Hg, which contributed the index value 32, 35, and 49, in the B-A, B-B, and C-C streams, respectively, despite the fact that there was no significant Hg contamination in the above-mentioned streams but, these values were due to the high *Unit Weight Value (Wi)* of Hg metal. Furthermore, similar findings were also noted in the degree of contamination index. According to this, all streams were classified as '*High Contamination*' because the index value was greater than 3. (5.5, 9.8, and 3.2 in the B-A, B-B, and C-C streams, respectively). In this case, too, the influential heavy metal was Pb in all the streams, and these findings also validated the HPI accuracy.

18.3.4 Assessment of Sediment Pollution

To assess sediment pollution, eight heavy metals (Pb, Cu, Cr, Co, Ni, Zn, Cd, and Hg) were used to calculate the Modified degree of contamination (mC_d) and pollution load index (PLI), and the concentrations of these metals were also compared to Sediment Quality Guidelines (SQGs) to establish the extent of contamination shown in Table 18.4. With reference to the mCd index, 9 samples (A1 to A6 and A9 to A12) of the B-A stream were classified to 'Moderate degree of contamination' with an index of $2 \leq mC_d < 4$. The highest mC_d index 3.6 was found in sample A1. However, A7 and A9 were classified as 'Low degree of contamination' with index values of 1.7, while A8 had the lowest mC_d

Table 18.4
Descriptive Concentration (mg/kg) of Sediment with Sediment Quality Guidelines

Stream/SQGs	Description/ Acronym	Pb	Co	Cr	Cu	Ni	Zn	Cd	Hg
B-A	Minimum	14	39	124	55	32	73	0.27	0.03
	Maximum	78	110	198	254	91	397	1.08	0.76
	Standard Dev.	19.95	20.90	23.46	60.14	21.43	83.39	0.28	0.22
	Average	**47**	**68**	**170**	**122**	**60**	**166**	**0.61**	**0.35**
B-B	Minimum	28	18	98	49	21	56	0.24	0.02
	Maximum	80	75	191	229	82	227	1.13	0.66
	Standard Dev.	18.78	19.90	29.07	59.37	14.73	62.21	0.33	0.20
	Average	**52**	**47**	**145**	**116**	**46**	**145**	**0.60**	**0.24**
C-C	Minimum	8	16	59	20	39	72	0.06	0.01
	Maximum	68	89	261	104	103	224	0.38	0.29
	Standard Dev.	13.78	15.71	51.29	14.99	13.18	44.35	0.06	0.06
	Average	**32**	**33**	**111**	**57**	**60**	**129**	**0.12**	**0.06**
Lowest effect level	LEL	31	NA	26	16	16	120	0.6	0.2
Threshold effect level	TEL	35	NA	37.3	35.7	18	123	0.596	0.174
Effect range—low	ERL	35	NA	80	70	30	120	5	0.15
Severe effect level	SEL	250	NA	110	110	75	820	10	2
Probable effect level	PEL	91.3	NA	90	197	36	315	3.53	0.0486
Effect range-median	ERM	110	NA	145	390	50	270	9	1.3
Toxic effect threshold	TET	170	NA	100	86	61	540	3	1

index of 1.3 and was classified as having a 'Nil to very low degree of contamination.' For the B-B stream, with an index 2≤mC_d<4, 5 samples (B1 to B3 and B8 to B9) were categorized as 'Moderate degree of contamination.' Sample B1 had the maximum mC_d index of 3.1, besides that 3 samples (B5 to B7) had an mC_d index of less than 1.5, hence these samples were classified as 'Nil to very low degree of contamination.' Whereas only one sample (B4) had an index of 1.5≤mC_d<2 with 1.8, hence it was considered in the 'Low degree of contamination' category. The lowest index in this stream was found in sample B6, which had an index of 1.1. Furthermore, in the C-C stream, only three samples (C1, C2, and C32) had an index of 2≤mCd<4, while four samples (C3, C4, C6, and C31) had an index of 1.5≤mC_d<2, hence putting them in the 'Moderate degree of contamination' and ' Low degree of contamination' categories, respectively. The rest of the 25 samples from these streams were classified as 'Nil to very low degree of contamination' with an index less than 1.5.

The PLI results revealed that only one sample (A8) in the B-A stream and three samples (B5 to B7) in the B-B stream had indexes less than 1, and the rest of the samples in both streams had indexes greater than 1, so these samples were classified as 'Polluted,' whereas samples with less than one were categorized in the 'Not Polluted' category. For the C-C stream, seven samples (C1 to C4, C6, and C31 to C32) had PLIs greater than 1 hence these samples were classified as 'Polluted,' category. However, similar to the mCd results, the remaining 25 samples of this stream were listed in the 'Not Polluted' category.

When metal concentrations in sediment were compared to empirically based Sediment Quality Guidelines (SQGs), it was concluded that all metals in all three streams exceeded the Threshold Effect Level (TEL), with the exception of Cd and Hg in the C-C stream. Whereas, the mean concentration of Cr and Ni in sediment samples from all streams crossed the Probable effect level (PEL). The Severe effect level (SEL) was observed to be higher for Cr from all streams and Cu from B-A and B-B streams sediment samples. The SQG values and comparison with the descriptive concentration of heavy metals from the sediment are shown in Table 18.4

18.3.5 Principal Components Analysis (PCA)

In the sample of B-A and B-B streams, three principal components (PCs) explain 77.03% and 85.23% variance, respectively. These percentages were adequate to give a detailed overview of the data structure. PC1 contributes 44.05% of total variance in B-A stream samples which consisted of more significant factors than others and characterized by high loadings of Hg (r=0.89), Co (r=0.89), Pb (r=0.85), Temp (r=0.80), Cu (r=0.75), Fe (r=0.74), Mn (r=0.71), Cd (r=0.70), and EC (r=0.59). PC2 contributes 25.94% of the total variance, with prominent loadings of Distance (r=0.88), pH (r=0.86), Cr (r=0.76), and Ni (r=0.64). Whereas, only one significant variable, Zn (r=0.58), was found in PC3. Consequently, in B-B stream samples, PC1 contributes 54.62% variance, which is higher than in the B-A stream of PC1. This was characterized by high loading EC (r=0.99), Mn (r=0.96), Hg (r=0.88), pH (r=0.87), Cr (r=0.86), Fe (r=0.85), Pb (r=0.84), Cd (r=0.84), and Temp (r=0.64). Furthermore, PC2 contributed 18.83% variance with two significant factors: Distance (r=0.96), and pH (r=0.82). Lastly, only one significant factor Cu (r=70) was observed in PC3. Likewise, four PCs account for 78.44% of the total variance in the C-C stream. PC1 and PC2 contribute nearly the same percentage of total variables, whereas PC3 and PC4 only have two and one variable respectively. PC1 attributed 31.08% of total variance with Cu (r=0.88), Zn (r=0.80), Ni (r=0.70), Cd (r=0.65), Co (r=0.64), Distance (r=0.58), Temp (r=0.57), and Hg (r=0.53), while PC2 contributed 28.93% of the total variance with major loadings of EC (r=0.82), Cr (r=0.61), and Pb (r=0.54). In PC3, Fe (r=0.63) and pH (r=0.41) whereas in PC4, Mn (r=0.68) was the significant variable accounting for 9.54% and 8.14% of total variance.

From the PCA results, as shown in Figure 18.2, it may be concluded that the influence of pH was altered for the different streams. In the B-B stream, pH had an influence on all metals except Zn and Cu, whereas in the B-A stream, it had an impact on Ni and Cr, and in the C-C stream, only Fe was the most influenced by pH. Electrical conductivity (EC) was most dominant in PC1 of the B-A

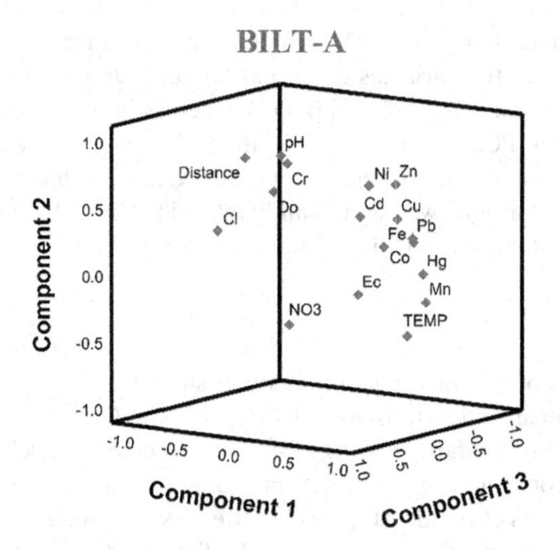

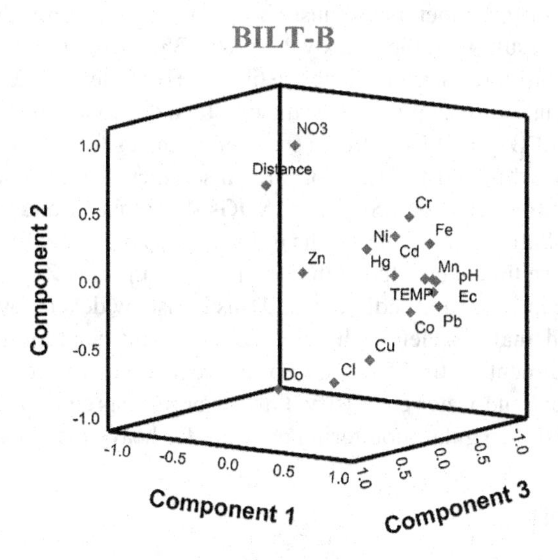

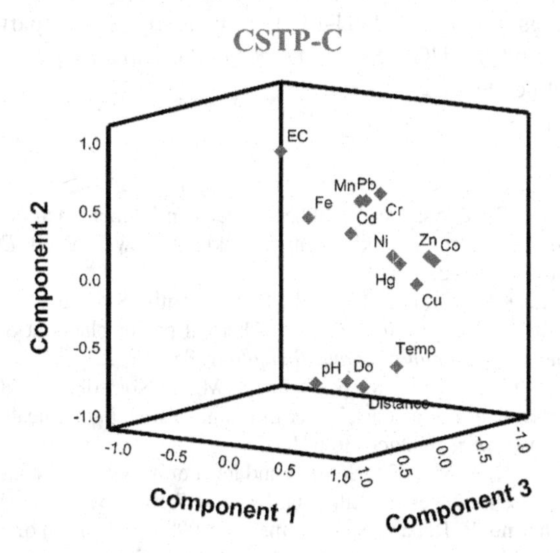

FIGURE 18.2 PCA loading plots.

and B-B streams and had an influence on all the metals in those streams, but in C-C it was significantly prevalent in PC2 with the variables of Cr and Pb only. Dissolved oxygen (DO) remarkably influenced Cu in the B-B stream as displayed in PC3, whereas in the B-A stream, Cr was impacted by it and this is reflected in PC2 of this stream. In the C-C stream, there was no significant influence of DO in any metals but, it was noticed that in PC2, EC and DO have an inversely proportional relationship. The Distance variable was significantly related to Zn in the B-B stream, while in B-A, it was directly proportional, and in C-C, it was indirectly proportional to pH.

18.4 CONCLUSION

According to the findings of this investigation, the B-B stream was overall more polluted than the other two streams. The mean concentrations of metals (Pb, Cu, Cr, Cd, Fe, and Mn) in this stream exceed the BIS standard limit, whereas the B-A stream was comparatively more polluted than the C-C stream, with mean concentrations of metals including Pb, Cu, and Cr exceeding the BIS standard limit. However, in terms of Fe and Mn, the C-C stream was more highly polluted than the other two streams; in addition, this stream also exceeded the BIS standard limits for Pb, Cu, and Cr, but the mean concentration of these metals was just above the standard limit threshold. Furthermore, HPI also confirmed this result, with the index value, B-B (257) >B-A (174) >C-C (160). Pb was the most prominent metal in all three streams, with the highest HPI index value. In context to sediment, the ideal tool for assessing sediment pollution is the modified degree of contamination (mC_d), which shows that the majority of B-A and B-B stream sediment samples fall into the 'Moderate degree of contamination' category, while most of the C-C stream sediment samples fall into the 'Nil to very low degree of contamination' category. Similarly, SQGs show that Cr and Cu have more contamination in the sediment of all streams. The results of the regression model for comparing metal concentrations in water and sediment concluded that Pb in all streams and Zn in the C-C stream have a strong correlation between water and sediment, and this is also evidenced by the log of distribution coefficient and statistical analysis. Hence, this study endorses the need for more research on heavy metal adsorption on sediment, with a focus on Pb, in order to determine the long-term effects of heavy metals on sediment and water chemistry. Furthermore, to protect these water bodies and to reduce environmental risk, a regular monitoring program for heavy metals is recommended.

ACKNOWLEDGMENT

The authors are thankful to RTM Nagpur University for funding under the University Research Project Scheme. Fundings under the DST-FIST scheme by the Department of Science and Technology, New Delhi, and the UGC-SAP scheme by the University Grants Commission, New Delhi, have also been acknowledged.

REFERENCES

Abrahim, G. M. S., Parker, R. J. 2008. Assessment of heavy metal enrichment factors and the degree of contamination in marine sediments from Tamaki Estuary, Auckland. New Zealand. *Environmental Monitoring and Assessment*. 136(1–3):227–238

Agrawal, P., Mittal, A., Prakash, R., Kumar, M., Singh, T. B., Tripathi, S. K. 2010. Assessment of contamination of soil due to heavy metals around coal fired thermal power plants at Singrauli Region of India. *Bulletin of Environmental Contamination and Toxicology*. 85:219–223.

Ali, M. M., Ali, M. L., Proshad, R., Islam, M. S., Rahman, M. Z., Kormoker, T. 2020. Assessment of trace elements in the demersal fishes of a coastal river in Bangladesh: a public health concern. *Thalassas: An International Journal of Marine Science*. 36:641e655.

Astatkie, H., Ambelu, A., Beyene, E. M. 2021. Sources and level of heavy metal contamination in the water of Awetu watershed streams, southwestern Ethiopia. *Heliyon*. 7:e06385

Backman, B., Bodis, D., Lahermo, P., Rapant, S., Tarvainen, T. 1998. Application of a groundwater contamination index in Finland and Slovakia. *Environmental Geology*. 36:55e64

Bakshe, P., Jugade, R., 2021. Distribution, association, and ecological risk evaluation of heavy metals and influencing factors in major industrial stream sediments of Chandrapur District, Central India. *Water Air and Soil Pollution.* 232:78.

Boyer, P., Wells, C., Howard, B. 2018. Extended Kd distributions for freshwater environment. *Journal of Environmental Radioactivity.* 192:128–142.

Bradl, H. B. (2004). Adsorption of heavy metal ions on soils and soils constituents. *Journal of Colloid and Interface Science.* 277(1):1–18.

EC, MENVIQ (Environment Canada and Ministere de l'Envionnement du Quebec). 1992. *Interim Criteria for Quality Assessment of St. Lawrence River sediment.* Environment Canada, Ottawa.

Hamlin, S. N. 1989. Preservation of samples for dissolved mercury. *Water Resources Bulletin.* 2:25.

Ideriah, T. J. K., David-Omiema, S., Ogbonna, D. N. B. (2012). Distribution of heavy metals in water and sediment along Abonnema Shoreline, Nigeria. *Resources and Environment.* 2(1):33–40.

Rodger, B., Bridgewater, L. 2017. *Standard Methods for the Examination of Water and Wastewater.* 23rd edition. American Public Health Association, Washington, DC.

Schwarzenbach, R. P., Egli, T., Hofstetter, T. B., Gunten, U., Wehrli, B. 2010. Global water pollution and human health. *Annual Review of Environment and Resources.* 35:109–136

Sedeño-Díaz, J. E., López-López, E., Mendoza-Martínez, E., Rodríguez-Romero, A. J., Morales-García, S. S. 2020. Distribution coefficient and metal pollution index in water and sediments: proposal of a new index for ecological risk assessment of metals. *Water.* 12:29.

Sheykhi, V., Moore, F. 2012. Geochemical characterization of kor river water quality, fars province, southwest Iran. *Water Quality Exposure and Health.* 4:25e38.

Shree, B. V., Nishikant, G. 2019. Examining the heavy metal contents of an estuarine ecosystem: case study from Maharashtra, India. *Journal of Coastal Conservation.* 23(6):977–984.

Smith, S. L., MacDonald, D. D., Keenleyside, K. A., Ingersoll, C. G., Field, J. 1996. A preliminary evaluation of sediment quality assessment values for freshwater ecosystems. *Journal of Great Lakes Research.* 22:624–638.

Srivastava, A., Chinnasamy, P. 2021. Developing village-level water management plans against extreme climatic events in Maharashtra (India)—a case study approach. In: Vaseashta, A., Maftei, C. (eds.) *Water Safety, Security and Sustainability. Advanced Sciences and Technologies for Security Applications.* Springer, Cham. https://doi.org/10.1007/978-3-030-76008-3_27

Srivastava, A., Chinnasamy, P. 2022. Assessing groundwater depletion in southern india as a function of urbanization and change in hydrology: a threat to tank irrigation in Madurai City. In: Kolathayar, S., Mondal, A., Chian, S. C. (eds.) *Climate Change and Water Security. Lecture Notes in Civil Engineering*, vol 178. Springer, Singapore. https://doi.org/10.1007/978-981-16-5501-2_24

Srivastava, A., Chinnasamy, P. 2023a. Tank cascade system in Southern India as a traditional surface water infrastructure: a review. In: Chigullapalli, S., Susha Lekshmi, S. U., Deshpande, A. P. (eds.) *Rural Technology Development and Delivery. Design Science and Innovation.* Springer, Singapore. https://doi.org/10.1007/978-981-19-2312-8_15

Srivastava, A., Chinnasamy, P. 2023b. Understanding declining storage capacity of tank cascade system of madurai: potential for better water management for rural, peri-urban, and urban catchments. In: Chigullapalli, S., Susha Lekshmi, S. U., Deshpande, A. P. (eds.) *Rural Technology Development and Delivery. Design Science and Innovation.* Springer, Singapore. https://doi.org/10.1007/978-981-19-2312-8_14

Srivastava, A., Khadke, L., Chinnasamy, P. 2021. Web application tool for assessing groundwater sustainability—a case study in rural-Maharashtra, India. In: Vaseashta, A., Maftei, C. (eds.) *Water Safety, Security and Sustainability. Advanced Sciences and Technologies for Security Applications.* Springer, Cham. https://doi.org/10.1007/978-3-030-76008-3_28

Srivastava, A., Khadke, L., Chinnasamy, P. 2022. Developing a web application-based water budget calculator: attaining water security in rural-Nashik, India. In: Kolathayar, S., Mondal, A., Chian, S. C. (eds.) *Climate Change and Water Security. Lecture Notes in Civil Engineering*, vol 178. Springer, Singapore. https://doi.org/10.1007/978-981-16-5501-2_37

Takata, H., Keiko, T., Tatsuo, A., Shigeo, U. 2014. Distribution Coefficient (Kd) of strontium and significance of oxides and organic matter in controlling its partitioning in coastal regions of Japan. *Science of the Total Environment.* 490:979–986.

Tariq, M., Ali, A. Z. Shah, Z. 2006. Characteristics of industrial effluents and their possible impacts on quality of underground water. *Soil & Environment.* 25(1):64–69.

Taylor, M., Governor, L. 2015. Surface Water Field Sampling Manual for Water Column Chemistry, Bacteria and Flows.

Thanh-Nho, N., Stardy, E., Nhu-Trang, T. T., David, F., Marchand, C. 2018. Trace metals partitioning between particulate and dissolved phases along a tropical mangrove estuary (Can Gio, Vietnam). *Chemosphere.* 196:311–322.

Tomlinson, D. L., Wilson, J. G., Harris, C. R., Jeffrey, D. W. 1980, Problems in the assessment of heavy-metal levels in estuaries and the formation of a pollution index. *Helgolander Meeresuntersuchungen.* 33(1–4):566–575.

USEPA, 2018. Edition of the Drinking Water Standards and Health Advisories Tables, Washington, DC.

Wagh, V. M., Panaskar, D. B., Mukate, S. V., Gaikwad, S. K., Muley, A. A., Varade, A. M. 2018. Health risk assessment of heavy metal contamination in groundwater of Kadava River Basin, Nashik, India. *Modeling Earth Systems and Environment.* 4(3):969–980.

Weber, A. A., Sales, C. F., de Souza Faria, F., Melo, R., Bazzoli, N., Rizzo, E. 2020. Effects of metal contamination on liver in two fish species from a highly impacted neotropical river: a case study of the fundao dam, Brazil. *Ecotoxicology and Environmental Safety.* 190:110165.

WHO (World Health Organisation). 2017. *Guidelines for Drinking-Water Quality*, 4th. WHO, Geneva.

Wijesiri, B., Liu, A., He, B., Yang, B., Zhao, X., Ayoko, G. A., Goonetilleke, A. 2019. Behaviour of metals in an urban river and the pollution of estuarine environment. *Water Research.* 164:114911.

Yaylali-Abanuz, G. 2011. Heavy metal contamination of surface soil around Gebze industrial area, Turkey. *Microchemical Journal.* 99:82–92.

Zabir, A. A., Zzaman, M. W. U., Hossen, M. Z., Uddin, M. N., Islam, M. S., Islam, M. S. 2016. Spatial dissemination of some heavy metals in soil adjacent to Bhaluka industrial area, Mymensingh, Bangladesh. *American Journal of Applied Scientific Research.* 2(6):38e47.

Zaid, M. H. 2015. *Correlation and Regression Analysis. The Statistical, Economic and Social Research and Training Centre for Islamic Countries (SESRIC).* Ankara, Turkey, 13–21 pp.

Zhang, C., Shan, B., Tang, W., Wang, C., Zhang, L. 2019. Identifying sedimentassociated toxicity in rivers affected by multiple pollutants from the contaminant bioavailability. *Ecotoxicology and Environmental Safety*, 171:84e91.

19 Water Waste Treatment and Reuse for Sustainability

Khilat Shabir, Rohitashw Kumar,
Munjid Maryam, and Falaq Firdous

19.1 INTRODUCTION

The sustainability of our planet depends on the conservation and management of water, a fundamental natural resource that supports life on earth. The demand for water is constantly rising as a result of population increase, industrialization, and urbanization, making effective and sustainable water management strategies more important than ever. The treatment and reuse of water waste are one of the most crucial components of water management (Black, 2016). Water waste treatment and reuse involve the process of treating wastewater and other forms of water waste to remove contaminants and impurities, making it safe for various beneficial uses such as irrigation, industrial processes, and even human consumption. This process not only provides a sustainable source of water but also helps to reduce the strain on natural water sources such as rivers, lakes, and groundwater. The management of water waste is essential for protecting the environment and human health. Untreated wastewater can contain a variety of harmful pollutants such as pathogens, heavy metals, and toxic chemicals, which can have detrimental effects on the environment and human health. The proper treatment and reuse of water waste can help to mitigate these negative impacts and contribute to sustainable water management.

In this chapter, we will explore the various aspects of water waste treatment and reuse, including the different treatment processes and technologies, the potential uses of treated wastewater, and the benefits and challenges of implementing water waste management programs. Through a comprehensive analysis of this topic, we can gain a deeper understanding of the importance of water waste treatment and reuse for sustainable water management and the protection of our environment and health.

In addition to the various treatment methods available, the degree of treatment required also depends on the intended use of the treated wastewater. For example, if the treated wastewater is to be reused for irrigation or industrial processes, a lower degree of treatment may be sufficient. However, if the treated wastewater is intended for discharge into water bodies, a higher degree of treatment is required to ensure that the wastewater meets the required quality standards (Vergine et al., 2017). The treatment of wastewater is not only important for protecting the environment and human health but can also have economic benefits. Treated wastewater can be reused for various purposes, such as irrigation, industrial processes, and even drinking water. This can help to reduce the demand for freshwater sources and provide a sustainable source of water for various uses. However, the implementation of wastewater treatment programs can also face challenges, such as high capital and operational costs, lack of technical expertise, and inadequate infrastructure. Therefore, it is important to consider the economic, environmental, and social factors when designing and implementing wastewater treatment programs (Zotos et al., 2009).

In conclusion, wastewater treatment and reuse are critical components of sustainable water management. The environment, human health, and a sustainable water source can all be protected with the help of wastewater treatment. We can assure the effective use of our water resources and support sustainable development by implementing efficient wastewater treatment procedures.

DOI: 10.1201/9781003303237-19

19.2 TYPES OF WATER WASTE

Several industries, such as agriculture, industry, and households, all produce wastewater. The sector, the activity, and the amount of water utilized all affect the kind and amount of water waste that is produced. The many types of water waste produced in these sectors are summarized as follows:

- **Agricultural waste**: Agricultural wastewater is a significant source of water waste, as agriculture is one of the largest industries in the world. This type of wastewater is generated during various agricultural activities, such as crop irrigation, livestock management, and crop processing.
    1. **Irrigation wastewater**: Irrigation is essential for crop production, but it generates a significant amount of wastewater. Irrigation wastewater may contain sediment, organic matter, nutrients, and pesticides. The amount and quality of irrigation wastewater vary depending on the type of crop, soil conditions, and irrigation method used.
    2. **Crop-washing wastewater**: During the harvesting and processing of crops, significant amounts of water are used for washing, cleaning, and sanitizing the crops. Crop-washing wastewater may contain pesticides, organic matter, and other contaminants. The amount and quality of crop-washing wastewater vary depending on the type of crop, processing methods, and cleaning techniques used.
    3. **Livestock wastewater**: Animal husbandry practices such as the washing of animal sheds, milking, and manure management generate a significant amount of wastewater. Livestock wastewater contains organic matter, nutrients, and pathogens that can pose a risk to the environment and public health. The amount and quality of livestock wastewater depend on the type and number of animals, management practices, and the type of manure management system used.
    4. **Crop processing wastewater**: Crop processing facilities generate wastewater during the processing of crops, such as canning, freezing, and juicing. Crop processing wastewater contains organic matter, nutrients, and chemicals used during the processing of crops. The amount and quality of crop processing wastewater vary depending on the type of crop, processing methods, and cleaning techniques used.
- **Industrial waste**: Depending on the production process, many industries produce wastewater with various properties, including manufacturing, mining, and energy generation. Heavy metals, hazardous compounds, and other pollutants that are dangerous to both human health and the environment can be found in the wastewater produced by these industrial activities.
    1. **Manufacturing wastewater**: Different manufacturing processes generate various types of wastewater. For example, metal-working processes generate wastewater that contains heavy metals, while the textile industry generates wastewater containing dyes and other chemicals.
    2. **Mining wastewater**: Mining activities generate wastewater that contains heavy metals, radioactive materials, and other harmful pollutants.
    3. **Energy production wastewater**: Power generation activities generate wastewater that contains chemicals such as sulfur dioxide, nitrogen oxides, and other pollutants.
- **Municipal waste**: Stormwater runoff, commercial wastewater, and residential sewage make up municipal waste. Domestic sewage, which often contains organic materials, detergents, and human waste, makes up the majority of municipal wastewater.
    1. **Domestic sewage**: Domestic sewage is the largest source of municipal wastewater. Domestic sewage contains human waste, detergents, and other organic matter.
    2. **Commercial wastewater**: Commercial activities such as restaurants, hotels, and shopping malls generate wastewater that contains detergents, oils, and other contaminants.
    3. **Stormwater runoff**: During heavy rain events, stormwater runoff can generate large quantities of wastewater that contains sediment, organic matter, and other pollutants.

- **Construction waste**: Wastewater is also produced during construction tasks like excavation, building deconstruction, and concrete pouring. Typically, suspended particles, silt, and other pollutants are present in this kind of wastewater.
  1. **Excavation wastewater**: Excavation activities generate wastewater that contains sediments, suspended solids, and other contaminants.
  2. **Concrete pouring wastewater**: Construction activities, such as concrete pouring, generate wastewater that contains cement, sand, and other materials.
  3. **Building demolition wastewater**: Building demolition generates wastewater that contains debris, metals, and other materials.
- **Healthcare waste**: Hospitals and clinics produce wastewater that is contaminated with diseases, chemicals, and medications. It is crucial to properly handle medical waste in order to stop the spread of illnesses, safeguard the environment, and safeguard public's health.
  1. **Hospital wastewater**: Healthcare facilities such as hospitals and clinics generate wastewater that contains various pathogens, chemicals, and pharmaceuticals.
  2. **Laboratory wastewater**: Laboratories generate wastewater that contains chemicals and other harmful substances.

Effective management of water waste is essential to protect the environment and public health while ensuring the sustainable use of water resources. Each sector generates water waste with unique characteristics, necessitating diverse treatment methods to ensure safe disposal or reuse. Agricultural activities, such as irrigation, crop washing, livestock management, and crop processing, generate significant amounts of water waste with varying degrees of pollution. Agricultural wastewater is particularly crucial to managing correctly due to its potential harm to the environment and public health. The composition and quantity of agricultural wastewater depend on the type of crop, livestock, and processing method used. For example, irrigation wastewater may contain sediments, nutrients, and pesticides, while livestock wastewater may contain organic matter, pathogens, and nutrients (Mateo-Sagasta et al., 2015).

To manage agricultural wastewater, various treatment methods can be utilized, including physical, chemical, and biological processes. The treatment method chosen depends on the characteristics of the wastewater and its intended reuse. For instance, irrigation wastewater may require primary treatment, while livestock wastewater may require secondary and tertiary treatment due to higher levels of pollutants. Overall, effective water waste management is essential to maintain environmental and public health while ensuring sustainable water resource use. Proper treatment of water waste generated by agriculture and other industries is critical in this regard.

19.3 GOALS OF WASTEWATER MANAGEMENT

The main goal of wastewater management is to promote sustainable development by safeguarding the environment and public health while wisely utilizing natural resources. The idea of sustainable development is to satisfy the requirements of the present without compromising the capacity of future generations to satisfy their own needs. Sustainable development in the context of water quality refers to the management of water resources in a way that does not jeopardize their availability for current or future generations. As a result, this principle should be taken into account when setting goals and objectives for the treatment, reclamation, reuse, and use of water. In summary, wastewater management's main objective is to support sustainable development by safeguarding public health and the environment and making appropriate use of natural resources without jeopardizing the needs of coming generations (Baud et al., 2001).

19.4 MANAGEMENT OF WATER QUALITY

Wastewater management is crucial for the sustainable development of natural resources and the protection of public health and the environment (UNDP, 2006). To achieve this, water quality management is necessary, where the intended use of water is determined, water quality requirements are delineated, and treatments or other management techniques are based on those requirements. Pollution can be caused by both humans and nature, and the philosophy of water pollution control has changed considerably since the passage of the Public Health Service Act of 1912, which primarily focused on investigating the spread of waterborne diseases (WHO, 2011). Today, water quality standards are usually limited by aquatic life considerations, and the cost of removal of a contaminant increases considerably as the percentage of removal increases.

Receiving surface waters have the innate ability to accept some contaminants without adverse environmental impact, known as the waste assimilative capacity. This capacity often dictates the wastewater treatment requirements imposed on a wastewater discharger. However, where sufficient assimilative capacity does not exist, regulations or standards may dictate at least a minimal degree of treatment obtainable by technology-based standards. Where assimilative capacity is not sufficient to accommodate wastewater effluents and maintain stream standards, compliance with water-quality-based standards is required.

Total maximum daily loads (TMDLs) are determined by state environmental agencies based on the designated use of the stream and results from ecotoxicological evaluations, considering contributions from non-point-source (diffuse) pollution, background conditions, and a factor of safety. A determination for each discharger is made as to acceptable point source effluent loadings, and discharge permit restrictions are then imposed and enforced. These actions typically require a high degree of treatment, thereby shifting focus to resource management, waste reduction, and water reuse options (Metcalf and Eddy, 2003).

The focus on wastewater treatment has shifted to one of wastewater management, which includes prevention instead of treatment/remediation and promotion of 'clean technologies'. Emphasis is now given by industry to waste minimization and product life cycle analysis to reduce raw materials, energy, and environmental releases, thereby conserving natural resources; reducing risks/liability; and at the same time, providing significant cost savings. In summary, wastewater management should be considered an essential element of water resources management to achieve the sustainable development of natural resources, protect public health and the environment, and ensure that the intended use of water is not impaired for future generations (UN, 2017).

19.5 GROUPING OF POLLUTANTS IN WASTEWATER

Water pollution is a significant environmental issue that affects the quality of water and the health of aquatic life. Pollutants can be classified into four categories: chemical, physical, physiological, and biological (Mallikarjuna et al., 2019). Each category poses different challenges to water quality, and methods for removal or treatment can vary depending on the specific pollutant.

Chemical pollutants can be inorganic or organic in nature, and their toxicity and effects on oxygen levels in the water can vary. Organic materials, which contain organic carbon, can be difficult to biodegrade and can result in a reduction of dissolved oxygen levels in the water. Aquatic life requires a certain level of dissolved oxygen to survive, and an excess of organic material can lead to oxygen depletion, creating harmful conditions for aquatic life. Inorganic chemicals, on the other hand, can be toxic, directly affecting aquatic life, and can change the pH of the water.

Physical pollutants include color, turbidity, temperature, suspended solids, foam, and radioactivity (Chowdhury et al., 2020). While color is not necessarily harmful, it can be aesthetically unacceptable for drinking water and some industrial uses. Temperature can affect biological activity and can increase the toxicity of heavy metals. Turbidity is caused by suspended solids and requires coagulation and filtration for removal. Suspended solids can reduce light penetration and inhibit photosynthesis, decrease benthic organism activity, and clog fish gills, negatively impacting aquatic life. Foam can cause aesthetic problems and reduce the rate of oxygen gas transfer into the water. Radioactive materials have special effects and as such they need to be regulated at the source. They can dissolve in water, be mixed into sludges, or affect biological life, and because of the unique effects of radioactive substances, they must be controlled at the source.

Physiological effects of pollution are primarily the result of taste and odor. Although taste and odor problems may be minor in effect, public reaction can result in magnification of problems and adverse publicity of the water purveyor. Taste and odor are particularly objectionable when present in drinking water or process water for food where palatability is important.

Biological pollutants include pathogens, algae, and other microorganisms. These pollutants can cause health problems in humans and aquatic life and can lead to algal blooms, creating harmful conditions for aquatic life.

Methods for removal or treatment of pollutants can vary depending on the specific pollutant. Some pollutants can be removed through physical processes like filtration or chemical processes like coagulation. Biological pollutants may require disinfection through UV light, ozone, or chlorination. Regardless of the method used, it is important to regulate the discharge of pollutants into water sources and ensure that water quality standards are maintained to protect aquatic life and public health.

In conclusion, understanding the different categories of pollutants and their effects on water quality is crucial in developing effective strategies for pollution control and prevention. By regulating the discharge of pollutants and implementing appropriate treatment and removal methods, we can protect the health of aquatic life and ensure that water sources remain safe and usable for human consumption.

19.6 CATEGORIZATION OF METHODS USED IN TREATING WASTEWATER

Wastewater treatment is a critical process that ensures the safe disposal of wastewater and prevents contamination of the environment and public health. The treatment process includes several stages, including source treatment, pretreatment, primary treatment, secondary treatment, and tertiary or advanced wastewater treatment.

Source treatment is used to remove toxins and other undesirable contaminants to prevent intermingling with other waste streams. This approach offers opportunities for the reuse of these constituents, such as metals. The process of source treatment is essential in industrial wastewater treatment, where specific pollutants are present at high concentrations. The applicable technologies for source treatment depend on the constituents in the process of wastewater targeted for removal. For example, if Volatile Organic Compounds (VOCs) and ammonia are to be removed, then air or steam stripping should be evaluated. If heavy metals are of concern, then oxidation/reduction, precipitation, filtration, ion exchange, and membrane processes may be investigated. Organic chemicals may require chemical oxidation, wet air oxidation, anaerobic treatment, granular activated carbon (GAC), polymeric resins, or reverse osmosis for effective removal (Wang et al., 2014).

Pretreatment is employed to render the raw wastewater compatible and/or amenable for subsequent treatment processes. Consideration is given to those constituents that pass through, interfere with, or accumulate in the sludge or are otherwise incompatible with following treatment processes. Equalization, spill retention, neutralization for pH adjustment, nutrient addition, toxics or inhibitory substance removal, oil and grease removal, and solids removal by flotation, sedimentation, or filtration are typical pretreatment processes. The goal of pretreatment is to minimize the load on downstream treatment processes and ensure that they operate efficiently.

Primary treatment is a subset of pretreatment methods and involves physical separation by screening, grit removal, and sedimentation. Depending on the number of organics contained in the solid material, primary treatment may remove a significant portion of the oxygen-demanding substances (biological oxygen demand (BOD)). A well-designed and operated primary plant may remove as much as 35%–40% of the BOD and as much as 60%–65% of the settleable solids for municipal wastewater. The primary treatment process separates large solids from the wastewater using screens or grates, and the wastewater then flows into sedimentation tanks, where the remaining solids are allowed to settle. The solids that settle are removed from the bottom of the tanks and are typically sent to a digester for further treatment.

The secondary treatment adds a biological process after primary treatment, which is commonly either activated sludge or trickling filtration for municipal wastewater. Typically, activated sludge or a modification of suspended growth treatment systems is used for industrial wastewater to achieve a high-quality effluent. These biochemical processes are typically aerobic and are the same as previously described as occurring in a river where organics are oxidized to carbon dioxide and water. A well-operated and designed secondary treatment plant can be expected to remove 85%–95% of both BOD and suspended solids. Under existing regulations in the United States, all discharges must be subjected to at least secondary treatment.

Conventional treatment includes pre- and primary treatment followed by secondary treatment processes. When necessary, tertiary treatment processes are included in the treatment sequence to remove specific constituents to very low residue levels. Conventional treatment is designed to remove conventional pollutants such as BOD, total suspended solids (TSS), etc., common to municipal wastewater. However, best conventional pollution control technology (BCT) does not effectively remove many constituents of present-day concern, especially those of industrial wastewater origin. These include many VOCs, toxins, non-biodegradable organics, persistent organic pollutants (POPs), nutrients, and emerging contaminants such as endocrine disrupting compounds (EDCs) and pharmaceuticals and personal care products (PPCPs). Hence, additional treatment technology is required.

Tertiary treatment is additional treatment in addition to primary and secondary processes. It may include precipitation, filtration, coagulation and flocculation, air stripping, ion exchange, adsorption, membrane processes, nitrification and/or denitrification, and other processes. These processes may be integrated into the secondary treatment plant or added to the secondary effluent. Tertiary or advanced wastewater treatment can attain virtually any removal efficiency desired. However, the cost increases as the percentage of contaminant removal increases. Tertiary treatment is employed to remove toxins, persistent organics, non-conventional pollutants, nutrients, etc., and is typically considered as best available technology economically achievable (BATEA).

In conclusion, wastewater treatment processes are classified into source treatment, pretreatment, primary treatment, secondary treatment, and tertiary or advanced wastewater treatment. Each method has its own unique techniques and processes that are employed to remove contaminants from wastewater. The selection of a particular method depends on the type of wastewater and the contaminants present in it. The ultimate goal of wastewater treatment is to produce high-quality effluent that is safe for discharge into the environment (Vesilind and Weiner, 2017).

19.7 CHARACTERIZATION OF WASTEWATER

Wastewater characterization is essential for designing and operating a WWTP. The composition and characteristics of wastewater must be known and understood to determine the appropriate treatment method. For example, the BOD measures the organic material present and its removal depends on its form (suspended, colloidal, dissolved, molecular). Solids' characteristics determine the sludge handling and transport facilities. pH may indicate metal speciation, toxicity, and/or the need for neutralization. Oxygen content and oxidation reduction potential (ORP) may demonstrate the need for odor control, as well as chemical speciation. Grease and oil may require special removal facilities as they cause operational problems (Worrell et al., 2016).

Wastes originating from industry may contain toxic or inhibitory pollutants to biological processes, and thus their discharge to a municipal treatment plant must be regulated. Pollutants may include heavy metals, VOCs, priority pollutants, grease, and oil. Pretreatment ordinances are used by municipalities to control discharges into municipal sewerage systems. The flow of wastewater varies considerably and may contain different contaminants in different concentrations at different times. Sampling of wastewater is often done in accordance with flow so that the mass of constituents can be determined on a weighted average basis (composite sampling).

Industrial waste characteristics and parameters of concern can change significantly. Important industrial waste parameters are defined by a standard industrial classification (SIC) grouping, which is used to regulate effluent discharge requirements under the national pollution discharge elimination system (NPDES) permit system. The strength and volume of industrial wastewaters are usually defined in terms of units of production (e.g., gal/bbl of beer and lb BOD/bbl of beer for a brewery) and variation in characteristics and flow by statistical distribution. Industrial waste of organic nature can be correlated to municipal waste loadings by the use of population equivalents. Treatability studies are often necessary to determine design parameters and potential pretreatment requirements due to the inherent variability in industrial waste characteristics.

Thus, wastewater characterization is critical to WWTP design and operation. Knowledge of wastewater composition and characteristics is essential to determine the appropriate treatment method and ensure compliance with regulations.

19.8 WASTEWATER OBJECTIVES

Wastewater treatment processes aim to achieve three main objectives. The first objective is to separate solids from the liquid fraction and concentrate the collected solids. The second objective is to remove or neutralize materials that could have adverse effects during the ultimate disposal of the effluent and residuals. The third objective is to maximize the potential for reusing treated wastewater and residuals. Domestic sewage contains only 0.1% solids, and less than 50% of the waste material in domestic sewage remains in suspension, which allows for separation through straining, skimming, or settling. Residuals from the separation process must be destroyed or rendered removable using biological, physical, or chemical methods. Chemical treatment involves combining coagulating chemicals with non-settleable material to form settleable flocs. In aerobic biological treatment, living organisms break down biodegradable substances and convert them into carbon dioxide and water, forming settleable films, slimes, or flocs. Physical treatment may involve adsorption processes, where contaminants with a high affinity for activated carbon can be removed (Howe et al., 2012).

19.9 WATER WASTE TREATMENT PROCESS

Wastewater is always produced whenever agricultural produce of any kind is handled, processed, packed, or stored. The amount of contamination in the wastewater will determine how it is handled. It could be just a preliminary treatment; a preliminary treatment followed by primary treatment; primary treatment then secondary treatment; or even a whole treatment that includes a preliminary, secondary, and tertiary treatment. The cost of treating wastewater depends on its unique qualities and requirements for discharge, either into bodies of water or municipal wastewater treatment plants (WWTP).

The treatment of wastewater is driven by physical forces as well as chemical and biological activities. Unit operations are medical procedures that use physical pressure. They might use flotation, filtration, sedimentation, or screening. Unit processes is a term used to describe treatment strategies based on chemical and biological processes. Precipitation, adsorption, and disinfection are examples of chemical unit processes. Microbial activity, which is in charge of the nutrient removal and decomposition of organic matter, is a component of biological unit processes (Englande et al., 2015).

The selection and organization of various unit operations are necessary for the design of a wastewater treatment facility. The integration of multiple procedures capable of treating various forms of wastewater is schematically depicted in Figure 19.1. The wastewater's properties, the desired effluent quality (including any future limits), prices, and land availability all play a role in the process selection. Pretreatment/primary treatment, secondary treatment, tertiary treatment, sludge treatment/stabilization, and final disposal or reuse of leftover treatment technologies are some of the numerous categories into which treatment methods can be divided. A treatment plant can successfully treat wastewater and create high-quality effluent while minimizing costs and land use by choosing the right combination of these methods.

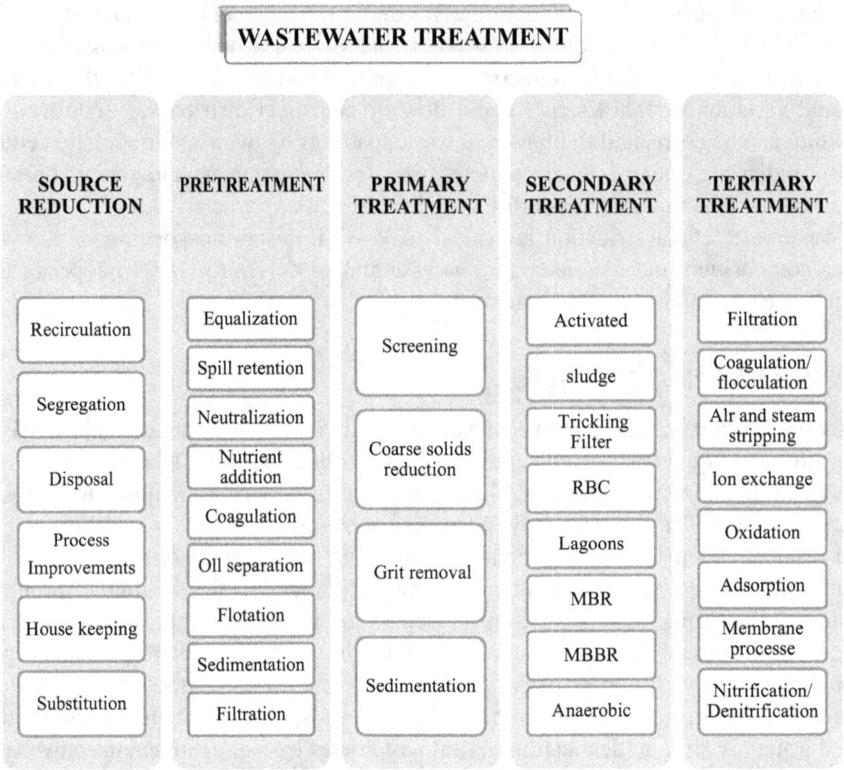

WASTEWATER TREATMENT

SOURCE REDUCTION
Recirculation
Segregation
Disposal
Process Improvements
House keeping
Substitution

PRETREATMENT
Equalization
Spill retention
Neutralization
Nutrient addition
Coagulation
Oll separation
Flotation
Sedimentation
Filtration

PRIMARY TREATMENT
Screening
Coarse solids reduction
Grit removal
Sedimentation

SECONDARY TREATMENT
Activated
sludge
Trickling Filter
RBC
Lagoons
MBR
MBBR
Anaerobic

TERTIARY TREATMENT
Filtration
Coagulation/ flocculation
Alr and steam stripping
lon exchange
Oxidation
Adsorption
Membrane processe
Nitrification/ Denitrification

FIGURE 19.1 Typical wastewater treatment processes.

19.9.1 Pretreatment/Primary Treatment

The purpose of pretreatment methods is to prepare wastewater for subsequent treatment processes by making it compatible and amenable. Such methods commonly include equalization, neutralization, and separation of oil and grease. On the other hand, primary treatment mainly involves the physical separation of municipal wastewater, comprising screening or comminution, followed by grit removal and sedimentation before secondary treatment.

- Screening/Comminution

  Screening and comminution are two methods used for the removal of large, solid matter from wastewater. Screening is preferred as it prevents flow obstruction and head loss, and the removed materials are usually buried or incinerated. Comminution is an alternative to screening but may contribute to the organic load of the treatment facility. Grit removal is usually done prior to comminution to protect the machinery.
- Grit Removal

  Grit is a mixture of small coarse particles of sand, gravel, or other mineral material found in wastewater. It is removed to prevent mechanical equipment damage and maintain tank volume capacity. This is done through an aerated chamber where the air is used to keep organic matter suspended, allowing heavier inorganic material to settle. Alternatively, controlling flow velocity in a chamber can cause the gritty material to settle while keeping organics suspended. Grit is typically washed and land disposed of.
- Oil and Grease Removal

  To avoid damage to primary sedimentation tanks or aeration basins, it is important to remove excessive oil, grease, and suspended solids from wastewater. This can be done

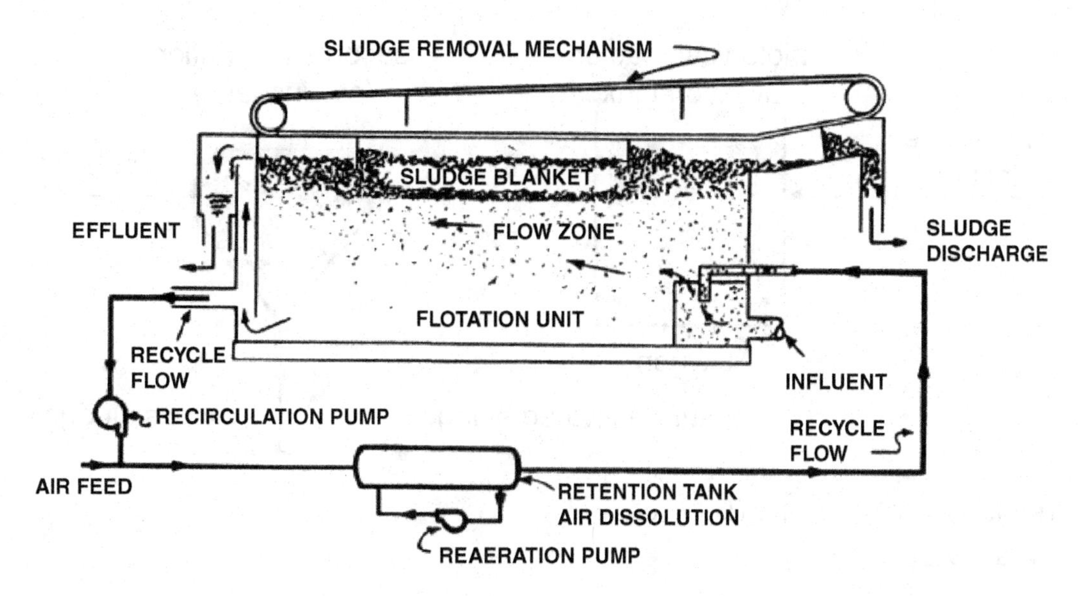

FIGURE 19.2 Dissolved air flotation unit. Courtesy of Komline–Sanderson Company.

using gravity separation, flotation, or a combination of both. Gravity separation can be achieved using an American Petroleum Institute (API) separator or a corrugated plate separator, while flotation involves introducing air under pressure to float oil globules, sludge flocs, and suspended solids to the surface. Chemical coagulants may be added to enhance the process. Flotation usually follows API separation, and a schematic diagram of a dissolved air flotation unit is shown in Figure 19.2. To minimize explosion hazards, the process can be conducted in covered units with methane or nitrogen purging.

- Equalization and Neutralization

  In some cases, an equalization basin may be needed to control fluctuations in pH and contaminant concentrations in incoming wastewater. Equalization helps maintain a steady flow and concentration through the treatment plant, which is important for effective treatment. Acidic and alkaline wastewater can be neutralized with various chemicals, and a pH of 6.5–8.5 is usually needed before biological treatment.
- Primary Sedimentation

  WWTPs use settling tanks to remove settleable solids from the wastewater. The primary settling tank removes solids and acts as a barrier for oil and grease. Solids are removed using a continuous belt device and pumped to a sludge treatment unit. The secondary settling tank follows biological treatment and removes solids without requiring scum removal. Chemicals can be added to enhance solids removal. The efficiency of solids removal depends on the overflow rate and liquid retention time. For industrial waste treatment, an equalization basin replaces the primary settling tank.

19.9.2 Secondary Treatment

Secondary treatment is an essential step in wastewater treatment that involves biological treatment to remove dissolved and colloidal organic matter. The process requires a sufficient number of acclimated microorganisms, oxygen supply, nutrients, intimate contact between microorganisms and wastewater, and a containment system. There are two commonly used methods for biological wastewater treatment: the activated sludge process and the trickling filtration process. In the activated sludge process, microorganisms are maintained in suspension in the tank, and air is introduced to supply oxygen and maintain turbulence to optimize the process. In contrast, the trickling filtration

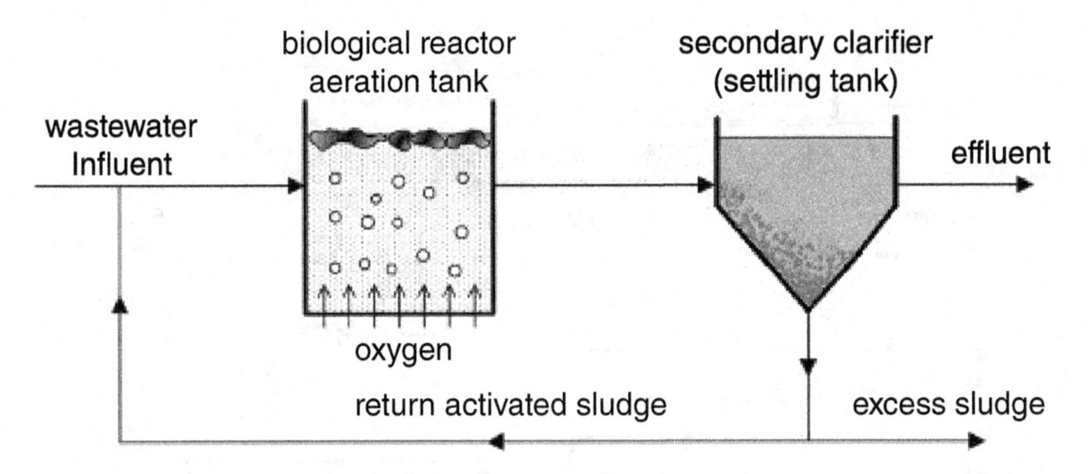

FIGURE 19.3 Complete mix biological reactor.

Source: Frącz (2016).

process involves microorganisms adhering to a suitable surface, and wastewater flows over the fixed film surface for the microorganisms to remove food and transfer end products to the liquid.

The processes in an aerobic biological treatment plant involve cell growth and death, depending on the food-to-microorganisms ratio. If the food supply is meager, bacteria will eat themselves, and bacterial cells will be converted into end products. The endogenous respiration process is similar to a starving human who uses its tissue to provide energy to sustain life processes. However, part of the waste material is non-biodegradable and remains in the system. Figure 19.3 schematically demonstrates a complete-mix biological wastewater treatment process. The mass of microorganisms per day wasted, divided into the inventory of microbes in the reactor, determines the sludge age or mean cell residence time (MCRT) used for system control to maintain the desired physiology of the biomass present. The retention time required may be hours to days and depends on treatability studies, wastewater complexity, temperature effects, and variation.

In summary, secondary treatment involves biological treatment to remove dissolved and colloidal organic matter from wastewater. The activated sludge and trickling filtration processes are commonly used, and the processes involve cell growth and death, with part of the waste material being non-biodegradable. A complete-mix biological wastewater treatment process involves several variables, including flow volume, the concentration of organics, volume of the tank, mass of microorganisms in the tank, concentration of organics in the tank, and fraction of return flow, with the sludge age used for system control.

- Activated Sludge Process

  The activated sludge process involves using a tank to cultivate a group of microorganisms that interact with incoming wastewater to break down organic compounds. This is typically done in a rectangular tank where air or pure oxygen is introduced to create optimal conditions for the microorganisms. The microbe concentration is maintained by returning a portion of settled material and disposing of the rest. This process creates new cell material that becomes a part of the activated sludge mass, which is necessary to maintain an active population of microorganisms that can effectively break down organic compounds in incoming wastewater (Figure 19.4). The design of the process depends on the food-to-microorganism ratio and the characteristics of the wastewater. Many modifications of the process exist, including high rate, step-aeration, tapered aeration, and contact stabilization processes. The extended aeration process, which involves long detention times, is used in small installations and for industrial waste treatment. The sequencing

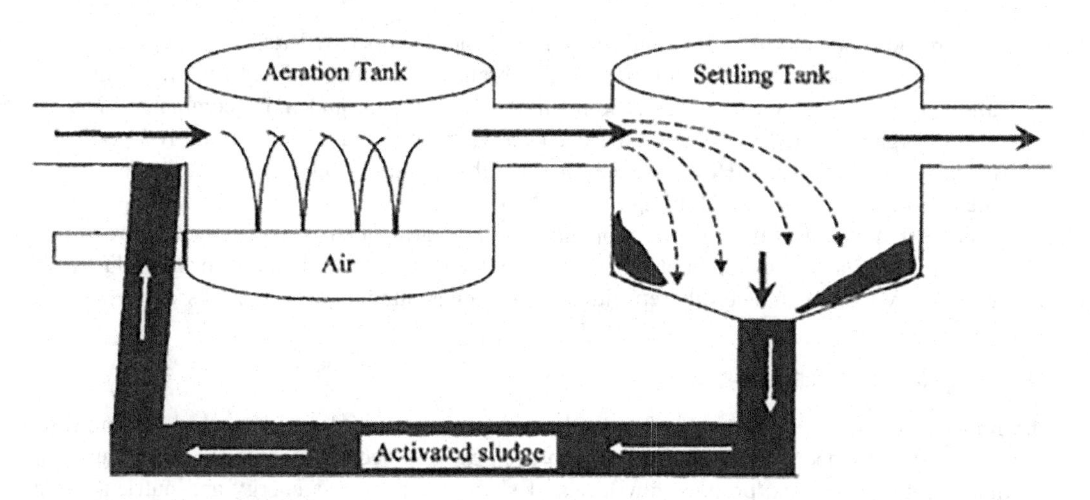

FIGURE 19.4 Activated sludge.

Source: MECC (2023).

batch reactor (SBR) is a combination of complete mix and plug flow operated on an intermittent basis, providing a good settling floc and high-quality effluent without the need for an external clarifier. The process can be modified with a selector to remove degradable organics and reduce the growth of filamentous microbes.

- Lagoons and Oxidation Ponds

  Stabilization ponds are an old but still used wastewater treatment method that can be used alone or with other processes depending on factors such as climate, land availability, purpose, and location. Four types of ponds are available, including facultative, aerated, aerobic, and anaerobic, with the most common being the facultative pond. Facultative ponds operate aerobically in the upper layers and anaerobically in the lower depths, and can have a detention time of 30–180 days, producing effluent with BOD values of 20–60 mg/L and suspended solids of 30–150 mg/L in warmer climates. Aerated lagoons are 6–20 ft deep and require large land areas, with detention times of 3–12 days, and can be used for pretreatment or designed in series. Aerobic ponds are 18–36 in. deep and require warm, sunny climates and have a detention time of 3–5 days. Anaerobic ponds are 8–15 ft deep and have a detention time of 20–50 days, but produce odor which can be reduced by the addition of sodium nitrate.

- Trickling Filtration

  The trickling filtration process is a type of biological wastewater treatment process that uses a fixed bed of media to support a biological film where microorganisms attach and grow to treat the organic waste in the wastewater. The media is usually made up of coarse, rough, and hard materials such as stones, gravel, or synthetic materials that provide surface area for the bacteria to attach and grow on. The wastewater is distributed over the bed through rotary nozzles and trickles down through the media. The microorganisms in the biological film consume and oxidize the organic matter in the wastewater as it passes through the media.

  The process requires minimal operator attention and uses less energy than other methods, but is highly temperature dependent and may not meet regulatory requirements. A biological contactor can be used to enhance effluent quality, and recirculation of effluent can reduce influent organic concentration. The process is commonly used as a pretreatment method for industrial wastewater to remove approximately 50% of the BOD. The filters may also be covered and off-gases treated to reduce or eliminate odor problems.

- Membrane Bioreactors (MBRs) and Moving Bed Bioreactors (MBBRs)

  MBRs use ultra- and micro-filtration membranes to remove solids from wastewater, allowing higher biomass concentration, smaller bioreactors, and more complete oxidation of organics. MBRs eliminate the need for secondary sedimentation and reduce the discharge of pathogens. However, they have high capital and energy costs and require membrane replacement and fouling control. MBBRs are an advanced secondary treatment process that increases the amount of biomass in the treatment basin using attached growth media. They have a smaller footprint and can enhance nitrification and priority pollutant removal. MBBRs can be easily retrofitted to existing aerated treatment processes.

19.9.3 Anaerobic Treatment

Anaerobic treatment is a process that breaks down organic waste into methane and carbon dioxide gases in the absence of oxygen. Anaerobic processes have become popular because they cost less than aerobic processes, produce low levels of sludge, require less energy and nutrients, and produce methane gas that can be used beneficially. Anaerobic treatment is commonly used to treat high-strength industrial waste from food industries and biological sludge. More than 850 anaerobic treatment systems are in operation globally, with 75% treating wastewater from the food industry. Chemical and petrochemical industries also use over 60 anaerobic treatment systems. Examples of anaerobic processes include the anaerobic filter reactor, anaerobic contact process, fluidized-bed reactor, upflow anaerobic sludge blanket, the ADI-BVF reactor is an anaerobic digestion system that offers very stable, effective wastewater treatment and biogas generation under a wide range of operating conditions (ADI-BVF) process, and expanded granular sludge bed process.

- Septic Tanks

  In discussing biological treatment, it's important to address the septic tank. Despite its primitive design, a large portion of the population still relies on this device, as depicted in Figure 19.5. Septic tank operates anaerobically and will accumulate solids. Without regular pumping (every 2–3 years), the tank will discharge unpleasant material. Typically, the discharge from the tank flows into a designated drain field and eventually into groundwater

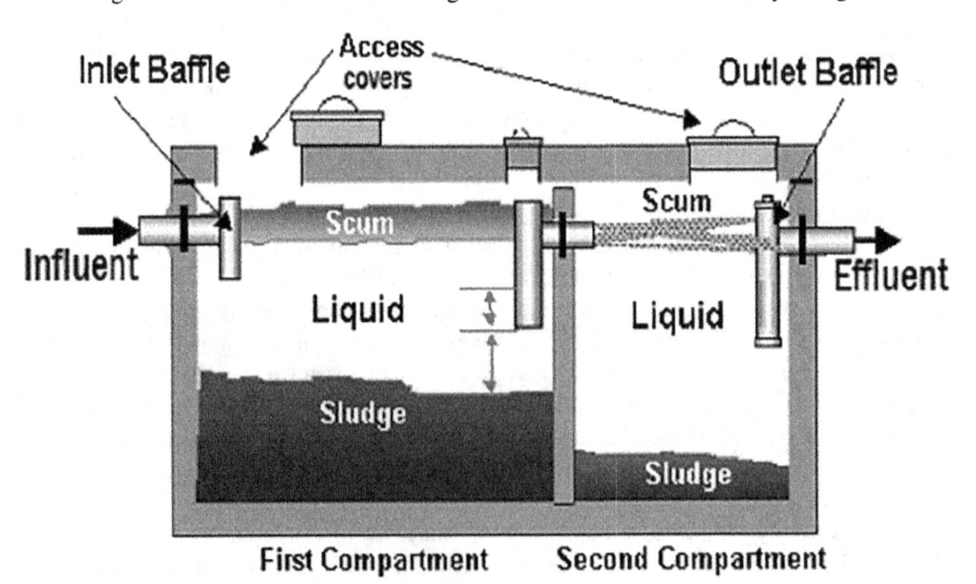

FIGURE 19.5 Septic tank.

Source: Alnahhal (2017).

or surface water. It's crucial to have a well-designed drainage system in appropriate soil, as the septic tank only functions as a settler, and its discharge remains high in soluble organic and microbial contaminants. Adequate hydrogeologic conditions are essential for effective drainage field treatment.

- Residuals (Sludge) Treatment

  Water pollution control faces a major challenge in disposing of the solids produced during separation processes. Chemical additives can alter the nature of the sludge and increase the volume produced, necessitating different treatment methods for ultimate disposal. The few choices for disposal include land application, incineration, or reuse, with incineration having air-quality impacts. Sludge treatment processes aim to reduce volume, remove water, control putrescibility, and stabilize organics to minimize transportation costs and environmental/public health effects. The typical sequence of unit processes for residual management includes thickening, stabilization, dewatering, and ultimate disposal or reuse.
- Thickening

  Sludge thickening aims to reduce the volume of sludge by concentrating the solids. Thickeners can increase the solids concentration by 2–5 times and produce a clear liquid effluent. This process can be achieved through either gravity or dissolved-air flotation methods. A dissolved-air flotation unit operates similarly to pretreatment, and the liquid that is decanted is returned to the treatment plant influent.
- Sludge Stabilization

  Sludge stabilization is necessary before reuse or ultimate disposal, typically achieved through aerobic or anaerobic digestion. Aerobic digestion oxidizes cellular organic matter using aeration and mixing, reducing volatile solids by 30%–50%. Nitrogen and phosphorus are released during this process, and the digested sludge can be disposed of without causing odor or nuisance conditions. Anaerobic digestion decomposes organic matter in the absence of oxygen, producing methane and carbon dioxide. Detention times may vary from 10 to 60 days, and the gas composition is about 70% methane and 30% carbon dioxide. The methane gas can be used to heat the digester or run machinery in the plant. Thermophilic digestion may be used to improve dewaterability and pathogen destruction but has higher energy requirements and lower process stability. Advantages of anaerobic digestion include no oxygen requirement, reduction in sludge volume, inactivation of pathogens, and production of a good soil conditioner. Disadvantages include high capital costs and the need for careful operation.
- Sludge Dewatering

  Various methods of sludge dewatering are available, including belt filter presses, pressure filters, and centrifugal dewatering. Chemicals such as lime, ferric chloride, or polymers are often used to enhance dewatering, and fly ash and coal fines can reduce cake compressibility. Sludge drying beds are suitable for small volumes and arid climates. Other methods include composting, thermal treatment, incineration, alkaline stabilization, and wet oxidation, but they may lead to air pollution and residual ash disposal issues. There are other less common methods, and treatment may be used before dewatering for stabilization and conditioning.

19.9.4 Ultimate Disposition of Solids

The United States Environmental Protection Agency (USEPA) regulates the disposal of municipal sewage sludge biosolids through 40 CFR Part 503 of the 1987 Clean Water Act Amendments. The EPA promotes the beneficial use of biosolids while protecting public health and environmental quality. Biosolids can be used as soil amendments or fertilizer supplements or converted to useful products. Processes that significantly reduce pathogens produce Class B sludge, which has restricted use. Processes that further reduce pathogens produce Class A sludge, which has unrestricted use. To be usable, the biosolids must be stable, non-infectious, and have low metal content (Marks, 1998). Heavy metals, nitrogen, and phytotoxic materials are of concern when using biosolids for edible crops. Nitrate in groundwater can cause methemoglobinemia in infants. Helminths, including Ascaris ova, pose a pathogen threat. Biosolids, when effectively treated and recycled, can conserve landfill capacity, improve soil fertility, reduce fertilizer use, and decrease energy consumption. Nationally, about 60% of biosolids are land-applied, and stabilized biosolids have been used for various applications, including landfill cover, land recovery, and wetlands reclamation.

19.9.5 Advanced Wastewater (Tertiary) Treatment

Conventional treatment methods (BCT) have been effective in reducing organic waste materials (BOD), suspended solids, and bacteria in wastewater with high efficiency. The treatment plant flow scheme can be designed with either activated sludge or trickling filtration. However, secondary treatment may not effectively remove other contaminants such as toxic organics, VOCs, phosphorus, nitrogen, heavy metals, refractory organics, and pathogens like *Giardia lamblia.* Therefore, additional removal techniques, such as tertiary treatment (BATEA), are necessary. Various processes, including coagulation and flocculation, filtration, ion exchange, nitrification, denitrification, membrane processes, air stripping, adsorption, and chemical oxidation, have been studied and applied successfully for the removal of these contaminants.

- **Coagulation**: By introducing a substance known as a coagulant, particles in a liquid are destabilized during coagulation. The coagulant causes the particles to cling together and neutralizes their charges, resulting in bigger clumps or flocs that are easily removed by settling or filtration. In order to remove suspended solids, turbidities, and some dissolved pollutants, such as metals and phosphates, coagulation is frequently used in the treatment of water and wastewater. Salts of aluminum and iron, such as alum and ferric chloride, are the most often used coagulants. Many variables, including pH, temperature, dose, and mixing vigor, affect how well coagulation occurs. For coagulation to work at its best, these parameters must be properly controlled.
- **Flocculation**: In order to encourage the production of bigger particles or flocs that can be easily settled or filtered, a liquid is flocculated. Flocculation is frequently employed after coagulation to maximize the removal of small particles and increase the effectiveness of settling or filtration. The coagulated particles undergo a gentle, moderate mixing process known as flocculation, which promotes the creation of larger, more stable flocs. The flocs are then either filtered out of the liquid or left to settle. The performance of flocculation may occasionally be improved by the addition of a polymer. The efficiency of flocculation is dependent on a number of variables, including the amount of mixing, the length of time, and the presence of impediments like oils or surfactants.
- **Filtration**: To eliminate suspended particles or other pollutants, a liquid must be filtered through a porous media or substance. Many water and wastewater treatment systems depend on filtration, which is used to remove contaminants to varying degrees. A variety of techniques, including sand, multimedia, membrane, and cartridge filters, can be used to achieve filtration. The size and type of pollutants, flow rate, and required level of treatment are just a few of the variables that must be taken into consideration when choosing the right filter type. In order to guarantee optimum performance and avoid clogging or fouling of the filter media, proper filter design, operation, and maintenance are essential. To remove contaminants at greater levels, filtration can also be combined with other treatment methods like flocculation and coagulation.

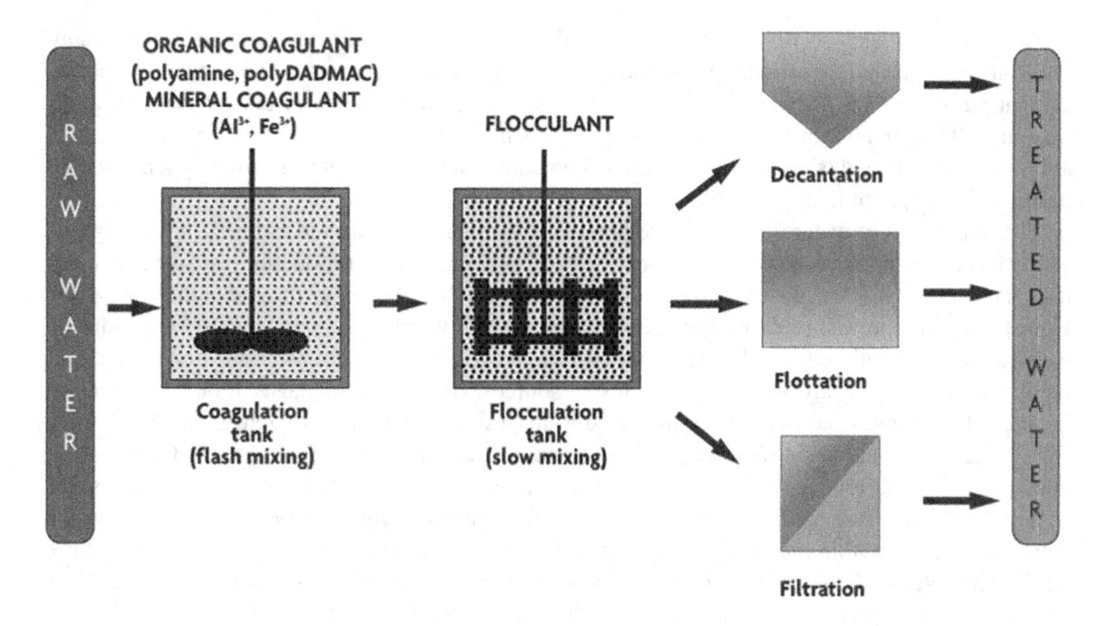

FIGURE 19.6 Physical-chemical process involved in coagulation-flocculation.

Source: SNF FLOERGER (2003).

19.10 WATER REUSE

Water scarcity and the uneven distribution of water resources across the globe have become strong drivers for water conservation and reuse. As the population continues to grow, conventional water supplies such as fresh surface water and groundwater are under extreme pressure, leading to the need to tap into unconventional supplies, such as brackish and sea water, which require significant energy resources for desalination (Tzanakakis et al., 2020). In areas such as the Middle East, water scarcity is an everyday concern, with limited renewable fresh groundwater supplies and high dependence on desalination, which is both energy-intensive and expensive. Water reuse has become an everyday practice in these regions, with treated sewage effluent (TSE) being reused for landscape irrigation. Furthermore, the cost of treating wastewater to high standards due to more stringent regulations, permitting requirements, public relations, and energy conservation has become a motivator for industries and communities to view wastewater as a valuable resource, rather than a waste product. Overall, the degree of treatment should be based on the quality requirements of the intended use of the water, and if wastewater can be used in a beneficial manner without causing adverse environmental or health effects, it should be considered a valuable resource and used accordingly.

19.11 WASTEWATER REUSE

Water reuse has been practiced for many years, with the Romans using a dual water system to distinguish between drinking water and other water supplies. In the early 1900s, numerous sewage farms were established around the world, which were primarily used for disposal, but some of the wastewater was used for crop production and other beneficial uses. Agricultural purposes have been the most widespread reuse of wastewater, with treated or untreated wastewater discharges applied to crops for irrigation. The use of dual distribution systems is also common, where one system is used for potable water, and the other is used for irrigation and other non-potable uses, reducing potable water demand (Englande et al., 2015).

As water shortages become more acute and the cost of wastewater treatment increases, many locations have started investigating the potential for reusing wastewater for non-potable purposes, alleviating some of the demand for potable water and giving an alternative to additional wastewater treatment. The demand for water is becoming increasingly acute, particularly in semi-arid and arid areas and when droughts occur. Thus, there is no doubt that reclaimed water will play a major role in the water supply of the future.

The major uses of water do not require potable water quality, such as in industry, where significantly large volumes of water are used for cooling, processing, steam production, and other general uses. The majority of that water evaporates (evaporative cooling) or leaves the site with the product. However, a good portion remains and ends up as processed water requiring disposal. The industry is driven more and more toward water conservation, recycling, and reuse as public and media, and regulatory scrutiny regarding how the industry sources, treats, and manages water continues to increase. Companies that treat water and wastewater as a commodity do so at great risk. Others who value water and wastewater as precious resources have a greater chance of sustaining their operations and improving their viability in the marketplace. From the perspective of industrial water users, water reuse can be approached from different angles. One approach is to conserve water and reuse effluent within the fence of their operation. A second approach is for the industry to look outside the fence of their operation for water conservation and reuse opportunities, addressing water management throughout their entire supply chain from production to final product use.

19.12 FACTORS AFFECTING REUSABILITY OF WASTEWATER

Wastewater from municipal and industrial sources can contain a wide range of contaminants that may make it unsuitable for reuse. Factors such as the presence of inorganic chemicals, organic chemicals, and microbiological agents must be taken into consideration to avoid adverse effects on public health and the environment. An economic analysis of the costs of treating and delivering the water as compared to the benefits of its beneficial use must be made to determine if advanced wastewater treatment is necessary. The parameters of concern are as follows.

- **Inorganic chemicals**: Inorganic chemicals are naturally present in water and can be contributed by various sources. Some ions, such as boron and heavy metals, may make the water unsuitable for irrigation or have potential toxicity to plants, humans, and animals. Advanced wastewater treatment can remove these ions to acceptable levels. Selected drinking and irrigation standards are shown in Table 19.1.
- **Organic chemicals**-: Organic chemicals in wastewater are numerous and difficult to list due to the numerous species existing, some of which may be formed by interactions with each other. Even with advances in analytical techniques, technology, and epidemiological studies, it cannot be concluded that many of these trace organics found in wastewater do not pose a potential health hazard.
- **Microbiological considerations**: Waterborne diseases can be brought on bÿ a variety of pathogenic microorganisms found in wastewater, including bacteria, viruses, protozoans, algae, and helminths. Indirect potable reuse may require advanced wastewater treatment even though conventional wastewater treatment can bring these pathogen levels down to tolerable levels. Human interaction with reclaimed water that might have any of these pathogens in it needs to be kept to a minimum.

It's important to take into account the possibility of the Legionnaires' illness (*Legionella*) spreading via cooling tower drift. With proper planning, implementation, monitoring, and maintenance of safety measures, there is very little risk associated with any application of recovered water. The Water Environment Federation states that no outbreak of infectious disease has been linked to water reuse.

TABLE 19.1
Selected Drinking and Irrigation Water Standards (mg/L)[a]

Constituent	Drinking Water	For Fine-Textured Soils	For any Soil	For Livestock
Aluminum		20		5
Arsenic	0.01	2	0.1	0.2
Barium	1			
Boron		0.75	2	5
Cadmium	0.01	0.05	0.01	0.05
Chromium	0.05	1	0.1	1
Copper	1	5	0.2	0.5
Iron	0.3	20	5	
Lead	0.05	10	5	0.1
Manganese	0.05	10	0.02	
Mercury	0.002			0.01
Nickel		2	0.2	
Selenium	0.01	0.02	0.02	0.05
Zinc	5	10	2	25

[a] *Source:* http://water.epa.gov/lawsregs/rulesregs/sdwa/regulations.cfm

19.13 ADDITIONAL WATER QUALITY PARAMETERS TO CONSIDER FOR WASTEWATER REUSE

With recovered wastewater, there are a number of additional water quality factors that must be taken into account. Excessive sediments can clog spray irrigation nozzles and settle out in the distribution system, among other issues. They can absorb heavy metals and clog soils, which lowers the rate of infiltration. Solids can also protect germs from disinfection, posing a health danger.

Salinity is yet another crucial element that must be taken into account. Typically, electroconductivity, which is correlated with total dissolved solids (TDS), is used to measure it. Because of plant evapotranspiration, salts that seep into the ground tend to concentrate in the root zone. This can have a significant impact on both plants and soils since plants must use more energy to compensate for osmotic effects in order to get water from the soil, which lowers productivity. As a result, good drainage is important for the long-term use of recycled water for irrigation, and it could be necessary to use more water than plants actually need to maintain a suitable salt balance in the soil.

Since it can alter soil structure, infiltration rates, and permeability rates, sodium in irrigation water is a significant problem. The high salt content in the soil that contains certain clays results in adverse conditions for water flow and plant development. There are ways to determine how sodium affects soils and plants, and its risk can be estimated by comparing it to calcium and magnesium.

As nutrients can be found in wastewater and have an impact on plant growth, they should also be taken into account. The most advantageous nutrient is nitrogen, which can substitute for an equivalent amount of nitrogen in fertilizer during the growing season. However, crops may suffer if there is too much nitrogen present; therefore, monitoring may be necessary. Because nitrate has a link to methemoglobinemia, it is vital to note that under aerobic circumstances, nitrogen in the wastewater that is currently present as organic or ammonia nitrogen will eventually be transformed into nitrate.

19.14 APPLICATIONS OF WATER REUSE

With water becoming an increasingly scarce resource, the utilization of reclaimed water is gaining momentum. A wide range of potential applications and limitations for reclaimed water are listed in Table 19.2.

TABLE 19.2
Categories of Wastewater Reuse and Potential Constraints[a]

No.	Wastewater Reuse Categories	Potential Constraints
1	Agricultural irrigation Crop irrigation Commercial nurseries	• Surface and groundwater pollution if not properly managed • Marketability of crops and public acceptance • Public health concerns related to pathogens (bacteria, viruses, parasites) • Effect of water quality, particularly on soils & crops
2	Landscape irrigation Parks Schoolyards Freeway medians Golf courses Cemeteries Greenbelts Residential	• Public health concerns related to pathogens • Use of area control including buffer zone; may result in high user cost
3	Industrial recycling and reuse Cooling Boiler feed Process water Heavy construction	• Constituents in wastewater related to scaling, corrosion biological growth, and fouling • Cross connection potable and reclaimed water lines • Public health concerns, particularly aerosol transmission of pathogens in cooling water
4	Groundwater recharge Groundwater replenishment Salt water intrusion control Subsidence control	• Organic chemicals in reclaimed wastewater and their toxicological effects • Possible contamination of groundwater aquifer used as a source of potable water • Total dissolved solids, nitrates, and pathogens in reclaimed wastewater
5	Recreational/environmental uses Lakes and ponds Marsh enhancement Streamflow augmentation Fisheries Snowmaking	• Health concerns related to presence of bacteria, viruses, and other pathogens (e.g., enteric infections and ear, eye, and nose infections) • Eutrophication due to P and N in receiving water • Toxicity to aquatic life
6	Non-potable urban uses Fire protection Air conditioning Toilet flushing	• Public health concerns on pathogens transmitted by aerosols • Effects of water quality on scaling, corrosion, biological growth, and fouling • Cross connections of potable and reclaimed water lines
7	Potable reuse	• Constituents in reclaimed water, especially trace organic chemicals and their toxicological effects • Health concern regarding pathogen transmission

[a] *Source*: Metcalf and Eddy (2003).

Each use requires specific treatment of wastewater to address concerns and fulfill requirements. As water-scarce regions, particularly those in arid and semi-arid areas, continue to grow, the demand for water reuse will likely increase. Although direct potable reuse is unlikely to be a common practice in the near future, as demonstrated in Namibia where no other source is available, direct use of reclaimed water is possible without any apparent adverse effects. Indirect potable reuse has also proven to be feasible, as evidenced by the use of Mississippi River water, which receives wastewater discharges, as a drinking water source for New Orleans. Advanced wastewater treatment has advanced significantly, and with experience and innovation, it can produce water quality that meets almost any desired standards.

When it comes to planned indirect potable reuse, the National Resources Council issued a statement in 1998 suggesting that any community contemplating potable reuse should proceed with caution, conducting a site-specific assessment that includes monitoring of contaminants, testing for health and safety, and evaluating the system's reliability. The careful assessment would enable the community to make informed decisions based on reliable data to ensure public health and safety.

To appreciate the viability and practicality of wastewater reuse and its potential to help alleviate severe water shortages, it is worth examining several successful water-reuse projects. These projects illustrate how wastewater reuse can provide a reliable and safe water supply for communities facing severe water scarcity. With careful planning, proper design, and reliable technology, water reuse can serve as a sustainable water management strategy to address the growing water demand.

19.15 REGULATIONS FOR WATER REUSE

The absence of federal standards for water reclamation and reuse has led several states to adopt their own regulations. As the need for water becomes more pressing, it is likely that states will develop more stringent guidelines for water reuse. Some states like Arizona, California, Colorado, Florida, Hawaii, Nevada, Oregon, and Texas have already enacted regulations that encourage water reuse. Other states may permit reuse on a case-by-case basis in the absence of any regulations. Additionally, foreign countries and the World Health Organization have published guidelines for water reuse (USEPA, 2004).

The primary objective of these regulations is to safeguard public health, with the main focus being on irrigation. However, there has been little attention paid to the possible adverse effects of certain wastewater constituents on plants and soil, which can be significant and should be considered. The regulations cover various categories of reuse with suitable restrictions and requirements, including unrestricted and restricted urban reuse, agricultural reuse on food and non-food crops, unrestricted and restricted recreational reuse, environmental reuse, and industrial reuse. Requirements may include water quality and treatment, monitoring, system reliability, storage for system or excess reclaimed water, application rates, groundwater monitoring, and a setback or buffer zones. The most stringent regulations are for indirect potable use via groundwater recharge by surface spreading or direct injection and augmentation of surface supplies. It is worth noting that public perception is a crucial consideration in indirect potable reuse, and several projects have been rejected due to public and/or political pressure, with health concerns being the primary reason for rejection.

19.16 CHALLENGES ON WASTEWATER TREATMENT AND REUSE

The growing human population and urbanization pose challenges to the availability of water resources and wastewater disposal. At present, wastewater is transported through collection sewers to a centralized WWTP located at the lowest elevation of the collection system near the point of disposal (Tchobanoglous et al., 2014). However, the centralized WWTPs are not suitable for dual distribution systems, thus limiting water reuse in urban areas. Moreover, the cost of infrastructure for storing and transporting reclaimed water to the point of use is high, making water reuse less economical. Therefore, decentralized wastewater management systems should be considered more seriously in the future, treating wastewater at or near the points of waste generation. Decentralized treatment at upstream locations, with localized reuse and/or the recovery of wastewater solids, is becoming a popular alternative to the conventional approach of transporting reclaimed water from a central WWTP.

The potential for water reuse to alleviate the strain on water resources is significant, but the challenge of biosolids disposal remains a concern, especially in densely populated urban areas. However, public perception is the primary obstacle in both water reuse and biosolids applications. Advanced technologies can improve energy efficiency and reliability, but the challenge of explaining emerging contaminants like pharmaceuticals and antibiotic-resistant bacteria to the public is significant. Public concerns over waterborne diseases, like cholera and cryptosporidiosis, reinforce the perception problem. To overcome these obstacles, advanced technologies such as online sensors, membranes, and advanced oxidation can help ease public concerns. Additionally, comparing engineered reused water to existing source waters can help to persuade the public (Angelakisand Snyder, 2015).

The issue of emerging chemical constituents has been further complicated by concerns over the toxicity of mixtures. Exposure to chemicals is not a discrete event, as chemicals exist as complex mixtures with widely varying compositions. The question of whether these mixtures are safe cannot be reasonably answered through animal testing alone, especially for mixtures, which can have an infinite number of combinations. As a result, there is growing interest in the use of rapid biological screening assays, primarily *in vitro*, to quickly and comprehensively evaluate the complex mixtures of chemicals in the water. High throughput bioassays are effective in identifying the chemicals present in a wide range of biological endpoints relevant to public health, both qualitatively and quantitatively. Given the constant introduction of new chemicals to the market and the potential for numerous transformation products, bioassays can help pave the way forward, enabling the public and regulators to make more informed decisions regarding water reuse projects.

With the growth of cities and the increasing strain on water resources, natural deposition alone cannot provide sufficient additional resources. Therefore, water reuse, desalination, and transportation are necessary to augment the available resources. Although water reuse offers several benefits, public acceptance remains a significant challenge. To overcome this challenge, scientists can play a vital role by improving the communication of complex data and ensuring that the quality of reused water is compared to that of existing urban water resources. This approach can help to move the field of water reuse forward.

19.17 CONCLUSION

In conclusion, water waste treatment and reuse are essential for maintaining a sustainable environment and conserving natural resources. With the ever-increasing demand for freshwater, it is becoming more important to implement measures that will reduce wastage and optimize the available resources. Water waste treatment technologies such as biological treatment, chemical treatment, and physical treatment play a significant role in removing contaminants from wastewater, making it safe for reuse. The benefits of water waste treatment and reuse extend beyond conserving water resources. Proper treatment and reuse of wastewater can reduce pollution in water bodies, improve public health, and provide alternative sources of water for agriculture and industrial processes. Additionally, reusing treated wastewater reduces the need for freshwater sources, thus conserving energy and reducing greenhouse gas emissions associated with water transportation. Despite the many benefits of water waste treatment and reuse, there are still challenges that need to be addressed. For instance, the cost of implementing wastewater treatment systems can be high, especially for small communities or low-income countries. Additionally, there is a need for proper regulation and monitoring of the quality of the treated water to ensure it meets the required standards for various uses.

Thus, water waste treatment and reuse are critical in maintaining sustainable water resources and protecting the environment. The implementation of appropriate wastewater treatment technologies can reduce water wastage, pollution, and greenhouse gas emissions, while also providing alternative sources of water for various uses. Governments, communities, and stakeholders should collaborate to develop sustainable solutions that promote water conservation and waste reduction. By doing so, we can ensure that future generations have access to safe and clean water resources.

REFERENCES

Alnahhal, S. 2017. Contribution to the development of sustainable sanitation in emerging countries. Ph. D. Thesis.

Angelakis, A.N. and Snyder, S.A. 2015. Wastewater treatment and reuse: past, present, and future. *Water*, 7(9), 4887–4895.

Baud, I. S. A., Grafakos, S., Hordijk, M. and Post, J. 2001. Quality of life and alliances in solid waste management: contributions to urban sustainable development. *Cities*, 18(1), 3–12.

Black, M. 2016. *The Atlas of Water: Mapping the World's Most Critical Resource*. University of California Press, Oakland, CA.

Chowdhury, S., Saha, S., Saha, B. and Mukherjee, A. 2020. Recent advancements in physical and chemical methods for removal of microplastics from water environment. *Journal of Environmental Management*, 2, 266.

Englande Jr, A. J., Krenkel, P. and Shamas, J. 2015. Wastewater treatment & water reclamation. Reference Module in Earth Systems and Environmental Sciences.

Frącz, P., 2016. Nonlinear modeling of activated sludge process using the Hammerstein-Wiener structure. In *E3S Web of Conferences*, Vol. 10.

Howe, K. J., Hand, D. W., Crittenden, J. C., Trussell, R. R. and Tchobanoglous, G. 2012. *Principles of Water Treatment*. John Wiley & Sons, New York.

Mallikarjuna, N., Sarkar, B. and Purakayastha, T. J. 2019. A review on water pollution, its causes, impacts and control. *International Journal of Latest Research in Science and Technology*, 8(4), 1–7.

Marks, H. 1998. Environmental regulations for land application of sewage sludge and municipal solid waste compost may not provide adequate protection against metals leaching. *Temple Journal of Science, Technology & Environmental Law*, 17, 123.

Mateo-Sagasta, J., Raschid-Sally, L. and Thebo, A. 2015. Global wastewater and sludge production, treatment and use. In: *Wastewater: Economic Asset in An Urbanizing World*. Springer, The Netherlands, pp. 15–38.

MECC. 2023. Lesson 6: Activated Sludge Treatment. Retrieved from https://water.mecc.edu/courses/ENV295WWII/lesson6.htm

Metcalf, L. and Eddy, H. 2003. *Wastewater Engineering: Treatment and Reuse*. McGraw Hill, New York, NY, pp. 384.

SNF FLOERGER. 2003. Physical-chemical process involved in coagulation-flocculation.

Tchobanoglous, G., Stensel, H. D., Tsuchihashi, R., Burton, F., Abu-Orf, M., Bowden, G. and Pfrang, W. 2014. *Wastewater Engineering: Treatment and Resources Recovery*. Metcalf and Eddy Inc, New York, NY.

Tzanakakis, V. A., Paranychianakis, N. V. and Angelakis, A. N. 2020. Water supply and water scarcity. *Water*, 12(9), 2347.

United Nations Development Programme. 2006. *Human Development Report: Beyond Scarcity: Power, Poverty and the Global Water Crisis*. Palgrave Macmillan, London.

United Nations. 2017. Sustainable Development Goal 6: Clean Water and Sanitation.

USEPA, U. 2004. Guidelines for water reuse EPA.

Vergine, P., Salerno, C., Libutti, A., Beneduce, L., Gatta, G., Berardi, G. and Pollice, A. 2017. Closing the water cycle in the agro-industrial sector by reusing treated wastewater for irrigation. *Journal of Cleaner Production*, 164, 587–596.

Vesilind, P. A. and Weiner, R. F. 2017. *Environmental Engineering: Principles of Practice*. CRC Press, Boca Raton, FL.

Wang, L. K., Chen, J. P. and Hung, Y. T. 2014. Physicochemical treatment processes. *Handbook of Environmental Engineering*, 4, 39–103.

World Health Organization. 2011. *Guidelines for Drinking-Water Quality*. WHO Press, Geneve, Switzerland.

Worrell, W. A., Vesilind, P. A. and Ludwig, C. 2016. *Solid Waste Engineering: A Global Perspective*. Cengage Learning, Singapore.

Zotos, G., Karagiannidis, A., Zampetoglou, S., Malamakis, A., Antonopoulos, I.S., Kontogianni, S. and Tchobanoglous, G. 2009. Developing a holistic strategy for integrated waste management within municipal planning: challenges, policies, solutions and perspectives for Hellenic municipalities in the zero-waste, low-cost direction. *Waste Management*, 29(5), 1686–1692.

20 Watershed Development – A Holistic Approach for Sustainable Agriculture

Falaq Firdous, Rohitashw Kumar, Munjid Maryam, and Khilat Shabir

20.1 INTRODUCTION

A watershed can mean a variety of things. It refers to a ridge line or a line with slopes in two opposite directions on either side in British English. A ridge line is a path that connects a terrain's greatest points of elevation. Hence, ridge line is sometimes referred to as a surface water divide or a watershed line. A path-breaking event is referred to as a "watershed" in everyday speech. When rainwater or storm water is gathered from an area that is surrounded by a ridge line, it is collected in a catchment or basin, which is referred to as a "watershed" in American English. This water gradually travels via different drainage systems, which eventually combine to form one or, in exceptional cases, more than one stream outfall. As a result, a watershed is described as the area bounded by a watershed line. The term "watershed" in this course refers to a small catchment or basin that represents a hydrological unit that empties all of its rainwater into a stream. In terms of its water in general and surface water in particular, it is so independent.

In order to distinguish a watershed—which typically denotes a small catchment or basin—Bali (1980) proposed a 2,000 km^2 maximum area threshold. The categorisation proposed by Rao (1975) for big river basins with an extent greater than 20,000 km^2; medium river basins with an area between 2,000 and 20,000 km^2; and small river basins, often known as watersheds, is extended by this classification. Bali's classification of watersheds was probably reflected in the 1990 All India Soil and Land Use Survey (AISLUS) classification of watersheds. Watersheds are further divided into five categories based on their sizes, including macro-watersheds, which have an area between 500 and 2,000 km^2; sub-watersheds, which have an area between 100 and 500 km^2, milli-watersheds, which have an area between 10 and 100 km^2, mini-watersheds, which have an area between 1 and 10, and micro-watersheds, which have an area of less than 1 km^2.

A watershed is a physical structure made up of the elements of nature, such as different kinds of plants of all sizes and shapes that develop over various types of soil or rock layers. Moreover, the watershed includes all man-made structures, including buildings, roads, bridges, tunnels, and animal burrows, all of which were mostly constructed by humans and occasionally by other animals. We will talk about the scope of watershed management in the section after this.

20.2 CHARACTERISTICS OF WATERSHEDS

The fundamental component of hydrological behaviour is a watershed. It is a geographic area where the hydrological cycle and its constituent parts can be studied. A watershed is typically described as an area that, based on geography, appears to contribute all the water that flows through a particular point of a stream. With all of its surface and subsurface elements, climate and weather patterns,

DOI: 10.1201/9781003303237-20

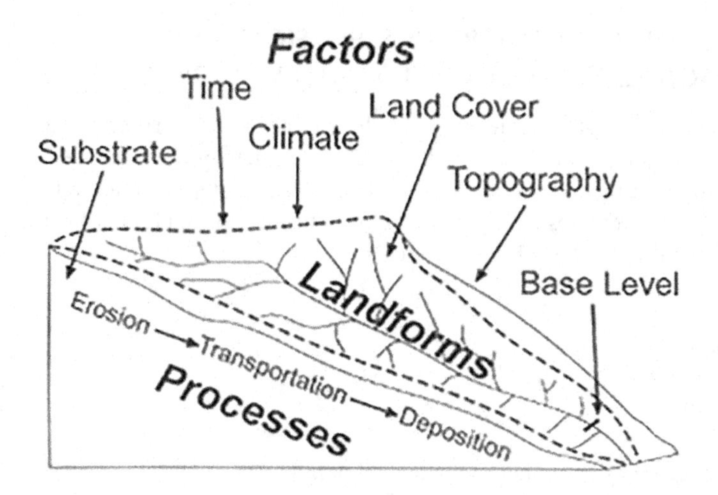

FIGURE 20.1 A Watershed illustration.

Source: Rees (1986).

geologic and topographic settings, soil and vegetation characteristics, and land use, a watershed encompasses both natural and man-made components (Figure 20.1). After rain and snow melt, water from the land is shed into a watershed. Water trickles through soils, groundwater, creeks, and streams, eventually reaching larger rivers and the ocean (Mauser et al., 2009).

20.3 CLASSIFICATION OF WATERSHED

Any measurable qualities in the area, such as size, shape, location, groundwater exploitation, and land use, can be used to categorise watersheds (Levner et al., 2007). Nonetheless, the primary division of a watershed is generally considered in terms of area and land use. If two watersheds of the same size do not exhibit similar land and channel phases, they may behave extremely differently. Here are descriptions of various watershed classifications.

1. **Size**: Watershed size has a major impact on the geographic heterogeneity of hydrological processes. Large watersheds have the highest geographic heterogeneity because size enhances the spatial variability of watershed parameters. Storage grows as the watershed size grows. The watersheds are classified into three types based on size
   1. Small Watersheds <250 km^2
   2. Medium Watersheds between 250 and 2,500 km^2
   3. Large Watersheds $>2{,}500$ km^2
2. **Land use**: The exploitation (natural and human interactions) characteristics of a watershed are defined by its land use, and these characteristics have an impact on the watershed's diverse hydrological processes. According to the type of land use, a watershed can be categorised as follows.
   1. Agricultural
   2. Urban
   3. Mountainous
   4. Forest
   5. Desert
   6. Coastal or marsh, or
   7. Mixed - a combination of two or more of the previous classifications.

20.4 WATERSHED CHARACTERISTICS: PHYSICAL AND GEOMORPHOLOGIC CHARACTERISTICS ASSOCIATED WITH WATERSHEDS

Geomorphology of a watershed is the study of the characteristics, arrangement, and development of landforms and qualities, as well as the physical development of the watershed. Within the watershed/basin boundary, it includes the characteristics of the land surface as well as the waterways. These watershed characteristics have a big impact on how runoff and other hydrological processes behave. The principal watershed characteristics are:

1. Basin Area
2. Basin Slope
3. Basin Shape
4. Basin Length

Basin shape is reflected by a number of watershed parameters as given below.

1. Form Factor
2. Shape Factor
3. Circularity Ratio
4. Elongation Ratio
5. Compactness Coefficient

Together with a watershed's surface features, channel characteristics play a crucial role in the movement of runoff water from the overland region to channels (streams) and from a primary channel to a higher-order channel (e.g. river stream). The most prevalent and significant watershed channel characteristics include:

1. Channel Order
2. Channel Length
3. Channel Slope
4. Channel Profile
5. Drainage Density

20.5 WATERSHED MANAGEMENT AND ITS OBJECTIVES

The study of a watershed's pertinent features with the goal of a sustainable distribution of its resources is called watershed management. In order to maintain and improve the watershed functions that have an impact on the plant, animal, and human communities within a watershed border, watershed management is a crucial component in developing and implementing plans, programmes, and projects (Wang et al., 2016).

The different objectives of watershed management programmes are:

1. To preserve soil and water by reducing destructive runoff and deterioration.
2. To control and put the runoff water to good use.
3. To safeguard, improve, and sustainably use the watershed's land for production.
4. To safeguard and improve the water resources coming from the watershed.
5. To stop soil erosion and lessen the watershed's impact from sediment yield.
6. To restore the land's failing conditions.
7. To reduce the heights of the floods in the downstream areas.
8. To increase rainfall infiltration.

9. To enhance and boost the output of animals, forage, and timber.
10. If appropriate, to improve groundwater recharge.

20.6 EFFECT OF PHYSICAL PROPERTIES ON WATERSHED MANAGEMENT

Hydrologic assessments are very interested in the physical characteristics of watersheds since they have a substantial impact on runoff characteristics. The following items list each physical characteristic's effects on watershed management (Dingman, 2015).

20.6.1 Size

The function of the watershed is significantly influenced by its size. The size of the watershed affects how much rain is captured, retained, and discharged (runoff). A tiny watershed is clearly defined by overland flow, which is the primary factor in the creation of a peak flow. A vast watershed has little-to-no overland flow; instead, channel flow is its primary attribute. Basin storage has an impact on significant watersheds as well. The size of the watershed is important because it interacts with the degree of land use changes as well as elements that influence weather and climate. In smaller watersheds, the interaction between runoff-causing events at the weather scale and the storm hydrograph predominates; in bigger watersheds, the interaction between the runoff-causing events at the climate scale and the yearly hydrograph predominates. Smaller, localised runoff-causing events have a tendency to produce more intense precipitation over constrained areas, which has a greater impact on the storm hydrograph in small watersheds or on small tributaries to larger watersheds, whereas large-scale events or land use changes may have an impact on small watersheds and even the storm hydrograph in large watersheds.

20.6.2 Shape

The common watershed might be circular, triangular, square, rectangular, oval, fern leaf-shaped, polygonal, long, or thin. More water will infiltrate, evaporate, or be used by the vegetation when the watershed is larger since there is more time for concentration. When the length of the watershed is smaller than the width, the situation is said to be reverse. Drainage patterns are significantly influenced by the shape of the land, which is determined by weather and geology. The ratio of overland runoff to infiltration depends on the number of streams in a watershed and their form. With a circular watershed, runoff would flow from several areas of the watershed to the outflow simultaneously. The runoff would be spread out over time in an elliptical watershed with the outlet at one end of the major axis and the same size as a circular watershed, resulting in a smaller flood peak than that of the circular watershed.

20.6.3 Topography

Topographic features, including slope, length, degree, and regularity of slope, have an impact on both soil loss and water disposal. Hence, the degree and length of the watershed's slope affect the amount of time that water concentrates and infiltrates.

20.6.4 Drainage

Drainage is regulated by topography. Topography affects drainage density (the number of drainage channels per unit area), major outlet size, and the length, width, and depth of the main and subsidiary channels. The duration of concentration is impacted by drainage pattern. Rapid reaction is an attribute of a high drainage density watershed. To assess the degree of flooding during high flows, more drainage cross-section data are required.

20.6.5 Area of the Watershed

The most significant watershed parameter for hydrologic study is the watershed area, commonly referred to as the drainage area. It displays the amount of water that a downpour can produce. An effective watershed management programme depends on identifying a practical watershed area size.

20.6.6 Length of Watershed

The hydrologic length is sometimes more appropriately used to describe this distance, which is conceptually the distance covered by the surface drainage. This distance is typically used to calculate a time parameter, which is a measurement of how long it takes water to pass through a watershed (time of concentration). Hence, the major flow path from the watershed outflow to the basin boundary is used to measure the length of the watershed. It is required to draw a line from the channel's terminus to the basin boundary since the channel does not go all the way to the basin boundary.

20.6.7 Slope of Watershed

The runoff's momentum depends on the watershed's slope. Channel slope as well as the watershed may be of interest. The slope of a watershed measures the pace at which elevation changes in relation to the length of the main flow corridor. It is typically calculated as the length divided by the elevation difference between the major flow path's endpoints. The point of highest elevation within the watershed may be located along a side boundary rather than at the end of the main flow line, so the height difference may not always be the greatest elevation difference within the watershed. Consider a number of sub-watersheds and calculate each one's slope if there is a considerable change in the slope along the main flow channel.

20.7 EFFECT OF GEOMORPHOLOGIC FACTORS AND ASSOCIATED PROCESSES ON WATERSHED MANAGEMENT

20.7.1 Geological Rocks and Soil

The degree of water erosion, the capacity of channels and hill slopes to erode, and sediment generations are all influenced by geological formation and rock types. While igneous rocks do not degrade, rocks like shale and phyllites do. The physical and chemical characteristics of the soil, particularly its texture, structure, and depth, affect how water is disposed of through infiltration, storage, and runoff (Khadri et al., 2015). The speed of water movement (both lateral and vertical) in the soil is influenced by the types of soil. For instance, relatively small voids exist between soil particles in fine-grained soils like clays, which prevent penetration and increase surface runoff. Contrarily, the bigger pore spaces of coarse soils, such sands, allow for higher rates of infiltration and less runoff. The runoff is influenced by a variety of factors, including surface roughness, soil properties like texture, soil structure, and wetness, and hydrologic soil groups. For instance, the ability for infiltration is influenced by soil qualities. Clay (d 0.002 mm), silt (0.002 d 0.02), or sand ($d > 0.02$ mm) are the three standard classifications for soil particles. Clay, silt, and sand particles are all present in a particular soil. In general, sandy soils have a greater infiltration capability compared to soils containing a considerable amount of tiny particles (Figure 20.2).

20.7.2 Climate

Climate variables have two effects on how a watershed operates and can be modified. Rain describes incoming precipitation in terms of both time and space, as well as other characteristics like strength and frequency. In addition to temperature, humidity, wind speed, and other variables, rainfall amount and these parameters control elements like soil and vegetation. Climate-related traits of the soil are reflected in the area. In the same way, a region's vegetation type is entirely dependent on its climate (Zucca et al., 2021).

20.7.3 Land Cover/Vegetation

This element controls how the watershed functions, such as infiltration, water retention, runoff generation, erosion, sedimentation, etc., depending on the kind and extent of vegetation. It prevents overland flow, catches rainfall, and encourages infiltration. Water is used by vegetation to grow as well. These variables collectively lessen the amount of runoff into streams. Erosion is less likely because of the binding and stabilising effects that vegetation has on soil. In addition to stabilising stream banks, vegetation creates habitat for both aquatic and terrestrial wildlife. In order to improve infiltration of rainfall into soils and groundwater recharge, which improves water storage for summer base flows, vegetation works to decrease runoff and reduce soil compaction. Moreover, vegetation patterns, sizes, and composition have an impact on how much soil erosion occurs. Raindrop splash is lessened by leaves and branches that divert falling rain. Dead leaves and branches from the vegetation build up an organic surface that shields the soil layer. Moreover, soil material is prevented from sliding down a slope by root systems.

20.7.4 Land Use

The main variables influencing watershed behaviour are the kind of land use, its extent, and management. Users' wise use of the land is crucial to the management and operation of the watershed. The collecting capacity and subsequent runoff behaviour of the watershed are substantially impacted by changes in land use, particularly in the variable source area (Price, 2011). Similar consequences to the association between areal storm extent and watershed size can be seen in the scope of land use change over the watershed. If the land use changes are small scale, the storm hydrograph will show their effects most clearly. Local characteristics dominate the storm hydrograph. The effects of land use changes on broader areas of the watershed can also be seen in the annual hydrograph.

20.8 SCOPE OF WATERSHED MANAGEMENT

Watersheds are small basins, as we previously saw in the previous section. A medium-sized or large river basin can be divided into a number of watersheds, each with a size of less than 2,000 km^2, by defining the ridgelines. Watersheds are always simpler to manage than river basins because of their smaller size. All of the natural resources, including soil, water, plants, etc., are preserved in a watershed that is well managed.

Plants and vegetation are essential for maintaining a watershed's natural resources, including its soil and water. Plants' subsurface parts, like their roots, disperse across the soil to fortify and stabilise it. Soil conservation is typically the result of this. Both the soil's voids and the point where the root surface and soil meet are ways that water seeps into the ground below. Plants' terrestrial parts, such as their stems, branches, and leaves, shield the soil under them from the sun's rays and the impact of rainfall. This causes/accelerates the downward passage of precipitation by stem flow and infiltration as a major portion of the velocity and energy in rainwater is absorbed. On the one hand, this process results in the formation of water bodies like rivers and groundwater reservoirs, which are good providers of the nutrients and water needed for plant growth. On the other hand, this procedure also significantly lowers stormwater surface flow velocity and soil erosion (Eubanks et al., 2002). Through the process of photosynthesis, there will also be a significant amount of oxygen released; the creation of vibrant, fragrant flowers; new leaves; and fruits. Because of this, all animals, including humans, migratory birds, and flying insects, enjoy the watershed very much. Besides being sustenance for humans and animals, the fruits and leaves are also edible.

A watershed is regarded as healthy if it has a lot of vegetation in it. A well-managed or green watershed is another name for it. It contains significant groundwater and surface water reserves, and there is little-to-no soil erosion. The majority of its natural resources are generally preserved.

Hence, all initiatives and programmes aimed at striking a balance between the use and preservation of natural resources within a watershed are included in the scope of watershed management. Via watershed management, it represents a sustainable method for resource conservation.

20.9 WATERSHED MANAGEMENT: INDIAN PERSPECTIVE

Despite having a well-developed monsoon precipitation pattern and an equally well-developed drainage system made up of 14 main river basins, 44 medium river basins, and hundreds of tiny river basins, India faces severe stress from ongoing overexploitation of its water and land resources. Numerous harmful hydro-meteorological effects have resulted from this, including extensive soil erosion, excessive water table lowering, extensive river/groundwater pollution due to municipal/industrial wastewaters, widespread loss of forests/grasslands/croplands/wetlands/water bodies, silting of existing water bodies, frequent occurrence of floods/droughts, alarming decrease in Himalayan glaciers, etc. Because of all these events, the Indian perspective on watershed management is often quite susceptible to climatic and human-made influences. Hence, in the Indian context, attaining sustainable water resource development and integrated watershed management are two of the biggest problems (Bansil, 2004).

In spite of this concerning situation, India's whole length and width are covered by hundreds of best management practices (BMPs), which have been the shining examples of effective management of water and land resources. These BMPs make use of traditional, modern, or a hybrid of the two technologies. These BMPs, which have been successfully used in various parts of India, include some of the following:

- The local government authorities in the northeastern state of Sikkim successfully implemented the prohibition on tree cutting, which led to an increase in the forest cover from 44% in 1995–1996 to 47.59% in 2009 (Hindustan Times, 2010).
- The 35 km Kali Bein River, which was severely polluted by industrial effluents and garbage, was nearly completely cleaned and rejuvenated between 2000 and 2006 thanks to the voluntary work of hundreds of people under the guidance of a spiritual saint near Jalandhar in the north Indian state of Punjab (The Times of India, 2007).
- Ralegan Siddhi, a village in the western Indian state of Maharashtra that was prone to severe drought, had changed during a 20-year period beginning in 1974 even with an annual rainfall of only approximately 200 mm into a village with plenty of food, drinkable water, and fodder. This was made possible by the use of social forestry, grassland development, continuous contour trenching, loose boulder structures, brushwood dams, nulla bunds, percolation tanks, underground dams, gabion bunds, check dams, farm ponds, staggered trenches for arresting soil erosion, and a ban on free grazing as part of the ridge to valley approach in watershed management (Padhye et al., 1994).

20.10 PROBLEMS AND CONSTRAINTS IN WATERSHED MANAGEMENT

a. One of the main issues is soil erosion brought on by runoff in places that receive their water from the rain. According to estimates, soil erosion in India increased by about twice as much in the 1990s as it did in the 1980s. Farmers in rain-fed areas are unable to invest because of the significant limitation of unpredictable rainfall and deplorable economic conditions. This results in poor management of the watershed.

b. Within farming communities and among the many watershed areas, equitable benefit sharing of watershed management is a major issue. In general, the activities associated with watershed management provide very little-to-no benefit to women, marginal farmers, and landless labourers. The overdevelopment of water harvesting facilities in the upstream portion of watersheds had drastically decreased the inflows into the downstream reservoirs, according to several case studies conducted in India's water-scarce Gujarat and Madhya Pradesh states. On the other hand, it is also observed that the construction of big reservoirs led to the submersion and hardship in the upstream sections of the same watershed or a neighbouring watershed, typically having an urban or an industrial sector, and benefits for residents in the downstream parts (Shiva, 1991).
c. Several watersheds with insufficient watershed management have had acute shortages of water, in general, and drinking water, particularly in the summer, which may lead to severe/recurrent droughts. It could frequently lead to brief and transitory increases in food yield.
d. Despite the watershed management efforts, common lands are frequently not handled properly, and re-vegetation does not occur as intended. Due to increased water withdrawals by other uses or as a result of overgrazing, domestic/ecosystem water needs and livestock water/fodder needs are either not sufficiently met or are made to suffer.
e. Issues with the management of watersheds arise from a poor understanding of the interactions between socio-economic and biophysical processes.
f. A conflict between different government ministries, including those related to agriculture (with an emphasis on food production), rural development (with an emphasis on employment generation and poverty alleviation), forests (with an emphasis on maintaining biodiversity & wildlife), as well as conflict between government bureaucracy and elected representatives in their zeal to control funds, is a major problem in watershed management programmes—one that needs to be resolved on a priority basis (Srinivasan, 2005).
g. Lack of baseline data for monitoring and comparing the current circumstances makes it difficult to undertake significant effect assessment studies on watershed management projects. Without adequately estimating water supply scenarios in drought, normal, or surplus years, as well as without effectively managing demand, particularly in drought years, the entire watershed management exercise is carried out.
h. Due to decreased forest yields, declining soil quality, a lack of tribal agriculture policy, and population pressure, large areas with a large tribal population lack facilities to capture water and stabilize their food, crop, and fodder production. As a result, the indigenous people experience ongoing suffering, sociopolitical turmoil, and rebellion.

20.11 NEW PROSPECTS AND OPPORTUNITIES ASSOCIATED WITH WATERSHED MANAGEMENT

Watershed management is related to fresh prospects and opportunities despite the aforementioned issues and limitations as well as some other issues and limitations. Below is a list of some of them:

a. More and better food must be produced without further harming the ecology and the environment, particularly the land, water, forests, wildlife, and atmosphere. This may involve implementing best management practices (BMPs) like organic farming, de-silting to increase crop productivity as well as reservoir capacity, sprinkler and/or drip irrigation to reduce excessive water use, a no-tree-falling policy, reforestation and arboriculture using high oxygen-yielding plants, etc.

b. The benefits of groundwater recharge must be protected from being lost to excessive groundwater extraction. Over-extraction of groundwater must be prevented in order to do this, and regulation and public awareness campaigns can help.
c. It is important to take into account how intensive upstream water conservation may affect downstream conditions. Watershed associations should be established for this purpose, including representation from each watershed stakeholder. These organisations have the authority to make decisions that are in everyone's best interests.
d. Increasing the moderate benefit–cost ratio while lowering the expenses at which the advantages are realised could open up new possibilities and opportunities for watershed management. To achieve this, low-cost technologies that may use readily available local resources; labour that is practically free; and conventional, tried-and-true technologies should be used to generate greater advantages that are distributed across the entire watershed among all the stakeholders.
e. A key priority should be to increase engagement from all groups of people after the project has been implemented in order to maintain sustainable watershed management. Only this can guarantee progress overcoming the hydro-geological, socio-political, and other uncertainties on a sustained basis.
f. Many villages in India have seen the implementation of numerous successful watershed management programmes on a modest scale, thanks to the combined efforts of government agencies, non-governmental organisations (NGOs), and research institutions. They stand for irregular BMPs. So, it is necessary to expand watershed management efforts over sizable areas, which may include remote and/or challenging terrain, in order to successfully address various issues facing our agricultural, rural, and forest sectors (Sahaand Mishra, 2017).
g. Given that there aren't many or any institutions dedicated to research and development on collective management of watersheds, there is a need to establish advanced learning institutions that make use of cutting-edge technologies like remote sensing, geographic information systems, decision support systems, computer-based planning tools, poverty and socio-economic analysis, etc.
h. The common pool resources (CPRs) of land, water, fodder, forests, fisheries, wildlife, and agriculture, which greatly support people's livelihoods, particularly in rural regions, need to be preserved and improved.
i. In order to reduce migration to urban areas, opportunities must be created in agriculture, natural disasters like floods and droughts, forest and mountain economies, and by stopping the decline in agricultural prices, the wage gap between urban and rural areas, and the employment opportunities between these areas.

20.12 LAND CAPABILITY AND ITS CLASSIFICATION

The ability of the ground surface to support artificial crop growth or human habitat, as well as natural plant growth and wildlife habitat, is known as land capability. The sort of land use (such as human habitation, agriculture, pastures, woods, wildlife habitat, etc.) that is appropriate over a specific type of land is thus indicated. Depending on the features of the land, such as slope, soil type, soil depth, and erosion conditions, different lands have varying capacities. It is desirable to use or ensure the continuity of that land area for other land uses, as discussed before, if certain land characteristics do not favour agriculture.

Complete soil conservation is the ultimate goal of the distribution of different land capacities throughout a large land area with a variety of characteristics. Comprehensive soil conservation entails optimal soil health and never-ending zero soil erosion. Moreover, it makes total water and total vegetation conservation easier. Long-term integrated watershed management is the end outcome of this.

The United States Department of Agriculture's (USDA) Soil Conservation Service (SCS) has pioneered the classification of land capacity (Klingbiel and Montgomery, 1961). Thus, the land capability is essentially divided into two groups based on the land's ability to support agriculture. Group 1 Lands are the first group, which consists of all the lands that are suitable for cultivation. Group 2 Lands are the remaining category, which consists of all the areas that are unfit for cultivation. These two groupings are each subdivided into a further four classes. In this way, cultivable land classes I through IV make up Group 1 Lands, while non-cultivable land classes V through VIII make up Group 2.

20.13 IMPACT ON WATERSHED DUE TO LAND USE

There is a strong relationship between the watershed and land use. Watersheds with a healthy biotic system—such as adequate flora and fauna—and an equally healthy aquatic system—such as appropriate streams and wetlands—are often sustainable systems. The land use changes substantially after they are subjected to extensive human interference and/or natural disasters. This has a significant impact on the hydrology, flora, and fauna of the watershed.

Large tracts of native grasslands and woods have been transformed into urban centres, agricultural land, or networks of roads and railways in various regions of the world. Riparian pathways have been altered as a result, as have wetlands that have been drained and natural river systems. The hydrologic changes in the watersheds, their stream systems, and their surface water–groundwater interactions are a result of these changes in land use. People and ecosystems can be impacted by changes in water quantity and quality in both upstream and downstream regions of watersheds (Ffolliott et al., 2013).

If these changes in watersheds are not adequately managed, they could eventually stop being sustainable. Thus, more focus is being placed on preserving or restoring natural stream channel systems, riparian communities, wetland ecosystems, and floodplains that can improve the hydrologic conditions of watersheds in order to minimize any unfavourable repercussions. Hey suggested a significant programme to maximise natural storage in wetlands and floodplains while reducing transportation in the upper Mississippi River Basin (Brooks et al., 2003). A programme like this would essentially undo some of the effects of the basin's past 200 years of levee building and other engineering techniques.

Watershed effects, or combined environmental effects of activities in a watershed, can negatively affect beneficial uses of land if watersheds are not managed sustainably (Sidle, 2000). These environmental influences might not seem important when considered separately. But over time and place, they might be substantial when considered collectively.

For instance, the flow of water and sediment may rise if cropland in one area of a watershed is converted from forest. Similar to how draining a wetland at another place might affect a watershed, road construction and drainage can likewise have an impact there. The removal of dense shrubs to boost forage production may also, in some situations, increase water yield, assist some wildlife species, and lessen the risk of fire. Yet, other kinds of species can suffer from the same shrub removal. Variations in the composition, density, age structure, and continuity of the vegetation across the landscape can have an impact on evapo-transpiration losses and, as a result, on the soil moisture conditions, timing of water yields, peak stream flow volumes, and other aspects of watersheds. Overgrazing lowers infiltration capacity and raises surface runoff because it causes excessive soil compaction and trampling in a watershed. Due to the exposure to erodible soil and subsoil during their construction, roads and paths may actually promote soil erosion. Because of the additional precipitation, there will be less infiltration and more overland flow, which will erode the side slopes of cuts and fills with higher gradients.

It is generally recognised that the development of finished impermeable surfaces, as well as the filling up of water bodies, particularly in metropolitan regions, which results in a sharp decline in infiltration or surface storage, both contribute to an increase in flooding. On the other side, wildfires frequently have an impact on watersheds of forests and wildlands. Due to the reduction in vegetation cover and the development of water-repellent soil layers, this causes both an increase in surface runoff and soil erosion.

Here are a few instances where altering land use has negatively affected watersheds and rendered them ecologically unsustainable. There are numerous other instances of land use changes that have an adverse impact on the watersheds' hydrologic, geomorphic, and water quality characteristics. Understanding and appreciating the impacts of cumulative watershed effects on water yield, other stream flow characteristics, and water quality are necessary for overcoming these undesirable effects. This requires an interdisciplinary approach involving hydrology, geomorphology, and ecology into watershed management and land-use planning. The planning of land use to guarantee sustainability in watershed management is covered in the next section.

20.14 PLANNING THE LAND USE

In order to choose and implement the optimal land-use options, land-use planning entails the methodical assessment of land and water potential, alternative land-use scenarios, and economic and social factors. Its goal is to identify and implement those land uses that will best serve the demands of the population while protecting the resources for the long term. Planning for land use is driven by the need for change, better management, or very diverse land-use patterns necessitated by shifting conditions (Arnold, 2010). Land-use planning takes into account all types of rural land use, including agriculture, pastoral areas, forestry, wildlife conservation, and tourism. By highlighting the sections of land that are most valuable when used for rural purposes, it also offers direction in situations when rural land use and urban or industrial expansion collide.

If the land-use planning is to be effective, the following two requirements must be satisfied:

1. The requirement for land use changes or the action to stop some undesirable changes, both of which must be agreed upon by the parties involved.
2. The plan must have the political will and capacity to be implemented.

It may be desirable to launch an awareness campaign or set up demonstration locations where these two requirements are not met but the problems are urgent in order to establish the circumstances required for efficient planning. We must rely on the land, which is in short supply, to provide for our fundamental needs of air, water, food, clothes, shelter, and fuel. The land becomes an increasingly limited resource as population and expectations rise.

Land must adapt to suit changing needs, which may result in new conflicts between the various competing uses of the land as well as between the interests of different land users and the greater good. Farming is no longer possible on the land that was taken for cities and industries. Similar to how forests, water resources, and wildlife may be in competition with new farmland development.

It is a well-established practice to plan for the greatest possible use of land. Farmers have formed plans over the years, determining what to cultivate and where to grow it season after season. Their choices have been based on their own requirements, understanding of the terrain and available technology, labour, and capital. Information and thorough techniques of analysis and planning are required as the scope of the problem, the number of parties involved, and the complexity of the issues grow.

Land-use planning is not only agricultural planning on a larger scale though. It also has another aspect, which is the general public's interest. Planning entails both anticipating and responding to the need for change. Its goals are established by social or political requirements that take into account the current environment. Since the land itself is being eroded, the current condition cannot persist in many places. Unwise land use examples include the following:

a. Clearing forests on sloping terrain or in areas with low soils, where there are currently no developed sustainable agricultural methods.
b. Overgrazing on pastures.
c. Polluting industrial, agricultural, and urban activities.

The degradation of land resources is mostly a result of consuming land today without making investments for tomorrow. It may also be attributed to human avarice, ignorance, unpredictability, or a lack of alternatives. By taking the following measures, land-use planning strives to make the best use of scarce resources:

1. Determine the present and future needs, and then methodically assess the land's capacity to meet those needs.
2. Recognise and resolve problems between overlapping uses, individual and collective needs, and requirements of the present generation and those of future generations.
3. Look for sustainable choices and select those that totally satisfy the needs that have been defined.
4. Develop a plan to implement the required improvements.
5. Get wisdom from knowledge.

20.14.1 Goals of Land-Use Planning

What is meant by the "optimal" use of the land is defined by the planning for land-use objectives. These ought to be mentioned right away in a specific land-use planning project. The following three categories—efficiency, equity & acceptability, and sustainability—can be used to group goals.

Efficiency: The projected land use must be financially viable. Making productive use of the land is thus one objective of development planning. Certain regions are more suitable than others for any given land use. Efficiency is attained by coordinating various land uses with the regions that will produce the highest advantages at the lowest cost, or maximum benefit–cost ratio.

Depending on who you ask, efficiency may mean different things. The highest value from the available land area, or the greatest return on cash and labour invested, for the specific land user. Increasing the foreign exchange situation by manufacturing for export or import substitution is one of the more challenging government goals.

20.14.1.1 Equity & Acceptability

Social acceptance must be ensured in land use. It should guarantee rural communities' access to food, jobs, and money. It is possible to enhance the land and redistribute it to combat absolute poverty or lessen inequality. Setting a minimum level of living to which the target groups should be increased is one strategy to do this. Income, dietary habits, food security, and housing conditions can all affect one's standard of living. Planning is required to meet these criteria, which includes allocating land for particular use as well as allocating funds and other resources.

Here is a good illustration of acceptance. The Ethiopian government first noticed the substantial soil deterioration in the highlands in the wake of the drought of 1973–1974 and the accompanying famine.

An intensive soil conservation effort was started, focusing on bunding and reforestation to preserve steep slopes. Although it significantly reduced soil erosion, it did little to boost agricultural productivity. Because it restricted the space available for cattle grazing and because forest conservation involved restricting access to the public for fuel wood collection, large-scale afforestation was also unpopular with the local population. If the public was to continue supporting soil conservation efforts without the use of incentives like the Food-for-Work Program, a balance between the conflicting demands of conservation and production was undoubtedly required.

The local population was found to be more accepting of a land-use plan that conserved steeper slopes by restoring good vegetative cover through closure, followed by regulated grazing, than large-scale afforestation used in isolation.

20.14.1.2 Sustainability

Sustainable land use is preserving resources for future generations while still providing for the needs of the present. Production and conservation must work together to achieve this. In order to ensure that production will continue in the future, it is necessary to combine the production of the things that people currently need with the conservation of the natural resources that support that production.

A community's future is lost if it destroys its land. Because individual land users frequently lack the capacity to conserve soil, water, and other land resources, land use must be planned for the community as a whole.

20.14.2 Trade-offs among Conflicting Goals of Land-Use Planning

These different land-use planning objectives obviously conflict with one another. Less efficiency could result from greater equity. The immediate requirements of the present might not be able to be met without using resources like burning oil or destroying natural forest areas. Decision-makers must take into account the trade-offs between various objectives. But for the system to continue as a whole, the use of natural resources must be offset by the growth of human or material resources of equal or greater value.

It is always important to have accurate information, such as knowledge about the requirements of the population; information about land resources; and information about the economic, social, and environmental effects of different course of action. The responsibility of the land-use planner is to make sure that decisions are made with an appropriate level of agreement or disagreement.

Planning the procedures, such as adopting the right new technology, can frequently minimise the costs in trade-off. By including the community in the planning process and by making the information and justifications for decisions public, it can also assist in resolving the issue.

20.14.3 The Focus of Land-Use Planning

The following points constitute the focus of land-use planning.

A. **Land-use planning is for the people:** The land-use planning process is driven by the requirements of the populace. Farmers in the area, other users of the land, and the larger community that depends on the land must acknowledge the necessity for a shift in land use because they will have to deal with its effects.

   Planning for land use must be constructive and beneficial to the populace. The planning team must learn about local knowledge, skills, labour, and capital as well as the demands of the populace. It must investigate the issues with current land-use strategies and look for alternatives while educating the public about the risks or drawbacks of sticking with the status quo as well as the possibilities for change.

   Rules intended to stop individuals from doing as they already do are most likely to fail for urgent reasons. Local involvement in land-use planning makes it simple to gain local acceptance. Local authorities' backing is crucial. In addition, it's critical that organisations that have the funding to carry out the plan participate.

B. **Land is not the same everywhere:** Land-use planning also has an emphasis on the land. Wherever it is needed, resources such as capital, labour, management expertise, and technology can be transported. However, as land cannot be transferred, different regions offer unique potential and management challenges. The climate and vegetation are clear indicators of how the land resources are usually changing. Nonetheless, instances like the depletion of water supplies or the loss of soil due to salinity or erosion serve as a reminder that resources can deteriorate, oftentimes irrevocably. Planning for land use therefore requires accurate knowledge regarding land resources.

C. **Technology:** Knowledge of technologies like agronomy, silviculture, livestock husbandry, and other methods of using land is a third component of land-use planning. The recommended technology must be relevant and compatible with the users' available resources, including money, knowledge, and other resources. The land-use planner should consider the social and environmental ramifications of new technologies.

D. **Integration:** Early attempts at land-use planning made the error of concentrating too intently on land resources without giving enough consideration to their potential uses. Excellent land for agriculture is typically also good for competing uses. Land use decisions are not solely based on the suitability of the land; they are also influenced by product demand and the degree to which a certain area's use is necessary for a given purpose. Planning must take into account data on the land's suitability, consumer demand for alternative goods or uses, and opportunities to meet such wants on the currently available land as well as in the future.

Land-use planning is not segmented as a result. Even in cases where a specific plan is sector-specific, such as small-holder tea development or irrigation, an integrated approach must be applied from the national level of strategic planning to the specifics of the individual projects and programmes at the district and local levels.

20.14.4 Land-Use Planning at Different Levels

The three main levels of land-use planning are national, district, and local. They don't have to be done in that order. These line up with the governmental tiers where land-use decisions are made. At each level, where planning techniques and plan types also vary, various judgements are made. Yet, a land use-strategy, policies that identify planning objectives, projects that address these goals, and operational planning are required at each level in order to complete the task efficiently, quickly, and affordably. The more the three layers of planning interact, the better it is for everyone. Both directions should be covered by the information flow. The amount of specificity required and the level of direct local participation rise with each subsequent planning level.

20.14.4.1 National Level Land-Use Planning

Land-use planning at the national level considers national objectives and resource distribution. National land-use planning frequently excludes the actual distribution of land among various purposes. It might set the order of importance for district-level projects in their place. A national land-use strategy might address:

1. **Land-use policy** aimed at balancing the competing demands for land from many economic sectors, including agriculture, tourism, the protection of wildlife, housing & public facilities, transportation, and industry.
2. **National development plans and budgets** which identify projects and allocate funds for development.
3. **Coordinating** the efforts of sectoral organisations concerned with land use.
4. **Laws** governing issues including water rights, logging, and land tenure.

National goals are intricate, and numerous people across a wide range of areas are impacted by legislative, legal, and fiscal decisions. It is impossible for decision-makers to be experts in every area of land use. So, it is the job of the planners to communicate the pertinent information in a way that the decision-makers can both comprehend and act on it.

20.14.4.2 District Level Land-Use Planning

District level refers to portions of territory that are between the national and local levels, not just administrative districts. The majority of development projects are in this stage, where planning first considers the diversity of the land and if it is suitable to achieve the project's objectives. National priorities must be translated into local plans when planning is started at the national level. It is important to reconcile conflicts between national and local interests. At this point, the following types of problems are addressed:

1. The placement of new construction, such as settlements, plantations, irrigation systems, etc.
2. The requirement for better infrastructure, such as greater access to water, roads, and marketing facilities.
3. Creating management policies to support better sorts of land use on each type of land.

20.14.4.3 Local Level Land-Use Planning

The hamlet, a cluster of settlements, a small watershed, or a catchment may all be considered local planning units. Using the expertise and contributions of the community at this level makes it very simple to adapt the plan to the people. Wherever district-level planning is started, local work must be done as part of the work programme to implement changes in land use or management. Alternately, this might be the initial stage of planning, with local residents setting the priorities. Planning at the local level involves determining what needs to be done where, when, and by whom on specific parcels of property.

Some instances of local land-use planning include:

1. Design of soil conservation, irrigation, and drainage systems.
2. **Design of infrastructure**: road alignment, location of facilities for milk collecting, veterinary care, or agricultural selling.
3. Planting particular crops on appropriate land.

Firm suggestions must be made in response to local requests, such as those for appropriate locations to introduce coffee or tobacco. Information at various scales and levels of generalisation are required for planning at these many levels. Maps might contain a lot of this information. The map scale that allows the entire country to fit on a single map sheet, which may require a scale ranging from 1:5 million to 1:1 million or larger, is the most appropriate for national-level land-use planning. Although some information may be summarised at smaller sizes ranging from 1:250,000 to 1:50,000, district-level land-use planning requires specifics to be mapped at around 1:50,000.

Maps with scales between 1:20,000 and 1:5,000 are proven to be the best for local level land-use planning. Reproductions of aerial photographs can serve as the foundation for local land-use planning since, according to field workers and experience, locals can identify their locations on the images.

20.14.5 Land Use in Relation to Sectoral and Development Planning

Land-use planning is by definition non-sectoral, but until a dedicated planning authority is established, sectoral agencies in agriculture, forestry, irrigation, etc., must implement a land-use plan. The different extension services will need to provide assistance with implementation.

The line separating land-use planning from other facets of rural development won't be obvious. For instance, the introduction of a cash crop may constitute a desired change in land usage. Using fertilisers may be necessary for effective management. This is impossible without local centres for fertiliser delivery, useful usage guidance, and a credit system for purchasing it.

Without a robust national distribution network and enough production of or allocation of foreign currency for imports, local services are useless. Although they are undoubtedly not a component of land-use planning, constructing a fertiliser factory and setting up nationwide distribution may be necessary for the achievement of planned land use. On the other side, a land-use planner may be responsible for locating local distribution centres in proportion to the population and available land.

As a result, there is a range of activities from those that primarily depend on the land-use planner's assessment of the physical characteristics of the property to those that require collaboration with other technical professionals. Furthermore, it is the planner(s)' responsibility to make it obvious when national policy issues, such as fair agricultural prices, are necessary conditions for effective land use.

20.14.6 People Involved in Land-Use Planning

Planning for land use entails bringing together a wide range of individuals to work towards shared objectives. Directly involved are the following three groups:

Those who reside in the planning area and whose livelihood is entirely or partially dependent on the land are referred to as land users. They comprise both individuals who actively use the land, such as farmers, herders, foresters, and others, as well as those who rely on their output, such as workers at sawmills and furniture companies, as well as those who prepare crops or raise livestock. All land users must be included in the planning process. They must ultimately implement the plan, so they must have faith in both its prospective advantages and the fairness of the planning process.

Although being the most valuable resource, the locals' expertise and tenacity in coping with their surroundings are typically the most underutilised. Individuals are more likely to take advantage of development possibilities that they themselves helped to plan than any other imposed plans. A plan will not likely be successful if local leaders are not on board.

It might be difficult to get the public to participate in planning effectively. The time and money needed to achieve involvement must be put out by planners through local talks, broadcasting, newspaper articles, technical workshops, and extension services. The more fruitful endeavours of the land users are marked by imagination, a true interest in people and the land, as well as a desire to experiment.

A. **Decision-makers:** Plan implementation is the responsibility of decision-makers. The majority of the time, they will be government ministers at the national and district levels. They will be representatives of other authorities or the local self-government.

   The planning staff often offers knowledge and professional guidance. The decision-makers choose whether to implement plans and, if so, which of the available options should be picked, while also advising the planning team on important concerns and objectives. The decision-maker(s) should be involved periodically, even though the planning team leader is in charge of the day-to-day planning activities. Through their willingness to make their decisions and the process by which they were reached public knowledge, decision-makers can also play a significant role in fostering public engagement.
B. **Land-use planning team:** The treatment of land and land use as a whole is a crucial component of land-use planning. Crossing borders between fields like engineering, agriculture, and the social sciences is required for this. Teamwork is therefore crucial. A team should ideally include individuals with a variety of specialised knowledge, including soil surveyors, land evaluation specialists, agronomists, foresters, range and livestock specialists, engineers, economists, and sociologists.

   A range like that might only be accessible at the national level. A more typical planning team at the local level might be made up of a land-use planner and one or two assistants. Each team member must handle a variety of tasks, necessitating specialised advice. Universities and the personnel of government agencies may be good places to get this guidance or help.

20.15 APPLICATIONS OF REMOTE SENSING AND GIS IN WATERSHED PLANNING

GIS and remote sensing are two crucial contemporary tools with numerous uses in watershed planning. This section first discusses the GIS applications in watershed planning before moving on to the remote sensing applications.

The improved meteorological gathering of information, such as wind speed and direction within weather systems, uses Doppler RADAR (i.e. radio amplification detection and ranging). Features on the seafloor can be mapped to a resolution of roughly a mile by monitoring the water bulges brought on by gravity. The altimeters assess wind speeds and directions as well as surface water currents and directions by measuring the height and wavelength of ocean waves. Airborne heights of objects and features on the ground can be assessed more precisely by light detection and ranging (LIDAR) than by radar technology. LIDAR is used to identify and measure the quantity of different substances in the atmosphere (Narumalani et al., 1997).

LIDAR's main use is for remote sensing of plant cover. Since the 1970s, simultaneous multispectral platforms have been in use, such as the photographs from the Landsat remote sensing satellite. Thematic mapping's maps of land cover and land use can be used to locate minerals, spot or track land use and deforestation, and assess the health of native flora and crops, including whole farming regions or forests.

In the context of the fight against desertification, remote sensing enables long-term follow-up and monitoring of risk areas, identification of desertification factors, assistance to decision-makers in defining the appropriate environmental management measures, and evaluation of the effects on watershed planning. India began developing an indigenous Indian remote sensing (IRS) satellite programme after the successful launch of its two remote sensing satellites, Bhaskara 1 and Bhaskara 2, in 1979 and 1981, respectively. This programme is intended to support the national economy in the areas of agriculture, water resources, forestry, ecology, geology, watersheds, marine fisheries, and coastal management.

The Department of Space (DOS) of the Government of India (GoI) is the nodal agency for the National Natural Resources Management System (NNRMS), which relies heavily on the IRS satellites and provides operational remote sensing data services. Received and distributed are data from the IRS satellites. High-resolution satellite technology has opened up new opportunities for mapping on a global scale, including planning for infrastructure and reducing urban sprawl. In the nation, NNRMS-administered remote sensing applications now cover a wide range of fields within the realm of watershed planning and management, including estimations of the pre-harvest crop area and production of major crops, drought monitoring and assessment based on vegetation condition, mapping of flood risk zones, etc.

In characterisation and assessment studies that call for a watershed-based methodology, GIS has been widely used. Digital elevation models (DEMs) and data from the National Hydrography Dataset (NHD) Program of the United States Geological Survey (USGS) are widely available and can be used to determine the basic physical properties of a watershed, such as the drainage network and flow routes. This helps to improve the creation of a watershed action plan and the identification of current and potential pollution issues in the watershed when combined with precipitation and other water quality monitoring data from sources like the USGS and the Environmental Protection Agency's (EPA) BASINS (i.e. better assessment science integrating point & non-point sources) database.

For a successful characterisation and assessment of watershed functions and conditions, data from environmental remote sensing systems and global positioning system (GPS) surveys can be used within a GIS.

- **Management Planning**

  Planning becomes crucial for dealing with problems affecting water quality and quantity brought on by both natural and human-induced risks (such as droughts, hazardous material spills, floods, and urbanisation) in order to lessen their effects and make the best possible use of the resources. To better comprehend the intricate linkages between natural and human systems as they relate to land and resource usage within watersheds, information from characterisation and assessment studies, typically in the form of charts and maps, can be merged with other datasets. A common framework [i.e. spatial location] is provided by GIS for data on watershed management that are gathered from many sources. GIS is a potent tool for understanding these processes and managing possible repercussions of human activity since watershed data and watershed biophysical processes have spatial dimensions (Mirchi et al., 2010).

Modern GIS's modelling and visualisation capabilities, along with the Internet's and the World Wide Web's fast growth, provide radically new tools for comprehending the processes and dynamics that affect the physical, biological, and chemical environment of watersheds. The integration of GIS, the Internet, and environmental databases is particularly beneficial for planning studies, where fast information interchange and input are essential—and even more so when several agencies and stakeholders are engaged.

- **Watershed Restoration (Analysis of Alternative Management Strategies)**

  Evaluation of multiple alternatives is a common component of watershed restoration projects, and GIS offers the ideal setting for doing so effectively and efficiently. GIS has been utilised for restoration studies involving everything from densely metropolitan landscapes to relatively small rural watersheds (Fullerton et al., 2009). GIS can help with unified source water assessment programmes, such as the total maximum daily load (TMDL) programme, when combined with hydrodynamic and spatially detailed hydrologic/water quality models. For instance, by making digital maps of the current conditions and contrasting them with maps that reflect the alternative scenarios, it is possible to study the various options for restoring a waterbody or a watershed. In addition, GIS can offer a platform for cooperation between academics, watershed stakeholders, and decision-makers, greatly enhancing consensus-building and presenting chances for cooperation on interdisciplinary environmental policy issues. The integrating capabilities of a GIS offer an interface to correctly and effectively translate and imitate the complexity of a real-world system within the boundaries of a digital environment.

20.16 WATERSHED POLICY ANALYSIS AND DECISION SUPPORT

The employment of computer-based simulation models in resource planning and management is significantly impacted by fundamental developments occurring in the field of watershed science, particularly watershed planning. On the one hand, GIS is becoming an indispensable tool for tasks related to watershed planning and management due to the greatly expanded availability of effective, affordable, and simple-to-use GIS software. But as usage has grown, so has the understanding that GIS alone cannot meet all the demands of planning and managing watersheds. Resource planners' interest in creating integrated decision support systems that integrate GIS, spatial and non-spatial data, computer-based biophysical models, knowledge-based (i.e. expert) systems, and advanced visualisation techniques has been rekindled by this realisation. These systems would support planning and policy analysis functions. As a part of a geographic decision support system, GIS offers extremely potent visualisation features for display and modification, providing quick, intuitive evaluation capabilities that a variety of non-technical users and decision-makers can understand.

With the help of flexible problem-solving settings and geoprocessing functions, GIS may help the decision-maker deal with challenging management and planning issues within a watershed. It is clear from a cursory review of the environmental and ecological scientific literature that there has been significant study into GIS-based watershed management and planning. The quick advancement of GIS into a diverse range of applications and implementations is evidenced by the enormous growth in its use for the aforementioned tasks.

20.17 CONCLUSION

In conclusion, watersheds play a vital role in supporting the environment and providing essential services to human society. The sustainable management of watersheds is necessary to maintain their ecological integrity and protect the livelihoods of communities that depend on them. The implementation of effective watershed management strategies requires an understanding of the complex interactions between the physical, biological, and socio-economic components of the watershed.

GIS have emerged as an essential tool for watershed management. GIS technology enables the integration and analysis of diverse data sources to provide a comprehensive understanding of the watershed's spatial and temporal dynamics. GIS-based models can be used to simulate and predict the impact of various management interventions on the watershed's health and inform decision-making processes.

The use of GIS technology in watershed management is not only limited to environmental conservation but also has important implications for land-use planning, infrastructure development, and disaster management. Overall, the incorporation of GIS in watershed management practices can help to ensure sustainable and equitable resource use, mitigate environmental degradation, and safeguard human well-being.

REFERENCES

Arnold, C.A. 2010. Adaptive Watershed Planing and Climate Change. *Environmental and Energy Law and Policy Journal*, **5, 417**.

Bali, Y. P. (1980). Selection of implementation area/ catchment, watershed delineation, classification and priority determination. National Seminar on Watershed Management Rainfed and Himalayan Development, New Delhi, India

Bansil, P.C. 2004. *Water Management in India*. Concept Publishing Company, New Delhi.

Brooks, K.N., Current, D. and Wyse, D. 2003. October. Restoring hydrologic function of altered landscapes: an integrated watershed management approach. In *Proceedings of the International Watershed Management Conference on Water Resources for the Future.* Porto Cervo, Sassari, Italy, pp. 101–114.

Dingman, S.L. 2015. *Physical Hydrology*. Waveland press, Long Grove, IL.

Eubanks, C. and Meadows, D. 2002. A soil bioengineering guide for streambank and lakeshore stabilization. *US Department of Agriculture Forest Service, Technology and Development Program.*

Ffolliott, P.F., Brooks, K.N., Neary, D.G., Tapia, R.P. and Chevesich, P.G. 2013. Soil erosion and sediment production on watershed landscapes: processes and control. *International Hydrological Programme for Latin America and the Caribbean*, 2013, 73.

Fullerton, A.H., Steel, E.A., Caras, Y., Sheer, M., Olson, P. and Kaje, J. 2009. Putting watershed restoration in context: alternative future scenarios influence management outcomes. *Ecological Applications*, 19(1), 218–235.

Hindustan Times, Daily Newspaper, Kolkata. (2010). Aug. 31, West Bengal, India.

Khadri, S.F.R. and Pande, C. 2015. Remote sensing based hydro-geomorphological mapping of Mahesh River Basin, Akola, and Buldhana Districts, Maharashtra, India-effects for water resource evaluation and management. *International Journal of Geology, Earth and Environmental Sciences*, 5(2), 178–187.

Klingbiel, A.A. and Montgomery, P.H. 1961. *Land Capability Classification, Agricultural Handbook*. Soil Conservation Service, United States Department of Agriculture, Washington, DC.

Levner, I. and Zhang, H. 2007.Classification-driven watershed segmentation. *IEEE Transactions on Image Processing*, 16(5), 1437–1445.

Mauser, W. and Bach, H. 2009. PROMET–Large scale distributed hydrological modelling to study the impact of climate change on the water flows of mountain watersheds. *Journal of Hydrology*, 376(3–4), 362–377.

Mirchi, A., Watkins Jr, D. and Madani, K. 2010. Modeling for watershed planning, management, and decision making. *Watersheds: Management, Restoration and Environmental Impact*, 2010, 354–392.

Narumalani, S., Zhou, Y. and Jensen, J.R. 1997. Application of remote sensing and geographic information systems to the delineation and analysis of riparian buffer zones. *Aquatic Botany*, 58(3–4), 393–409.

Nguyen, T.T., Ngo, H.H., W. Guo, X.C., Wang, N., Ren, G., Ding, J. and Liang H. 2019. Implementation of a specific urban water management-Sponge City. *Science of the Total Environment*, 652, 147–162.

Padhye, A.A., Pathak, A.A., Katkar, V.J., Hazare, V.K. and Kaufman, L. 1994. Oral histoplasmosis in India: a case report and an overview of cases reported during 1968–92. *Journal of Medical and Veterinary Mycology*, 32(2), 93–103.

Price, K. 2011. Effects of watershed topography, soils, land use, and climate on baseflow hydrology in humid regions: a review. *Progress in Physical Geography*, 35(4), 465–492.

Rao, K.L. 1975. *India's Water Wealth. Orient Longman*, Hyderabad, Andhra Pradesh, India.

Saha, D. and Mishra, S. K. 2017. Watershed management in India: a review of approaches and challenges. *Environmental Science & Policy*, 79, 1–13.

Shiva, V. 1991. Equity Issues in Watershed Development: A Review of Literature and Practices. *Economic and Political Weekly*, 26, 51–52.

Sidle, R.C., 2000. Watershed challenges for the 21st century: a global perspective for mountainous terrain. *Land Stewardship in the 21st Century: The Contributions of Watershed Management*, U.S. Department of Agriculture, Forest Service, Rocky Mountain Research Station, Fort Collins, CO, pp. 45–56.

Srinivasan, S. 2005. Integrated watershed management in India: constraints and directions for future research. *Agricultural Economics Research Review*, 18, 1–18.

The Times of India, Daily Newspaper, Kolkata. (2007). Jan. 20, West Bengal, India.

Wang, G., Mang, S., Cai, H., Liu, S., Zhang, Z., Wang, L. and Innes, J.L. 2016. Integrated watershed management: evolution, development and emerging trends. *Journal of Forestry Research*, 27, 967–994.

Zucca, C., Middleton, N., Kang, U. and Liniger, H. 2021. Shrinking water bodies as hotspots of sand and dust storms: the role of land degradation and sustainable soil and water management. *Catena*, 207, 105669.

Index

Printed in the United States
by Baker & Taylor Publisher Services